Wind and Solar Energy Applications

This book examines the recent advances, from theoretical and applied perspectives, addressing the major issues associated with renewable energy systems, with each chapter covering fundamental issues and latest developments. This book covers important themes, including solar energy equipment, wind and solar energy systems, energy storage and bioenergy applications, hybrid renewable energy systems, as well as the measurement techniques that are used for these systems. Further, it focusses on original research outcomes on various technological developments and provides insights to taxonomy of challenges, issues, and research directions in renewable energy applications.

Features:

- Covers research and technological developments in wind and solar energy applications.
- Proposes resolution of limitations and performance issues of existing system models and design.
- Incorporates the challenges of adoption of renewable energies system.
- Provides hypotheses, mathematical analysis, and real-time practical applications to practical problems.
- Includes case studies of implementation of solar and wind systems in remote areas.

This book is aimed at researchers, professionals, and graduate students in electrical and mechanical engineering and renewable energy.

Wind and Solar Energy Applications
Technological Challenges and Advances

Edited By
Satish Kumar Peddapelli and Peter Virtic

CRC Press
Taylor & Francis Group
Boca Raton London New York

CRC Press is an imprint of the
Taylor & Francis Group, an **informa** business

Designed cover image: © Shutterstock

First edition published 2023
by CRC Press
6000 Broken Sound Parkway NW, Suite 300, Boca Raton, FL 33487–2742

and by CRC Press
4 Park Square, Milton Park, Abingdon, Oxon, OX14 4RN

CRC Press is an imprint of Taylor & Francis Group, LLC

© 2023 selection and editorial matter, Satish Kumar Peddapelli and Peter Virtic; individual chapters, the contributors

Reasonable efforts have been made to publish reliable data and information, but the author and publisher cannot assume responsibility for the validity of all materials or the consequences of their use. The authors and publishers have attempted to trace the copyright holders of all material reproduced in this publication and apologize to copyright holders if permission to publish in this form has not been obtained. If any copyright material has not been acknowledged please write and let us know so we may rectify in any future reprint.

Except as permitted under U.S. Copyright Law, no part of this book may be reprinted, reproduced, transmitted, or utilized in any form by any electronic, mechanical, or other means, now known or hereafter invented, including photocopying, microfilming, and recording, or in any information storage or retrieval system, without written permission from the publishers.

For permission to photocopy or use material electronically from this work, access www.copyright.com or contact the Copyright Clearance Center, Inc. (CCC), 222 Rosewood Drive, Danvers, MA 01923, 978–750–8400. For works that are not available on CCC please contact mpkbookspermissions@tandf.co.uk

Trademark notice: Product or corporate names may be trademarks or registered trademarks and are used only for identification and explanation without intent to infringe.

ISBN: 978-1-032-28846-8 (hbk)
ISBN: 978-1-032-34390-7 (pbk)
ISBN: 978-1-003-32189-7 (ebk)

DOI: 10.1201/9781003321897

Typeset in Times
by Apex CoVantage, LLC

Contents

Editors' Biographies vii

Chapter 1 Investigation of Offshore Wind Energy Potential across Three Southern Coastal Regions in India 1

Saravana Venkatesh G, Amutha N, and N. Kumutha

Chapter 2 Power Quality Enhancement of Fixed- and Variable-Speed WEGS Using HSAPF Based on 5-Level Cascaded Multilevel Inverter and Fuzzy Logic Controller 13

Seema Agrawal and Mahendra Kumar

Chapter 3 Forecasting of Wind Power Using Hybrid Machine Learning Approach 27

Mahaboob Shareef Syed, Ch.V. Suresh, B. Sreenivasa Raju, M. Ravindra Babu, and Y. S. Kishore Babu

Chapter 4 Improving Power Quality of Modern Hybrid Polygeneration SOFC- and PMSG-Based WES Using ANN-Controlled UPQC 35

Ch. Siva Kumar

Chapter 5 Review on Reconfiguration Techniques to Track Down the Maximum Power Under Partial Shadings 47

V. Ramu, P. Satish Kumar, and G. N. Srinivas

Chapter 6 Electric Vehicles – Past, Present, and Future 59

A. Jagadeeshwaran and H. Shree Kumar

Chapter 7 Onboard Electric Vehicle Charger in G2V and V2G Modes Based on PI, PR, and SMC Controllers with Solar PV Charging Circuit 81

Premchand Mendem and Satish Kumar Gudey

Chapter 8 Experimental Investigation on Hybrid Photovoltaic and Thermal Solar Collector System 101

P Narasimha Siva Teja, S. K. Gugulothu, and B. Bhasker

Chapter 9 Concentrated Solar Integrated Hydrothermal Liquefaction of Wastes and Algal Feedstock: Recent Advances and Challenges 109

Namrata Sengar, Matthew Pearce, Christopher Sansom, Xavier Tonnellier, and Heather Almond

Chapter 10 Integrated PV-Wind-Battery-Based Single-Phase System 121

B. Mangu, P. Satish Kumar, and A. Jayaprakash

Chapter 11 Modeling of Power Management Strategy Using Hybrid Energy Generating Sources 137

Hinal Surati

Chapter 12 Photovoltaic Transformerless Inverter Topologies for Grid-Integrated High-Efficiency Applications 151

Ahmad Syed, S. Tara Kalyani, Xiaoqiang Guo, and Freddy Tan Kheng Suan

Chapter 13 Performance Analysis of Rooftop Grid-Connected Solar PV System Under Net Metering System: A Case Study 169

T. Bramhananda Reddy, G. Sreenivasa Reddy, and Y. V. Siva Reddy

Chapter 14 Isolated Bidirectional Dual Active Bridge (DAB) Converter for Photovoltaic System: An Overview 175

Nishit Tiwary, Venkataramana Naik N, Anup Kumar Panda, Rajesh Kumar Lenka, and Ankireddy Narendra

Chapter 15 Sustainable Energy Management in Lighting Urban Public Places 189

Melita Rozman Cafuta and Peter Virtič

Chapter 16 A Review on Multiobjective Control Schemes of Conventional Hybrid DC/AC Microgrid 197

S. Mamatha and G. Mallesham

Chapter 17 Recent Advancements in Solar Thermal Technology for Heating and Cooling Applications 205

Sunita Mahavar

Chapter 18 Developments in Wide-Area Monitoring for Major Renewables: Wind and Solar Energy 227

S. Behera and B. B. Pati

Chapter 19 Solving Issues of Grid Integration of Solar and Wind Energy Models by Using a Novel Power Flow Algorithm 251

R. Satish, K. Vaisakh, and Almoataz Y. Abdelaziz

Chapter 20 Multifunctional PV-Integrated Bidirectional Off-Board EV Battery Charger Targeting Smart Cities 267

Rajesh Kumar Lenka, Anup Kumar Panda, Venkataramana Naik N, Laxmidhar Senapati, and Nishit Tiwary

Chapter 21 Integration of Wind, Solar, and Pumped Hydro Renewable Energy Sources in Rayalaseema Region: A Case Study 281

Y. V. Siva Reddy, T. Bramhananda Reddy, and E. Sanjeeva Rayudu

Chapter 22 Photovoltaic-Based Hybrid Integration of DC Microgrid into Public Ported Electric Vehicle 287

S. Pragaspathy, V. Karthikeyan, R. Kannan, N. S. D Prakash Korlepara, and Mr. Bekkam Krishna

Chapter 23 Battery Packs in Electric Vehicles 305

Antonio Peršić

Chapter 24 Alternative Wind Energy Turbines 323

Andrej Predin, Matej Fike, Marko Pezdevšek, and Gorazd Hren

Chapter 25 MPPT Controller for Partially Shaded Solar PV System 333

M. Subashini and M. Ramaswamy

Chapter 26 Adaptive Control of Smart Microgrid Using AI Techniques 349

Krishna Degavath and Mallesham Gaddam

Index 363

Editors' Biographies

Satish Kumar Peddapelli is professor in the Department of Electrical Engineering, University College of Engineering, Osmania University, Hyderabad, India. He has completed his BTech in EEE from JNTU, obtained his MTech in power electronics from JNTUH, and got his doctorate in the area of multilevel inverters in the year 2011 from JNTUH. His areas of interests are power electronics, drives, power converters, multilevel inverters, special machines, and renewable energy systems. He is a senior member of IEEE, life member of ISTE and FIE(I), and member of SSI. He completed two major research projects as the principal investigator funded by UGC and SERB, government of India. He is currently implementing Indo–Sri Lanka joint research project funded by DST, government of India. Under his supervision, four research scholars were awarded and eight are pursuing their PhD degrees. He established "Research Lab for Multilevel Inverters" at EED, UCE, OU. He has over 24 years of teaching and research experience and published more than 98 publications in international journals and international conferences. He received a "certificate of merit" for his research paper which was presented in an international conference at the University of California, USA. He has delivered keynote addresses in international conferences held at Singapore and Paris. He applied for three patents on cascaded multilevel inverters, neutral point clamped multilevel inverters, and power conditioners. He received the "Best Teacher Award" from the state government of Telangana, the "Award for Research Excellence," the "Global Teacher Role Model Award," and the "Fast Track Scheme for Young Scientist Award." Dr. P. Satish Kumar visited various countries like the United States of America, France, Switzerland, Japan, Hong Kong, Singapore, Bangkok, and Sri Lanka to present his research papers in various international conferences, to engage in collaborative research, and to deliver expert lectures and keynote addresses.

Peter Virtic was born in Slovenj Gradec, Slovenia, in 1979. He received a BS degree in electrical engineering in 2004 and a PhD degree in electrical engineering in 2009 from the University of Maribor, Faculty of Electrical Engineering and Computer Science. From 2004 to 2009, he was a researcher with the Development Centre for Electrical Machines. Since 2009, he has been an assistant professor with the Faculty of Energy Technology, University of Maribor, Slovenia. His research interests include permanent magnet machines, axial-flux machines, analytical and numerical modeling, and developing analytical tools for the analysis of electrical machines. His bibliography comprises 238 bibliographic units. He has national patent on "the system for smart home control using artificial intelligence." He received the award for the Best Student in Secondary School of Electrotechnics and Computing in Velenje; the Rector's Award for the best student in the generation of the Faculty of Electrical Engineering and Computer Science, University of Maribor; the award for the "Best Graduate Student" of study program in electrical engineering at the Faculty of Electrical Engineering and Computer Science; the award of the Polish Society of Theoretical and Applied Electromagnetics; the award of the President of the Polish Association of Applied Electromagnetics; the award for the "Best Poster Paper" at the 13th Conference of the IEEE CEFC 2008; the award at the Day of the Faculty of Electrical Engineering and Computer Science, University of Maribor; the award at the University of Maribor for extremely important successes and achievements in scientific research and educational field in the development of professional and personal contribution to the reputation of the Faculty of Energy Technology, University of Maribor; the award for top-rated teacher (the best grades in the student survey) at the Faculty of Energy Technology, University of Maribor, in the academic year 2013/2014 and 2014/2015; the award for top-rated teacher at the Faculty of Energy Technology, University of Maribor; and the ranking on top publications list (Metina lista) for published paper in applied energy. He is a member of IEEE and a member of the editorial board of the *Journal of Energy Technology*. He has worked on several EU-funded and national projects, such as the "L2–1180": National Agency for Research and Development (1.2.2008–30.1.2010) high-efficient hybrid synchronous permanent magnet motor; the "EN-DIFF": Energy2B_Inteligent Energy Europe Programme; the "ENERSCAPES": Territory, landscape, and renewable energy sources – MED Programme; the "MANERGY": Paving the way for self-sufficient regional energy supply based on sustainable energy concepts and renewable energy sources, Central Europe Programme; the "PV-NET": Promotion of PV energy through net metering optimization, MED Programme; the "TREND": Training for Renewable Energy Network Development, Erasmus+ Programme; the "ERESPLAN": Innovative Educational Tools for Energy Planning, Erasmus+ Programme; the "TOGETHER": Towards a Goal of Efficiency Through Energy Reduction, Central Europe Programme; the "STORES": Promotion of higher penetration of distributed PV through storage for all, MED Programme; the "IQ HOME," financed by the Ministry of Education, Science, and Sport; the "OSCI-GEN," financed by the Ministry of Education, Science, and Sport; and the "PAKT," smart devices and models in power grid, by SPIRIT Slovenia.

1 Investigation of Offshore Wind Energy Potential across Three Southern Coastal Regions in India

Saravana Venkatesh G, Amutha N, and N. Kumutha

CONTENTS

1.1 Introduction 1
1.2 Site Description 2
1.3 Weibull Distribution Function 3
    1.3.1 Weibull Distribution Function 3
    1.3.2 Most Probable and Most Energy-Carrying Wind Speeds 4
    1.3.3 Estimation of Weibull Shape k and Scale c Parameters 4
    1.3.4 WAsP Method 4
    1.3.5 Graphical Method 4
    1.3.6 Maximum Likelihood Method 4
    1.3.7 Prediction Model for Weibull Distribution Model 4
1.4 Results and Discussions 5
    1.4.1 Wind Speed Frequency Distribution 5
    1.4.2 Wind Rose Analysis 5
    1.4.3 Wind Rose Analysis 5
    1.4.4 Wind Power Class and Wind Power Density Analysis 5
    1.4.5 Wind Turbine Net Power Output (kW) 10
1.5 Conclusion 10
1.6 References 11

1.1 INTRODUCTION

Due to globalization, energy demand is increasing day by day in developed countries as well as in emerging economic and developing countries [1]. In recent years, due to limited usage of fossil fuel, importance has been given to renewable energy sources, like wind, solar, and tidal energy. Among these different kinds of energy sources, wind energy is one of the most popular renewable energy sources because of the many positive factors associated with it, such as its cleanliness, abundance, inability to produce any harmful gases, and also producing power at very attractive prices [2].

Utilizing wind sources for generating energy can be done with wind turbines. Locating the wind turbines either in onshore or offshore regions for estimating wind energy potential is the next challenging issue. The major cities of the world are located near coastal regions, so in order to prevent the requirement of longer transmission lines and provide feasibility in generating larger amount of energy in an economic manner, offshore wind turbines are suitable in such cases. Thus, in recent years, significant importance has been given to offshore wind energy potential across the world. Many research works have been carried out to estimate offshore wind energy potential across the world.

Normally, wind resources at offshore locations have lesser turbulence and low wind shear; this lesser turbulence leads to harvesting more energy from wind turbines effectively and also reducing fatigue loads on the turbines, which ultimately results in an increase in the lifetime of offshore wind turbines [3–5]. Also, rather than installing and locating onshore wind turbines near highly densely populated areas, offshore wind energy is an alternative choice [6].

India is the world's fourth largest in wind energy market, and the country's installed wind energy capacity is 32.8 GW at the end of 2017. Though the Ministry of New and Renewable Energy (MNRE), India, predicted that the onshore wind power installation at the end of year 2022 will be around 60 GW approximately [7], India is now preparing for offshore wind energy potential road map with the support of the Global Wind Energy Council (GWEC).

Murthy et al. [8] carried out a wind power potential assessment over the coastal region of Bheemunipatnam in northern Andhra Pradesh, India, and observed that the variations in mean, minimum, and maximum wind speed values are in the range of 4.41–5.61 m/s, 0.57–1.64 m/s, and 9.1–15.03 m/s, respectively. Serhat et al. [9] conducted a wind resource assessment of Izmit in the west Black Sea coastal region of Turkey and analyzed its turbulence intensity and economic analysis for constructing wind turbines in this region. Cumali et al. [10] studied the offshore potential of wind power in the coastal areas of Turkey. Sidi Mohammed Boudia et al. [11] evaluated the wind power potential at Oran, northwest of Algeria, and found the annual mean wind power density to be 129 (W/m^2). Allouhi et al. [12] investigated the wind energy potential in coastal locations in the Kingdom of Morocco and found the capacity factor, dominant wind direction, and wind speed variation with respect to height.

The main objective of this chapter is to evaluate the offshore wind energy potential from the historical time series data for offshore coastal regions across India, namely, Andaman, Kanayakumari, and Chennai. For this research work, 11 years' (2005–2015) time series wind speed (MEERA) datasets have been downloaded from the Windographer software. The MEERA datasets consist of the time stamp of each record, air temperature at 10 m, and wind speed and wind direction at 50 m hub height (m/s). In this research work, the crucial parameters related to wind characteristics, like Weibull probability distribution function, variation of Weibull parameters, wind speed and wind power density at different hub heights, wind direction, and wind power class, are determined.

This chapter is ordered as follows. Section 2 describes the offshore location. Section 3 describes Weibull distribution function and its parameter estimation methods. Section 4 presents wind speed frequency distribution analysis, wind rose analysis, and wind turbine power output analysis. Finally, Section 5 is the conclusion.

1.2 SITE DESCRIPTION

The three offshore regions of Andaman, Kanyakumari, and Chennai are considered for the study. These regions are chosen because Andaman (island) is surrounded by water in all its sides, while Kanyakumari and Chennai are surrounded by three sides and one side of water, respectively. Table 1.1 describes the site information in terms of their latitudes, longitudes, and altitudes. Figures 1(a), 1(b), 1(c) show their corresponding geographical locations (*source*: Google Maps).

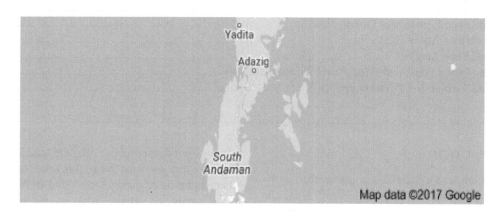

FIGURE 1.1(A) Andaman Google Map location.

FIGURE 1.1(B) Kanyakumari Google Map location.

FIGURE 1.1(C) Chennai Google Map location.

TABLE 1.1
Site Description Details

S. No.	Offshore Locations	Latitude (Degree)	Longitude (Degree)	Altitude (Meter)	Measurement Period
1	Andaman	11.5	92.5	79	2005–2015
2	Kanyakumari	8	77.5	30	2005–2015
3	Chennai	13	80	14	2005–2015

1.3 WEIBULL DISTRIBUTION FUNCTION

The Weibull distribution function is widely used in various engineering fields and statistical applications. It was discovered by Swedish mathematician Waloddi [15]. The accurate estimation of wind energy potential at a particular location for studying the wind speed frequency distribution is a vital parameter. If the wind speed frequency distribution for a particular site is known, then the wind power potential and its economic feasibility of installing the wind turbines can be evaluated easily [16]. Wind speed is an intermittent, random variable, and variation of wind speed over a period of time can be expressed by probability density functions [17]. There are numerous probability density functions, such as gamma, lognormal, three-parameter beta, Rayleigh, and Weibull distribution function [18]. Among these different distribution functions, the Weibull distribution function is the suitable and accepted function for a wide variation of wind speed data [19]. The Weibull distribution model has been recommended to find the wind characteristics at offshore location and is defined as [20]:

$$f(v) = \left(\frac{k}{c}\right)\left(\frac{v}{c}\right)^{k-1} \exp\left(\frac{-v}{c}\right)^k \quad v>0, k>0, c>0 \quad (1.1)$$

Where $f(v)$ is the probability of observed wind speed, v 'k' is the Weibull shape parameter (dimensionless), and 'c' is the Weibull scale parameter (m/s).

Weibull shape parameter 'k' is the width of the distribution, and scale parameter 'c' (m/s) is the average wind speed [14].

The Weibull cumulative density function can be derived from the probability density of Weibull distribution and can be expressed as [21]:

$$F(v) = 1 - \exp\left(\frac{-v}{c}\right)^k \quad (1.2)$$

Where $F(v)$ is the corresponding distribution function wind speed (v).

The mean wind power density can be derived from the probability density of Weibull distribution function and can be expressed as [22]:

$$WPD = \frac{1}{2}\rho c^3 \Gamma\left(1 + \frac{3}{K}\right) \quad (1.3)$$

Where ρ is the air density in Kg/m^3 and Γ is the gamma function.

1.3.1 WEIBULL DISTRIBUTION FUNCTION

The variation of wind speed over a height is called wind shear. Wind shear can be calculated by two methods, namely, the log law model and the power law model. When using the log law model, aerodynamic roughness is essential for calculating the wind speed and wave characteristics;

however, it is very difficult to calculate aerodynamic roughness [13]. For this reason, the simple power law model is applied for offshore-location wind power potential estimation. The power law model can be expressed mathematically as [23]:

$$\left(\frac{V_2}{V_1}\right) = \left(\frac{H_2}{H_1}\right)^{\alpha} \quad (1.4)$$

Where V_1 and V_2 are wind speeds for corresponding heights H_1 and H_2 in meters. The value of $\alpha = 0.14$ is recommended for offshore location [24].

1.3.2 Most Probable and Most Energy-Carrying Wind Speeds

Probable wind speed and maximum energy-carrying wind speed can be calculated using the Weibull shape 'k' and scale 'c' parameters. The most probable wind speed is a most frequent wind speed, and the most energy-carrying wind speed is wind speed which should have maximum wind energy. The most probable wind speed can be expressed as [25]:

$$V_{mp} = c\left(1 - \frac{1}{k}\right)^{1/k} \quad (1.5)$$

The maximum energy-carrying wind speed can be expressed as [25]:

$$V_{max} = c\left(1 + \frac{2}{k}\right)^{1/k} \quad (1.6)$$

1.3.3 Estimation of Weibull Shape k and Scale c Parameters

Several methods for estimating the shape parameter 'k' and scale parameter 'c' of Weibull distribution functions include the graphical method, maximum likelihood method, empirical method of Lysen, energy pattern factor method, equivalent energy method, probability weighted moments based on power density method, and Wind Atlas Analysis of Application Program (WAsP) method [26]. In this chapter, the Weibull parameters 'k' and 'c' are determined by the WAsP method, maximum likelihood method, and graphical method.

1.3.4 WAsP Method

The WAsP method is based on the wind flow model. The WAsP algorithm is used to fit the measured wind speed data. It can be expressed as [27]:

$$-\ln X = \Gamma\left(\frac{1}{K} + 1\right)^k \quad (1.7)$$

$$c = \sqrt[3]{\frac{\sum V_i^3}{N\Gamma\left(\frac{3}{K} + 1\right)}} \quad (1.8)$$

Where v_i is the mean wind speed.
Γ is the gamma function.

1.3.5 Graphical Method

The Weibull parameters 'k' and 'c' are determined by using the concept of least square regression principle. A straight line is fitted to time series wind speed data. Data are sorted into bins. It can be represented by [27]:

$$k = a; c = e^{\left(\frac{-b}{k}\right)} \quad (1.9)$$

Where a and b are the slope and intercept.

1.3.6 Maximum Likelihood Method

This method is used to determine the Weibull shape 'k' and scale 'c' parameters iteratively using the following expression [27]:

$$k = \left(\frac{\sum_{i=1}^{N} v_i^k \ln(v_i)}{\sum_{i=1}^{N} v_i^k} - \frac{\sum_{i=1}^{N} \ln(v_i)}{N}\right)^{-1} \quad (1.10)$$

$$c = \left(\frac{1}{N}\sum_{i=1}^{N} v_i^k\right)^{\frac{1}{k}} \quad (1.11)$$

Where N is the total number of wind speed measurement, V_i the measured wind speed value for the ith measurement.

1.3.7 Prediction Model for Weibull Distribution Model

The correlation coefficient R^2 value has been determined from equation 12 to find the best Weibull parameter method for calculating the wind speed frequency distribution data. The highest R^2 value determines the best method. It can be described as [28]:

$$R^2 = \frac{\sum_{i=1}^{N}(y_i - Z)^2 - \sum_{i=1}^{N}(x_i - Z)^2}{\sum_{i=1}^{N}(y_i - Z)^2} \quad (1.12)$$

Where y_i is the ith actual data, x_i is the predicted data, Z is the mean of actual data, and N is the number of observations.

1.4 RESULTS AND DISCUSSIONS

1.4.1 WIND SPEED FREQUENCY DISTRIBUTION

In this analysis of 11 years (2005–2011), MEERA datasets are taken into consideration. These datasets are downloaded from a wind navigator through the Windographer software. The three different methods, namely, WAsP algorithm, maximum likelihood method, and graphical method, are applied for measuring wind speed data for various locations that are tabulated. From the calculated values shown in Tables 1.2, 1.3, and 1.4, the WAsP algorithm shows a high correlation coefficient R^2 value, so it is considered the best method for measuring wind speed data. Also, wind speed frequency distribution analyses carried out at the hub height of 50 m are shown in Figures 1.2(a), 1.2(b), and 1.2(c) for the 11-year wind direction taken at (a) Andaman, (b) Kanyakumari, and (c) Chennai.

1.4.2 WIND ROSE ANALYSIS

The wind rose diagram graphically shows the distribution of wind speed and wind direction for a particular location in a certain period of time. Figures 1.3a, 1.3b, 1.3c show the prevailing wind direction for the hub height of 50 m at Andaman, Kanyakumari, and Chennai offshore locations to be 225°; diagram (b) shows the overall 11-year Kanyakumari offshore location data (2005–2015) and also represents wind direction at the hub height of 50 m. The dominant wind directions indicated in this diagram is 270° and 45°, respectively.

1.4.3 WIND ROSE ANALYSIS

Table 1.6 shows the annual variation of mean wind speeds, Weibull 'k', and Weibull 'c' parameters at three different offshore locations. The highest yearly value of mean wind speed is recorded at the Andaman location, as shown in Table 1.6 (2012, at 7.304 m/s), and the lowest value at the Chennai location (2010, at 4.082 m/s).

1.4.4 WIND POWER CLASS AND WIND POWER DENSITY ANALYSIS

Wind power classes are categorized into seven types of 1 to 7 rating from poor to superb with respect to wind power density and wind speed [29]. Table 1.7 represents a wind power class classification table. Each wind power class describes the various range of wind power density (W/m^2) and corresponding mean wind speeds. Wind power class 4 or higher is the most comfortable of wind turbine applications. Wind power class 3 has been recommended for wind power generation using taller wind turbines. Class 2 is described as marginal for wind power generation. Class 1 denotes unfit for wind power generation [30]. Wind power density is

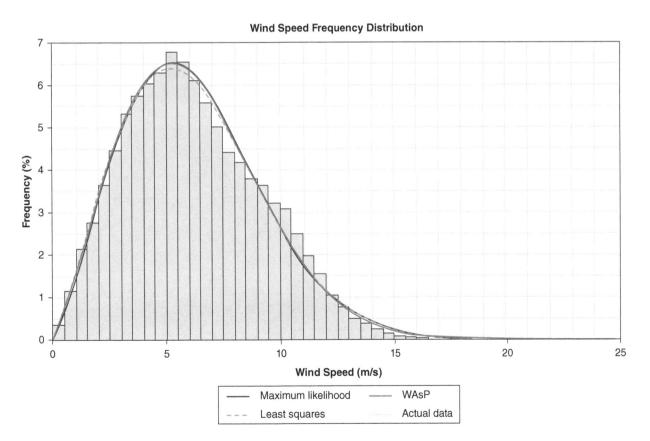

FIGURE 1.2A Wind speed vs. frequency distribution, Andaman location.

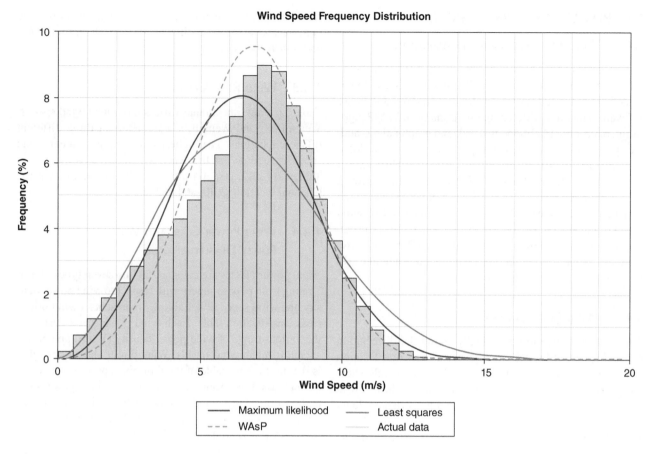

FIGURE 1.2B Wind speed vs. frequency distribution, Kanyakumari location.

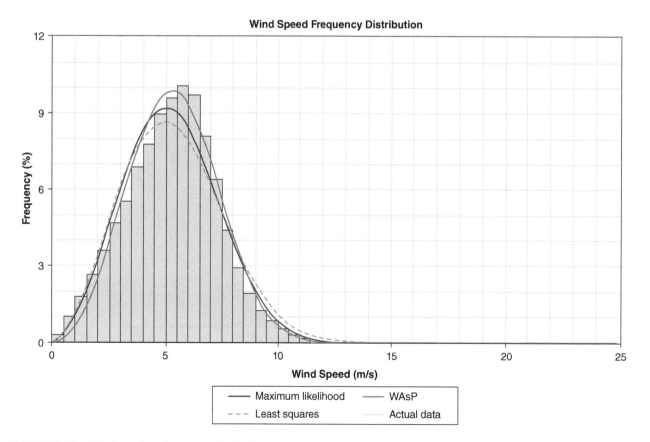

FIGURE 1.2C Wind speed vs. frequency distribution, Chennai location.

TABLE 1.2
Wind Speed Characteristics, Andaman Location, at 50 M Hub Height

Algorithm	Weibull k	Weibull c (m/s)	Mean Wind Speed (m/s)	Wind Power Density (W/m^2)	Correlation Coefficient (R^2)	Most Probable Wind Speed (V_{mp})	Max Energy-Carrying Wind Speed (V_{max})
WAsP algorithm	3.178	6.959	6.77	252.6	0.99318	5.24	15.03
Graphical method	3.126	7.008	6.206	261.5	0.99206	5.23	15.23
Maximum likelihood method	3.199	6.993	6.193	254.2	0.99167	5.3	15.05

TABLE 1.3
Wind Speed Characteristics, Kanyakumari Location, at 50 M Hub Height

Algorithm	Weibull k	Weibull c (m/s)	Mean Wind Speed (m/s)	Wind Power Density (W/m^2)	Correlation Coefficient (R^2)	Most Probable Wind Speed (V_{mp})	Max Energy-Carrying Wind Speed (V_{max})
WAsP algorithm	3.751	7.475	6.55	238.2	0.93926	6.88	13.23
Graphical method	3.034	7.462	6.624	277.4	0.88734	6.13	15.17
Maximum likelihood method	3.03	7.308	6.529	238.1	0.93341	6.4	*13.92*

TABLE 1.4
Wind Speed Characteristics, Chennai Location, at 50 M Hub Height

Algorithm	Weibull k	Weibull c (m/s)	Mean Wind Speed (m/s)	Wind Power Density (W/m^2)	Correlation Coefficient (R^2)	Most Probable Wind Speed (V_{mp})	Max Energy-Carrying Wind Speed (V_{max})
WAsP algorithm	2.608	6.024	4.43	132.9	0.98805	4.97	11.44
Graphical method	2.589	5.999	5.327	142.5	0.96554	5.29	12.12
Maximum likelihood method	2.761	5.932	5.28	132.9	0.97827	5.04	11.69

TABLE 1.5
Comparative Analysis of Wind Speed Characteristics at 50 M Hub Heights in 11-Year Time Period

S. No.	Wind Speed Characteristics	Andaman	Kanyakumari	Chennai
1	Wind power density (W/m^2)	252.6	238.2	132.9
2	Mean wind speed m/s	6.77	6.55	4.43
3	Most probable wind speed m/s	5.24	6.88	4.97
4	Max energy carrying wind speed m/s	15.03	13.23	11.44

wind power available per unit area and also determines wind resource available at a particular site. In this work, wind power density has been calculated during an annual time period at different hub heights at three coastal locations (Andaman, Kanyakumari, and Chennai), as tabulated in Table 1.8. Wind power densities calculated for hub height of 50 m are 252.6 (W/m²), 238.2 (W/m²), and 132.9 (W/m²), respectively, and according to wind power class analysis (Table 1.9), the corresponding wind power classes are identified as 2 (marginal), 2 (marginal), and 1 (poor), respectively.

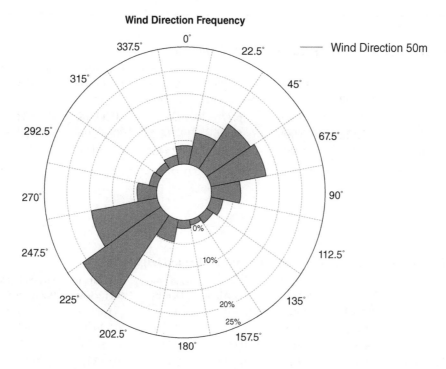

FIGURE 1.3A Wind rose diagram, Andaman.

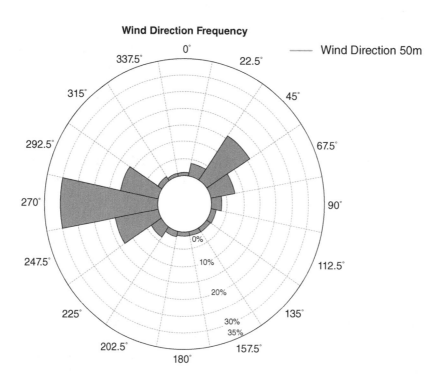

FIGURE 1.3B Wind rose diagram, Kanyakumari.

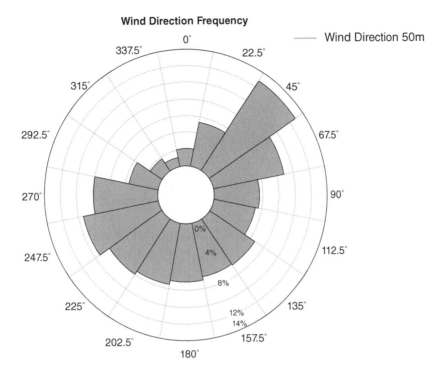

FIGURE 1.3C Wind rose diagram, Chennai.

TABLE 1.6
Annual Variation of Mean Wind Speed, Weibull k, and Weibull c Parameters

		Andaman			Kanyakumari			Chennai		
S. No.	Year	Mean Wind Speed (m/s)	Weibull k	Weibull c (m/s)	Mean Wind Speed (m/s)	Weibull k	Weibull c (m/s)	Mean Wind Speed (m/s)	Weibull k	Weibull c (m/s)
1	2005	6.825	3.285	7.611	6.505	5.852	6.994	4.454	2.701	5.007
2	2006	7.064	4.514	7.736	6.112	5.015	6.636	4.666	3.157	5.203
3	2007	6.319	3.253	7.052	6.292	4.832	6.837	4.372	2.861	4.896
4	2008	6.384	2.89	7.144	6.147	4.745	6.693	4.244	3.053	4.74
5	2009	7.126	3.927	7.885	6.629	4.805	7.216	4.686	2.898	5.247
6	2010	5.685	3.286	6.348	6.43	3.745	7.097	4.082	2.85	4.576
7	2011	7.262	4.699	7.914	6.805	5.562	7.344	4.568	3.186	5.095
8	2012	7.304	4.548	7.984	6.843	6.044	7.361	4.301	2.802	4.821
9	2013	7.026	4.698	7.67	7.14	4.979	7.763	4.692	2.879	5.251
10	2014	6.906	3.654	7.654	6.798	4.464	7.438	4.511	2.999	5.048
11	2015	6.627	3.406	7.389	6.349	4.541	6.943	4.163	2.557	4.683

TABLE 1.7
Classification of Wind Power Classes [31]

Wind Power Class	Description	Wind Power Density (W/m²)	Wind Speed (m/s)
1	Poor	<200	<5.6
2	Marginal	200–300	5.6/6.4
3	Fair	300–400	6.4/7.0
4	Good	400–500	7.0/7.5
5	Excellent	500–600	7.5/8.0
6	Outstanding	600–800	8.0/8.8
7	Superb	>800	>8.8

TABLE 1.8
Wind Power Density (W/m^2) at Different Offshore Locations

Height (m)	Andaman	Kanyakumari	Chennai
10	120	114	63
30	191	181	100
50	252.6	238.2	132.9
80	288	273	150

TABLE 1.9
Wind Power Class Variation of 10 M and 50 M Height at Different Offshore Location

Height (m)	Andaman	Kanyakumari	Chennai
10	1 (Poor)	1 (Poor)	1 (Poor)
50	2 (Marginal)	2 (Marginal)	1 (Poor)

TABLE 1.10
Offshore Wind Turbine Characteristics

S. No.	Wind Turbine Name	Rated Power (kW)	Cut-In Wind Speed (m/s)	Rated Wind Speed (m/s)	Cut-Out Wind Speed (m/s)	Rotor Diameter (m)	Swept Area (m²)
1	Gamesa G128–5 MW (80m)	4500	1	12	27	128	12,860
2	RE power 5M (85m)	5000	3.5	14.5	30	126	12,469
3	Gamesa G132–5 MW (95m)	5000	1.5	13	27	132	13,685
4	Vestas V112–3.0 MW (100m)	3450	3	13	25	112	9,852
5	Sinovel SL6000/128 (110m)	6000	3.5	13	25	128	12,868

TABLE 1.11
Annual Wind Turbine Output Power (kW) Generation at Different Offshore Locations

Wind Turbine	Andaman	Kanyakumari	Chennai
Gamesa G128–5 MW (80m)	1,295.30	1,164.80	676.50
RE power 5M (85m)	1,325.20	1,207.50	698.60
Gamesa G132–5 MW (95m)	1,465.20	1,294.30	774.20
Vestas V112–3.0 MW (100m)	1,051.10	902.90	563.40
Sinovel SL6000/128 (110 m)	1,419.90	1,301.40	708.70

1.4.5 Wind Turbine Net Power Output (kW)

In this chapter, five offshore commercial wind turbines are considered to estimate net power (kW). Table 1.10 shows the different offshore wind turbine characteristics. For the selected 11-year (2005–2015) time period in the Andaman offshore location, Table 1.11 shows that the Gamesa G 132–5 MW offshore wind turbine (95 m) has generated the highest net power of 1465.20 kW, and for the Chennai offshore location, Vestas V112–3.0 MW has produced the lowest amount of annual generated net power of 563.40 (kW).

1.5 CONCLUSION

In this chapter, 11-year historical time series wind speed and wind direction data, from 2005 to 2015, have been taken into consideration for analyzing three offshore coastal regions across India. Mean value of wind speed, Weibull wind speed frequency distribution, and Weibull shape k and scale c parameters have been determined. Prevailing wind direction has been analyzed, and dominant wind direction was plotted through wind rose diagrams. The results are summarized as following:

- Accordingly, the highest correlation coefficient (R^2) value WAsP algorithm best fit the measured wind speed time series data.
- Wind power density analysis at the height of 50 m at the Andaman location was recorded at 252.6 W/m^2, the Kanyakumari location at 238.2 W/m^2, and the Chennai location at 132.9W/m^2.
- Mean wind speed analysis at the height of 50 m was registered at the Andaman location at 6.77 m/s, Kanyakumari at 6.55 m/s, and Chennai at 4.43 m/s.
- Wind power class analysis at the height of 50 m wind power class represented were 2, marginal, for both Andaman and Kanyakumari locations, and 1, worst, for the Chennai location.
- Accordingly, wind power density analysis, mean wind speed analysis, and wind power class analysis at both the Andaman and Kanyakumari offshore locations were identified for offshore wind turbine installation.
- Most probable wind speed and maximum energy-carrying wind speed were calculated at all three offshore locations.
- Yearly mean wind speed analysis registered highest in Andaman at 7.304 m/s and lowest in Chennai at 4.082 m/s.
- The Gamesa G 132–5 MW offshore wind turbine (95 m) has generated the highest amount of annual net energy output (1,465.20 kW) at the Andaman offshore location.

1.6 REFERENCES

[1] www.iea.org.
[2] Waewask J, Landry M, Gagnon Y. Offshore wind power potential of Gulf of Thailand. *Renewable Energy* 2015, 81; 609–626.
[3] Bilgili M, Yasar A, Simsek E. Offshore wind power development in Europe and its comparison with onshore counterpart. *Renewable and Sustainable Energy Reviews* 2011, 15; 905–915.
[4] Estban MD, Diez JJ, Lopez JS, Negro V. Why offshore wind energy? *Renewable Energy* 2011, 36; 444–450.
[5] Sethu Raman S, Raynor GS. Comparison of mean wind speeds and turbulence at a coastal site and offshore location. *Applied Meteorological* 2011, 36; 15–21.
[6] Nagababu G, Simha R, Naidu NK, Kachhwaha SS, Savsani V. Application of OSCAT satellite data for offshore wind power potential assessment of India. *Energy Procedia* 2016, 90; 89–98.
[7] www.gwec.net.
[8] Murthy KSR, Rahi OP. Preliminary assessment of wind power potential over the coastal region of Bheemunipatnam in northern Andhra Pradesh, India. *Renewable Energy* 2016, 99; 1137–1145.
[9] Kucukali S, Dinckal C. Wind energy resource assessment of Izmit in the West Black Sea Coastal Region of Turkey. *Renewable and Sustainable Energy Reviews* 2014, 30; 790–795.
[10] IIkilic C, Aydin H, Wind power potential and usage in the coastal regions of Turkey. *Renewable and Sustainable Energy Reviews* 2015, 44; 78–86.
[11] Boudia SM, Guerri O. Investigation of wind power potential at Oran, northwest of Algeria. *Energy Conversion and Management* 2015, 105; 81–92.
[12] Allouhi A, Zamzoum O, Islam MR, Saidur R, Kousksou T, Jamil A, Derouich. Evaluation of wind energy potential in Morocco's coastal regions. *Renewable and Sustainable Energy Reviews* 2017, 72; 311–324.
[13] Shu ZR, Li QS, Chan PW. Investigation of offshore wind energy potential in Hong Kong based on Weibull distribution function. *Applied Energy* 2015, 156; 362–373.
[14] Karthikeya BR, Prabal Negi S, Srikanth N. Wind resource assessment for urban renewable energy application in Singapore. *Renewable Energy* 2016, 87; 403–414.
[15] Nedaei M, Assareh E, Walsh PR. A comprehensive evaluation of the wind resource characteristics to investigate the short term penetration of regional wind power based on different probability statistical methods. *Renewable Energy* 2018, 128; 362–378.
[16] Coelingh JP, Van Wijk AJM, Hoitslag AAM. Analysis of wind speed observations over the North Sea. *Wind Engineering Industrial Aerodynamics* 1996, 61; 362–373.
[17] AL-Yahyai S, Charabi Y, Gastli A, Al-Alawi S. Assessment of wind energy potential locations in Oman using Data from existing weather stations. *Renewable and Sustainable Energy Reviews* 2010, 14; 1428–1436.
[18] Li J, Yu Xiong (Bill). Lidar technology for wind energy potential assessment: Demonstration and validation at a site around Lake Erie. *Energy Conversion and Management* 2017, 144; 252–261.
[19] Bagiorgas HS, Giouli M, Rehman S, Al-Hadhrami LA. Weibull parameters estimation using four different methods and most energy-carrying wind speed analysis. *International Journal of Green Energy* 2011, 8; 529–554.
[20] Akdag SA, Guler O. A novel energy pattern factor method for wind speed distribution parameter estimation. *Energy Conversion and Management* 2015, 106; 1124–1133.
[21] Akdag SA, Dinler A. A new method to estimate Weibull parameters for wind energy applications. *Energy Conversion and Management* 2009; 50:1761–1766.
[22] Khahro SF, Tabbassum K, Soomro AM, Dong L, Liao XZ. Evaluation of wind power production prospective and Weibull parameter estimation methods for Babaurband, Sindh Pakistan. *Energy Conversion and Management* 2014, 78; 956–967.
[23] Safari B, Gasore J. A statistical investigation of wind characteristics and wind energy potential based on the Weibull and Rayleigh models in Rwanda. *Renewable Energy* 2010, 35; 2874–2880.
[24] Islam MR, Saidur R, Rahim NA. Assessment of wind energy potentiality at Kudat and Labuan, Malaysia using Weibull distribution function. *Energy* 2011, 36; 985–992.
[25] Ozay C, Celiktas MS. Statistical analysis of wind speed using two-parameter Weibull distribution in Alacati region? *Energy Conversion and Management* 2016, 121; 49–54.
[26] Baseer MA, Meyer JP, Rehman S, Mahbub Alam Md. Wind power characteristics of seven data collection sites in Jubali, Saudi Arabia using Weibull parameters. *Renewable Energy* 2017, 102; 35–49.
[27] Solyali D, Altunc M, Tolun S, Aslan Z. Wind resource assessment of Northern Cyprus. *Renewable and Sustainable Energy Reviews* 2016, 55; 180–187.
[28] Kaplan YA. Determination of the best Weibull methods for wind power assessment in the southern region of Turkey. *IET Renewable Power Generation* 2017, 11; 175–182.
[29] Irwanto M, Gomesh N, Mamat, Yusoff MRYM. Assesment of wind power generation potential in Perlis. *Malasiya. Renewable and Sustainable Energy Reviews* 2014, 38; 296–308.

[30] Ouammia A, Dagdougui H, Sacile R, Mimet A. Monthly and seasonal assessment of wind energy characteristics at four monitored locations in Liguria region (Italy). *Renewable and Sustainable Energy Reviews* 2010, 14; 1959–1968.

[31] Jang J.-K., Yu B.-M., Ryn K.-W., Lee J.-S. Offshore wind resource assessment around Korean peninsula by using QuikSCAT satellite data. *Journal of the Korean Society for Aeronautical & Space Sciences* 2009, 37(11); 1121–1130.

2 Power Quality Enhancement of Fixed- and Variable-Speed WEGS Using HSAPF Based on 5-Level Cascaded Multilevel Inverter and Fuzzy Logic Controller

Seema Agrawal and Mahendra Kumar

CONTENTS

2.1 Introduction 13
2.2 Proposed System Configuration 14
    2.2.1 The Stand-Alone Wind Energy Generation System 14
        2.2.1.1 Wind Turbine and Drivetrain Modeling 14
        2.2.1.2 PMSG Dynamic Modeling 15
        2.2.1.3 Switch-Mode Rectifier MPPT Algorithm...16
    2.2.2 Shunt Active Power Filter and Its Architecture 16
        2.2.2.1 SAPF Control Strategy 16
            2.2.2.1.1 Reference Current Generation 16
            2.2.2.1.2 Cascaded MLI and Switching Pulse Generation 17
            2.2.2.1.3 Intelligent DC Bus Voltage Regulation...17
2.3 Results and Discussions..19
2.4 Conclusion 25
2.5 References...25

2.1 INTRODUCTION

The proliferation of atmospheric gases' intensity as blazing of fossil fuel is an accredited global warming fact. Remote-area wind energy generation systems (WEGS) turn out to be the best emulating and nature-friendly proficient elucidations for the electrification of distant community consumers [1].

Variable-speed WEGS are mostly centered on doubly fed induction generators (DFIGs) or permanent-magnet synchronous generators (PMSGs) with power electronics interface. The weighty gearbox requires frequent upkeep of WEGS due to faults with DFIGs [2–3]. Therefore, WEGS with direct-drive PMSGs gained more deliberation due to its self-excitation, high power factor, and proficient operation [4–7]. Optimum wind energy utilization from the wavering wind under variable-speed operation is accomplished by numerous best tracking schemes, for example, maximum torque/power tracking (OT&PSF), perturb and observe (P&O), etc. [8]. However, switch-mode rectifier (SMR) is more superior as it is a simple and economic output power-boosting scheme without the necessity of wind speed sensor for small-scale wind turbine [9].

Remote-area power system control under unexpected conditions such as wind speed variations in the vast era of penetrating power electronics–based nonlinear load discrepancies is a demanding problem [10–11]. Such discrepancies may increase power quality problem like voltage and frequency variations. Additionally, solid-state power electronics–based load crops harmonics contain current that indirectly add disturbance as voltage and current disruption at PCC, capacitor blustering, intense neutral currents, destabilization of DC link voltage. Therefore, SAPF is strongly used in the power system to cancel out adverse impacts of harmonics produced by power electronic converters–based nonlinear load [12–15].

Inverter control with a balanced source signal is a key question in a stand-alone-mode operation. Likewise, blackout, random variations in source voltage, flickers, noise, notches, harmonic, and voltage unbalance are key power quality concern issues that are obtained in WECS. Undesirable harmonics signals (voltage/current) are produced due to power electronics interface as controlled/uncontrolled rectifier, inverter, and DC–DC converter between generator and load. Power quality issues cannot be accepted by user and so need remedy.

Conventional passive harmonic filters can be applied to eliminate upper-level harmonics with slow response; presently, active harmonic filters have the capability to perform harmonic filtering over a broad operating range of fundamental frequency (50Hz). The hybrid active power filtering has become more popular in the present scenario due to a substantial price fall in power electronics semiconductors and signal-processing devices.

After seeing the limitations of the converters and higher power demands, there was requirement for a new type of inverters. The device which came up to meet the demands was an inverter which has various levels in their output; these are known as multilevel inverters. These inverters utilize medium-powered semiconductor switches and capacitor voltage sources. They are arranged in an array structure generating an output which is a stepped waveform. The more the number of steps, the more sinusoidal the waveform will look [16–18].

MLIs are diode-clamped, flying capacitor (FC), and cascaded H-bridge multilevel inverter (CHB-MI). The cascaded multilevel inverter is the newest member of the MLI family. The cascaded multilevel inverter is an ideal solution for power circuit topology because of lower-order harmonic elimination in its output voltage signal without enhancing the switching frequency or diminishing the inverter power output [19–20]. With an increase in the number of H-bridge level, the output voltages have more steps, thus resembling a sinusoidal wave and reducing harmonics.

Owing to the advantages of cascaded multilevel inverter over conventional two-level inverters, such as source current with low distortion, better harmonic reduction, and low voltage stress on load, the proposed work is based on a shunt active power filter using CMLI topology. The cascaded five-level inverter is controlled through the linear PWM technique [21–22]. The multilayered output waveform resembles a sinusoidal waveform, hence reduces the harmonic content in the system [23–24].

APF performance is affected by precision and time required in harmonic compensation. Literature survey includes compensation methods as Id–Iq instantaneous current concept, P–Q active reactive power scheme, SRF technique, and self-tuning filter [25–27]. One other scheme for reference current generation via unit vector formation for shunt APF using synchronizing PLL is applied in this book chapter [28–31].

Sudden fall/increase exists in a DC-side inverter voltage. It is the sign of ripples and babbling as a slow dynamic response of DC link voltage regulation with conventional controllers-based SAPF. As a result, controlling the DC-side inverter voltage and upholding it steadily have a major role in SAPF current harmonics elimination ability. Hence, DC link voltage management via nonconventional soft computing algorithm–based controllers as an artificial neural network, sliding mode, imprecise inputs–based fuzzy logic controller (FLC), genetic algorithm, and heuristic algorithm–tuned conventional controller are utilized for better gating pulse generation to lessen ripple and switching losses [32–34].

This chapter will explore and analyze harmonic content level in PMSG-based WECS with hybrid active power filter assembly. It is applicable to mitigate current harmonics and indirectly plays a role in the life span improvement of PMSG. Lower active power losses will show more electrical power extraction, improving the overall WECS competence as a result. The implementation of HSAPF will enhance WEGS functionality over a wide wind speed range. Positive sequence detection using a phase-locked loop is used to reference current generation. Work with imprecise inputs property of FLC controller is applied for DC link voltage stabilization in this chapter. In this system, various nonlinear loads are connected, like uncontrolled converter with R–L load, controlled converter, and AC–DC–DC converter generating current harmonic, which is then reduced by a compensating current produced from CMLI SAPF, and hence increasing the efficiency and providing better current harmonic reduction as compared to conventional VSI as per IEEE standards.

2.2 PROPOSED SYSTEM CONFIGURATION

The suggested assembly has a wind energy generation system (WEGS) based on the permanent magnet self-excited generator (PMSG) with hybrid SAPF for harmonics elimination, owing to being thereof three-phase diode-based uncontrolled and controlled rectifier feeding (R, L) load. A series of passive components–based static filter to compensate for voltage harmonic distortion exists in the wind sinusoidal AC voltage output. The cascaded five-level MLI-based shunt active power filter enhances current signal quality by bringing in the same magnitude and reversing current signal. The SAPF control assembly is based on unit vector creation via fundamental frequency-locking property-based phase-locked loop for reference current production and carrier wave PWM method for pulse-making fed cascaded MLI with self-supported bus. Artificial controller handled nonlinearity without mathematical modeling-based fuzzy logic controller (FLC) is utilized as DC bus voltage management by elimination of active power losses.

2.2.1 THE STAND-ALONE WIND ENERGY GENERATION SYSTEM

The wind energy generation system is connected to controlled and uncontrolled rectifier feeding with resistive and inductive load. It is depicted in Figure 2.1. WEGS is made by a variable-speed wind turbine, which drives PMSG. This segment is devoted to WEGS system mathematical modeling.

2.2.1.1 Wind Turbine and Drivetrain Modeling

WEGS converts available kinetic energy from the wind into mechanical energy on the PMSG shaft. Power is taken out by wind turbine rotor, expressed as:

$$P_{wt} = \frac{1}{2} A \rho C_P(\beta,\lambda) v_w^3 \quad (2.1)$$

Where, $A, \rho, v_w,$ and C_p are swept area (m²), air density (kg/m³), wind velocity (m/s), and power coefficient, respectively.

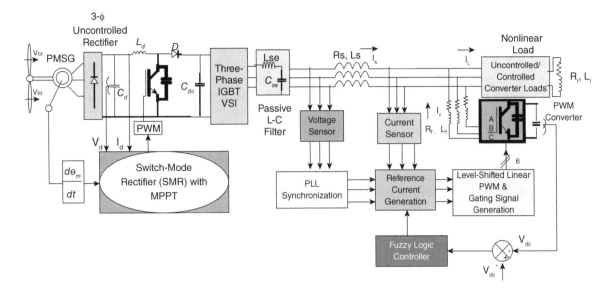

FIGURE 2.1 Suggested architecture.

The tip speed ratio (λ) defines the ratio of blade tips' linear speed to the wind speed as:

$$\lambda = \frac{\omega_{wt} R}{v_w} \quad (2.2)$$

Where ω_{wt} and R are mechanical turbine rotor (rad/s) and blade radius (m), respectively.

In accumulation, the turbine rotor coefficient $C_p(\lambda, \beta)$ shows power-capturing capability of wind turbine, which is a nonlinear function of pitch angle of blades (β) and tip speed ratio (λ). In this study, we assume fixed pitch rotor as angle (β) is considered zero, then the relation between C_p and λ is depicted in Figure 2.2 and expressed by using the following empirical equation:

$$C_p(\beta, \lambda) = 0.5 \left(\frac{116}{\lambda_x} - 0.4\beta - 5 \right) \exp(-21/\lambda_x) \quad (2.3)$$

Where $\lambda_x = \left(\frac{1}{\lambda + 0.08\beta} - \frac{0.035}{\beta^3 + 1} \right)^{-1}$.

A more efficient two-mass drivetrain model senses wind turbine dynamics. The following numerical equations govern the model:

$$2H_{wt} \frac{d\omega_{wt}}{dt} = \Gamma_{wt} - \Gamma_{sh}$$

$$\frac{1}{\omega_{elb}} \frac{d\theta_{sta}}{dt} = \omega_{wt} - \omega_r$$

$$2H_g \frac{d\omega_r}{dt} = \Gamma_{wt} - \Gamma_g$$

Where H_{wt} and H_g are wind turbine and PMSG inertia constants, respectively. θ_{sta}, ω_{wt}, ω_r, and ω_{elb} are shaft twisting angle, wind turbine angular speed in p.u, PMSG rotor speed in p.u, and electrical base speed, respectively.

Shaft torque Γ_{sh} is written as follows:

$$\Gamma_{sh} = K_{sh} \theta_{sta} + D_{wt} \frac{d\theta_{sta}}{dt} \quad (2.4)$$

Here, K_{sh} is shaft stiffness and D_{wt} is wind turbine damping constant.

The maximum wind turbine can be received when turbine functions at maximum $C_p(C_{p\_opt})$. So it is compulsory to correct the rotor speed at the optimum value of the tip speed ratio (λ_{opt}) with different values of wind speeds.

Then, the wind turbines' optimized power can be written as follows:

$$P_{wt\_opt} = \frac{1}{2} A\rho C_{p\_opt} \left(\frac{\omega_{wt\_opt} R}{\lambda_{opt}} \right)^3 = \alpha_{opt} \left(\omega_{wt\_opt} \right)^3 \quad (2.5)$$

Where $\alpha_{opt} = \frac{1}{2} A\rho C_{p\_opt} \left(\frac{R}{\lambda_{opt}} \right)^3$.

Additionally, the optimum torque can be calculated as:

$$\Gamma_{wt\_opt} = \alpha_{opt} \left(\omega_{wt\_opt} \right)^2 \quad (2.6)$$

2.2.1.2 PMSG Dynamic Modeling

The dynamic model of PMSG is generally used, in which stator voltages of PMSG are described in d–q frame by the following dynamic equations:

$$V_{sd} = R_s i_{sd} + p(L_d i_{sd} + \psi_m) - P\omega_{wt}(L_q i_{sq}) \quad (2.7)$$

$$V_{sq} = R_s i_{sq} + p(L_q i_{sq}) + P\omega_{wt}(L_d i_{sd} + \psi_m) \quad (2.8)$$

Where R_s is the stator winding resistance, i_{sd} and i_{sq} are the d–q axis currents respectively, L_d and L_q are the d–q axis inductances respectively, p is the differential operator, P is the pole pairs, ω_{wt} is the rotor speed (mech. rad/s), and ψ_m is the permanent magnet generated flux linkage.

If PMSG is surface-mounted, then we can assume $L_d = L_q$. Therefore, the electromagnetic torque is given as:

$$\Gamma_e = \frac{3}{2} P \psi_m i_{sq} \qquad (2.9)$$

2.2.1.3 Switch-Mode Rectifier MPPT Algorithm

The generated power output using variable-speed PMSG is not appropriate to use because of consistent changes in wind speed, which affect its magnitude and frequency. Therefore, reduced switch three-phase uncontrolled bridge rectifier is put into application to transform variable output generator voltage into a fixed-DC voltage, and after that, it is converted to AC-voltage waveform by a power electronics inverter.

The reduced SMR comprises a three-phase uncontrolled bridge rectifier and DC–DC converter. The SMR output is controlled by controlling an active switch (IGBT) through its duty cycle to take out the highest power at any value of wind speed from the wind turbine so that it can be supplied to loads.

The structure of the proposed MPPT algorithm is presented through the control block, as shown in Figure 2.2, and generator speed v/s torque is depicted in Figure 2.3.

2.2.2 SHUNT ACTIVE POWER FILTER AND ITS ARCHITECTURE

The shunt APF can be applied to execute current harmonics compensation present in the source current. The shunt APF compensation principle is equal amplitude but reverse load current harmonic introduction at PCC, which lies between the source and the load. A positive sequence detection scheme via PLL is applied in a 5L-CMLI-based SAPF for reference current generation. Self-supported DC bus voltage profile regulation is obtained by a fuzzy logic controller. The subsequent subsection explains the detailed description of the proposed SAPF control system constituents.

The shunt active power filter (SAPF) is coupled parallel at the point of common coupling (PCC) between WEGS and the controlled and uncontrolled converter-based nonlinear load. The APF architecture is shown in Figure 2.1. It is composed of a current-controlled cascaded multilevel voltage source inverter as a power circuit, owing to the advantages of a cascaded multilevel inverter over conventional two-level inverters, such as source current with low distortion, better harmonic reduction, and low voltage stress on load. So in the proposed work, the shunt active power filter based on MLI topology is modulated and simulated. Five-level CMLI is controlled through linear PWM; this multilayered output waveform resembles a sinusoidal waveform, hence reduces the harmonic content in the system. This inverter comprises of IGBT/diode switches and capacitors, which are connected in a way to give the required results. The inverter used for simulation in the proposed work is a five-level cascaded MLI which is connected in a star connection. Like any other inverter or converter, this MLI also has two sides, that is, one side is the AC side, from where the output is supplied at PCC to reduce the harmonics, while the other side is the DC side, having a capacitor which is used for the generation of PWM. CMI utilizes two separate DC sources per phase to generate an output voltage with five levels.

2.2.2.1 SAPF Control Strategy

Compensation performance is dependent on the reference current generation technique and DC self-supported bus voltage profile (V_{dc}) regulation techniques.

2.2.2.1.1 Reference Current Generation

The reference current signal for the proposed system is generated using an indirect current control methodology. A synchronizing vector waveform is achieved by a phase-locked loop (PLL), as illustrated in Figure 2.1. FLC output is measured as amplitude of 3-φ reference current signal.

The three-phase unit vector waveform signals are U_{sa}, U_{sb}, U_{sc}, then unit vector (U) is defined as follows:

$$U = \begin{bmatrix} U_{sa} \\ U_{sb} \\ U_{sc} \end{bmatrix} = \begin{bmatrix} \sin(\omega t) \\ \sin(\omega t - 120^0) \\ \sin(\omega t + 120^0) \end{bmatrix} \qquad (2.10)$$

The 3-φ reference source current signal is expressed as:

$$\begin{bmatrix} i_{sa}^* \\ i_{sb}^* \\ i_{sc}^* \end{bmatrix} = \begin{bmatrix} U_{sa} \\ U_{sb} \\ U_{sc} \end{bmatrix} \begin{bmatrix} i_{sp} \end{bmatrix} \qquad (2.11)$$

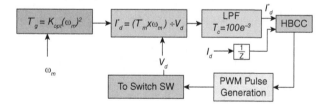

FIGURE 2.2 Block diagram of MPPT.

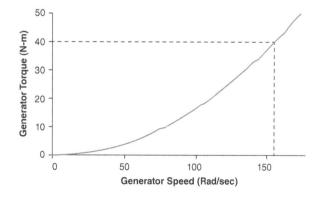

FIGURE 2.3 Graph between generator speed and torque.

2.2.2.1.2 Cascaded MLI and Switching Pulse Generation

CMLI is controlled through linear PWM. This multilayered output waveform resembles a sinusoidal waveform, hence reduces the harmonic content in the system. The three-phase five-level cascaded MLI is illustrated in Figure 2.4.

Switching gating pulse for cascaded multilevel voltage source inverter is generated via comparing actual and reference source currents' waveform. Actuating error signals flowing through levels shifted in-phase disposition (IPD) linear PWM produce the needed gating signals, owing to quick prevails over rapid current transitions. Pulses produced by the hysteresis band are always vibrating in nature, which, as a result, produces gate circuitry losses, so linear PWM is preferred. If the amplitude of the reference wave is more than the carrier wave, then the PWM will be generated, as shown in Figure 2.5. In in-phase disposition linear PWM, all carriers are in-phase. The five-level MLI requires four carrier signals. The five-level CMLI is controlled through linear PWM; this multilayered output waveform resembles a sinusoidal waveform, hence reduces the harmonic content in the system. The v_{ref} represents reference wave, while $v_{cr1}, v_{cr2}, v_{cr3}$ and v_{cr4} represent the triangular carrier signals, as shown in Figure 2.5. Similarly, the switching signals for other leg b and c of the five-level CMI are created via their corresponding actual and reference filter currents.

2.2.2.1.3 Intelligent DC Bus Voltage Regulation

Soft computing artificial intelligent FLC is utilized as a self-supported DC bus voltage profile regulation by compensating active power system losses. The DC bus capacitor voltage profile is observed and matched with the DC-link reference voltage signal for sensed error (E) and change in error (CE). Now, these are defined in the following equations:

$$E(t) = V_{dc}^* - V_{dc}(t)$$
$$CE(t) = E(t) - E(t-1) \quad (2.12)$$

The error $E(t)$ and change in error $CE(t)$ at t sampling time are applied as inputs for the FLC inference system. The intelligent controller response is I_{sp} and expressed as:

$$I_{sp} = I_{sp}^*(t-1) + \delta I_{sp}^*(t) \quad (2.13)$$

Real power load demand and losses in VSI are looked out by the aforementioned current I_{sp}. The gating waveforms for CMI are produced via variation of existent and source currents reference waveform. Error signal is gone with level

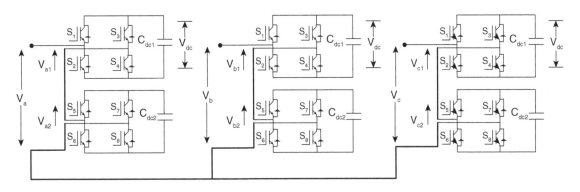

FIGURE 2.4 Five-level CMLI.

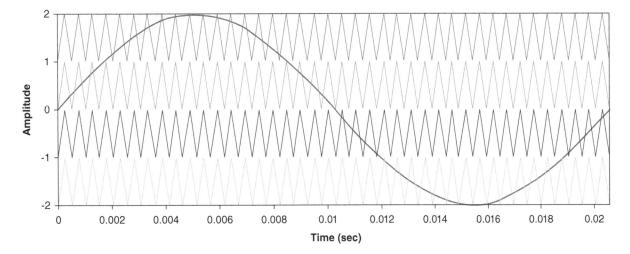

FIGURE 2.5 Level shifted (PD) PWM for 5-level cascaded MLI.

shifted (IPD) linear PWM to create the required gating pulses.

Designed FLC rules are described in fuzzy rule matrix. It is shown in Table 2.1. The FLC inputs membership degree (E, CE) are illustrated in Figure 2.6(a) and 2.6(b). Its output membership degree is demonstrated in Figure 2.6(c). MATLAB/Simulink rule editor is shown in Figure 2.7 as per designed rule base. Figure 2.8 shows the complete block of SAPF control strategy via fuzzy logic controller.

TABLE 2.1
Fuzzy Logic Rule Matrix

E/CE	NB	NM	NS	ZE	PS	PM	PB
NB	NB	NB	NB	NB	NM	NS	ZE
NM	NB	NB	NB	NM	NS	ZE	PS
NS	NB	NB	NM	NS	ZE	PS	PM
ZE	NB	NM	NS	ZE	PS	PM	PB
PS	NM	NS	ZE	PS	PM	PB	PB
PM	NS	ZE	PS	PM	PB	PB	PB

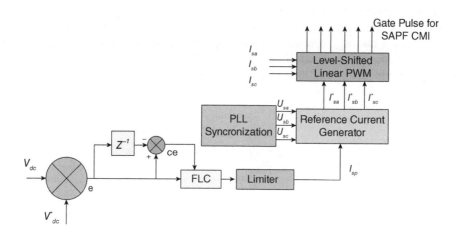

FIGURE 2.6 Block diagram of SAPF control scheme using FLC.

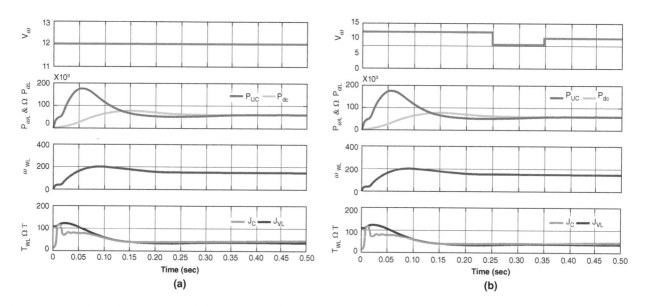

FIGURE 2.7 Dynamic response of wind energy generation system for fixed and variable speed.

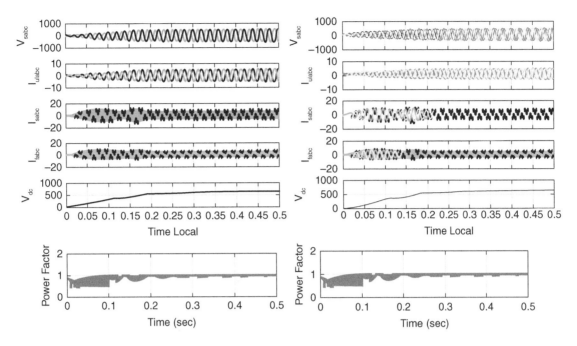

FIGURE 2.8 Performance of SAPF based on VSI with fixed and variable speed with uncontrolled converter load.

2.3 RESULTS AND DISCUSSIONS

System architecture of WECA-HSAPF using PLL is shown in Figure 2.1. The simulation study has been done to show the effectiveness of the proposed small-scale fixed-/variable-speed WEGS using HSAPF CMI based on positive sequence detector PLL strategy and DC voltage regulation using FLC. Maximum power from WECA can be extracted using SMR MPPT. HSAPF is formed using series passive (L-C filter) and shunt active power filter (SAPF). A series passive filter eliminates voltage harmonics, and SAPF eliminates current harmonics injected by the nonlinear load. Additionally, it can be seen that the positive sequence detection method based on PLL is used as unit vector generation for reference current generation of SAPF gating signals. The robustness of the system is represented in order to reduce the harmonic currents produced by uncontrolled and controlled converter feeding with (R_l, L_l) load under fixed- and variable-speed consideration.

The overall system architecture is simulated using MATLAB/Simulink 2015(a) platform. The system data used in the simulation study are given in Table 2.2. A hysteresis band current controller generates pulses after 0.1 sec. Before the shunt APF operation between $T_1 = 0$ sec and $T_2 = 0.1$ sec, the source current is highly distorted and rich on harmonics, out of phase with the source voltage, and the power factor is poor, with high consummation of reactive power. The SAPF starts the compensation process instantly ($T_2 = 0.1$ sec) when it is connected to the nonlinear load. The source current after compensation is practically sinusoidal and in phase with the corresponding source voltage.

Figures 2.9(a) and 2.9(b) give the simulation response of WECA for fixed wind speed $v_w = 12$ m/s and step change in wind speed from $v_w = 12$ m/s to 8 m/sec to 10 m/sec to 12 m/sec, respectively. It can be seen that generator electrical power output follows turbine mechanical input at a level of 6 kW. The turbine rotor rotates at a speed of 152.89 rad/sec, and T_e follows T_m quite well at a level of 40 N-m, that is, the DC current regulates turbine torque to extract peak power from wind turbines.

Simulation results for SAPF based on VSI under fixed and variable speed with uncontrolled converter feeding R–L load are shown in Figures 2.10(a) and 2.10(b). The three supply voltage (V_{sabc}), load current (I_{ulabc}), source current (I_{sabc}), compensation current (I_{fabc}), DC link voltage (V_{dc}), and power factor waveforms are depicted from top to bottom.

The APF is switched on at $T_1 = 0.1$ sec; the load current is balanced, having a magnitude of 6.11 A with 28.32% THD, as shown in Figure 2.11(a). It can be observed from Figure 2.11(b) that supply current is sinusoidal due to the compensation provided by the SAPF after 0.1 sec. The level and THD of supply current are 7.134A and 3.95%, as shown in Figure 2.11(b).

Current amplitude levels with performance parameter THD for variable speed are also illustrated in Figures 2.12(a) and 2.12(b) before and after filter insertion.

Simulation performances for SAPF based on VSI under fixed and variable speed with controlled converter feeding R–L load are depicted in Figures 2.13(a) and 2.13(b). The three supply voltage (V_{sabc}), load current ($I_{conlabc}$), source current (I_{sabc}), compensation current (I_{fabc}), DC link voltage (V_{dc}), and power factor waveforms are shown from up to down.

The APF is switched on at $T_1 = 0.1$ sec; the load current is balanced, having a magnitude of 6.096 A, with 28.56% THD, as shown in Figure 2.14(a). It can be observed from Figure 2.14(b) that supply current is sinusoidal due to the compensation

TABLE 2.2
System Simulation Parameters

Source	Parameter/Nomenclature	Value
Wind turbine	Density of air (ρ)	1.225 kg/m^3
	Area swept by rotor blades (A)	1.06 m^2
	Optimum coefficient (K_{opt})	1.67* 10^{-3} nm/ (rad/sec)^2
	Base wind speed	12 m/sec
Drivetrain	Turbine inertia constant (H_{wt})	4 sec
	Generator inertia constant (H_g)	0.4 sec
	Shaft stiffness coefficient (K_{sh}) Damping coefficient (D_t)	0.3 p.u./ el. rad
		0.7 p.u./el. rad.
PMSG	Pole pairs (p)	5
	Nominal speed (wg)	153 rad/sec
	Nominal current	12 A
	Armature phase resistance (R_s)	0.425 Ω
	Flux linkages	0.433 Wb
	Stator inductance ($L_d = L_q$)	8.4 mH
	Nominal torque (T_e)	40 N/m
	Nominal power	6 KW
HSAPF/PLL	Series passive filter (L_{se}, C_{se})	60 mH, 440μF
	DC link capacitance (C_{dc})	1500 μF
	Reference DC link voltage (V_{dc}^*)	700 volts
	Filter resistance and inductance (R_f, L_f)	0.5Ω, 3.35 mH
	PLL parameters (K_P, K_I)	0.18, 1
Loads	Three-phase uncontrolled and controlled converter load (R_l, L_l)	100 Ω, 100 mH
Simulation time	0 to 0.5 sec	

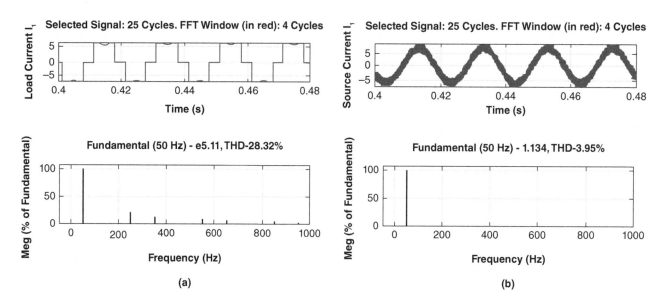

FIGURE 2.9 THD analysis before and after compensation for phase 'a' fixed speed with uncontrolled converter load.

Power Quality Enhancement of Fixed- and Variable-Speed WEGS

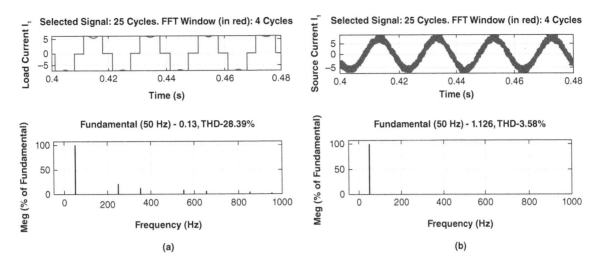

FIGURE 2.10 THD analysis before and after compensation for phase 'a' variable speed with uncontrolled converter load.

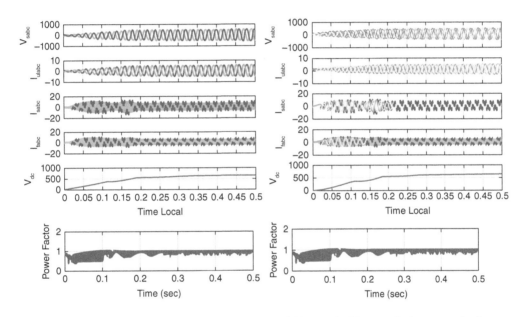

FIGURE 2.11 Performance of SAPF based on VSI with fixed and variable speed with controlled converter load.

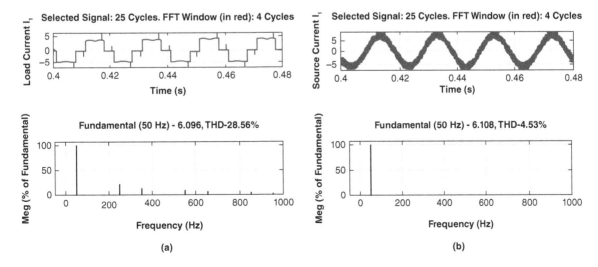

FIGURE 2.12 THD analysis before and after compensation for phase 'a' fixed speed with controlled converter load.

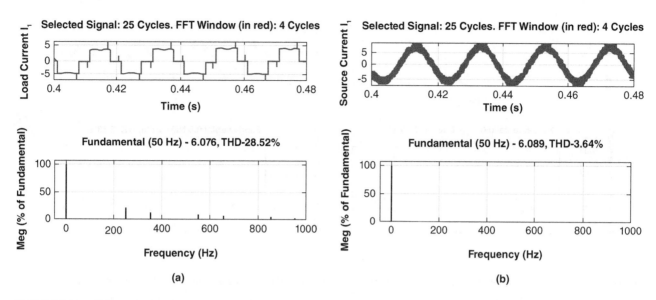

FIGURE 2.13 THD analysis before and after compensation for phase 'a' variable speed with controlled converter load.

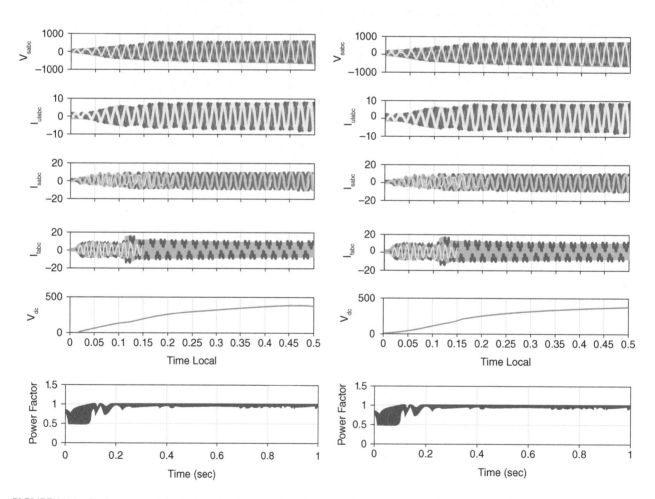

FIGURE 2.14 Performance of SAPF based on 5L-CMLI with fixed and variable speed with uncontrolled converter load.

provided by the SAPF after 0.1 sec. The level and THD of supply current are 6.168A and 4.53%, as shown in Figure 2.14(b) for fixed speed. Current amplitude level with performance parameter THD for variable speed is also examined in Figures 2.15(a) and 2.15(b) before and after filter application. THD is lying as per IEEE standard 3.64% for variable speed.

Simulation performances for SAPF based on 5L-CMI under fixed and variable speed with uncontrolled converter

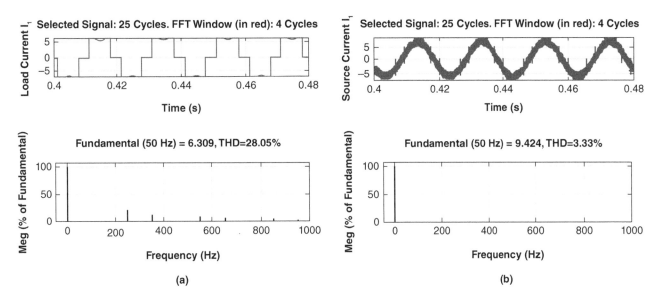

FIGURE 2.15 THD analysis before and after compensation for phase 'a' fixed speed with uncontrolled converter load.

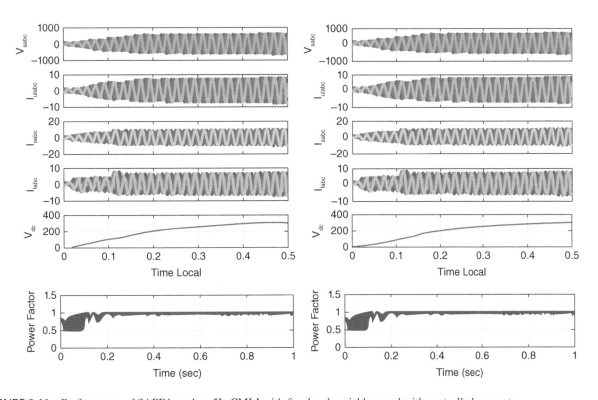

FIGURE 2.16 Performance of SAPF based on 5L-CMLI with fixed and variable speed with controlled converter.

feeding R–L load are depicted in Figures 2.12(a) and 2.12(b). The three supply voltage (V_{sabc}), load current (I_{ulabc}), source current (I_{sabc}), compensation current (I_{fabc}), DC link voltage (V_{dc}), and power factor waveforms are represented in Figures 2.13(a) and 13(b).

The APF is switched on at the same time as before case; the load current is balanced, having a magnitude of 6.309 A, with 28.05% THD. It can be observed from Figure 2.14(a) that the supply current is nonsinusoidal. It becomes sinusoidal after 0.1 sec compensation provided by the SAPF. Supply current THD is 3.33%, as examined in Figure 2.15(b), for fixed speed, and 3.39% for variable speed, as in Figure 2.16(b).

Simulation performances for SAPF based on 5L-CMI under fixed and variable speed with controlled converter feeding R–L load are depicted in Figures 2.16(a) and 2.16(b). The three supply voltage (V_{sabc}), load current (I_{ulabc}), source current (I_{sabc}), compensation current

(I_{fabc}), DC link voltage (V_{dc}), and power factor waveforms are represented in Figure 2.16(a) and 2.16(b) from top to bottom.

This control strategy is very effective for current harmonic compensation while maintaining 0.99 power factor.

The power quality index (THD) parameter before compensation is 28.05%, as illustrated in Figure 2.20(a). After compensation, supply current THD is 2.09%, as examined in Figure 2.17(b) for fixed speed, and 1.81% for variable speed, as in Figure 2.18(b)

Figures 2.19(a) and 2.19(b) show % THD of voltage before and after compensation. It also stands under IEEE 519 standards, and obtained voltage is perfectly sinusoidal.

Table 2.3 illustrates THD for the different case studies. The system performance in terms of eliminating harmonics is very acceptable. The THD values obtained in all case studies for current compensation are under IEEE standards 519–1992 (THD ≤ 5%). Fuzzy controller maintained DC link voltage constant with zero overshoot. Power factor and THD performance are better for 5L-CMI-based SAPF as compared to VSI-based SAPF.

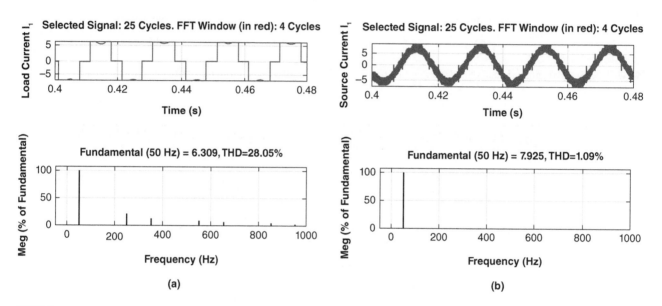

FIGURE 2.17 THD analysis before and after compensation for phase 'a' fixed speed with controlled converter load.

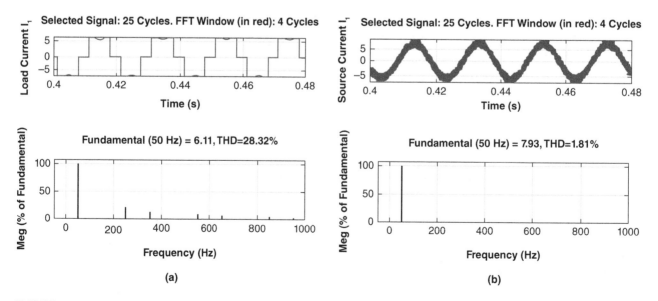

FIGURE 2.18 THD analysis before and after compensation for phase 'a' variable speed with controlled converter load.

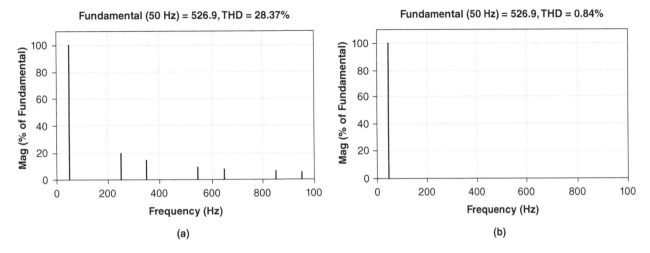

FIGURE 2.19 FFT voltage spectrum analysis before and after filter.

TABLE 2.3
Result Summary (% THD) of Various Case Studies

Case Study	Before Compensation (%THD)		After Compensation (%THD)	
Speed	Fixed Speed	Variable Speed	Fixed Speed	Variable Speed
VSI uncontrolled converter load	28.32	28.39	3.95	3.58
VSI controlled converter load	28.56	28.52	4.53	3.64
Cascaded 5L-MLI with uncontrolled converter load	28.05	28.01	3.33	3.39
Cascaded 5L-MLI with controlled converter load	28.05	28.32	2.09	1.81

2.4 CONCLUSION

In this chapter, the five-level H-bridge inverter-based shunt active filter was designed and verified using MATLAB software. Level shifted PWM techniques were used to give the gating signals. It was found that as the number of levels increased, the output was found to have fewer harmonics and its shape more resembled a perfect sine wave. The unit vector current control using PLL control strategy was used to produce the reference currents for compensation. Maintaining the DC-link capacitor voltage constant was the main task in designing the filter and getting a harmonic-free source current. An intelligent fuzzy controller is employed to get the filter working without any overshoot. Using a comparator, all the reference compensation currents were compared with those of actual filter currents. Errors obtained were subjected to PWM using a triangular carrier signal. The THD was measured for all types of connected load, which are the R–L load feeding uncontrolled converter load, thyristor-based controlled converter load with 45° switching. Without a filter in the system, the total harmonic distortion generated is very high. However, with the implementation of cascaded multilevel inverter-based shunt active power filter, the total harmonic distortion in the system was reduced to as per IEEE 919 standard. This decrease in the harmonic content shows that the C5LI-based SAPF is working efficiently and is a useful tool to increase the efficiency of the system. It has been shown that a single filter can be designed for all the loads, controlled as well as uncontrolled. Voltage performance in terms of quality index parameter THD has been improved using filter.

2.5 REFERENCES

[1] "Global wind report: Annual market update", Published by global wind energy council (GWEC). Accessible from www.gwec.net, accessed on May 2019.
[2] H. Polinder, F. F. Van, G. J. De Vilder, and P. J. Tavner, "Comparison of direct-drive and geared generator concepts for wind turbines", *IEEE Transactions on Energy Conversion*, vol. 21, no. 3, pp. 725–733, 2006.
[3] A. Beainy, C. Maatouk, N. Moubayed, and F. Kaddah, "Comparison of different types of generator for wind energy conversion system topologies", *IEEE International Conference on Renewable Energies for Developing Countries (REDEC)*, pp. 1–6, 2016.
[4] M. Fatu, L. Tutelea, I. Boldea, and R. Teodorescu, "Novel motion sensor less control of standalone permanent magnet synchronous generator (PMSG): Harmonics and negative sequence voltage compensation under nonlinear load", *IEEE European Conference on Power Electronics and Applications*, pp. 1–10, 2007.

[5] C. N. Bhende, S. Mishra, and S. G. Malla, "Permanent magnet synchronous generator-based standalone wind energy supply system", *IEEE Transactions on Sustainable Energy*, vol. 2, no. 7, pp. 361–373, 2011.

[6] K. R. Sekhar, R. Barot, P. Patel, and N. V. Kumar, "A novel topology for improved DC bus utilization in PMSG based wind energy generation system", *IEEE International Conference on Renewable Energy Research and Applications (ICRERA)*, pp. 525–530, 2015.

[7] A. Nasr and M. El-Hawary, "Harmonics reduction in a wind energy conversion system with a permanent magnet synchronous generator", *International Journal of Data Science and Analysis*, vol. 3, no. 6, pp. 58, 2017.

[8] J. P. Ram, N. Rajasekar, and M. Miyatake, "Design and overview of maximum power point tracking techniques in wind and solar photovoltaic systems: A review", *Renewable and Sustainable Energy Reviews*, vol. 73, pp. 1138–1159, 2017.

[9] D. M. Whaley, W. L. Soong, and N. Ertugrul, "Investigation of switched mode rectifier for control of small-scale wind turbines", *IEEE Conference Annual Meeting*, pp. 2849–2856, 2005.

[10] X. Liang, "Emerging power quality challenges due to integration of renewable energy sources", *IEEE Transactions on Industry Applications*, vol. 53, no. 2, pp. 855–866, 2017.

[11] N. Golovanov, G. C. Lazaroiu, M. Roscia, and D. Zaninelli, "Power quality assessment in small scale renewable energy sources supplying distribution systems", *Energies*, vol. 6, no. 2, pp. 634–645, 2013.

[12] Y. Oguz, I. Guney, and H. Çalik, "Power quality control and design of power converter for variable-speed wind energy conversion system with permanent-magnet synchronous generator", *The Scientific World Journal*, pp. 1–14, 2013.

[13] J. Tsai and K. Tan, "H APF harmonic mitigation technique for PMSG wind energy conversion system", *IEEE Australasian Universities Power Engineering Conference (AUPEC)*, pp. 1–6, 2007.

[14] P. Acuna, L. Moran, M. Rivera, J. Dixon, and J. Rodriguez, "Improved active power filter performance for renewable power generation systems", *IEEE Transactions on Power Electronics*, vol. 29, no. 2, pp. 687–694, 2014.

[15] S. H. Qazi and M. W. Mustafa, "Review on active filters and its performance with grid connected fixed and variable speed wind turbine generator", *Renewable and Sustainable Energy Reviews*, vol. 57, pp. 420–438, 2016.

[16] J. Rodriguez, J. S. Lai, and F. Z. Peng, "Multilevel inverters: A survey of topologies, controls and applications", *IEEE Transactions on Industrial Electronics*, vol. 49, no. 4, pp. 724–738, 2002

[17] M. Grath, B. Peter, and D. G. Holmes, "Multicarrier PWM strategies for multilevel inverters", *IEEE Transactions on Industrial Electronics*, vol. 49, pp. 858–867, 2002.

[18] S. Kathalingam and P. Karantharaj, "Comparison of multiple carrier disposition PWM techniques applied for multilevel shunt active filter", *Journal of Electrical Engineering*, vol. 63, pp. 261–265, 2012.

[19] S. X. Du, J. J. Liu, and J. L. Lin, "Hybrid cascaded H-bridge converter for harmonic current compensation", *IEEE Transactions on Power Electronics*, vol. 28, pp. 2170–2179, 2013.

[20] Y. Suresh and A. K. Panda, "Investigation on hybrid cascaded multilevel inverter with reduced dc sources", *Renewable Sustainable Energy Review*, vol. 26, pp. 49–59, 2013.

[21] A. Hoseinpour, S. M. Barakati, and R. Ghazi, "Harmonic reduction in wind turbine generators using a shunt active filter based on the proposed modulation technique", *International Journal of Electrical Power & Energy Systems*, vol. 43, pp. 1401–1412, 2012.

[22] P. Palanivel and S. S. Dash, "Analysis of THD and output voltage performance for cascaded multilevel inverter using carrier pulse width modulation techniques", *IET Power Electronics*, vol. 4, no. 8, pp. 951–958, 2011.

[23] S. Jain and V. Sonti, "A highly efficient and reliable inverter configuration based cascaded multilevel inverter for PV systems", *IEEE Transactions on Industrial Electronics*, vol. 64, no. 4, pp. 2865–2875, 2017.

[24] M. Malinowski, "A survey on cascaded multilevel inverters", *IEEE Transactions on Industrial Electronics*, vol. 57, no. 7, pp. 2197–2206, 2010.

[25] S. Chourasiya and S. Agrawal, "A review: Control techniques for shunt active power filter for power quality improvement from non-linear loads", *International Electrical Engineering Journal*, vol. 6, no. 10, pp. 2028–2032, 2015.

[26] M. A. M. Moftah, G. E. S. A. Taha, and E. N. A. Ibrahim, "Active power filter for variable-speed wind turbine PMSG interfaced to grid and non-linear load via three phase matrix converter", *Eighteenth International Middle East Power Systems Conference (MEPCON)*, pp. 1013–1019, 2016.

[27] A. Oruganti and S. R. Prasad, "Real-time control of hybrid active power filter using conservative power theory in industrial power system", *IET Power Electronics*, vol. 10, pp. 196–207, 2017.

[28] S. Agrawal, D. K. Palwalia, and M. Kumar," Performance analysis of Ann based three-phase four-wire shunt active power filter for harmonic mitigation under distorted supply voltage conditions", *IETE Journal of Research*, vol. 68, no. 1, pp. 566–574, 2019.

[29] Seema Agrawal, Vijay Kumar Gupta, D. K. Palwalia, and R. K. Somani, "Power quality improvement of standalone wind energy generation system for non-linear load," *2nd IEEE International Conference on Power Electronics, Intelligent Control and Energy Systems (ICPEICES-2018)*, October 2018.

[30] S. Agrawal, C. Sharma, and Dinesh Birla, "Enhanced PLL (EPLL) Synchronization and HBCC Controlling of Grid-Interactive (PV-SOFC) hybrid generating system," *Indian Journal of Science and Technology*, vol. 14, no. 2, pp. 154–169, 2021. DOI:10.17485/IJST/v14i2.2079

[31] K. G. Masoud, S. A. Khajehoddin, P. K. Jain, and A. Bakhshai, "Derivation and design of in-loop filters in phase-locked loop systems," *IEEE Transactions on Instrumentation and Measurement*, vol. 61, no. 4, pp. 930–940, 2012.

[32] S. Mikkili and A. K. Panda, "Real-time implementation of PI and fuzzy logic controllers based shunt active filter control strategies for power quality improvement", *International Journal of Electrical Power & Energy Systems*, vol. 43, no. 1, pp. 1114–1126, 2012.

[33] Seema Agrawal, Seemant Chourasiya, and Dheeraj Kumar Palwalia, "Performance measure of shunt active power filter applied with intelligent control technique", *Journal of Power Technologies*, vol. 100, no. 3, pp. 272–278, 2020.

[34] M. Suleiman, "Modified synchronous reference frame based shunt active power filter with fuzzy logic control pulse width modulation inverter", *Energies*, vol. 10, p. 758, 2017.

3 Forecasting of Wind Power Using Hybrid Machine Learning Approach

Mahaboob Shareef Syed, Ch.V. Suresh, B. Sreenivasa Raju, M. Ravindra Babu, and Y. S. Kishore Babu

CONTENTS

3.1 Introduction: Background and Driving Forces 27
3.2 Wind Power 27
3.3 Forecasting Methods 28
3.4 Persistent Extreme Learning Machine Algorithm (PELM) 28
3.5 Methodology for Forecasting Wind Speed 29
3.6 Results and Analysis 30
3.7 Summary 34
3.8 References 34

3.1 INTRODUCTION: BACKGROUND AND DRIVING FORCES

In the present electric world, most of the power is generated through fossil fuels only. The main disadvantages of these fuels are, the resources may get exhausted at any time, the cost involved in procuring and processing these fuels is huge, and these also lead to pollution of the environment.

At this situation, power producers globally are looking for an alternate source of energy. The only acceptable solution they have arrived at is utilization of nonconventional sources of energy, that is, renewable energy sources (RES). The advantages which can be drawn from these sources are, these are available free of cost, are inexhaustible, are available in abundance, do not at all pollute the environment, and finally, are a clean form of energy. But the main problem that arises with RES is their uncertain and intermittent nature of power generation. In such condition, it is not suggestible to integrate RES with the electric grid. Therefore, it is very essential to know the actual available generation very accurately by the RES prior to the integrating of the grid. For the proper planning and a secure operation of power system, it is inevitable to have information of generation by RES.

In this context, the forecasting of available wind speed in a particular area has gained much attention of power producers. It necessitates a sophisticated algorithm for accurate forecasting. Therefore, in this chapter, a hybrid machine learning algorithm is proposed for forecasting of wind speed.

Combining the features of conventional persistent algorithm and extreme learning machine (ELM) algorithm, a new topology is developed in this chapter. This hybrid machine learning algorithm is known as persistent extreme learning machine algorithm (PELM).

At first, the developed PELM network is trained with the obtained historical data. After testing the accuracy of training, the same is used to forecast wind speed. For measuring the accuracy of the training, two metrics, root mean square error (RMSE) and mean absolute error (MAE), are used. Later, forecasting is carried out over short-term time periods, which involves the forecasting duration of some minutes to hours, and also long-term time periods, which involves forecasting duration of some days or weeks to a few months. Forecasting is also carried out with respect to the seasons. In summer, winter, and rainy seasons, forecasting is done with the proposed hybrid machine learning algorithm.

3.2 WIND POWER

Wind turbines generate electrical power with the help of energy stored in wind. The kinetic energy (KE) in wind is primarily converted to rotational KE and then to electrical energy.

Mathematically, the KE stored in a mass m (kg) moving with a velocity v (m/s) is given by:

$$KE = \frac{1}{2} m v^2 \qquad (3.1)$$

The rate of change of KE is power stored in the wind (P_{Wind}).

$$P_{Wind} = \frac{d(KE)}{dt} = \frac{1}{2} v^2 \frac{dm}{dt} \qquad (3.2)$$

But

$$\frac{d(m)}{dt} = \rho A \frac{d(l)}{dt} \qquad (3.3)$$

Where ρ is air density = 1.225 kg/m³, A is area swept by the turbine blade in m², and l is the distance moved in meters.

$$\frac{d(m)}{dt} = \rho A v \quad (3.4)$$

$$P_{Wind} = \frac{1}{2}\rho A v^3 \quad (3.5)$$

Betz's law [1] demonstrates that the maximum power that can be extracted from the wind is independent of the design of a wind turbine in open flow. It states that no wind turbine can extract more than 59.3% of the KE. This is denoted with a factor C_p. Wind power expression can be modified as:

$$P_{Wind} = \frac{1}{2}\rho A v^3 C_p \quad (3.6)$$

From equation 1.6, it can be observed that the wind power generated is directly proportional to cube of the velocity of wind. Any small change in velocity also reflects huge considerable change in wind power generated; therefore, it is very much essential to estimate wind velocity very accurately to obtain the available wind power.

3.3 FORECASTING METHODS

To overcome the uncertainty that arises due to the chaotic behavior of wind speed, sophisticated forecasting methodologies are indeed necessary. It makes a very comfortable and secure operation of the power system. Various researchers have thrown a light on accurate forecasting of wind by reviewing enormous literature. The authors of [2–6] have presented a review of various forecasting algorithms. These algorithms have gotten much attention in terms of forecasting time horizons that depend on the considered electrical application. Broadly, these methods are categorized into persistence method, physical methods, statistical methods, artificial intelligence–based approaches, and hybrid approaches.

The persistence approach is a very basic and naive approach. It assumes that the situation at $(t + 1)^{th}$ time instant is similar to the tth instant. It's very much simple to implement and mostly suitable for forecasting in very short periods of time scales. But this is not suitable for longer periods of forecasting time horizons. The accuracy of forecasting is not guaranteed.

In physical methods, the forecasting procedure involves the calculation of wind speed or power based on weather parameters, such as temperature, pressure, terrain effect, smoothness of the surface, and obstacles on the way of passage of wind. The only drawback with these methods is, it involves high computational procedures.

The statistical method of forecasting involves using historical data stored in metrological stations. The fundamental technique is time series, and later, regression methods are also proposed. These methods comprise auto regressive (AR), auto regressive moving average (ARMA), and auto regressive integrated moving average (ARIMA) methods, which find their application in various engineering problems. The problem faced with these methods is, forecasting will not be accurate as prediction time horizons enlarges.

In the present engineering world, artificial intelligence (AI) methods find their contribution in achieving accurate solutions in various complicated fields. With the advancements in AI, various new algorithms such as artificial neural networks (ANN), fuzzy logic, adaptive neuro-fuzzy inference systems (ANFIS), and support vector machines are proposed, developed, and proved to be very accurate in forecasting wind power. An enormous volume of work has been carried out for forecasting wind speed using AI techniques.

The hybrid methods are formulated by combining any of these algorithms. The benefits obtained through hybrid methods are, they retrieve the advantages of individual algorithm and develop a most suitable algorithm which leads to a globally acceptable solution. Hybrid algorithms can be developed by combining the features of physical and AI methods, statistical and AI, or any other advancement.

In this chapter also, a novel hybrid algorithm is proposed to forecast the available wind speed. A persistent extreme learning machine (PELM) algorithm is developed by combining the features of both persistence and extreme learning machine algorithm (ELM).

3.4 PERSISTENT EXTREME LEARNING MACHINE ALGORITHM (PELM)

The persistent algorithm is a fundamental and basic technique which is most suitable for short-term time horizons rather than long-term time periods. The ELM algorithm is an AI-based algorithm. Traditionally, it is one input, one output, with a single hidden layer feed-forward neural network. The ELM was developed to overcome the drawbacks of overfitting, overtraining, and converging at local optima, which very often occur in ANN.

To make use of the prominent features of both persistent and ELM algorithms, a hybrid algorithm is proposed in this chapter by combining the aforementioned two, and it is named persistent extreme learning machine algorithm (PELM). The structure of PELM is similar to that of ELM, but the process of training the network is a little bit different. The structure of PELM is demonstrated in Figure 3.1.

There are S number of samples considered, and each sample of pattern will be formatted as $(x_i, x_{i+1})_{i=1}^{S-1}$. It indicates that for the input x_i, x_{i+1} is the output for that particular input. Where $x_i \in R^S$, $x_i = [x_{i1}, x_{i2}, x_{i3}, \ldots, x_{i(S-1)}]^T$, and $x_{i+1} = [x_{(i+1)1}, x_{(i+1)2}, x_{(i+1)3}, \ldots, x_{(i+1)(S-1)}]^T$, these matrices consist of a collection of attributes. The mathematical equation which describes the PELM with h number of neurons in hidden layer and the activation function $F(.)$ is presented as:

$$E(x_j) = \sum_{i=1}^{h}\beta_i F(\alpha_i \bullet x_j + b_i) \quad (3.7)$$

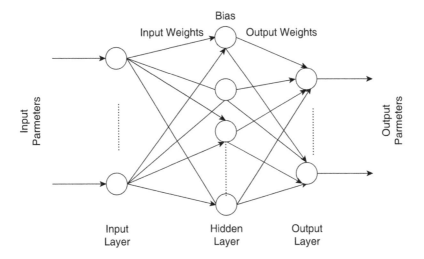

FIGURE 3.1 Architecture of PELM.

Equation 1.7 is denoted in the matrix notation as:

$$F * \beta = E \quad (3.8)$$

The activation function F in the matrix notation can be presented as:

$$F = \begin{bmatrix} F(\pm_1 x_1 + b_1) & F(\pm_2 x_1 + b_2) & \cdots & F(\pm_h x_1 + b_h) \\ \cdot & \cdot & \cdot\cdot & \cdot \\ \cdot & \cdot & \cdot\cdot & \cdot \\ \cdot & \cdot & \cdot\cdot & \cdot \\ F(\pm_1 x_{S-1} + b_1) & F(\pm_2 x_{S-1} + b_2) & \cdots & F(\pm_h x_{S-1} + b_h) \end{bmatrix}_{(S-1)\S h} \quad (3.9)$$

And E is the target matrix, which can be represented as:

$$E = [x_2, x_3, x_4, \ldots, x_s]^T$$

After randomly initializing input weights and bias factors for the network, the output layer weights are analytically calculated. The cost function to be minimized is:

$$E_{BP} = \sum_{j=1}^{S} \left[\sum_{i=1}^{h} \beta_i F(\alpha_i * x_j + b_i) - x_{j+1} \right]^2 \quad (3.10)$$

Finally, the output weight matrix will be computed with the Moore–Penrose method as:

$$\beta = F^+ X \quad (3.11)$$

3.5 METHODOLOGY FOR FORECASTING WIND SPEED

The present work is carried out on real-time Andhra Pradesh state system. A few areas are identified which are highly windy in nature. These are the possible areas for wind plant installation. For the installation and production of wind power, there should be a minimum wind speed required. Depending on the data provided by New and Renewable Energy Development Corporation of Andhra Pradesh (NREDCAP) [7], a state government–owned body, the areas are selected for future wind speed forecasting. The general steps to be followed for forecasting wind speed are given here:

Step 1: Identify the areas to be forecasted using the data provided by NREDCAP.
Step 2: Collect the historical weather data of the selected areas.
Step 3: Segregate the data into testing and training clusters.
Step 4: Train the developed network model with the training data.
Step 5: After confirming the accuracy, forecast the wind speed for short-term and as well as long-term time period horizons.
Step 6: The forecasting needs to be carried out in all the selected areas.

To validate the accuracy of the proposed PELM methodology, two metrics are considered: root mean square error (RMSE) and mean absolute error (MAE).

$$MAE = \frac{1}{S} \sum_{i=1}^{S} (V_{Wi}^{Forecasted} - V_{Wi}^{Actual}) \quad (3.12)$$

$$RMSE = \sqrt{\frac{1}{S} \sum_{i=1}^{S} (V_{Wi}^{Forecasted} - V_{Wi}^{Actual})^2} \quad (3.13)$$

Where, $V_{Wi}^{Forecasted}$ is the forecasted value of wind speed, V_{Wi}^{Actual} the actual value of wind speed of ith sample.

3.6 RESULTS AND ANALYSIS

Analysis is carried out by forecasting the wind speed in the area of Guntur, located in the Andhra Pradesh state of India. At first, wind speed is forecasted for short-term time duration; later, it is extended to long-term time horizons. Wind speed is strongly dependent on weather parameters, which include temperature and pressure. Therefore, for the winter season, meteorological historical data was collected on the date of January 1, 2020. Data patterns were collected for 24 hours a day; each sample at 2 minutes results into 720 samples. Within this, 70% of data, that is, 504 samples, are utilized to train the proposed network, and the remaining 216 data samples were used for testing. Each data sample contains pressure in mbar, temperature in °C, and wind speed for the present instant, which are given as the input parameters for the network. The output parameters contain temperature, pressure, and wind speed for the next instant. The same thing was carried out for the summer season by considering data on the day of May 1, 2020. Similarly, for the rainy season, October 1 data was considered.

To validate the forecasted wind speed, the support of hardware setup also was considered. Figure 3.2 shows the hardware setup, which includes an anemometer and a data logger. The forecasted values with the proposed algorithm are compared with that of the values recorded by the hardware setup. The validated results for all the three seasons are shown in Figure 3.3 for summer, winter, and rainy seasons.

From the obtained results of forecasting, it is evident that the proposed algorithm predicts very nearby values which can be compared with that of hardware results obtained. Further, the obtained forecasting results are also compared with that of the algorithms which were already proposed in the previous literature. For measuring accuracy, the mathematical metrics RMSE and MAE are calculated and compared, which are shown in Table 3.1.

From Table 3.1, MAE and RMSE obtained from the proposed PELM are 0.0484 and 0.0491, respectively. As compared with that of the algorithms ANFIS, ARIMA, and the hybrid algorithms, the proposed PELM yields reduced errors by 67% and 70%, and also, it happened in all the seasons.

After the accuracy of the proposed algorithm is validated, it was used to forecast wind speed in the selected locations of the Andhra Pradesh state depending on the feasibility of wind power installation. As per the data provided by NREDCAP, Ananthapur, Kadapa, Kurnool, and Chittoor are the windiest locations and are most suitable for installation of wind power plant as compared to the remaining areas. Historical data of the selected locations were collected, and the proposed network is trained with the same. Figure 3.4, Figure 3.5, and Figure 3.6 show forecasting of wind speed in the considered areas for winter, summer, and rainy seasons.

For the winter season, the forecasting time considered was January 15, 2020; for summer, it was May 15, 2020; and for the rainy season, it was October 15, 2020. The time duration for representing testing accuracy considered was 16:80 to 24:00. In all the three seasons, the proposed algorithm forecast the wind speed as accurately as the actual value in all the selected regions.

In this chapter, forecasting is also carried out on long-term time horizons in the same selected areas for short-term. Wind speed forecasting was carried out for a period of 15 years. For training the proposed network for long-term, data from the year 2007 to 2017, totaling 10 years' data, was acquired from the Indian solar resource data website. For training in winter season, data for the month of January over 10 years, that is, a total of 120 training patterns, were formulated, and 24 patterns were considered for testing purpose. Every pattern consists of first three successive years'

FIGURE 3.2 Hardware module.

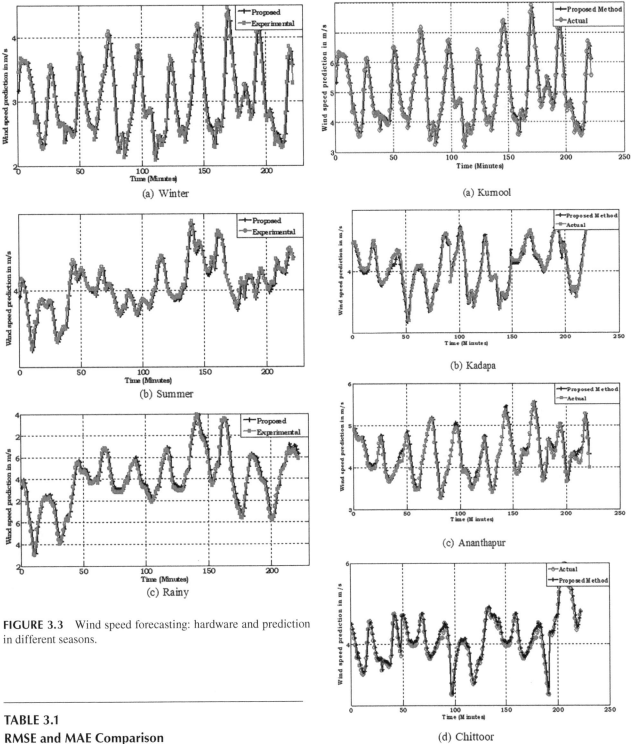

FIGURE 3.3 Wind speed forecasting: hardware and prediction in different seasons.

FIGURE 3.4 Short-term wind speed forecasting in the month of January.

TABLE 3.1
RMSE and MAE Comparison

Algorithms	MAE	RMSE
ANFIS [8]	0.1499	0.168
SARIMA [8]	0.1399	0.1800
EEMD + ANFIS + SARIMA [8]	0.0491	0.0599
Proposed PELM	0.0484	0.0491

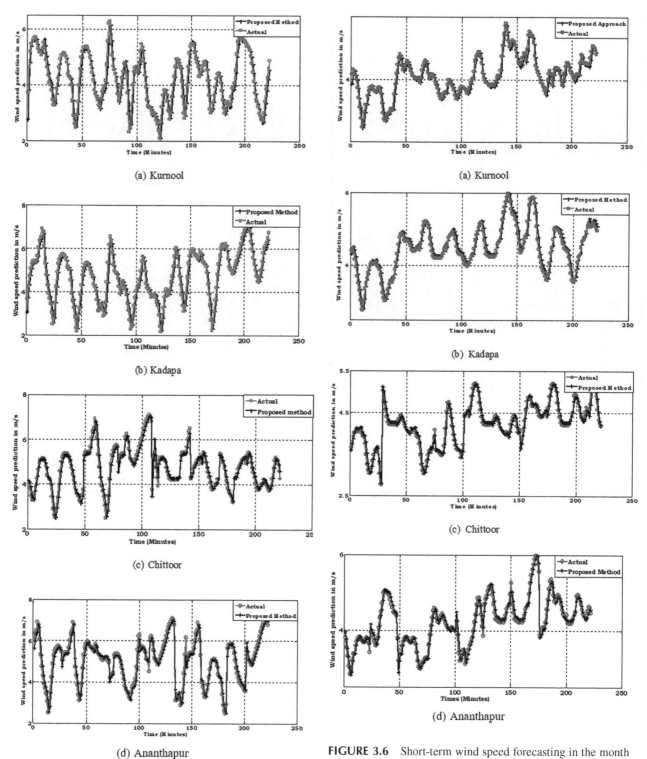

FIGURE 3.5 Short-term wind speed forecasting in the month of May.

FIGURE 3.6 Short-term wind speed forecasting in the month of October.

daily average wind speed, in which the first two samples are for training and the last sample is the target.

Figure 3.7, Figure 3.8, and Figure 3.9 show forecasting of wind speed in the considered areas for winter, summer, and rainy seasons in long-term time periods.

Upon observing the obtained results from long-term time periods, it is proved that the proposed PELM algorithm forecasts very accurately in the selected locations in all the three seasons.

Forecasting of Wind Power Using Hybrid Machine Learning

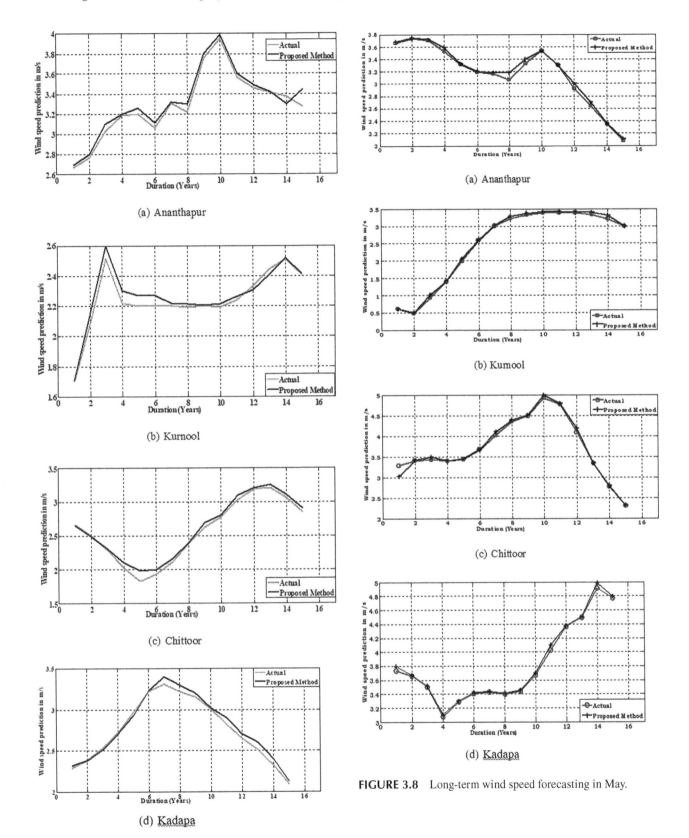

FIGURE 3.7 Long-term wind speed forecasting in January.

FIGURE 3.8 Long-term wind speed forecasting in May.

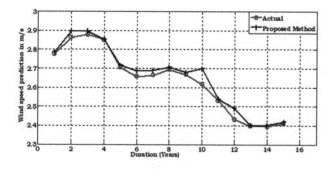

(a) Ananthapur

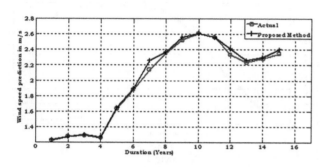

(b) Kurnool

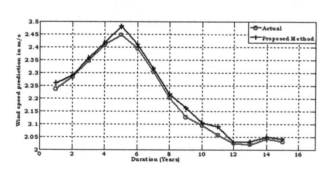

(c) Chittoor

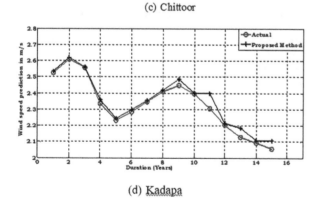

(d) Kadapa

FIGURE 3.9 Long-term wind speed forecasting in October.

3.7 SUMMARY

In this chapter, a new hybrid PELM algorithm is developed by putting together the properties of persistent and ELM algorithms. Based on the data provided by the NREDCAP, the windiest locations of the AP state are identified as Ananthapur, Kadapa, Kurnool, and Chittoor. With the help of historical data obtained from the Indian solar resource data website, PELM network is trained and wind speed is forecasted in the identified areas for three different seasons for short-term as well as long-term time horizons. From the obtained analytical results, it is observed that the proposed PELM forecasts much better when compared with that of existing algorithms.

3.8 REFERENCES

[1] *Wind Power in Power Systems*, Thomas Ackermann, 2nd Edition, 23-Apr-2012, John Wiley & Sons Ltd.
[2] Harsh S. Dhiman and Dipankar Deb, "A Review of Wind Speed and Wind Power Forecasting Techniques", *Signal Processing*, 2020, Vol. 1, pp: 1–11.
[3] Shahram Hanifi, Xiaolei Liu, Zi Lin and Saeid Lotfian, "A Critical Review of Wind Power Forecasting Methods – Past, Present and Future", *Energies*, 2020, Vol. 13, pp: 1–24.
[4] Wen-Yeau Chang, "A Literature Review of Wind Forecasting Methods", *Journal of Power and Energy Engineering*, 2014, Vol. 2, pp: 161–168.
[5] Temitope Raphael Ayodele, Richard Olarewaju, and Josiah Lange Munda, "Comparison of Different Wind Speed Prediction Models for Wind Power Application", paper presented in the 2019 Southern African Universities Power Engineering Conference/Robotics and Mechatronics/Pattern Recognition Association of South Africa during 28–30 Jan. 2019.
[6] Saurabh S. Soman, Hamidreza Zareipour, Om Malik, and Paras Mandal, "A Review of Wind Power and Wind Speed Forecasting Methods with Different Time Horizons", paper presented in Annual North American Power Symposium, 2010, pp: 1–8.
[7] http://nredcap.in/GuidelinesWindPowerProjects.aspx
[8] J. Zhang, Y. Wei, Zhong-fu Tan, W. Ke, and Wei Tian, "A Hybrid Method for Short-Term Wind Speed Forecasting", *Sustainability*, 2017, Vol. 9, No. 4: 596. https://doi.org/10.3390/su9040596.

4 Improving Power Quality of Modern Hybrid Polygeneration SOFC- and PMSG-Based WES Using ANN-Controlled UPQC

Ch. Siva Kumar

CONTENTS

4.1 Introduction: Background and Driving Forces 35
4.2 Carbon Emission Reduction Using Modern Hybrid Polygeneration SOFC- and PMSG-Based WES 36
4.3 Modeling Using Simulation 36
    4.3.1 Wind Turbine System and Wind Profile 36
        4.3.1.1 Wind Turbine System 36
        4.3.1.2 Wind Models 37
            4.3.1.2.1 Fluctuating Wind Model 37
            4.3.1.2.2 Random Noise Wind Model 38
            4.3.1.2.3 Gusty Wind Model 38
    4.3.2 PMSG-Based Electrical Generator 38
    4.3.3 Fuel Cell System: SOFC 39
    4.3.4 ANN-Based Controller for UPQC 41
4.4 Simulation Studies 41
    4.4.1 Impacts of Fluctuating Wind 42
    4.4.2 Improvement of Power Quality Issue: Sag 42
    4.4.3 Improvement of Power Quality Issue: Swell 43
4.5 Conclusion 45
4.6 References 46

4.1 INTRODUCTION: BACKGROUND AND DRIVING FORCES

Day by day, due to an increasing population, changing living standards, and growth in the industrial sector, the demand for electrical energy is increasing. To meet this demand, the consumption rate of fossil fuels, natural gas, is increasing. Along with these conditions, deforestation and transportation lead to a change in temperature across the globe by 1.5°C [1–2]. Low-carbon-emission-based green energy systems, like wind, solar, and fuel cells, have the capability to lower down carbon emission. Regarding the technological changes in the automobile sector, fuel cell in vehicles meets the requirement and expectation of automobiles. Moreover, electric vehicles have attractive characteristics, like a quiet operation with precise controls and high efficiency, fast charging technology, and wireless charging technology. This technology is also extending towards a combination with internal combustion engines as hybrid vehicles with rechargeable battery storage systems. With an increase in awareness and utilization of electrical vehicles, plug-in installations are encouraged by governments across the globe.

But these green sustainable energy systems cannot dominate and replace fossil fuel–based systems in today's existing technology and infrastructure. But it can be used directly for smaller applications/loads and as a cogenerating system with the combination of fossil fuel systems and renewable energy systems, as a hybrid system to meet base as well as peak load demands, depending on the availability and the amount of electrical energy power generation capability. With the developments in the control strategies of power electronic systems and their applications in modern electrical power systems, the concept of hybrid generation is increasing across the globe [3]. A hybrid system may be a combination of a fossil fuel–based electrical power generating system with a green energy system or a combination of different green energy systems.

Working with different green energy systems as a hybrid system is a hectic task for engineers to control individually and collectively as energy source in a modern power system, as these green energy systems have their own limitations, like amount of available energy, duration of availability, nature of availability, continuous and discontinuous mode of availability, etc., and have an impact on the distribution system. This chapter concentrates on the impacts of WES in the existing distribution systems [4].

The generating sources cannot control independently the power quality issues that arise due to the operation of fluctuating and uncertain-natured green energy systems. The demand for a state-of-the-art controller is needed to maintain the modern power system electrical parameter values within. The most promising devices at distribution level are custom power devices [5]. Different control strategies and methods have been proposed by researchers for custom power devices. An H∞ matching model is proposed to analyze the coupling effect of both the controllers of UPQC by PengLi et al. [6], while H. Akagi et al. controlled reactive power without using any storage components [7]. Jian et al. have proposed online VA loading for effective control of series and shunt controller circuits by adjusting the displacement angle to reduce manufacturing cost [8], for fast extraction of reference voltages using wavelets [9], to vary effective impedance by thyristor switching arrangements of the converters [10], and for instantaneous reactive power compensators comprising switching devices without energy storage components [11]. For this work, a custom power device, unified power conditioner (UPQC), is used to improve the power quality of a modern hybrid power system.

In this chapter, to address the impact of greenhouse emissions, it is suggested to build a hybrid system with air pollution–free wind energy system along with polygeneration SOFC. To address power quality issues, an ANN controller–based UPQC should be built.

4.2 CARBON EMISSION REDUCTION USING MODERN HYBRID POLYGENERATION SOFC- AND PMSG-BASED WES

The long-term impacts of greenhouse gas emissions are a rise in sea level, leading to the submerging of islands; the melting of sea ice; an increase in extreme climate conditions; and the risk to plant and animal species. This impact mainly depends on its chemical nature and its relative concentration in the atmosphere, especially in industrialized countries, where the per capita consumption of fossil fuel, gas-based generation of electrical energy, is more. To reduce greenhouse gases, especially in industrialized countries, the Kyoto Protocol [12] was operationalized. For the total power generating capacity of 21.61 MW, the amount of reduction in carbon emission is by 21.19941 t [13].

4.3 MODELING USING SIMULATION

The practical design, control, and implementation of real-time systems are expensive and time-consuming when it comes to understanding and checking performance and doing the corresponding modifications. Simulation platforms are best suited for design, control, repeated modifications, and getting an optimized system. The complete work is carried out in a MATLAB/Simulink environment. The following sections describe building simulation models of the system study.

4.3.1 Wind Turbine System and Wind Profile

4.3.1.1 Wind Turbine System

It is a collective unit consisting of rotating blades and low-speed shaft connected to a gear mechanism to high-speed shaft and electrical generator. The main function of the wind turbine is to convert the dynamic motion of the wind to rotational mechanical motion, which in turn rotates the electrical generator to produce electrical energy. For low- and medium-speed winds, the better-suited wind turbine is the horizontal-axis wind turbine over the vertical-axis wind turbine [14]. The energy that is available in wind cannot be transferred completely to electrical energy due to the area swept by the blades that is missed out. As per the theory developed by Albert Betz, only about 59.3% KE of wind is converted to mechanical power at ideal conditions. Power coefficient (C_p) is a measure of wind turbine efficiency, which includes the complete wind energy system components (turbine blades, bearings, gear mechanism, electrical generator, and power electronic systems), as shown in Figure 4.1.

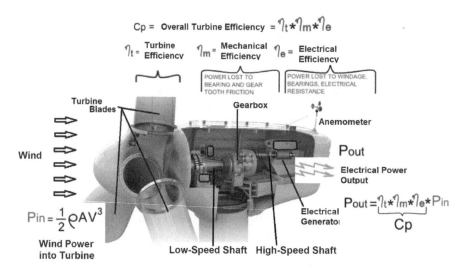

FIGURE 4.1 Power flow in a typical wind turbine.

The power in the wind (P_m) is the function of the dimensionless power coefficient (C_p), air density (ρ), blade-swept area (A), and wind speed (v), and the power coefficient in terms of tip speed ratio and angle of attack of wind turbine blades is shown in equation 4.1 and equation 4.2, respectively.

$$P_m = \frac{1}{2}\rho A v_w^3 C_p \quad (4.1)$$

$$C_p = f(\lambda,\beta); \frac{16}{27} = 0.593 \quad (4.2)$$

The power characteristics of a turbine are shown in Figure 4.2. As it clearly indicates, the available power and maximum power availability depend on wind speed. The study proposes building a wind turbine with mechanical output of (MW) 21.6e6. Suzlon Model: S52/600 specifications are considered for the simulation studies.

4.3.1.2 Wind Models

The power output of a wind turbine depends on wind speed, as the wind is fluctuating in nature and varies with height and is dependent on environmental conditions. For this work, considered is a three-component aggregate wind model consisting of fluctuating wind, random wind noise, and gusty wind. This wind model is defined as:

$$V_t = V_f + V_r + V_g \quad (4.3)$$

Where V_t is the aggregate wind velocity in m/s, V_f is the base fluctuating wind velocity in m/s, V_r is the random wind velocity in m/s, and V_g is the gust wind velocity in m/s.

4.3.1.2.1 Fluctuating Wind Model

There is no ideal wind model in existence as wind is fluctuating in nature with respect to time. To consider these real-time conditions, the sinusoidally fluctuating wind model is shown in Figure 4.3. It is assumed that this fluctuating wind

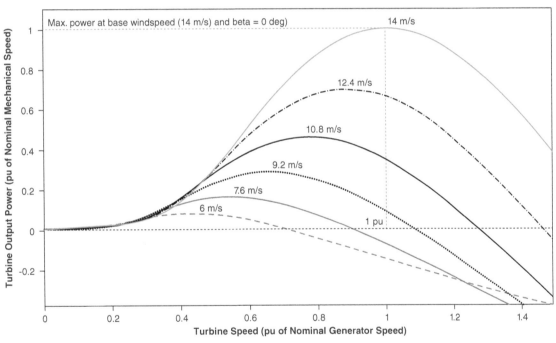

FIGURE 4.2 Wind turbine characteristics.

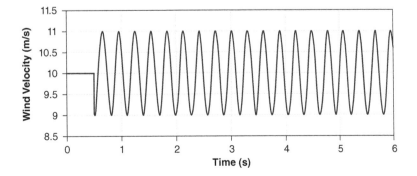

FIGURE 4.3 Fluctuating wind model.

component is always present in studies. The output of the sine wave is represented by:

$$W_f = \text{Amplitude} \times \sin(2 \times \pi \times (k + \sigma)/p + b) \quad (4.4)$$

Where *amplitude* is the amplitude of the sine wave (10.8 m/s); p is the number of time samples/sine wave period (50E-6); k is a repeating integer value in the range of 0 to p-1; σ is the signal phase; and b is the signal bias.

4.3.1.2.2 Random Noise Wind Model

Variations in wind speeds due to different objects in its flow create the noise. The magnitude of the noise changes with respect to time in small magnitudes. These noises disturb the wind flow pattern. For this model, generated are normally distributed random numbers. The correlation time of the noise is given by:

$$Tc = 1/100 \times 2\pi / f_{max} \quad (4.5)$$

Where f_{max} is the bandwidth of the system in rad/sec. Figure 4.4 shows the random noise wind model.

4.3.1.2.3 Gusty Wind Model

Another aspect of wind disturbance is gusty wind. The property of a gusty wind is wind, over its mean value, suddenly and briefly rising in its speed. This is due to uneven solar heating of the ground. The impact of the gust suddenly changes the wind profile, speed, and direction.

$$u(t) = \begin{cases} u - 037 u_{gust} \sin(3\pi/T)(1-\cos(2\pi t/T)) \\ u \end{cases} \quad (4.6)$$

$$for \ 0 \leq t \leq T$$

Figure 4.5 shows the built gust wind model.

An aggregate wind model or the wind profile of fluctuating wind model, random noise wind model, and gusty wind model is shown in Figure 4.6.

4.3.2 PMSG-Based Electrical Generator

For the conversion of mechanical energy to electrical energy at the high-speed shaft end, an electrical device

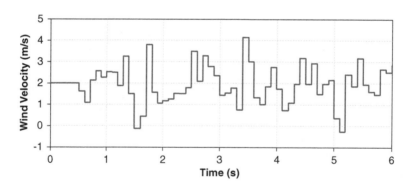

FIGURE 4.4 Random noise wind model.

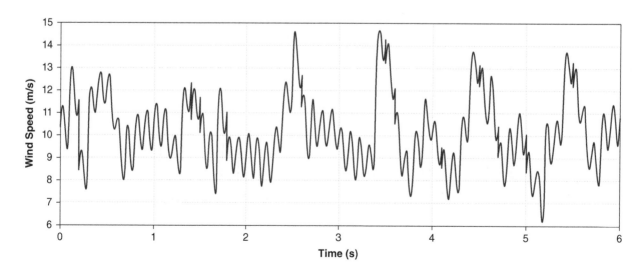

FIGURE 4.5 Gust wind model.

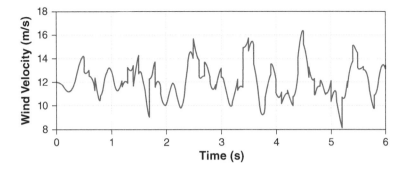

FIGURE 4.6 An aggregate wind model.

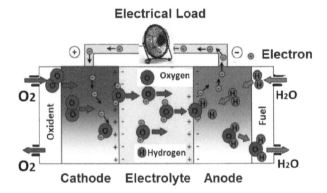

FIGURE 4.7 Solid oxide fuel cell (SOFC) components and working principle.

called electrical generator is used. Induction generator is used to operate at other-than-synchronous speeds. For nearly synchronous speeds, synchronous generators are used. The concept of direct-driven PMSG is used in many industrial manufactures [15–17] due to its better efficiency and reliability. In this work, simulation models of PMSG are built to generate 21.6 MW (0.6 MW × 36). The torque for the machine is represented with the rotational damping factor (D), equivalent inertia (J_{eq}), and mechanical rotational speed (ω_m):

$$T_e = J_{eq}\frac{d\omega_m}{dt} + D\omega_m + T_\omega \tag{4.7}$$

4.3.3 Fuel Cell System: SOFC

The modern power system is the combination of fossil fuel systems, nonconventional energy systems, microgrids, and energy storage systems to meet the electrical demands of commercial, industrial, and residential requirements, along with the recent addition of eco-friendly electrical charging stations. With the recent developments in infrastructures of gas and biofuel systems, future electrical vehicle technology and hybrid electrical vehicle technology are playing a major role around the world in reducing the impact of greenhouse gases in the transportation sector [18]. It is triggering the building of a number of charging stations. In building charging stations, nonconventional energy sources will play a major role. Fuel cell technology is one of the promising technologies in building electrical vehicle charging stations. A number of fuel cells have been developed by many researchers to date. Out of all the different types of fuel cells, the solid oxide fuel cell is used to generate electrical power as an independent source and as a combined source [19–22]. Figure 4.7 shows the basic components of solid oxide fuel cell (SOFC) and its working principle. The corresponding chemical equations are shown in the following.

Chemical reaction at cathode:

$$O_2 + 4\bar{e} \rightarrow 2O^{2-} \tag{4.8}$$

Chemical reaction at anode:

$$H_2 + O^{2-} \rightarrow H_2O + 2\bar{e} \tag{4.9}$$

or

$$CO + O^{2-} \rightarrow CO_2 + 2\bar{e} \tag{4.10}$$

Overall equation:

$$H_2 + O_2 + CO \rightarrow H_2O + CO_2 \tag{4.11}$$

The expression for stack output voltage V according to [21, 23, 24] is represented by:

$$V = N_0\left(E_0 + \frac{RT}{2F}\left[\frac{\ln pH_2 p_{o2}^{0.5}}{pH_2O}\right]\right) \tag{4.12}$$

For this work, build a 100 kW dynamic model of a solid oxide fuel cell, as shown in Figure 4.8. The SOFC model parameters are specified in Table 3.1.

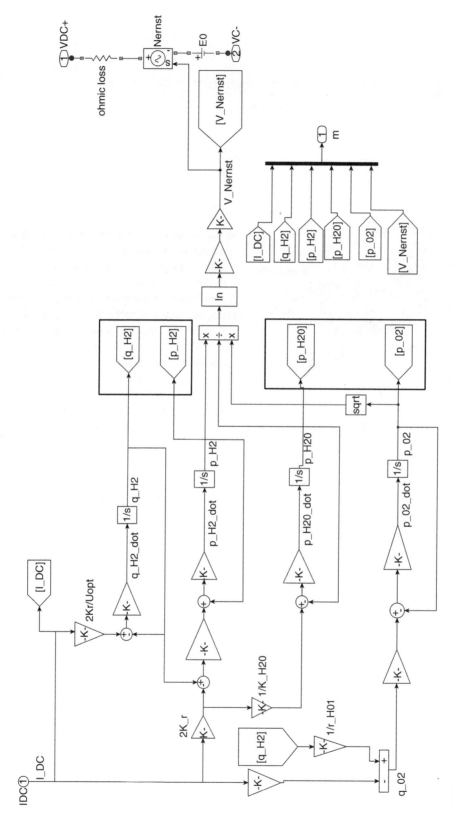

FIGURE 4.8 Solid oxide fuel cell Simulink model.

TABLE 4.1
SOFC Model Parameters

SOFC: V_{dc} rated fuel cell terminal voltage	690 V	KH_{2O}. water valve molar constant	0.281 mol/(s.atm)
T, the operating temperature	1273 k	K_{O2} oxygen valve molar constant	2.25 mol/(s.atm)
E_0, the ideal standard potential-	1.18 V	τH_2 hydrogen flow response time	26.1 sec
N_0, number of series connected cells in SOFC stack	723	τH_2O water flow response time	78.3 sec
K_r modeling constant	$0.993 \times 10e^3$	oxygen flow response time	2.91 sec
Fuel utilization factor setting	0.8	r ohmic losses	0.126 ohms
K_{H2}, hydrogen valve molar constant	0.843 mol/(s.atm)		

4.3.4 ANN-Based Controller for UPQC

Day by day, power quality issues in the power sector are increasing due to generating sources, especially green energy systems and nonlinear loads. The main responsibility of the suppliers is to maintain the quality of power in the system with the specified magnitudes and frequencies to the consumers. But maintaining the quality of the power under different operating conditions is a difficult task. Researchers have proposed different custom power devices, as discussed in the introductory part. The block diagram representation of UPQC shown in Figure 4.9 consists of series and shunt controller with a common storage device [22]. Initially simulated is the network with instantaneous power theory-based PI controller circuit. Instantaneous active power (p) and reactive power (q) are used to determine voltage and current quantities:

$$p_{\alpha\beta} = v_\alpha i_\alpha + v_\beta i_\beta \quad (4.13)$$

$$q_{\alpha\beta} = v_\alpha i_\alpha - v_\beta i_\beta \quad (4.14)$$

$$v_{\alpha\beta 0} = \begin{pmatrix} v_\alpha \\ v_\beta \\ v_0 \end{pmatrix} = [T_{\alpha\beta 0}] \begin{pmatrix} v_a \\ v_b \\ v_c \end{pmatrix} \quad (4.15)$$

$$i_{\alpha\beta 0} = \begin{pmatrix} i_\alpha \\ i_\beta \\ i_0 \end{pmatrix} = [T_{\alpha\beta 0}] \begin{pmatrix} i_a \\ i_b \\ i_c \end{pmatrix} \quad (4.16)$$

Three instantaneous powers corelated as:

$$\begin{pmatrix} p_0 \\ p \\ q \end{pmatrix} = \begin{pmatrix} v_0 & 0 & 0 \\ 0 & v_\alpha & v_\beta \\ 0 & v_\beta & -v_\alpha \end{pmatrix} \begin{pmatrix} i_0 \\ i_\alpha \\ i_\beta \end{pmatrix} \quad (4.17)$$

$$P_{shunt}(t) = \frac{3}{2} V_d^{PCC}(t).I_d^{shuC}(t) \quad (4.18)$$

$$Q_{shunt}(t) = -\frac{3}{2} V_d^{PCC}(t).I_q^{shuC}(t) \quad (4.19)$$

In real-time world, many of the practical problems are very much complex, nonlinear, with complex relationships between the variables. To solve these kinds of problems, the brain processing concept of training of neurons,

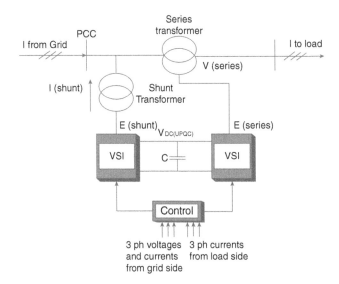

FIGURE 4.9 Block diagram representation of UPQC.

called an artificial neural network (ANN), is best suited as it could be trained to predict when it comes across a new situation without any restrictions on input variables. It may be single-layer, as show in Figure 4.10, or multilayer perceptions, as shown in Figure 4.11. Each node uses nonlinear activation function and is trained by backpropagation technique.

In this work, an ANN-based shunt controller of a UPQC is modeled. The input data is the simulation input parameters of PI controller used in an ANN controller. The ANN controller's fundamental flow diagram is shown in Figure 4.12.

The built Simulink model of shunt controller is shown in Figure 4.13, with the three phase of voltages, currents, and the DC-link voltage as inputs to generate the input pulses to the shunt converter to operate in different operating conditions. The number of hidden layers in the system is considered: 3, number of epochs; 500, number of inputs and outputs; 1, number of the error voltages of PCC- and DC-link voltage. The output of ANN is given as input to the PWM generator.

4.4 SIMULATION STUDIES

The distributed network presented in [25] is used in a hybrid system as shown in Figure 4.14, with the following notations used to represent the bus numbers: BB(i), Zij,

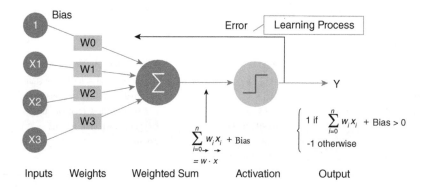

FIGURE 4.10 Single-layer perception.

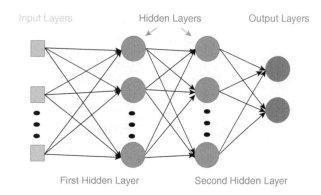

FIGURE 4.11 Multilayer perception.

and ZTR – bus bar numbers, buses, and line numbers – are represented by i and j. UPQC is connected at BB6 in the loads with switch (SW). Studied are the power quality issues generated by wind energy systems with the impacts of sag and swells.

4.4.1 Impacts of Fluctuating Wind

An aggregate wind model, shown in Figure 4.6, clearly demonstrates that the wind generally does not flow constantly during the span of a few seconds, minutes, hours, and days. But it continuously fluctuates due to atmospheric weather conditions. Equation 4.1 shows the directly proportional relationship between the mechanical power out of a wind turbine and the v^3. This, in turn, creates voltage and power fluctuations on the electrical generator, which in turn impacts the connected equipment and other generating sources. These wind fluctuations create the fluctuating output voltages and power of the wind generator. These fluctuations have an impact on consumers, like industries and domestic appliances, but the major impact is on electronics-based equipment and IT equipment. This is one of the power quality issues that arises in the power system due to wind energy systems. Figures 4.15(a), (b), (c), and (d) show the impact of voltage fluctuations. The simulation outputs shown in Figure 4.16, Figure 4.17, and Figure 4.18 are P and Q fluctuations at grid, wind terminal, and fuel cell system due to fluctuating wind. This is one of the power quality issues related to wind energy system injecting into the grid. Custom

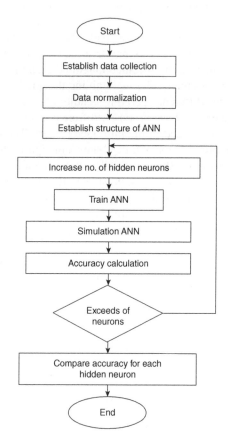

FIGURE 4.12 Fundamental flowchart to implement ANN.

power devices like UPQC are the best solutions to mitigate these kinds of power quality issues at distribution level. In this work, an ANN controller–based UPQC started its controlling operation at 2.5 sec. Figure 4.18(a), (b), (c), and (d) and Figure 4.19, Figure 4.20, and Figure 4.21 demonstrate the effectiveness of the built ANN controller–based UPQC.

4.4.2 Improvement of Power Quality Issue: Sag

The electrical network, either in the transmission or distribution, is continuously subjected to variations in loads; switching the lines and disconnecting the lines and abnormal operating conditions are the causes of power quality issues like sag and swell and harmonics into the system.

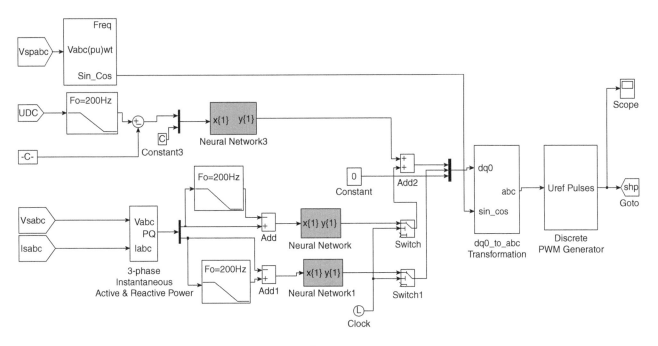

FIGURE 4.13 Simulink model of UPQC ANN shunt controller.

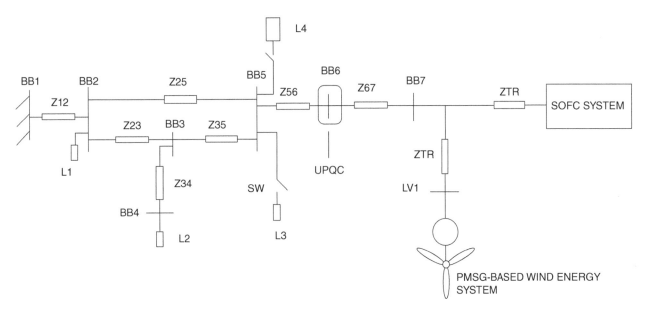

FIGURE 4.14 SOFC- and PMSG-based modern power system.

In this work, these power quality disturbances are studied. The first case study, sag, is generated by connecting different loads at different switching time periods, as presented in Table 4.2.

Figure 4.18(a) demonstrates the voltage fluctuations at PCC are effectively nullified within the IEEE 1159–1995 international standards by ANN-based UPQC.

4.4.3 Improvement of Power Quality Issue: Swell

As the power system is complex and dynamic in nature, when it is subjected to a sudden opening of any line and any unsymmetrical fault, there is a possibility of system voltages rising. This change in the voltage has an impact on the life span of the devices connected to the system, failure of insulations, etc. Initially, the system is loaded with two loads which are similar to the previous case study, then after that, a swell is created for a time period of 4.6 to 4.8 sec by creating a three-phase fault with a resistance of 0.001 ohms and a ground resistance of 0.001 ohms on the secondary side of the transformer, as shown in Figure 4.19(b). In a similar manner, created are two swells at different instances of time with L–L fault with 9 ohms, 0.001 ground resistance, as show in Figure 4.19(c), and L–G fault with 0.001 ohms, as shown in Figure 4.19(d). The built ANN-based UPQC effectively mitigated the swells in the system to improve

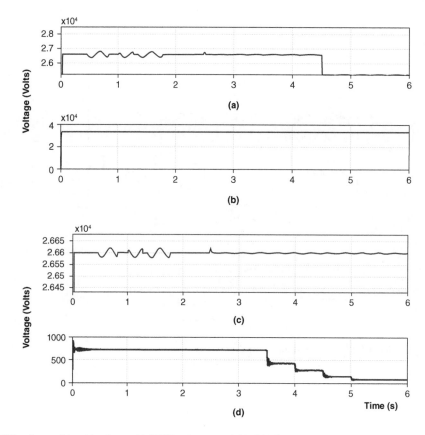

FIGURE 4.15 (a) PCC voltage, (b) grid voltage, (c) WES voltage, and (d) SOFC voltage.

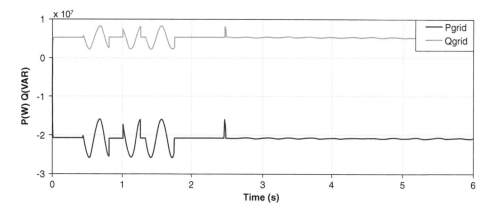

FIGURE 4.16 Grid active power and reactive power.

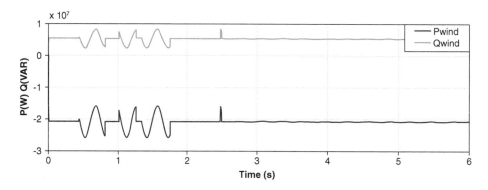

FIGURE 4.17 WES active power and reactive power.

TABLE 4.2
Switching Time Periods for Loads

Load	Switching Time in Seconds
15 MW + j 6.31 MVAr	3.6
12 MW − j 2.4 MVAr	4.2
9.2 MW + j 1.85 MVAr	4.6
20 MW + j 150 MVAr	5.2

the power quality of the system, as shown in Figure 4.19(a), within IEEE 1159–1995 standards [26].

4.5 CONCLUSION

This chapter discussed the importance of minimizing global carbon emissions and the solutions for fossil fuel–based electrical generating systems: hybrid generating systems – SOFC- and PMSG-based WES. The mathematical model

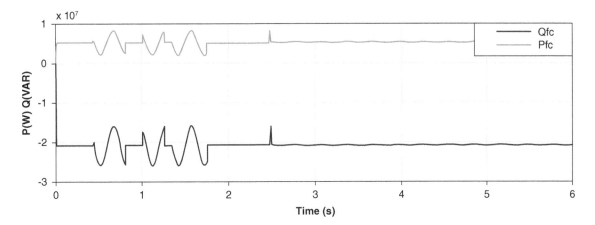

FIGURE 4.18 SOFC active power and reactive power.

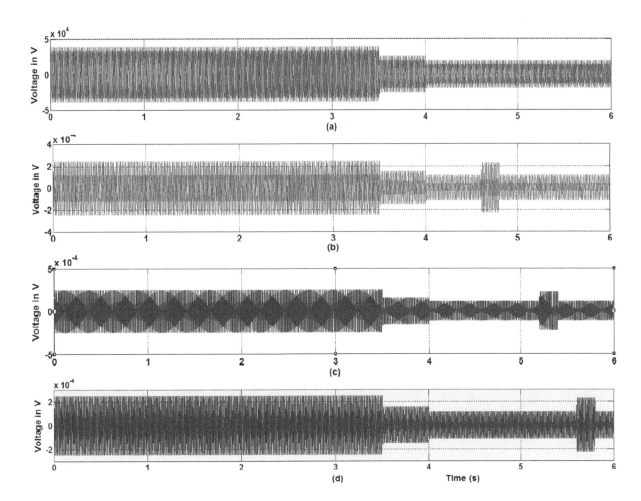

FIGURE 4.19 Voltage swells due to R_g:0.001 Ω (a) 3-phase fault, (b) 3-phase fault with R_f, (c) L–L fault with R_f, and (d) L–G fault.

of hybrid systems was built in Simulink. Discussed were the causes of power quality issues due to wind green energy system for variations in wind models and their impacts. Further, to mitigate power quality issues in the network for different operating conditions. At the end, an ANN controller–based UPQC was built to bring power quality issues within IEC 61000–4–15 and IEEE 1159–1995 standards.

4.6 REFERENCES

[1] P. J. First, "Global warming of 1.5 can IPCC special report on the impacts of global warming of 1.5 c above pre-industrial levels and related global greenhouse gas emission pathways, in the context of strengthening the global response to the threat of climate change," *Sustainable Development, and Efforts to Eradicate Poverty*. www.ipcc.ch/sr15/. Accessed, vol. 1, 2019.

[2] R. Jones, B. Haley, G. Kwok, J. Hargreaves, and J. Williams, "Electrification and the future of electricity markets: Transitioning to a low-carbon energy system," *IEEE Power and Energy Magazine*, vol. 16, no. 4, pp. 79–89, July 2018.

[3] S. Guo, Q. Liu, J. Sun, and H. Jin, "A review on the utilization of hybrid renewable energy," *Renewable and Sustainable Energy Reviews*, vol. 91, pp. 1121–1147, 2018.

[4] L. Fan and Z. Miao, "An explanation of oscillations due to wind power plants weak grid interconnection," *IEEE Transactions on Sustainable Energy*, vol. 9, no. 1, pp. 488–490, 2018.

[5] E. Hossain, M. R. Tur, S. Padmanaban, S. Ay, and I. Khan, "Analysis and mitigation of power quality issues in distributed generation systems using custom power devices," *IEEE Access*, vol. 6, pp. 16 816–16 833, 2018.

[6] P. Li, Y. Li, and Z. Yin, "Realization of UPQC h coordinated control in microgrid," *International Journal of Electrical Power & Energy Systems*, vol. 65, pp. 443–452, 2015.

[7] H. Akagi, Y. Kanazawa, and A. Nabae, "Instantaneous reactive power compensators comprising switching devices without energy storage components," *IEEE Transactions on Industry Applications*, vol. IA-20, no. 3, pp. 625–630, May 1984.

[8] J. Ye, H. B. Gooi, and F. Wu, "Optimal design and control implementation of UPQC based on variable phase angle control method," *IEEE Transactions on Industrial Informatics*, vol. 14, no. 7, pp. 3109–3123, 2018.

[9] M. Forghani and S. Afsharnia, "Wavelet based control strategy for UPQC control system used for mitigating voltage sag," *IECON 2006–32nd Annual Conference on IEEE Industrial Electronics. IEEE*, pp. 2003–2008, 2006.

[10] L. Gyugyi, "Reactive power generation and control by thyristor circuits," *IEEE Transactions on Industry Applications*, no. 5, pp. 521–532, 1979.

[11] H. Akagi, Y. Kanazawa, and A. Nabae, "Instantaneous reactive power compensators comprising switching devices without energy storage components," *IEEE Transactions on Industry Applications*, no. 3, pp. 625–630, 1984.

[12] N. Maamoun, "The Kyoto protocol: Empirical evidence of a hidden success," *Journal of Environmental Economics and Management*, vol. 95, pp. 227–256, 2019.

[13] CIPS, "Co2base line data base for the Indian power sector-user guide version13.0 June 2018," Available: www.cea.nic.in/reports/others/thermal/tpece/cdm/CO2/user/guide/ver13.pdf.

[14] V. Nelson and K. Starcher, *Wind energy: Renewable energy and the environment*. CRC Press, 2018.

[15] S. M. Tripathi, A. N. Tiwari, and D. Singh, "Grid-integrated permanent magnet synchronous generator-based wind energy conversion systems: A technology review," *Renewable and Sustainable Energy Reviews*, vol. 51, 2015 November 1, pp. 1288–305.

[16] D. Bang, H. Polinder, G. Shrestha, and J. Ferreira, "Review of generator systems for direct-drive wind turbines," in: *Proceedings of European Wind Energy Conference Exhibition*; March/April 2008, pp. 1–11.

[17] A. Uehara, A. Pratap, T. Goya, T. Senjyu, A. Yona, N. Urasaki, and T. Funabashi, "A coordinated control method to smooth wind power fluctuations of a PMSG-based WECS," IEEE Transactions on energy conversion, vol. 26, no. 2, pp. 550–558, 2011.

[18] A. Adib, K. K. Afridi, A. Amirabadi, F. Fateh, M. Ferdowsi, B. Lehman, L. H. Lewis, B. Mirafzal, M. Saeedifard, M. B. Shadmand, and P. Shamsi, "E-mobility – Advancements and challenges," *IEEE Access*, vol. 7, pp. 165226–165240, 2019.

[19] F. Ramadhani, M. A. Hussain, H. Mokhlis, and S. Hajimolana, "Optimization strategies for Solid Oxide Fuel Cell (SOFC) application: A literature survey," *Renewable and Sustainable Energy Review*, vol. 76, pp. 460–484, 2017.

[20] Y. Inui, T. Matsumae, H. Koga, and K. Nishiura, "High performance SOFC/GT combined power generation system with CO2 recovery by oxygen combustion method," *Energy Convers Manage*, vol. 46, pp. 1837–1847, 2005.

[21] Y. Zhu and K. Tomsovic, "Development of models for analyzing the load-following performance of microturbines and fuel cells," *Electric Power Systems Research*, vol. 62, no. 1, pp. 1–11, 2002.

[22] F. Ramadhani, M. A. Hussain, H. Mokhlis, M. Fazly, and J. M. Ali, "Evaluation of solid oxide fuel cell based polygeneration system in residential areas integrating with electric charging and hydrogen fueling stations for vehicles," *Applied Energy*, 238, pp. 1373–1388, 2019.

[23] N. Brandon, *Solid oxide fuel cell lifetime and reliability: Critical challenges in fuel cells*. Academic Press, 2017.

[24] V. Khadkikar, "Enhancing electric power quality using UPQC: A comprehensive overview," *IEEE Transactions on Power Electronics*, vol. 27, no. 5, pp. 2284–2297, May 2012.

[25] L. Barelli, G. Bidini, and A. Ottaviano, "Solid oxide fuel cell modelling: Electrochemical performance and thermal management during load-following operation," *Energy*, vol. 115, p. 107e9, 2016.

[26] J. Padulles, G. W. Ault, and J. R. McDonald. "An integrated SOFC plant dynamic model for power systems simulation," *Journal of Power Sources*, vol. 86, p. 495e500, 2000.

5 Review on Reconfiguration Techniques to Track Down the Maximum Power Under Partial Shadings

V. Ramu, P. Satish Kumar, and G. N. Srinivas

CONTENTS

5.1 Introduction 47
5.2 Literature Survey 48
5.3 MPPT Control Methods for PV System 49
    5.3.1 Conventional MPPT Strategies 49
        5.3.1.1 Parameter Selection–Based Controllers 49
        5.3.1.2 Sampled Data–Based Direct MPPT Controllers 50
    5.3.2 Smart MPPT Control Techniques 50
    5.3.3 Order and Outline for MPPT Control Strategies 51
5.4 MPPT Control Strategies Under PSCs 51
    5.4.1 Array Reconfiguration–Based Hardware Control Methods 51
    5.4.2 Control Methods Based on Artificial Intelligence Algorithms 53
5.5 Research Challenges, Gaps, and Limitations 55
5.6 Problem Identification and Statement 55
5.7 Objectives 55
5.8 Conclusion 56
5.9 References 56

5.1 INTRODUCTION

Solar energy is a freely accessible asset excessively significant for decreasing the reliance on regular sources. Photovoltaic (PV) frameworks produce electricity by changing the limitless energy into power. Generally significant expense, low transformation efficiency of electric power, reliance on ecological conditions, and nonlinearity of the P–V and I–V quality of PV (as shown in Figure 5.1) exhibits are the fundamental difficulties in PV cluster usage.

Following worldwide pinnacle of a PV, it is altogether essential to ensure the maximum power transformation. Numerous extreme power extraction techniques are proposed in the literature [1–3], with mainstream MPPT strategies like P&O, hill climbing (HC) (as shown in Figure 5.2), and IC techniques, which are compelling under irradiance condition of PV modules. Since the following turns out to be more muddled under fractional shading condition, that is, when all the modules don't get uniformity in radiation, these techniques simply neglect to follow the max power extractions. In uniform based conditions, the P–V plots of PV cluster has only one global peak, yet under PSCs, the P–V trademark shows various peaks. Henceforth, a few MPPT techniques are proposed which are material in PSCs. These strategies can be classified into two gatherings: equipment-based strategies and programming-based strategies [4].

The investigation and evaluation of interconnected PV cells under various concealing conditions and different concealing examples are considered. The incomplete concealing conditions because of the different elements diminish the power yield of PV clusters, and its attributes have various tops because of the crisscrossing misfortunes between PV boards. The chief target is to demonstrate, examine, mimic, and assess the presentation of PV cluster topologies [5], for example, series parallel (SP), honeycomb (HC), total cross-tied [6] (TCT), ladder (LD), and bridge-linked (BL) under various concealing conditions, to deliver the power

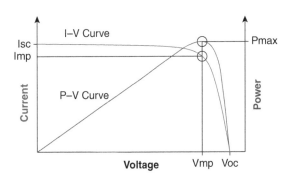

FIGURE 5.1 P–V and I–V curves of PV cells. (a) Block diagram; (b) flowchart of MPPT control system.

DOI: 10.1201/9781003321897-5

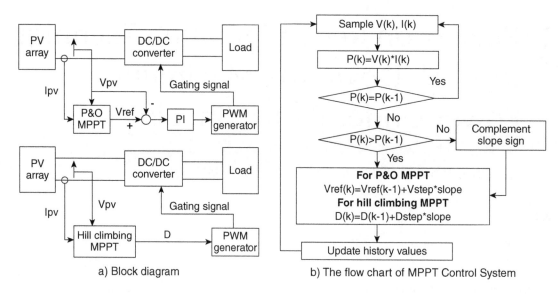

FIGURE 5.2 P&O and hill climbing MPPT system.

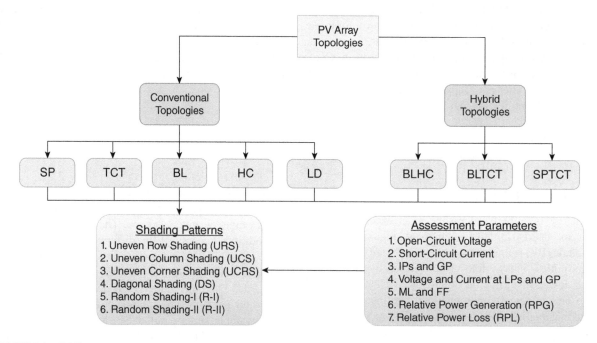

FIGURE 5.3 Different photovoltaic array (PV) topologies.

by decreasing the crisscrossing misfortunes. Alongside the ordinary PV clusters, the exploration manages the crossover PV exhibits, for example, bridge-linked honeycomb (BLHC), bridge-linked total cross-tied (BLTCT), and series parallel total cross-tied (SPTCT), as shown in Figure 5.3. The present investigation of the customary PV exhibit topologies alongside the mixture is done during static and dynamic concealing examples by contrasting the different boundaries, for example, the global peak (GP), local peaks (LPs), comparing voltage and current at GP and LPs, fill factor (FF), and ML. Furthermore, the voltage and current conditions of the HC setup under two concealing conditions are determined. The different boundaries of the PV module and its conditions are utilized for PV displaying and recreation in MATLAB software. Consequently, the derived results give valuable data to the exploration for solid activity and power amplification [5] of PV framework.

5.2 LITERATURE SURVEY

Renewable energy sector growth in India has been significant, even for electricity generation. Renewable energy is energy generated from natural resources, such as sunlight, wind, rain, tides, and geothermal heat, which are naturally replenished. Even for the decentralized systems, the growth for solar power deployment capacity has increased by

11 times in the last five years, standing in fifth position, crossing Italy, with 30 GW generation in July 2019. The installation of solar home lighting systems has been 300%, solar lanterns 99%, and solar photovoltaic water pumps 196%. The smart lighting market at 14.23 billion USD in 2019 is expected to peak to 58.57 billion USD at the end of 2025, as per CAGR of 29.30% in forecast period 2020–2025.

This is a phenomenal growth in the renewable energy sector mainly for applications that were supplied only through major electricity utilities. Some large projects have been proposed, and an area of 35,000 km^2 in the Thar Desert has been set for solar power projects to generate 700 to 2,100 GW. Renewable energy systems are also being looked upon as a major application for electrification of 20,000 remote and unelectrified villages and hamlets in the upcoming five-year plans of India. Our country has put fourth an international solar alliance with 121 countries in 2015, with motive of "One Sun, One World, One Grid" and "World Solar Bank," to harness adequate power on a global scale.

The following chart presents an idea of installed solar capacity as of March 31, 2020.

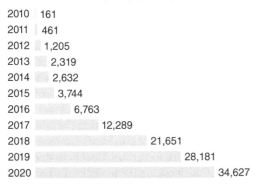

Year Cumulative Capacity (in MW)

Year	Capacity
2010	161
2011	461
2012	1,205
2013	2,319
2014	2,632
2015	3,744
2016	6,763
2017	12,289
2018	21,651
2019	28,181
2020	34,627

The preceding bars indicate the revolutionary increment in the solar power generation in the world's green energy revolution. However, even if there have been many technological advancements in the solar power sector, it still lacks the touch of efficient usage of resources and electrical elements, such as the solar PV modules and materials used in the cell configuration, maximum power generation techniques, MPPT strategies, efficient chopper conversions for power transmission, and switching technologies in the power converters.

The great amount of research manuscripts from IEEE, Springer, Elsevier, and a web of science journals has been surveyed to assess the scenario of solar power generation right from its past roots to choices in the future. The chapter here depicts the modeling of the solar harnessing power with the PV system incorporated with the PV reconfiguration techniques, PV generation adopting various MPPT algorithms, various switching sequences and algorithms in the power converter operations, energy storage elements, etc.

The characteristics of a PV module are nonlinear, affected by the irradiance and temperature solar patterns. For PV array even under irradiance, there is a unique peak point on the PV curve. Some traditional MPPT techniques include VK tracking, Voc tracking, Isc tracking, P&O, IC, FLC, and other intelligent controllers. However, a PV installation can encounter many environmental factors, such as dust, clouds, bird droppings, leaves, buildings, etc., which cast nonuniform shadows on solar surfaces. Under these partial shading conditions, many peaks occur on PV plots where the conventional MPPT methods may fail to distinguish local peaks from global peaks. Hence, it is of great deal to study control algorithms of PV systems under PSCs for enhancing their efficiency and reliability.

The first of all the PV array reconfiguration techniques was introduced in 1990, experimenting the MPPT strategies under PSCs; thereafter, many more researchers have introduced reconfiguration structures, such as SS, PS, TCT, SP, BL, HC, etc. Similarly, many such control techniques are being studied to switch relays quickly during reconfiguring the PV modules. AI algorithms are also developed and can be classified as bionic and evolutionary algorithms. In contrast to evolutionary algorithms that are genetic algo and differential evolution, next-level bionic algorithms are investigated by researchers, which include PSO, ACO, ABC, SSA, GWO, etc. However, many prefer to combine AI algorithms with multiple intelligent algorithms to achieve MPPT.

In the following sections, the main MPPT methods are reviewed and classified in terms of their control theory and their adaptations in the solar PV systems. Comparisons of existing methods are classified into three categories, as follows:

    i. Conventional MPPT techniques
    ii. Intelligent control–based MPPT strategies
    iii. MPPT strategies under PSCs

The chapter is organized into the following sections:

1. MPPT control methods are reviewed under normal irradiation conditions, including traditional and AI methods.
2. Secondly, a comparison of different categories of MPPT control methods under PSCs is presented.
3. Finally, conclusions are drawn.

5.3 MPPT CONTROL METHODS FOR PV SYSTEM

5.3.1 Conventional MPPT Strategies

Conventional MPPT methods are of the following two types:

    a. Parameter selection–based controllers
    b. Sampled data–based direct MPPT controllers

5.3.1.1 Parameter Selection–Based Controllers

The control technique is acheived by defining the PV panel parameters by tracking down the maximum power [7, 8] from the VI, PV plots measured under different operating conditions.

The intrinsic boundaries of a PV board with estimations of the solar temperatures make the foundation of an upgraded numerical model. Such models empower the P–V and I–V benchmarks of the board under given working conditions and thus decide the MPP. The control techniques in this classification principally incorporate Vk tracking, Voc tracking, Isc tracking, and so on, and delegate models are summed up beneath. Vk tracking, which is the least-complex boundary-selective MPPT strategy, directs the module terminal voltage dependent on its open-circuit esteem, although the technique cannot adequately accomplish MPPT perfectly, especially under enormous temperature variations.

Voc tracking is similar to Vk tracking, but in contrast, it tracks a constant voltage; this method starts from an open circuit to steps of varying voltage. It is clear from the PV array output that when the open-circuit voltage Voc changes under different operating conditions, the MPP voltage Vm will also change proportionally. An approximate linear relationship between Vm and Voc is calculated by Vm = Kv × Voc, where *Kv* is proportionality constant (<1) and *Voc* can be known by disconnecting PV array from load. In [9], Voc tracking was used for irradiation factor less than 350 W/m^2.

The results obtained from [10] are directed by sweeping the array current and determining the peak power point voltage. In contrast to all other methods, the main downfall of current scanning method is, its tracking speed is lesser than actually required in reality.

5.3.1.2 Sampled Data–Based Direct MPPT Controllers

This category of methods tracks the MPP by sampling data such as voltage, current, and power from PV array. These direct methods do not rely on any model of the PV and are simple to implement in practice and hence are widely used [11, 12]. These methods include P&O, IC, PC, RCC, PAC, AM, VF, PF, etc., which are often combined to further improve accuracy and speed of tracking the power. Some of these methods are summarized as follows. The principle of P&O [13], one of the most commonly used implementing MPPT in a PV system, is to increment or decrement the voltage or the duty cycle of the PV array at regular intervals and observe power deviation to determine the next control signal. P&O is solid and simple to execute. The calculation tracks the MPP by continually changing the terminal voltage of the PV cluster, which can undoubtedly yield power swaying. Moreover, with the difference in the natural conditions, the technique can bring about power loss of the PV framework. To beat this issue, numerous alterations of this technique [14–17] are introduced.

IC [18, 19] is another normally utilized for calculations of MPPT control which works by observation of conductance *I/V* with the negation of the *-dI/dV*. At the point when the natural conditions change, the technique can easily follow the changes in MPP, paying little attention to attributes and boundaries of the PV module. The control cycle of IC is unpredictable, and precision will influence the presentation of the following partially. Simultaneously, the progression size of the voltage augmentation is additionally relative to the following mistake. An improved IC strategy has been proposed (i), in which the reasonable mistake is taken inside a specific precision range that fulfills the MPPT. When contrasted with conventional strategy, the altered technique can rapidly recognize an expansion or abatement in illumination and settle on the correct choice, the consistent state wavering is disregarded, and the deficiency of power is decreased.

PC [20] is comparable to IC, which considers the parasitic capacitance of the PN crossing points in the cell and stray inductance between PV cells. The introduction of parasitic capacitance can diminish the slipup sign when not at MPP value; besides, it improves the counterimpedance of the PV system. The accuracy of PC is in helpful applications. Anyway, the technique is difficult to realize. PC procedure resembles P&O preserves trading factor with converter to trouble the voltage and current of the PV cluster for empowering MPPT. The trouble occurs in the PV structure, as there is no convincing motivation to pester the voltage and current. Moreover, the procedure makes it pointless to get the limits of the PV show early. Thusly, some fundamental circuits can be used to execute relevant limits. By using this technique, it requires the converter to work in CCM [21]. ASP [22] is a procedure which changes the game of PV groups, as demonstrated by different weights, so the PV can meet assorted weight requirements and work at the MPP.

Regardless of all the aforementioned merits, this technique has insufficiency in that its progressing introduction is poor, as it can't change the working reason for the PV group quickly enough as the external atmosphere changes. AM [23] uses a hoax PV group module to develop a reference model of the working PV to depict a comparison. This strategy can avoid the insufficiency of light in developing the PV cell; the issue of different most prominent characteristics furthermore should be considered for a powerful structure. PF [24] learns current to yield power in programming, by yielding voltage and currents of the PV. It changes the voltage, as demonstrated by the power relation between the current and past cycles. The realized voltage may not be exceptional at a comparative generated power; the controller should be arranged as a single-value control mode, that is, only one side of the P–V curve is used.

5.3.2 Smart MPPT Control Techniques

The P–V plot of a PV module is a curvy plot and will change with respect to the temperature and illumination patterns, which will then display various plotting patterns. Consequently, the traditional MPPT strategies may fail. To counter this issue, MPPT strategies dependent on astute calculations are proposed and effectively applied; these techniques incorporate FLC, NN, SMC, and so on. Some normal astute MPPT control techniques are dissected as follows.

FLC is a conscious calculation, with a bunch of fuzzy principles. Its usage can be partitioned into the three stages of fuzzification, control rule assessment, and defuzzification. The vital component of FLC is consolidating experiences and information into language rules to control the framework. Also, FLC can follow the MPP rapidly to arrive to the MPP, that is, it has better, powerful, and consistent execution. But the fundamental drawbacks of FLC are that various irradiance patterns can cause float and the execution is intricate. A few specialists have attempted to join the conventional MPPT calculation with FLC to improve the intermingling speed and decrease consistent state wavering and usage intricacy. In [3], the beta halfway factor was presented as the third channel, which streamlines the work and diminishes the reliance on the experience. [25] proposed a PV board–sourced support converter, with FLC, under 25–60°C temperature and 700–1,000 W/m^2 irradiance. This framework was associated with a buck converter with PI control to work as a battery charging regulator. What's more, FLC has likewise been proposed for an MPPT conspire in which four advancement calculations are introduced for streamlining fuzzy and producing the legitimate duty cycle.

[26–28] is another sort of data handling innovation. The mostly utilized neural organization structure has three layers: input layer, concealed layer, and yield layer. At the point when applied in a PV framework, numerous examples of information and yield information are utilized to prepare factors of variation. As of now, numerous scientists have consolidated NN with other keen calculations to acquire better outcomes. One model proposed is the consolidated PSO-RBFNN calculations in [26] to decrease preparing time; the PSO is utilized to improve the RBFNN boundaries, change the weight boundaries and speed to accomplish the impact of actualizing. In [28], the regulator boundaries are enhanced and tuned utilizing a prescient neural organization regulator. It predicts the control boundaries by matrix flow and DC-transport voltage and killing these blunders in a short time.

5.3.3 Order and Outline for MPPT Control Strategies

Summarizing the preceding conversation, some MPPT control strategies are assessed from the perspective of unpredictability level, velocity, productivity, and so on. From Table 5.1, it very well may be seen that the different MPPT control strategies have their own points of interest and inescapable deficiencies. These MPPT control techniques are not great and have wide exploration possibilities and advancement space, particularly on account of PSCs.

5.4 MPPT CONTROL STRATEGIES UNDER PSCS

To beat the issue of the yield power drop due to the multitop property of the trademark bend of the PV framework under PSCs, a few analysts have proposed strategies which have indicated expanded precision of the MPPT and thus expanded the yield intensity of the PV exhibit under PSCs.

As current control hypothesis and AI consciousness grow, some more reasonable techniques following have opened up. These arising keen control strategies give more alternatives to MPPT control of PV frameworks under PSCs. Following a survey of current writings [29–34], the current GMPPT strategies under PSCs can be gathered into two classes, as follows:

i. Array reconfiguration–based hardware control methods
ii. Artificial intelligence algorithms–based control methods

5.4.1 Array Reconfiguration–Based Hardware Control Methods

These methods are based on PV array reconfiguration using switches and sensors to dynamically change the array connections to increase the system efficiency under PSCs. The PV array reconfiguration concept was first proposed in 1990 by Salameh and Dagher [35], which were widely developed and applied [36, 37]. As shown in Figure 5.4, the PV array interconnection structures [38] include SS, SP, BL, TCT, HC, etc. The output characteristics of all these connections have been analyzed through simulation and experiment [39]. Moreover, the performance of common PV array joint structures under PSCs has been evaluated [38, 40]. It can be seen that changing the exhibit association structure properly can reduce the concealing impact and improve the power output somewhat. Moreover, lately, a few strategies joining smart control with equipment exhibit reconfiguration have been proposed to upgrade the resistance of PV frameworks under PSCs.

In [39], the PV exhibit utilizes the TCT association structure, and another riddle-based actual reconfiguration conspire is proposed. This is named the predominant square strategy and prompts actual migration of the modules inside the TCT interconnection. This assists accomplishing uniform current in each line of the PV exhibit. In contrast to this, another diverse zigzag strategy was proposed by [41] which exhibits its viability just for 3 × 3 PV cluster.

Another actual game plan conspire for the PV exhibit has been introduced [42], which is primarily founded on expanding the mathematical distance between modules that are topologically neighboring inside the PV cluster. Furthermore, an advanced exchanging set (SWS) geography for PV modules is created in [43]. Numerous PV exhibit setups have been accounted for until this point, and the regularly utilized arrangements depend on the TCT and SP designs. The SS, parallel, BL, and HC designs are seldom utilized because of low proficiency, muddled wiring, and so forth. For the TCT design, the most testing part is to interface PV modules with comparable irradiance levels in each column, which is known as irradiance adjustment.

For SP association, the point of reconfiguration is to fabricate a series of arrangement-associated modules with comparative irradiance levels. A rundown of dynamic PV

TABLE 5.1

Dynamic PV Array Reconfiguration Methods

Structure	Strategy	Required Switches	Acquired Parameters	Advantage	Disadvantage
SS	Series	-	0	High applicability	Low efficiency and high power losses
parallel	Parallel	-	0	High applicability and high output current	Low efficiency and low output voltage
TCT	z-z	-	Irradiance	High efficiency and high reliability	Limited to 3 × 3 array
	IE	24 DPST	V, I, irradiance	High applicability and high efficiency	High complexity and acquired three parameters
SP	Dynamic PV array	15-SPST, 5-DPDT	V, irradiance	High reliability	Complexity and low reliability
	RPV	6-SPDT, 5-DPST, 4-DPDT	V, irradiance	High applicability	Poor versatility
	SWS	6-switches for each SWS	I, irradiance	High convergence	Only two transition modes of connection
	Adaptive	6NFST + 3NFMIM + (NFST − 1) + (NFMIM − 1)	I, irradiance	High compatibility	Complexity and many switches
	IE	-	Irradiance	High convergence	High randomness
BL	IE	-	Irradiance	High applicability and low cost	High complexity and not widely used
HC	IE	-	Irradiance	High stability	High complexity and not widely used

exhibit reconfiguration techniques detailed in the writing is introduced in Table 5.1. These techniques are summed up in terms of system, required switches, and other distinctive boundaries. A basic assessment of the favorable circumstances of these techniques is given. In the table, DPST, SPDT, and SPST signify twofold-shaft single-toss, single-post double toss, and single-shaft single-toss sorts of switches individually. NFST is the quantity of adaptable strings, and NFMIM is the quantity of adaptable miniature inverter modules in the PV framework.

The pick of rightful arrangement of PV configuration under PSCs is essential since the power loss is dependent upon PV concealing state by PV cluster design. Among the previously mentioned, PV cluster arrangement assumes an essential part in deriving maximum power output. There are numerous PV exhibit arrangements noted in the writing to cut down the inability to relate losses brought by concealing, for example, "series arrangement, parallel, series parallel (SP) arrangement, total-cross-tied (TCT), bridgelink (BL), and honeycomb (HC)" [5, 44]. Among all, PV cluster arrangements SP, TCT, and BL are respected for their unwavering quality utilizing probabilistic methodology, which will cut down the inability to relate impacts because of production. The survey expressed that over SP-designed PV exhibit, TCT and BL PV setups are more dependable. In [8], distinctive PV exhibit arrangements, for example, "SS, SP, TCT, BL, and HC," with halfway concealing examination are introduced. The results show TCT PV cluster has cut down the losses just as that TCT PV exhibit is indicating less weakness to PSCs and creating the maximum power over different arrangements [11, 12]. The significant issue with the TCT cluster design is, if the quantity of PV modules are concealed in succession, the output is restricted in the exhibit [9]. All things considered, numerous researchers have proposed reconfiguration methods to settle this issue for TCT PV cluster to disperse concealing impacts from one column to various lines to cut down confound losses under PSCs [10, 45].

These strategies are ordered into dynamic and static reconfigurations. As of survey, a unique method of reconfigured PV modules helps in getting maximized power output under PSCs. A strategy in [13, 14] is created to change the associations between the PV modules, where fuzzy regulator is applied to pick the electrical array reconfiguration (EAR) to driving position. In [16], another versatile PV cell exhibit comprises of fixed part, versatile part, and exchanging network to diminish PSCs. Here framework assumes a fundamental job to associate versatile PV cells into the set cells to take care of any issues with irradiance drop in each column. The modules are changed progressively, as indicated by a changing framework to the flow to the most extreme degree of the single string in case of concealing by an electrical reconfiguration method [18], where power loss is brought somewhere near a few concealed conditions yet the strategy is cost-inadequate. So to make savvy, even now different papers [16, 19, 45] manage dynamic reconfiguration which deals with most extreme power under fractional concealing. Writing expressed that the dynamic reconfiguration procedure requires an observing framework, reconfiguration calculation, and exchanging grid, which results in the expense that the dynamic reconfiguration strategy would increment.

In static reconfigurations, as the name itself demonstrates, the module position is fixed for all concealing conditions with fixed interconnection plot, which will help the power output under fractional concealing condition, since this method even doesn't need any sensors, reconfiguration calculation, and switch grid. However, the issue with this procedure is, reconfigurable example for remastering the PV modules to appropriate concealing impacts over the cluster should be viable. The novel interconnection plot [21] is proposed, where the electrical association between PV modules are made subsequent to renumbering, which assumes an imperative part than SP-, TCT-, and BL-designed PV clusters by executing them in 3 × 3 exhibit. Similarly for 4×4 TCT PV exhibit to disperse PSCs, an enchantment square [23] PV cluster plan is created, which makes the proposed course of action, boosting the output power and cutting down the power loss as contrasted and existing PV cluster configurations. Meanwhile [26, 36, 46], dependent on the numerical methodology, they stretched out different example courses of action to disseminate PSCs, which made them across a onetime physical arrangement [37] of PV modules associated in a TCT PV exhibit, which is proposed to support the power age under PSCs. The modules are truly positioned without changing their electrical association, dependent on the 9×9 SuDoKu puzzle design, which was made to help the PV energy generated. The cluster design depends on arbitrary riddle design, and the improvement in power output is drawn for different shading conditions, for example, line losses which are relying upon the length of the wire needed for the association of SuDoKu game plan and the scattering of shade haven't been considered for the investigation. The proportion of concealing scattering relies upon the capacity of the example to scatter any shade to the most extreme conceivable number of lines to limit the concealing impact.

Ideal SuDoKu [37] example will chop down the line losses and lift conceal scattering for improved energy creation without changing their electrical associations, dependent on ideal SuDoKu. The actual area of PV modules in the TCT cluster is revamped; the proposed course of action will be additionally reviewed with TCT and SuDoKu PV exhibit plans by over the worldwide greatest power point (GMPP), befuddle losses, efficiency (η), fillfactor (FF) under different concealing conditions utilizing MATLAB-Simulink.

SuDoKu course of action utilized in [37] appears in Figure 5.1. For instance, keeping the electrical association stay unaltered module 42 (fourth line, second section) is genuinely migrated to the first column, in the fourth line, according to this game plan. Concealing scattering is by moving the PV modules to particular columns according to the SuDoKu structure, which diminishes the probability of concealed modules in a similar line and, in this manner, disperses them all through the cluster. It is to be noticed that this design is a fixed one; it stays as before for all concealing examples.

5.4.2 Control Methods Based on Artificial Intelligence Algorithms

AI control–based calculations are well-known because of their main techniques, including PSO [47], ABC [48], ACO [49], SSA [50], GA [51], DE [52], so on numerous applications to PV, clusters have been proposed and have indicated their viability for MPPT in PV framework under PSCs. PSO calculation is a multiextraordinary capacity worldwide streamlining strategy created by reproducing winged creatures' scavenging conduct [47].

The exhibition for taking care of multi-issues in multivariable framework has been generally recognized by specialists in various fields. The goal of the PSO is to locate the best framework which speaks to ideal arrangement GMPP of the PV cluster. As of now, PSO calculations have been improved and applied to PSCs by numerous specialists. Ishaque in [53] utilized procedures to actualize the PSO calculation even under various irradiance patterns, which improved the execution. Hamdi in [26] incorporated a PSO with ANN and RBF to improve the calculations. In [29], Li proposed another general dispersion PSO calculation which sets the conceivable pinnacle guide voltage toward the underlying situation by OD-MPPT, along these lines guaranteeing that the PSO calculation just necessitates discovering MPP rapidly and proficiently in an exceptionally little pursuit zone. Saad [54] proposed a novel crossbreed control of environmentally friendly framework. An improved PSO, which goes about as a regulator, is utilized to control the power of the sources depending on interlinking control. Contrasted with the standard PSO calculation, numerous arbitrary and meddling factors are taken out, thus making the structure more fundamentally improved. The ideal position is that it can adjust to the ongoing changes and run ceaselessly [55–58]. Socha and Dorigo in [56] introduced an expansion of ACO to persistent domains. Another biomotivated MPPT regulator depends on the subterranean insect province improvement calculation with another system, saving calculation time with high exactness, zero motions, and high strength [57]. Even though the ACO calculation shows incredible execution in managing power control issues, there aren't many papers on MPPTs utilizing the ACO calculation; rather, it mainly upgrades different regulators to improve MPPT execution. Lekshmi and Umamaheswari [58] zeroed in on the acknowledgment of a productive-based power taking care of DC–DC single-finished essential inductor converter (SEPIC). Another proposition by Karaboga (2005) [48] in light of the honeybee's searching attributes recreates the pioneer to search for the nectar source, while a cycle of ceaseless emphasis is utilized to locate the ideal source. This is appropriate for ideal estimation of dynamic boundaries. When contrasted with the other traditional calculations, ABC is better in gathering participation measure and thus gives quicker assembly and better results. In [59], Belhaouas, ABC joined with P&O to improve calculation dependability. Padmanaban

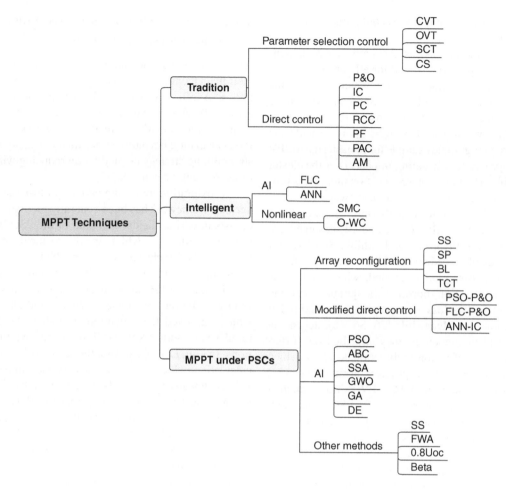

FIGURE 5.4 Different classification of MPPT control algorithms.

in [60] presented a novel controlled MPPT for a lattice PV framework. A crossover with neuro-fuzzy framework and ABC calculation was utilized to improve the overall work.

Also, Yang in [61] proposed a bio-enlivened technique named swarm calculation, creating the first of its type with various free salp chains; in this manner, it can carry out more extensive investigation. The presentation of SSA, GWO, and different calculations was broken down by Mohamed [28]. GA is an arbitrary pursuit calculation that re-enacts the choice of the organic world, and the ideal arrangement of the issue is developed through hereditary qualities and replication. Particularly for nonlinear and multimodular target work issues, it has high relevance; in any case, GA isn't entirely steady and thus still finds the nearby ideal arrangement. In [62], Huang proposed calculation of GA with the FA and its computation cycle, which additionally improved calculation. Among them, the transformation cycle and the assimilation cycle consolidate to mitigate aftereffects of the hereditary calculation. The genuine coded improved hereditary calculation and controls the DC-interface voltage, as indicated by a reference DC-constant voltage, thereby improving the dynamic conduct of framework. Differential evolution (DE) calculation is a transformative calculation utilized for worldwide streamlining [52]. The calculations are utilized for tackling these issues that are nonceaseless, nonlinear, or having numerous neighborhood minima and constraints. In contrast to PSO, there are just two boundaries that are needed to be set in DE; along these lines, it is necessary to accomplish precise MPPT. The standard DE calculation is utilized to perform MPPT under PSCs [63]. The different classification of MPPT control algorithms is as shown in Figure 5.4.

By adopting the preceding algorithms, the MPPT strategies have been designed to extract the maximum power from the solar PV systems. Even though they are efficient in their own way, they have their respective pros and cons, whose data are as follows:

According to the preceding summary of MPPT techniques, it is evident that research on MPPT methods for PV system has been an active topic, but further extensive research is needed on the balance of rapidity and stability of MPPT, especially under complex, fast-changing irradiation conditions. Based on this, possible future research directions in MPPT methods can be laid.

TABLE 5.2

Performance of Some MPPT Methods under Partial Shading Conditions

Category		MPPT	MPPT Performance Indicator					
			Complexity	Tracking Speed	Cost	Efficiency	Accuracy	Hardware implementation
MPPT methods under PSCs	Array reconfiguration methods	SS, SP, BL TCT. etc.	High	Slow	High	<90%	Medium–low	Difficult
	Improve direct control methods	ABC-P&O, ACO-P&O. PSO-O, etc.	High	Fast	High	>99%	High	Difficult
	intelligence methods	PSO	Medium-high	Fast	High	>98%	High	Medium
		ACO	Medium-high	Fast	High	>983%	High	Medium
		ABC	Medium-high	Fast	High	>99%	High	Medium
		FA	Medium-high	Fast	High	>98.5%	High	Medium
		S5A	Medium	Fast	High	>99%	High	Medium
		CA	Medium-high	Fast	High	>98%	Medium-high	Easy
		DE	Medium	Fast	High	>98%	Medium-high	Easy
	Other methods	SS. RS. FLS,	Medium-high	Fast	High	>98%	Medium-high	Medium
		FWA, etc. Bate method	Medium	Fast	High	>98%	High	Easy
		0.8Uoc	Medium-low	Medium	Medium	<90%	Medium-low	Easy

5.5 RESEARCH CHALLENGES, GAPS, AND LIMITATIONS

To break down the power improvement among various calculations, a near report was performed. According to the studies and analysis presented, the following are the gaps and limitations for future research:

1. The reconfiguration methods developed must be incorporated for high-rated PV plants.
2. Array-level switching is to be implemented instead of module-level, because module-level will not exhibit much improvement in power in a high-rated plant.

   More examination on working of exchanging grid is required, which works effectively significantly under quick concealing conditions. Furthermore, the fractional concealing conditions are unusual and dynamic; in a brief period, there is an extension to acquaint effective reconfiguration strategies with fulfilling the states of concealing.
3. Investigating the adjustment of MPPT is another significant component which needs to be executed.
4. Temperature having effect on power production, it is critical to consider while shaping the conditions for the reconfiguration.

5.6 PROBLEM IDENTIFICATION AND STATEMENT

Although the literature stated that the different connection structures reduced the impact of PSCs and improved the power output of the array, TCT has one major drawback of limiting the output array current based on the number of modules shaded in a row.

From the scripts, it is seen that, for the reconfiguration of 4×4 PV cluster, it's necessary to have 48 double-pole double-throw (DPDT) switches. By presenting the twofold post double-pole four-throw (DP4T) switches, the switch number can be diminished to 16.

Along these lines, reconfiguration with DP4T switches has extraordinary effect on diminishing exchanging losses and its center subject to work for future improvement.

In the vast majority of the static techniques, the first section of the exhibit will be the equivalent. Accordingly, the framework can't work effectively under a quick change in the illumination conditions. Along these lines, future research can counter this issue.

Dynamic reconfiguration approaches are considered as ongoing choices for the static reconfiguration methods. In this way, it is normal to present new and effective meta heuristic procedures to give more efficient method of exchanging lattice under a few shading conditions.

Solar PV power enhancement strategies with most efficient AI-based algorithms befit the future scope of PV plant–based electrical grids. These are to be implemented with cost optimization, power storage elements, efficient algorithms in maintenance in both hardware and software operating loops.

5.7 OBJECTIVES

The reconfiguration is a significant strategy for giving the most extreme execution on a PV framework. In playing out the technique, a significant advancement is deciding the ideal framework. This progression should be exact and as fast as under various circumstances, particularly in huge PV frameworks.

In this examination, a proficient calculation can be applied to continuous research on frameworks. The ideal structure is then decided, and the outcome is applied to the PV exhibits, which concludes the main objective of the research as follows:

- To reconfigure the best PV array structure for maximum power generation and with best cost optimization.
- To develop a better algorithm by assessing the cons of the previously introduced strategies for PV systems in solar power generation.
- To validate the developed solar-to-electrical-energy conversion system in hardware model to derive the actual results as proposed in the research.

5.8 CONCLUSION

From the past to the present, many MPPT strategies have been developed to extract maximum power from the PV system by adopting new control algorithms and PV reconfiguring techniques, which are based on the traditional MPPT topologies. But for the smart future, the conventional structures fail to possess the maximum potential of power extraction to meet the exponentially increasing demand. So there's a need for solar PV systems to adopt to new bionic technologies, such as genetic algorithms, ANN strategies, PSO, etc., which favor continuous research in this segment.

Therefore, this segment is making optimum use of PV clusters to extract maximum power under uniform and irradiance patterns, though lag at the unwanted elemental disturbances from environment. Therefore, there is enough research gap to further improve the solar PV system, by adopting to ongoing developments in control algorithms, which can help control algos and MPPT strategies to extract max power from PV system with ease and speed.

5.9 REFERENCES

[1] Esram, T., Chapman, P. L., 2007. Comparison of photovoltaic array maximum power point tracking techniques. *IEEE Trans. Energy Convers.* 22 (2), 439–449, Jun.

[2] Ishaque, K., Salam, Z., 2013. A review of maximum power point tracking techniques of PV system for uniform insolation and partial shading condition. *Renew. Sustain. Energy Rev.* 19, 475–488, Mar.

[3] Liqun, L., Meng, X., Liu, C., 2016. A review of maximum power point tracking methods of PV power system at uniform and partial shading. *Renew. Sustain. Energy Rev.* 53, 1500–1507, Jan.

[4] Kouchaki, A., Iman-Eini, H., Asaei, B., 2013. A new maximum power point tracking strategy for PV arrays under uniform and non-uniform insolation conditions. *Solar Energy* 91, 221–232, May.

[5] Pilawa-Podgurski, R. C., Perreault, D. J., 2013. Submodule integrated distributed maximum power point tracking for solar photovoltaic applications. *IEEE Trans. Power Electron.* 28 (6), 2957–2967, Jun.

[6] Darussalam, R., Pramana, R. I., Rajani, A., 2017. Experimental investigation of serial parallel and total-cross-tied configuration photovoltaic under partial shading conditions. *2017 International Conference on Sustainable Energy Engineering and Application (ICSEEA), Jakarta*, 140–144, doi: 10.1109/ICSEEA.2017.8267699.

[7] Chiang, Mao-Lin, Hua, Chih-Chiang, Lin, Jong-Rong, 2002. Direct power control for distributed PV power system. *Proceedings of the Power Conversion Conference-Osaka 2002 (Cat. No.02TH8579), Osaka, Japan* 1, 311–315, doi: 10.1109/PCC.2002.998566.

[8] Maiti, A., Mukherjee, K., Syam, P., 2016. Design, modeling and software implementation of a current-perturbed maximum power point tracking control in a DC-DC boost converter for grid-connected solar photovoltaic applications. *2016 IEEE First International Conference on Control, Measurement and Instrumentation (CMI), Kolkata*, 36–41, doi: 10.1109/CMI.2016.7413706.

[9] Veerapen, S., Wen, H., Du, Y., 2017. Design of a novel MPPT algorithm based on the two-stage searching method for PV systems under partial shading. In: *Proceedings of the IEEE 3rd International Future Energy Electronics Conference and ECCE Asia, 2017*, IEEE.

[10] Noguchi, T., Togashi, S., Nakamoto, R., 2002b. Short-current pulse-based maximum power point tracking method for multiple photovoltaic and converter module system. *IEEE Trans. Ind. Electron.* 49 (1), 217–223.

[11] Murtaza, A. F., Chiaberge, M., Spertino, F., Ahmad, J., Ciocia, A., 2018. A direct PWM voltage controller of MPPT & sizing of DC loads for photovoltaic system. *IEEE Transactions on Energy Conversion*, vol. 33, no. 3, pp. 991–1001, Sept., doi: 10.1109/TEC.2018.2823382.

[12] Patel, A., Tiwari, H., 2017. Implementation of INC-PIMPPT and its comparison with INC MPPT by direct duty cycle control for solar photovoltaics employing zeta converter. *2017 International Conference on Information, Communication, Instrumentation and Control (ICICIC), Indore*, 1–6, doi: 10.1109/ICOMICON.2017.8279173.

[13] Alik, R., Jusoh, A., 2018. An enhanced P & O checking algorithm MPPT for high tracking efficiency of partially shaded PV module. *Sol. Energy* 163, 570–580.

[14] Manickam, C., Raman, G. P., Raman, G. R., Raman, G. R., Ganesan, S. I., Chilakapati, N., 2017. Fireworks enriched P & O algorithm for GMPPT and detection of partial shading in PV systems. *IEEE Trans. Power Electron.* 32 (6), 4432–4443.

[15] Ghamrawi, A., Gaubert, J. P., Mehdi, D., 2018. A new dual-mode maximum power point tracking algorithm based on the perturb and observe algorithm used on solar energy system. *Sol. Energy* 174, 508–514.

[16] Abdel-Salam, M., Ookawara, S., 2017. Adaptive reference voltage based MPPT technique for PV applications. *IET Renew. Power Gener.* 11 (5), 715–722.

[17] Sundareswaran, K., Vigneshkumar, V., Sankar, P., Simon, S. P., Srinivasa Rao Nayak, P., Palani, S., 2016. Develofpment of an improved P & O algorithm assisted through a colony of foraging ants for MPPT in PV system. *IEEE Trans. Ind. Inform.* 12 (1), 187–200.

[18] Yatimi, H., Ouberri, Y., Aroudam, E., 2019. Enhancement of power production of an autonomous PV system based on robust MPPT technique. *Procedia Manufa* 32, 397–404.

[19] Motahhir, S., Ei, G. A., Sebti, S., Derouich, A., 2018. Modeling of photovoltaic system with modified incremental

[19] conductance algorithm for fast changes of irradiance. *Int. J. Photoenergy* 3286479.
[20] Pahari, O. P., Subudhi, B., 2018. Integral sliding mode-improved adaptive MPPT control scheme for suppressing grid current harmonics for a PV system. *IET Renew. Power Gener.* 12 (16), 1904–1914.
[21] Esram, T., Kimball, J. W., Krein, P. T., Chanpman, P. L., Midya, P., 2006. Dynamic maximum power point tracking of photovoltaic arrays using ripple correlation control. *IEEE Trans. Power Electron.* 21 (5), 1282–1290.
[22] Nguyen, D., Lehman, B., 2008. An adaptive solar photovoltaic array using model-based reconfiguration algorithm. *IEEE Trans. Ind. Electron.* 55 (7), 2644–2654.
[23] Rabinovici, R., Frechter, Y. B., 2010. Solar cell single measurement maximum power point tracking. In: *Proceedings of 2010 IEEE 26-th Convention of Electrical and Electronics Engineers*. Israel.
[24] Hua, C., Lin, J., Shen, C., 1998. Implementation of a DSP-controlled photovoltaic system with peak power tracking. *IEEE Trans. Ind. Electron.* 45 (1), 99–107.
[25] Yilmaz, U., Kircay, A., Borekci, S., 2018. PV System fuzzy logic MPPT method and PI control as a charge controller. *Renew. Sustain. Energy Rev.* 81, 994–1001.
[26] Hamdi, H., Ben Regaya, C., Zaafouri, A., 2019. Real-time study of a photovoltaic system with boost converter using the PSO-RBF neural network algorithms in a MyRio controller. *Sol. Energy* 183, 1–16.
[27] Farajdadian, S., Hosseini, S. M. H., 2019b. Optimization of fuzzy-based MPPT controller via metaheuristic techniques for stand-alone PV systems. *Int. J. Hydrog. Energy* 44 (47), 25457–25472.
[28] Mohamed, M. A., Zaki, D. A. A., Rezk, H., 2019b. Partial shading mitigation of PV systems via different meta-heuristic techniques. *Renew. Energy* 130, 1159–1175.
[29] Hu, K., Cao, S., Li, W., Zhu, F., 2019. An improved particle swarm optimization algorithm suitable for photovoltaic power tracking under partial shading conditions. *IEEE Access* 7, 143217–143232.
[30] Belhachat, F., Larbes, C., 2019. Comprehensive review on global maximum power point tracking techniques for PV systems subjected to partial shading conditions. *Sol. Energy* 183 (2019), 476–500.
[31] Kumari, P. A., Geethanjali, P., 2019. Parameter estimation for photovoltaic system under normal and partial shading conditions: A survey. *Renew. Sustain. Energy Rev.* 84 (2018)
[32] Kihal, A., Krim, F., Laib, A., Talbi, B., Afghoul, H., 2019. An improved MPPT scheme employing adaptive integral derivative sliding mode control for photovoltaic systems under fast irradiation changes. *ISA Trans.* 87, 297–306.
[33] Eltamaly, A. M., Farh, H. M. H., Othman, M. F., 2018. A novel evaluation index for the photovoltaic maximum power point tracker techniques. *Sol. Energy* 174, 940–956.
[34] Fernandez-Guillamon, A., Gomez-Lazaro, E., Muljadi, E., 2019. Power systems with high renewable energy sources: A review of inertia and frequency control strategies over time. *Renew. Sustain. Energy Rev.* 115, 109369.
[35] Salameh, Z. M., Dagher, F., 1990. The effect of electrical array reconfiguration on the performance of a PV-powered volumetric water pump. *IEEE Trans. Energy Convers.* 5 (4), 653–658.
[36] Sai, K. G., Moger, T., 2019. Reconfiguration strategies for reducing partial shading effects in photovoltaic arrays: State of the art. *Sol. Energy* 182, 429–452.
[37] Solodovnik, E., Liu, S., Dougal, R. A., 2004. Power controller design for maximum power tracking in solar installations. *IEEE Trans. Power Electron.* 19 (5), 1295–1304.
[38] Woyte, A., Nijs, J., Belmans, R., 2003. Partial shadowing of photovoltaic arrays with different system configurations: Literature review and field test results. *Sol. Energy* 74 (3), 217–233
[39] Dhana, L. B., Rajasekar, N., 2018. Dominance square based array reconfiguration scheme for power loss reduction in solar photovoltaic (PV) systems. *Energy Convers. Manage.* 156, 84–102.
[40] Karatepe, E., Boztepe, M., Colak, M., 2007. Development of a suitable model for characterizing photovoltaic arrays with shaded solar cells. *Sol. Energy* 81 (8), 977–992.
[41] Vijayalekshmy, S., Bindu, G. R., Iyer, S. R., 2016. A novel zig-zag scheme for power enhancement of partially shaded solar arrays. *Sol. Energy* 135, 92–102.
[42] Belhaouas, N., Cheikh, M. S. A., Agathoklis, P., Oularbi, M. R., Amrouche, B., Sedraoui, K., Djilali, N., 2017. PV array power output maximization under partial shading using new shifted PV array arrangements. *Appl. Energy* 187, 326–337.
[43] Iraji, F., Farjah, E., Ghanbari, T., 2018. Optimisation method to find the best switch set topology for reconfiguration of photovoltaic panels. *IET Renew. Power Gener.* 12 (3), 374–379
[44] Woei-Luen, C., Tsai, C., 2015. Optimal balancing control for tracking theoretical global MPP of series PV modules subject to partial shading. *IEEE Trans. Ind. Electron.* 62 (8), 4837–4848, Aug.
[45] Ahmad, R., Murtaza, A. F., Sher, H. A., 2019. Power tracking techniques for efficient operation of photovoltaic array in solar applications – a review. *Renew. Sustain. Energy Rev.* 101, 82–102.
[46] Velasco-Quesada, G., Guinjoan-Gispert, F., Pique-Lopez, R., Roman-Lumbreras, M., Conesa-Roca, A., 2009. Electrical PV array reconfiguration strategy for energy extraction improvement in grid-connected PV systems. *IEEE Trans. Ind. Electron.* 56 (11), 4319–4331
[47] Kennedy, J., Eberhart, R., 1995. Particle swarm optimization. In: *Proceedings of IEEE Computational Intelligence Society Conference on Neural Networks*. Perth, Australia, 1942–1948.
[48] Karaboga, D., 2005. *An Idea Based on Honey Bee Swarm for Numerical Optimization*. Technical Report-TR06, Erciyes University.
[49] Dorigo, M., Gambardella, L. M., 1997. Ant colony system: A cooperative learning approach to the traveling salesman problem. *IEEE Trans. Evol. Comput.* 1 (1), 53–66.
[50] Mirjalili, S., Gandomi, A. H., Saremi, S., Faris, H., 2017. Salp swarm algorithm: A bio-inspired optimizer for engineering design problems. *Adv. Eng. Softw.* 114, 163–191.
[51] Holland, J. H., 1992. Genetic algorithms. *Sci. Am.* 267, 66–72. Hu, K., Cao, S., Li, W., Zhu, F., 2019. An improved particle swarm optimization algorithm suitable for photovoltaic power tracking under partial shading conditions. *IEEE Access* 7, 143217–143232.
[52] Storn, R., Price, K., 1997. Differential evolution-a simple and efficient heuristic for global optimization over continuous spaces. *J. Global Optim.* 11, 341–359.
[53] Ishaque, K., Salam, Z., 2013. A deterministic particle swarm optimization maximum power point tracker for photovoltaic system under partial shading condition. *IEEE Trans. Ind. Electron.* 60 (8), 3195–3206.
[54] Saad, N. H., El-Sattar, A. A., Mansour, A. E. A. M., 2018. A novel control strategy for grid connected hybrid renewable energy systems using improved particle swarm optimization. *Ain Shams Eng. J.* 9 (4), 2195–2214.

[55] Sundareswaran, K., Peddapati, S., Palani, S., 2014. Application of random search method for maximum power point tracking in partially shaded photovoltaic systems. *IET Renew. Power Gener.* 8 (6), 670–678

[56] Socha, K., Dorigo, M., 2008. Ant colony optimization for continuous domains. *European J. Oper. Res.* 185 (3), 1155–1173.

[57] Titri, S., Larbes, C., Toumi, K., Benatchba, K., 2017. A new MPPT controller based on the ant colony optimization algorithm for photovoltaic systems under partial shading conditions. *Appl. Soft Comput.* 58, 465–479.

[58] Lekshmi, S. B., Umamaheswari, M. G., 2018. A Hankel matrix reduced order SEPIC model for simplified voltage control optimization and MPPT. *Sol. Energy* 170, 280–292.

[59] Belhaouas, N., Cheikh, M. S. A., Agathoklis, P., Oularbi, M. R., Amrouche, B., Sedraoui, K., Djilali, N., 2017. PV array power output maximization under partial shading using new shifted PV array arrangements. *Appl. Energy* 187, 326–337.

[60] Pathy, S., Subramani, C., Sridhar, R., Thamizh Thentral, T. M., Padmanaban, S., 2019. Nature-inspired MPPT algorithms for partially shaded PV systems: A comparative study. *Energies* 12 (8), 1451.

[61] Tafti, H. D., Sangwongwanich, A., Yang, Y., Pou, J., Konstinou, G., Blaabjerg, F., 2019. An adaptive control scheme for flexible power point tracking in photovoltaic systems. *IEEE Trans. Power Electron.* 34 (6), 5451–5463.

[62] Huang, Y. P., Chen, X., Ye, C. E., 2018. A hybrid maximum power point tracking approach for photovoltaic systems under partial shading conditions using a modified genetic algorithm and the firefly algorithm. *Int. J. Photoenergy* 7598653.

[63] Shi, J. Y., Xue, F., Qin, Z. J., Ling, L. T., Yang, T., Wang, Y., Wu, J., 2016. Tracking the global maximum power point of a photovoltaic system under partial shading conditions using a modified firefly algorithm. *J. Renew. Sustain. Energy* 8 (3), 033501.

6 Electric Vehicles – Past, Present, and Future

A. Jagadeeshwaran and H. Shree Kumar

CONTENTS

6.1 Introduction 59
6.2 The Fascinating Story of the Electric Vehicle Battery 61
6.3 The Tesla Motor – An Innovation That Changed the Perception of EVs Forever 64
6.4 The Electric Vehicle Charging Problems 69
6.5 Alternatives to the Lithium-Ion Battery and Emerging Battery Technologies 76
6.6 Conclusion 78
6.7 Bibliography 79

6.1 INTRODUCTION

An electric vehicle (EV) is a road vehicle which runs with electric propulsion. With this basic definition, EVs may include only battery as a source of power and are called battery electric vehicles (BEV). Vehicles involving both battery as well as internal combustion engine in the same vehicle are called hybrid electric vehicles (HEVs). Vehicles that use fuel cells are called fuel cell electric vehicles (FCEV). Electric vehicles use electric motor drives as a medium of propulsion, compared to internal combustion engines used on vehicles, which use fossil fuels such as petrol, diesel, etc. Electric vehicles are a multidisciplinary subject which covers broad and complex aspects, including chassis, body technology, propulsion technology, and energy source technology. The characteristics of different types of electric vehicle are given in the table that follows.

The concept of electric vehicles is not a twenty-first-century idea, because electric-powered motors came up at the same time when petroleum-driven engines were being developed. Almost two centuries ago, in 1828, a Hungarian engineer named Anyos Jedlik invented the first prototype of the electric motor and used it to power a small model car. In 1834, an American blacksmith, Thomas Davenport, created a similar device that could be driven for a short distance using an electric track. In the Netherlands, Professor Sibrandus Stratingh built a tiny electric car powered by nonrechargeable batteries. The major problem faced during this era was that the primary cell batteries were required in very large numbers to drive the motor for a very short range, making such vehicles nonpracticable for daily use. In 1857, French physicist Gaston Plante invented the lead-acid battery, which changed the world of electric cars for the next many years. Many countries started producing three-wheeled electric cars, until the United States made a huge breakthrough. In 1891, they created the first electric vehicle for commercial purposes, which was a six-passenger wagon with a top speed of 23 km/hr. After this, the electric car market thrived, and the late-1890s' electric-powered taxis filled the streets of London. At that time, electric vehicles had many advantages over gas-guzzling IC engines, because they did not vibrate nor did they give out awful "burning gasoline" smell, and most importantly, they did not require much effort to start. By the early 1900s, almost one-third of cars in the United States were electric-powered. However, this situation changed in the late 1920s, after infrastructure in the United States improved and vehicles needed to go farther more efficiently. So fossil fuel–powered cars took the lead because they could cover longer distance with higher speed compared to their electric vehicle counterparts. The electric vehicle of that era had a maximum speed of 24 to 32 km/hr, with a maximum range of 50 to 80 km in the best possible situation on single charge, which means they required frequent charging for many hours. Things were about to take a turn for the worse, along with improvement in the infrastructure, when the electric starter was invented and gasoline cars started using them along with mufflers, which reduced the noise of the engine. Electric cars took a final hit in 1910, when Henry Ford started mass production of gasoline-powered vehicles, which made them very cheap compared to electric vehicles of that era. This trend continued till 1935, when automobile companies realized that there was no future scope for electric vehicle and stopped their production completely. The era between 1935 and 1955 is called the dark age of electric vehicles because during this time, IC engines had completely replaced electric engines and no research was going on for further development of electric vehicles anywhere in the world. After 1955, once again technical interest was revived in electric vehicles. By the mid 1960s, solid-state electronics allowed the development of new ideas for electric vehicles driven by innovative three-phase induction motors. The discovery of the induction motor was done independently by Galileo

TABLE 6.1
Comparison of Different Types of EVs

Types of EVs	Battery EVs	Hybrid EVs	Fuel Cell EVs
Propulsion	• Electric motor drives	• Electric motor drives • Internal combustion engines	• Electric motor drives
Energy System	• Battery • Supercapacitor	• Battery • Supercapacitor • ICE generating unit	• Fuel cells
Energy Source and Infrastructure	• Electric grid charging facilities	• Gasoline stations • Electric grid charging facilities (optional)	• Hydrogen • Methanol or gasoline • Ethanol
Characteristics	• Zero emission • Independence from crude oils • 100–200 km short range • High initial cost • Commercially available	• Very low emission • Long driving range • Dependence on crude oils • Complex • Commercially available	• Zero emission or ultralow emission • High energy efficiency • Independence from crude oils • Satisfied driving range • High cost now • Under development
Major Issues	• Battery and battery management • High-performance propulsion	• Managing multiple energy sources • Dependent on driving cycle	• Fuel cell cost • Fuel processor • Fueling system

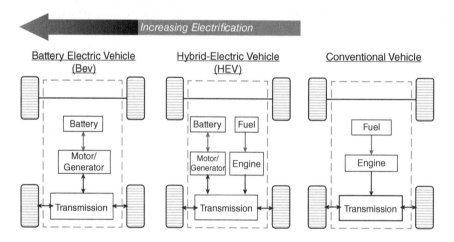

FIGURE 6.1 Working mechanism of different types of vehicles.

Ferraris, an Italian professor, and Nikola Tesla, who was a Serbian American inventor, between 1885 and 1887. A linear outfit and independent company in the United States developed electric vehicles in the mid-1960s along with General Motors, which built high-performance electric vehicles using silver zinc batteries. This electric vehicle from GM was called Electrovire II and was launched in 1966. During this time, GM launched another vehicle, which was a van, called the GM Electrovan, and used a hydrogen oxygen fuel cell, thus becoming the first vehicle to do so in 1966. The fuel cell was supplied by a company called Union Carbide at a cost of $6,000. In the 1970s, geopolitical instability in the Middle East, especially in major OPEC countries, such as Iran, Iraq, and Kuwait, led to a steep rise in crude oil prices. This created the urgency to develop alternate-fuel vehicles. GM converted an IC engine vehicle into electric vehicle using nickel zinc batteries in 1977, and it was called the GM Electrovette. However, for a short duration, the crude oil prices fell and interest in electric vehicles also reduced. Probably the most famous electric car of the 1970s was the LRV (Lunar Roving Vehicle), which was sent to the moon on Apollo missions thrice, and it was powered by 36-volt silver zinc potassium hydroxide batteries which were nonrechargeable. Each rover had a theoretical range of 57 mi. The major focus in 1960s and 1970s was to use electricity to reduce air pollution. With the 1973 oil crisis, which began in October 1973 during the Yom Kippur War, Western countries started developing an interest in electric vehicles and many prototypes were developed, but none of it was practical in nature. This was due to limited range and speed of the electric vehicles of that era, because the battery technology hadn't developed much. It was only in 1994 when General Motors decided to mass-produce an electric car. This decision was inspired by the California Air Resources Board, which passed a mandate that made the production and sale of zero-emission vehicles a requirement for major automakers selling cars in the United States. The electric vehicles were made available through lease-only agreements initially to residents of the cities of Los Angeles, California, Phoenix, Tucson, and Arizona. While customer reaction to the EV1 was positive, GM believed that electric cars occupied an unprofitable niche of the automobile market and ended up crushing most of the cars, regardless of protesting customers. The EV1 program was subsequently discontinued in 2002, and all cars on the road were taken back by the owner, under the terms of the lease. Lessees were not given the option to purchase their cars from GM, which cited parts, service, and liability regulations. The majority of the EV1s taken back were crushed, with about 40 delivered to museums and educational institutions, with their electric power trains deactivated, under the agreement that the cars were not to be reactivated and driven on the road. Approximately 20 units were donated to overseas institutions. The EV1's discontinuation remains controversial, with electric car enthusiasts, environmental interest groups, and former EV1 lessees accusing GM of self-sabotaging its electric car program to avoid potential losses in spare parts sales (sales forced by government regulations), while also blaming the oil industry for conspiring to keep electric cars off the road. The electric vehicles once again made a strong comeback in the first decade of the 2000s, with many companies such as Tesla, General Motors, Toyota, Lucid Motors, etc. developing their EVs which are comparable and, in some parameters, exceeding the capabilities of their internal combustion engine counterparts. This trend is still continuing with the latest innovations in battery technology, and the future of EVs looks pretty bright.

6.2 THE FASCINATING STORY OF THE ELECTRIC VEHICLE BATTERY

The present electric vehicles use lithium-ion batteries, and the most famous among them is the Tesla battery, such as the Tesla 2170 battery. Around 2,976 of these batteries combine together to form the battery pack of the Model 3 Tesla car. However, this cell originated through a long evolution process to become the highest-energy-density cell in the world. These batteries represent the first hopeful steps in transitioning the society towards a new standard in practical and economical transportation through electric vehicles. The electrochemical battery has been around for more than 200 years. Using them practically in the manner comparable with IC engine–based vehicles was very challenging. It's only within the last twenty years of innovation that batteries have developed up to a point where it is feasible to use them on a practical purpose in electric vehicles.

Alessandro Volta, an Italian physicist, had developed the first electrochemical battery, known as the voltaic cell, in 1799. It uses silver and zinc as electrodes, with salt water as electrolyte. The zinc plate would react with the salt water, producing an accumulation of electrons forming an anode. An electrode that resolves in the production of electrons is known as an anode. Meanwhile, in the silver plate, a simultaneous reaction with the salt water enables it to accept electrons, and it is known as the cathode. The liquid with which both the anode and the cathode react, which in this case is salt water, is called an electrolyte. The electrolyte also functions as a pathway for the transfer of positively charged ions to balance the flow of electrons from the anode to the cathode, keeping the reaction running. In order to prevent the ions of the more noble metal from plating out the other electrode, a semipermeable membrane is sometimes used to divide the reaction happening in the electrolyte. This type of chemical reaction is known as reduction–oxidation reaction, or redox reaction. The entire reaction can be split into two half reactions, where one half reaction occurs at the anode, while the other half reaction occurs at the cathode. Each of these reactions possesses a particular standard potential. The difference in standard potential between the electrodes becomes the cell's overall electrochemical potential, or its voltage. The greater the difference, the greater will be the voltage. Individual cells can be combined into a configuration that can increase both the total voltage and the current capacity; this is known as a battery. On a primary battery, the electrodes become depleted as they release their positive or negative ions into the electrolyte, and the buildup of the reaction products on the electrodes prevents the reaction from further continuing, which results in a onetime-use battery. In secondary batteries, the chemical reaction that occurs during the discharging process can be reversed. However, the process isn't perfect, because during each charge cycle, the battery loses performance over time, leading to battery wear.

In 1859, French physicist Gaston Plante invented the lead-acid battery, which was the truly rechargeable battery. A lead-acid cell consists of a lead anode and a lead dioxide cathode immersed in a sulfuric acid electrolyte. While discharging, both electrodes react with the sulfuric acid to produce lead sulfate, while the electrolyte loses its dissolved sulfuric acid. This chemical reaction could be reversed by passing a reverse current through the battery, thus recharging it.

By the 1880s, the the lead-acid battery took on a more practical aspect that made them easier to mass-produce. Because the electrodes and electrolytes are not 100% conductive in nature, they all have internal resistance. Lead-acid batteries have low internal resistance, making them ideal for producing large surge current. This property makes them ideal for powering large current-intensive load, such as in electric motors. By the end of 1880s, the lead-acid battery brought the rise of the first practically viable electric vehicles in Europe.

Before the IC engines took over, electric vehicles held many speed and range records. Among the most notable of these records was the breaking of the 100 km/hr speed on

April 29, 1899. In the early 1900, electric vehicles began to gain popularity in the United States after thriving in Europe for over 15 years. By 1905, in the United States 40% of automobiles were powered by steam, 38% by battery, and the rest, 22%, by gasoline. However, the golden age of electric vehicles soon ended, as gasoline-powered vehicles could travel at higher speed for longer distances. The lack of sufficient infrastructure for recharging batteries, as well as very long charging duration, also hindered the adoption of electric vehicles. Henry Ford's mass-produced Model T IC engine–based vehicles were the final blow for electric vehicles. Introduced in 1908, the Model T made gasoline-powered vehicles widely available as well as affordable. These cars could be purchased for as low as $650, which was half the price of most electric vehicles of that era.

As a result, most electric vehicle manufacturers ceased production of their EV models. Most electric vehicles of that era had a top speed well below 30 km/hr and generally had a range of about 40 km. Despite these limitations, lead-acid batteries would still remain a part of the automotive industry, as their high current capacity makes them ideal for powering electric starters on gasoline vehicles. The pead-acid batteries would evolve over the next century, becoming a key part of electromechanical and electronic systems in cars. The energy storage characteristics of a battery's chemistry can be compared to a handful of key parameters which are very essential design parameters. *Specific energy* is the amount of energy which can be stored per mass unit (Wh/kg). *Energy density* (Wh/L) is the specified amount of energy that can be stored per unit of volume. *Specific power* (W/kg) is the quantity of power it can generate per unit of mass. The parameters that determine a battery's suitability for application are:

1. Self-discharge rate (% per month)
2. Cycle durability (number of charge cycles)
3. Energy efficiency (%)
4. Nominal cell voltage (volts)
5. Cost per unit of energy (Wh/USD)

A typical modern AA alkaline primary battery, when compared to lead-acid battery, has four and a half times specific energy, five and half times the energy density, but only one-third of the specific power. They also have very low self-discharge rate, losing 0.17% per month, compared to 3–20% for lead-acid batteries. Lead-acid batteries have a huge advantage in cost per watt-hour compared to alkaline batteries, and lead-acid batteries can store 40% more energy at a reduced cost compared to alkaline batteries. Over the next 60 years, electric vehicles entered a dark age, and there was not much development in battery technology due to the availability of cheap, abundant gasoline. During the oil crisis of 1973, enthusiasm in EVs once again developed, but it was short-lived. It should be noted that during this period in history, the most famous EV of today, that is, LRV (Lunar Roving Vehicle), made its first drive on the moon in 1971. Though it helped to raise the profile of EVs, from a technical point of view, it did not contribute much to the development of electric vehicles.

Fundamentally, very little had changed on battery technology from the golden age of electric vehicles because, still, lead-acid batteries were the only available option, seriously limiting their speed and range. In the late 1960s, research had begun in the global community, such as by COMSAT Laboratories, to develop a new battery technology based on nickel hydrogen, designed specifically for use on satellite and spacecraft. These batteries use hydrogen stored at about 82 bars, with nickel oxide hydroxide as cathode and platinum catalyst as anode, with potassium hydroxide as electrolyte. As the battery discharges, hydrogen is consumed by the anode, producing water, which is simultaneously consumed in the nickel oxide hydroxide reaction. The pressure of the hydrogen would decrease as the cell discharges, indicating the charge status of the cell. These batteries offered slightly better energy storage capability than lead-acid batteries, as well as service life, exceeding 15 years, and had cycle durability exceeding 20,000 cycles; as well, they were resilient to overcharging and overheating. By the early 1980s, the batteries were used on many famous space missions, such as with the Hubble Space Telescope and the International Space Station (ISS).

Nickel hydrogen cell chemistry was based on nickel cadmium, which was one of the first rechargeable alkaline cells ever developed. First created by Waldemar Junger of Sweden in 1897, nickel cadmium cells also use a nickel oxide hydroxide cathode with potassium hydroxide electrolyte. They use the toxic metal cadmium as an anode.

They offer slightly better energy density and specific power than lead-acid batteries. They offer more than six times cycle durability, making them ideal for consumer products and portable electronic tools such as cameras, flashlights, cordless phones, etc. However, they are less ideal for use in electric vehicles because of their complex charging requirements, lack of a lustier energy capacity, and high cost. They also suffer from a problem called memory effect, wherein a battery loses its maximum charge capacity over a period of time if they are recharged after being only partially discharged. Simultaneously, other methods were also being developed for achieving better results, including the nickel metal hydride. This new chemistry relied on metal hydrides, which are a class of material containing metal or metalloid bonded to hydrogen to function as an anode. Over the next two decades, research took place in nickel metal hydrides and was funded by companies such as Volkswagen and the Audi Group, resulting in batteries having storage capability similar to nickel hydrogen though with a fivefold increase in specific power. However, these batteries suffered from electrode alloy instability within the alkaline electrolyte, thus suffering from low-charge-cycle durability, typically around 500 charge cycles. Finally, in 1987, a breakthrough in research led to an anode material consisting of a mixture of neodymium, lanthanum, nickel, cobalt, and silicon that allowed the cell to retain 84% charge capacity after 4,000 charge–recharge cycles. This breakthrough led to the first consumer-grade nickel metal hydride batteries to be commercially available in 1989. In the nineties, more alloys, such as titanium and nickel modified with chromium, cobalt, and manganese, would result in batteries with specific energy almost two and a half times higher than that of lead-acid cells. This new anode material offered energy density five and half times greater than that of lead-acid batteries. Thus, new battery technology had evolved as a viable successor to the lead-acid technology in electric vehicles. Nickel metal hydride batteries did suffer from some drawbacks, such as having a higher self-discharge rate than lead-acid batteries, memory effect issues, as well as cost, which was two to three times higher than that of lead-acid batteries. With the advancement of microprocessor, advanced charging and battery monitoring technology could be developed, which could negate many of the problems faced by these batteries. These batteries became quite famous in the consumer market, replacing nickel cadmium batteries; almost 100 years after the first golden age of electric vehicles had come to an end, a new era for electric vehicles was about to begin. In the 1990s, the California Air Resource Board began to push major automakers to produce zero-emission vehicles to minimize the level of air pollution. This, along with the recent development of nickel metal hydride battery technology, made practical electric vehicles possible.

By the late 1900s, mass production of electric vehicles had once again been started by major automakers based on available battery technology. Some notable all-electric vehicles include the Honda EV hatchback, Ford Ranger EV, S10 EV pickup, and the controversial EV1 from General Motors. While other battery technologies were explored, these first-generation electric vehicles all used nickel metal hydride batteries. Due to the limitation of vehicle size and target price, most of these vehicle had battery pack of around 26.4 kWh, with range of around 160 kilometers, and could be charged using 6.6 Kw Magne Charge inductive converter with a curb weight of 1,319 kg. For the first time, electric vehicles were now a practical option outside the city as well. These vehicles, however, lacked mass appeal because they were still slow, low powered, expensive, and range-limited, with very few charging stations available. In the United States, companies such as General Motors leased their electric vehicles to common people and started collecting daily data from customers to bring further improvement in their EVs. However, it would still take more than a decade for the electric vehicles to finally arrive in the market and compete comprehensively with their internal combustion engine counterparts.

During the mid-1970s, Exxon Mobil started backing a breakthrough made by British scientist Stanley Whittinghan, who had discovered a way to make electrode from a layer material that could store lithium ions within sheets titanium sulfur and could be moved from one electrode to other, creating a battery from the highly reactive property of lithium. In lithium-based cells, the electrolyte doesn't take part in the reaction but only mediates in the movement of lithium ions, controlling the cell's characteristics. This method was known as intercalation, and it permitted the addition of lithium ions into a post material without significantly changing its structure.

Because lithium was used in these batteries, they were initially called lithium metal batteries. Compared to other battery chemistry, they have higher energy density, with some as high as ten times that of lead-acid battery, and have less self-discharge rate. However, these batteries were very expensive because of costly synthesis process. They were also very dangerous, as pure lithium would instantly react with water, releasing flammable hydrogen gas. By the late 1970s, new cathode materials were developed, which made commercial production of lithium batteries possible, and they found use in a variety of electronic equipment, such as pacemakers, and computer equipment, such as motherboard, where long life of battery was essential. These new lithium cells found in consumer applications use manganese dioxide as the cathode, with a salt of lithium dissolved in organic solvent as the electrolyte. Despite the advantages offered by lithium, a practical rechargeable lithium-ion battery remained a challenge. The electrochemical reaction that allowed the lithium cells to function so efficiently also made them prone to explosion when overcharged. The cathode would also quickly erode from repeated charge and discharge cycles. The problems of Whittinghan's lithium cells would be solved by American material scientist John B Goodenough. He had become very much familiar with the family of materials known as metal oxides.

These compounds combined oxygen and metal to form metal oxides. He proposed that metal oxide would allow charging and discharging a higher voltage than Whittinghan's lithium cells, which would result in greater energy storage. His team experimented various metal oxides and had concluded that cobalt was the most stable, allowing lithium ions to be extracted at 04 volts without eroding the electrodes. Completed in 1980, the lithium cobalt oxide cathode–based lithium ion cell became a massive breakthrough in the field of rechargeable batteries. The world's first lithium ion battery had energy density unmatched by anything else ever developed, offering a specific energy and energy density almost seven times that of lead-acid batteries, because lithium cobalt oxide was such a stable material which could be used with a negative electrode material other than lithium metal. In the same year, Moroccan engineer Rachid Yazami demonstrated the reversible chemical interpolation of lithium in graphite, inventing the far safer and more stable lithium graphite anode, eliminating the need for purer metallic lithium. During recharging, when a voltage is applied, positively charged lithium ions from the cathodes migrate to the graphite anode and become lithium metal.

Because lithium has a strong electrochemical driving force to be oxidized upon discharging, it moves back to the cathode, becoming lithium ions while giving up its electrons to the cobalt. In 1991, Sony combined Goodenough's cathode and carbon anode for its commercial lithium-ion battery. The result was a huge commercial success, as these batteries found wide use in consumer goods like cameras and handheld video cameras.

However, these batteries had a major problem of thermal runaway due to overcharging. This made them unsuitable for large-capacity mobile use. By the late 1990s, Goodenough once again made a huge leap in battery technology by introducing a far more stable lithium-ion cathode–based lithium iron phosphate battery. The cathode material was thermally stable, and it allowed the formation of crystalline LiFePO$_4$ (phospho-olivine) that allowed the ions a better pathway to move through them. This phospho-olivine cathode has similar structure to mineral olivines.

Lithium ferro phosphates came as a successor to a similar group of cathodes made from manganese spinel, which also utilized the crystal structure to get thermal stability. However, manganese spinel suffers from poor cyclic stability due to the tendency of manganese to dissolve in the electrolyte, and they are thermally less stable compared to lithium ferro phosphate. However, they were very cheap to construct due to the low cost of manganese. Lithium-ion cells can now be made safe into a large format that can undergo rapid charge and discharge cycles. This new cathode material finally opened up lithium batteries to new high-demand applications, such as in electric vehicles. In December 1997, Nissan introduced the Nissan Ultra EV. It was the first commercial vehicle to use lithium-ion batteries in its power train. It had a 33 kWh battery and weighed 360 kg, with a range of 200 km. Despite the success of the lithium-ion battery, the later part of 1990s and early part of 2000s saw EVs powered by nickel metal hydride batteries partly because of the high cost of lithium-ion batteries.

It would take a small Silicon Valley start-up, Tesla Motors, starting with their announcement of producing a luxury electric vehicle that could go more than 200 mi on a single charge, to ignite interest in lithium-ion batteries–powered electric vehicles by 2010. Tesla Motors established a manufacturing plant in California and became the largest automobile industry in California. Tesla Motors is unique, because unlike other car manufacturers, it produces only electric vehicles. The success of Tesla Motors inspired other major car manufacturers as well as start-ups to venture in the field of electric vehicles. By 2020, virtually every major automaker has started manufacturing electric vehicles, with vast majority of them using lithium-ion battery technology with ranges between 200 km to 450 km, with recharge time between half an hour and ten hours based on the method of recharging. Countries such as Denmark and Netherlands have set up a complete network of charging stations both in rural and urban areas, eliminating the concern of range anxiety among electric vehicle owners. In 2010, cathode materials of lithium-ion battery would once again evolve with the use of lithium nickel manganese cobalt oxide (NMC) cells. These cells have the lowest self-heating rate among lithium cells and have been used by almost every major electric vehicle makers, with the only exception being Tesla Motors, who uses lithium nickel cobalt aluminum oxide (NCA) cells. These cells have been in use since 1999 for mobile applications such as cellphone, offering high specific energy and specific power with longer lifetime. NCA batteries are not safe and require a cooling material to be circulated through the outer body of the cell to reduce heating. However, Tesla claims that these cells are much cheaper for mass production. These batteries all rely on two key materials, namely, lithium and cobalt, both of which are rare elements, with estimates suggesting that their supply could become critical by 2050. So research is being carried out to develop better battery technology compared to lithium-ion battery technology.

6.3 THE TESLA MOTOR – AN INNOVATION THAT CHANGED THE PERCEPTION OF EVS FOREVER

There are many but mainly five different types of electric motors which are being used in the current generation of electric vehicles. They are:

1. Brushless DC motors (BLDCM)
2. Permanent magnet synchronous motor (PMSM)
3. AC induction motor (ACIM)
4. Interior permanent magnet motor (IPMM)
5. Permanent magnet switched reluctance motor (PMSRM)

However, before the type of motor is chosen, software simulation is very much necessary. MATLAB simulation is commonly preferred for research in this field. This is preferred because of the following reasons:

1. Explore electric power train architecture
2. Tune regenerative braking algorithms
3. Modify a suspension design
4. Optimize vehicle performance
5. Develop active chassis control
6. Validate ADAS (advanced driver-assistance system) algorithms
7. Test using hardware-in-the-loop (HIL)

Using software simulation, around 95% accuracy can be obtained in comparison with the physical model. The power required to move the vehicle depends on four forces, which are:

1. Push force, which is generated by the motor
2. Aerodynamic drag force generated in opposite direction to the push force due to the friction offered by the body of the vehicle
3. Frictional force due to the friction of wheels with the road surface
4. Gravitational force, which is simply the gravitational pull of the Earth

It is obvious that the power generated by the motor is the product of total tractive force and velocity of the vehicle.

Forces on the Vehicle

Total Tractive Force

1. Rolling Resistance Force (F_{rr})
2. Aerodynamic Drag Force (F_{aero})
3. Hill Climbing Force (F_{hc})
4. Acceleration Force (F_{rr})
   1. Linear Acceleration Force
   2. Angular Acceleration Force

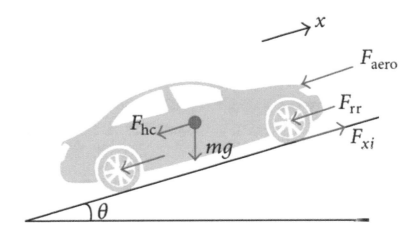

FIGURE 6.2 Various forces on a vehicle.

The rolling resistance force (Frr) is one of the forces that act to oppose the motion of the vehicle. The coefficient is a function of the tire material, structure, temperature, pressure, thread geometry, road roughness, material, and presence and absence of liquids on the road. Its value varies with the speed of the vehicle, and as the tire pressure increases, its vale decreases.

$$\text{Rolling Resistance Force} \left(\text{Frr}\right) = \mu_{rr} m.g \quad \text{Newton} \quad (6.1)$$

Where μ_{rr} is rolling resistance constant, m is the mass of the vehicle, and g is the gravitational acceleration constant.

The value of rolling resistance force (Frr) is around 0.013 for car tires on concrete/asphalt roads, while on rolled gravel, its value is 0.02, and on unpaved roads, the value becomes 0.05. This means in Earth gravity, a car of 1,000 kg on asphalt will need a force of around 130 Newton for rolling (1,000 kg × 9.81 m/s² × 0.013 = 127.53 N).

The aerodynamic drag force (F_{aero}) is another force that acts on the vehicle, and it resists the motion of the vehicle through the air, which is also called as drag. Drag is a force which acts parallel to and in the same direction as the airflow. Aerodynamic drag increases with the square of the speed.

$$\text{Aerodynamic Drag Force} \left(F_{aero}\right) = 0.5 \rho C_d A v^2 \quad \text{Newton} \quad (6.2)$$

Where ρ is the air density kg/m³, A is the frontal area m², C_d is the drag coefficient, and V is the velocity of the vehicle in m/s.

The value of drag coefficient (C_d) varies between 0.3 and 0.7, depending on the type of vehicle, as shown in Figure 6.3.

The hill climbing force (F_{hc}) is another force acting on a vehicle in motion and is due to the ups and downs in the road surface. This force acting on a vehicle is given as follows:

$$\text{Hill Climbing Force} \left(F_{hc}\right) = m.g.Sin\theta \quad (6.3)$$

Where m is the mass of the vehicle, g is the gravitational acceleration constant, and $Sin\theta$ is the angle between surface plane and rise plane.

So the total power required for the vehicle from the motor is given by:

$$\text{Total Power } P = F_{te}.V \quad (6.4)$$

Where F_{te} is the total tractive force and V is the velocity of the vehicle in m/s.

$$\text{The Total Tractive force} \left(F_{te}\right) = Frr + F_{aero} + F_{hc} + F_{la} + F_{wa} \quad (6.5)$$

Where Frr is the rolling resistance force, F_{aero} is the aerodynamic drag force, F_{hc} is the hill climbing force, F_{la} is the linear acceleration force, F_{wa} is the angular acceleration force.

$$\text{The Linear Acceleration Force} \left(F_{la}\right) = m.a \quad (6.6)$$

Where m is the mass of the vehicle and a is the acceleration, which is simply the change in velocity (Δv) over the change in time (Δt), given by the formula a = $\Delta v/\Delta t$.

$$\text{The Angular Acceleration Force} \left(F_{wa}\right) = IG^2 a/r^2 \eta_q \quad (6.7)$$

Where I is the moment of inertia, G is the gear ratio, A is the acceleration, r is the radius of the tire, and η_q is the efficiency of the gear.

For ease in calculations during software simulation, we can use the values given in Table 6.2. The standard speed of the vehicle has been set at 40.0 kmph as a safe limit.

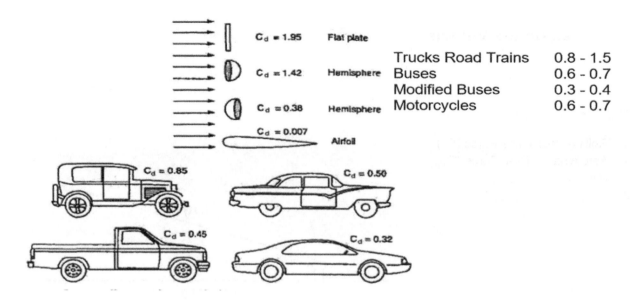

FIGURE 6.3 Drag coefficient values of vehicle body.

TABLE 6.2
Values of Various Parameters for Simulation

Parameters	Notations	Value	Units
Weight	M	1,000	kg
Width	W	1.7	m
Height	H	1.5	m
Rolling resistance	μ_{rr}	0.013	
Air density	P	1.225	kg/m³
Drag coefficient	C_d	0.4	
Gravitational force	G	9.8	m/s²
Speed of vehicle		40	kmph
Frontal area	A	2.55	m²
Velocity	V	11.11	m/sec

Based on these formulas, if we calculate the power required to run a vehicle at 40 kmph, the value comes to around 2,563.580 watts. The same value reaches 11,472.00 watts for the vehicle running at the speed of 100 kmph. Figure 6.4 shows the graph between vehicle speed and power consumption, which clearly indicates that as the speed of the vehicle increases, its power consumption also increases, and accordingly, the motor as well as the battery for EVs have to be chosen. Based on these calculation and formulas, we can choose the power of the motor and its type.

Brushless DC motors (BLDCM) are being used in most lightweight two-wheeler and three-wheeler EVs, like electric scooters and electric motorcycles. Permanent magnet synchronous motors (PMSMs) are used by EV manufacturers for their high-performance electric motorcycles, electric cars, and electric buses. AC induction motor (ACIM) is being used by some EV manufacturers in two-wheelers and four-wheelers and are also found in Indian EVs, like Mahindra e-Verito, Mahindra e2o, Mahindra e-Supro, TATA Tigor (EV), and TATA Tiago (EV). Interior permanent magnet motors (IPMMs) are also used by some two-wheeler EV manufacturers for their high-performance electric motorcycles. Permanent magnet switched reluctance motors (PMSRMs) are used by Tesla Motors for their EVs. Most EV manufacturers modify and combine the characteristics of two different motors to get superior results. Tesla Motors engineers made a stunning design choice when they developed the Tesla Model 3. They abandoned the conventionally used and well-proven induction motors and replaced them with a new kind of motor, that is, IPMSynRM. These motors have a totally different design, making use of both magnetic and reluctance action. The big news is that Tesla Motors has started replacing induction motors in the Model S with these new models as well. How does it work, and what is so special about IPMSynRM? We first need to understand the Model S electric motor, which is an induction motor. Clearly, as you can see, the rotating party here is an arrangement of conducting bars.

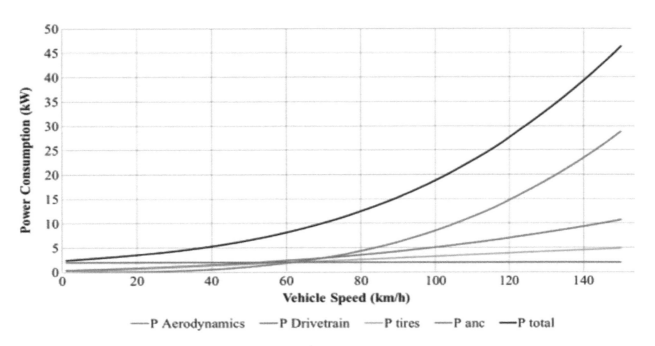

FIGURE 6.4 Graph between vehicle speed and power consumption.

Alternating currents from the battery pack into the motor windings would create a rotating magnetic field; the fluctuating field interacts with the rotor bars and generates electromotive force on them, which in turn generates currents in the rotor bars. The interaction between these induced currents and the RMS imposes force on the rotor bars, and the rotor starts to spin. These motors are more efficient, but not up to the mark. For instance, long drives at cruise speeds lose 3% to 4% of energy to generate currents in the rotor bars, which is definitely not efficient. For the most crucial performance parameter is its starting torque even though the induction motor has a better starting torque than IC engines, there is a motor technology that provides even better starting torque from the same volume of motor technology based on permanent magnets. The permanent motors work on the basis of the attraction between two magnetic fields and produce a good starting torque when you operate them using a controller, and they do not experience energy loss in the rotor. An efficient permanent magnet motor can be made by placing permanent magnets around the solid iron cylinder, so why not replace the screw gauge–type rotor with a permanent magnet one? These four permanent magnets produce a combined magnetic field, and the shape of this combined field is quite important for further analysis.

Now, we need to analyze the interaction between the RMF and the combined magnetic field. Analyzing the force interaction between two magnetic fields is simple. Just observe how the cells in the north poles interact with each other. For simplification, let's hide the magnetic field produced by the permanent magnet and keep only the north and south poles.

At this angle, the RMF definitely produces a torque on the rotor. Now, let's rotate the RMF to 45°. Interestingly, at this angle the rotor experiences maximum torque because the attractive and repulsive forces are passing almost tangentially to the rotor and they are producing the torques in the same direction.

Thus, this is the perfect and go-to start for your electric car; maintaining this angle or another angle regulation is the smart controller's job. In this design, the rotor has no induced current, which reduces the input energy required and its higher efficiency than those of induction motors. Along with higher starting torque, the permanent motor also runs at a synchronous speed, but search for the perfect electric motor is not over yet. A permanent magnet motor produces good torque when you start the car or ride up the hill; however, when the car cruises down the road at high speeds, permanent magnet motors have terrible performance. The villain here is the back EMF. The magnetic field lines produced by the permanent magnet link with the stator windings and generate an EMF there. This EMF is called back EMF, which is clearly reverse voltage to the stators; supply voltage. The higher the speed, the more it produces back EMF. This phenomenon is why permanent magnet motors performed terribly in high-speed applications. Moreover, these high-strength magnets of the resultant magnetic eddy current loses, which increases the heat in the machine. How can we modify this design so that it would work efficiently even at high-speed operation? For

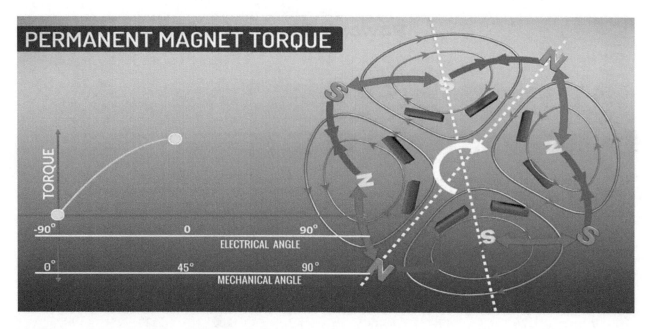

FIGURE 6.5 Maximum torque at permanent magnet.

high-speed operations, Tesla's engineers made use of iron's reluctance property. The ability to oppose magnetic fields is known as reluctance. Iron is a good keeper of magnetic field lines, but air is not. If the rotor was turned by 45°, it will face a very low reluctance. The rotor always has a tendency to attain a low reluctance state, and therefore, if the magnetic field rotates, the rotor will rotate along with it so that the rotor can always be in a low reluctance state. The rotor's rotation speed will be the same as the RMF speed. The torque produced by this phenomenon is known as reluctance torque, and such motors are called synchronous reluctance motors. Synchronous reluctance motors (SynRM) are highly efficient, and they don't have back EMF issues. The permanent magnet motor is good for low speed, while SynRM are quite good for high-speed operations, if we can integrate the SynRM technology into the permanent magnet motor technology. We can easily achieve this design integration by placing the permanent magnets into the slotted cuts of the SynRM deep within the iron core. This placement further reduces the magnetic effect on stator winding and thus reduces back EMF. This design is the internal permanent magnet synchronous reluctance motor (IPMSynRM) used in the Tesla Model 3 electric car to achieve superior performance.

The relative permeability of the magnet is almost the same as that of the air, so they will oppose the field to pass through it just as the air did earlier, thereby generating the reluctance torque. To analyze this motor properly, we first need to observe the resultant magnetic field that permanent magnet arrangement producers. The FEA simulation software EMWorks is used to study this in detail. The resultant magnetic field produced by this arrangement would be shown.

Using this, let's do further analysis. The interesting thing about this design is that the permanent magnet and reluctance part of this motor have totally different behavior in regard to the position of the RMF. Let's analyze them separately from latest IPSynRM design. Let us remove the iron core and keep only the permanent magnets. At this RMF position, the permanent magnets do not experience torque as there is no tangential component for these four forces and the torque; the remaining forces produced cancel each other out. If the RMF is rotated by 45°, a torque acts upon the magnets in a clockwise direction due to the effect of the RMF. At this angle, we get maximum torque at a permanent magnet.

Let's see what happens when we turn by another 45°. The torque the rotor produces goes to zero again, allowing us to obtain the permanent magnet's torque curve.

The iron part of the rotor has an opposite effect on the same technique. At the initial angle, torque production would be zero because it is a perfectly misaligned and symmetrical case. When we slightly offset the RMF in a clockwise direction, the rotor will experience negative and maximum torque. As the RMF reaches 45°, the torque becomes zero again. Since this is a perfectly symmetrical case, again as we further rotate the RMF, the reluctance torque produced will be positive. Now, let's analyze the Tesla Model 3's motor or the combined permanent magnetic and reluctance effect together on the motor. It is clear from the total torque graph that if RMF angle is around 50°, we'll get maximum torque from the motor. So Tesla Motors engineers made sure that when you start the car, the RMF angle is around 50°, which will guarantee maximum torque production. We know that as the motor speed increases, the permanent magnets induce a back EMF on the stator coils. To overcome this issue, Tesla engineers used a solution that is quite simple for turning at high speeds. They aligned the RMF opposite to the permanent magnetic field, thus weakening or canceling the permanent magnetic field. This

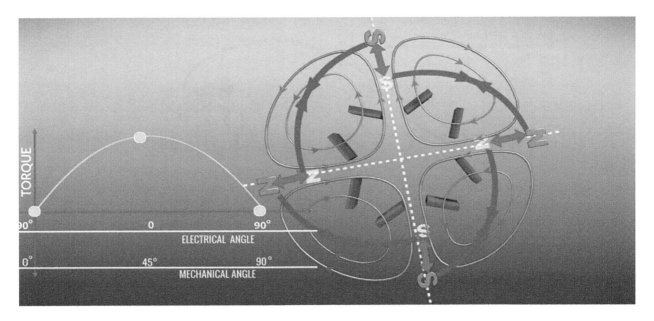

FIGURE 6.6 Minimum torque at permanent magnet.

way, even at high speed, such motors won't produce much back EMF. Obviously, at this stage, torque production work mostly comes from the reluctance effect. The Tesla Model 3 uses a six-pole design which is no different from a four-pole design apart from getting higher torque.

The latest models of Tesla Model 3 have IPSynRM motors in the front of the car. They have efficiencies in the range of 96% compared to the traditional induction motors having efficiency up to 94%. These motors have set a new performance standard in the world of electric vehicles.

6.4 THE ELECTRIC VEHICLE CHARGING PROBLEMS

The largest problem towards mass adoption of EVs presently is battery technology used, which has range concerns as well as the time required to recharge these batteries, which is called the charging problem. The alternatives to Li-ion battery are being developed but still in experimental stage and will be discussed in the next section, but the problem of recharging the batteries can be solved to some extent by fast charging. However, the infrastructure to establish fast chargers is not cheap, and large-scale installation of fast chargers isn't a feasible solution, given the fact that EVs are still used only in very small numbers throughout the world when compared with internal combustion vehicles.

As we can see from the previous figure, the EVs are much simpler in their construction compared to their IC engine counterparts. The only source of power to run the motor comes from the batteries, and since batteries provide direct current, an inverter is required to convert direct current to alternating current. All EVs, regardless of the battery technology used, have a limited source of power and need their battery to be recharged in due course. The major components of EVs are:

1. Electric traction motor
2. DC/DC converter
3. Battery pack
4. Onboard charger
5. Power electronics controller
6. Battery (auxiliary)
7. Transmission channel or drive train
8. Charging port
9. Thermal system of cooling

The electric traction motor is the engine of the EV, and we have already seen the various types of motors used in EVs in the previous section. The DC/DC converter is present along with the battery pack, which has the rechargeable batteries, about which we have already discussed in the previous sections. Onboard chargers are present, and we will discuss them in this section. An auxiliary battery, which is a lead-acid battery, is also used for meeting the requirements of the electrical systems, such as for headlights and taillights. The drive train is the group of components which deliver power to the driving wheels and excludes the engine or motor that generates the power. The charge port is used for charging the battery pack, which can be done by different types of chargers, which we will discuss in this section. A thermal cooling system is used which circulates coolant through the battery pack as well as other parts which helps in reducing the heat buildup within the system. The general EV configuration has three subsystems, which are:

1. Electric propulsion subsystem
2. Energy source subsystem
3. Auxiliary subsystem

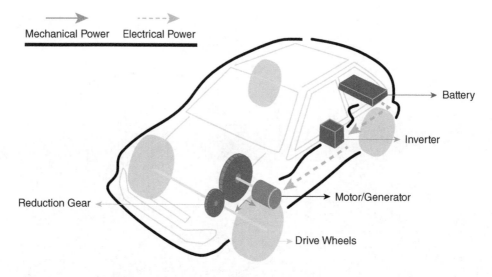

FIGURE 6.7 A simple concept of EV with its parts.

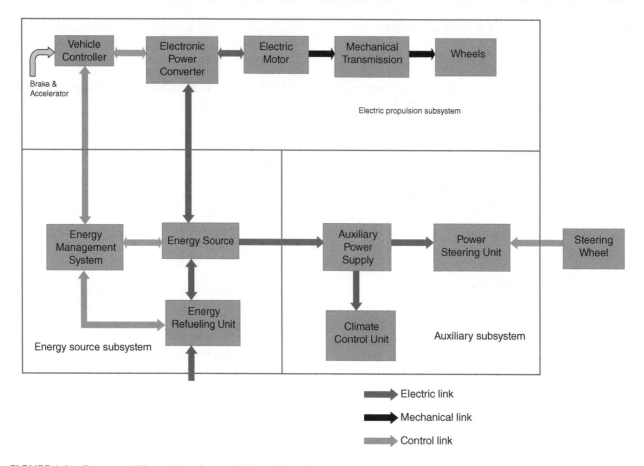

FIGURE 6.8 Conceptual illustration of general EV configuration.

The electric propulsion subsystem consists of the vehicle controller, which controls the power flow to the electric motor. The electronic power converter is connected to the electric motor and receives power from the battery pack. The electric motor is connected to the mechanical transmission, which drives the wheels by transferring power to them. The energy source subsystem consists of the energy source, which in this case is the battery pack and is mostly lithium-ion-based rechargeable batteries. The energy refueling unit consists of onboard chargers which are used to charge the battery pack. The energy management unit is the most important part of this subsystem and helps in efficient energy management, which plays a critical role in determining the range of the EV.

Software-based solutions are also used in EVs, such as in the Tesla Model S.

The auxiliary subsystem consists of the auxiliary power supply for fulfilling the electrical requirements of the EV, such as in headlights, taillights, power windows, etc. The hotel climate control unit is responsible for meeting the air-conditioning requirement of the EV. It also has the power steering unit at the driver end. Figure 6.8 shows the conceptual illustration of general EV configuration in which the three subsystems, along with their parts, are shown. The mechanical links are shown with double solid lines, while electric links are shown with thick solid lines. The control link is shown with thin arrows. The electric motor and the electronic power converter have the bidirectional electric link, which means that whenever there is regenerative braking applied, the electric motor becomes a generator and provides power, which is fed back to the battery pack. Brake and accelerator are connected to the vehicle controller, which helps in controlling the EV.

Since the present generation of EVs use lithium-ion battery technology, let's first compare the energy storage characteristics of these batteries. Here we have compared the lithium cobalt oxide ($LiCoO_2$), lithium nickel manganese cobalt oxide ($LiNiMnCoO_2$), lithium manganese oxide ($LiMn_2O_4$), lithium iron phosphate ($LiFePO_4$), and lithium titanate (Li_2TiO_3).

We observe that among the present Li-ion battery technology, the lithium cobalt oxide ($LiCoO_2$) and lithium nickel manganese cobalt oxide ($LiNiMnCoO_2$) appear to be the most promising due to their higher energy density and longer life cycle, with lithium nickel manganese cobalt oxide ($LiNiMnCoO_2$) appearing to be the best available option with the present technology. However, the problem of charging these batteries is a big concern. A typical Tesla Model S has an 85 kWh battery pack which weights around 540 kg and contains 7,104 AA lithium-ion battery cells in 16 modules wired in a series. Each module contains 6 groups of 74 cells wired in parallel, and then the 6 groups are wired in a series within the module. The charging of these battery packs can be done in three ways:

1. Level 1 charging
2. Level 2 charging
3. DC fast charging

TABLE 6.3
Comparison of Various Lithium-Ion Battery Technologies

Li Technology	$LiCoO_2$	$LiNiMnCoO_2$	$LiMn_2O_4$	$LiFePO_4$	Li_2TiO_3
Parameters	NCA	NMC	LMO	LFP	LTO NMC
Nominal Voltage	3.6 V	3.6 V	3.7 V	3.2–3.3 V	2.4 V
Operating Range	3.0–4.2 V	3.0–4.2 V	3.0–4.2 V	2.5–3.65 V	1.8–2.85 V
Charge Rate	0.7 C–1 C/3Hr	0.7 C–1 C/3Hr	0.7 C–1 C/3Hr	1 C/3Hr	1 C/3Hr
Discharge Rate	1 C, 2.5 C	1 C, 2 C	1 C, 10 C	1 C, 25 C	10 C
Life Cycle	500–1,000	1,000–2,000	300–700	1,000–2,000	3,000–7,000
Thermal Runaway	150 C full promote runaway	210 C	250 C	270 C	Safest
Energy Density	150–200 Wh/Kg	150–200 Wh/Kg	110–130 Wh/Kg	95–130 Wh/Kg	70–80 Wh/Kg
Characteristics	Risky when damaged, low discharge rates, high specific energy and power, long life	Longer life and inherent safety, less prone to heating, low cost, high specific energy	Longer life and inherent safety, used in hybrid vehicles, high discharge rates	Reduced risk of heating/fire, much less volumetric capacity, higher self-discharge, long life span	Operates at low temperature (-40°C), rapid charge and discharge, high cost, long cycle life
Supplier	Panasonic	Samsung, Litech	AESC, LGChem, LiEnergy Jpn	A123, ATL, Calb, Lishen Tianjin, Saft, BYD	Toshiba, Altaimano, Tiankang
Applications	Tesla Model S	German EV, BMW i3, Fiat	US + Japan EV, Nissan Leaf, Renault	Chevrolet Spark, Coda, eBus China	Grid Storage, EV buses and ferries

TABLE 6.4
Electric Vehicle Charging Levels

Charging Type	Level 1	Level 2	DC Fast
Charging Time (Hours)	20–22	06–08	0.2–0.5
Charger Location	Onboard (1-phase)	Onboard (1- or 3-phase)	Off-board (3-phase)
Voltage Supply (Volts)	120	240	208–600
Range Added to the Battery Pack	2–5 miles per hour of charging	10–20 miles per hour of charging	80–100 miles per hour of charging
Primary	Residential charging	Residential and public charging	Public charging

Level 1 and level 2 charging can be done at home with compatible chargers supplied by the manufacturer, while DC fast charging is done in charging stations. Level 1 chargers use standard 120 V connection, which is the same as a standard household outlet in the United States. While most of the EV manufacturers refer to them as chargers, the reality is that they are mere extension cords, because in level 1 and level 2 charging, the AC-to-DC conversion is done on board the car. So these two levels of charging are very slow and are best useful for short-distance uses and overnight charging of EVs.

The most important thing to remember about EV charging is the charging standards. The following charging standards are used worldwide.

Level 2 uses a much higher charging voltage, and like level 1 it, uses onboard charger of EV to charge the battery pack. Normally, a 240 V 80 amp circuit is used in level 2 charging. These types of chargers are also called destination chargers. They deliver around 10–20 kWh of charge and are used in homes, hotels, and even parking lots and follow the same charging standards as level 1 chargers.

The DC chargers are the fastest and most convenient means of charging EVs and are also the most expensive to build. They can charge up a battery pack up to 80% in 20–25 min using fast charging, and they don't rely on the onboard charger of EV.

DC chargers are off-board chargers designed to charge the battery of the EVs with an electrical output between 50 Kw and 350 Kw. With higher-power operation, the AC-to-DC converter, DC-to-DC converter, and power control circuits become larger and more expensive. This is the reason these types of chargers are implemented as off-board chargers rather than as onboard chargers, so that it doesn't take up place within the EV and the fast charger can be shared between many vehicles. Let us now analyze the power flow of the DC charger from the charger to the battery pack using the block diagram that follows.

In the first step, the alternating current is first converted into DC using a rectifier inside the DC charging station. Then the power control unit appropriately adjusts the voltage and current of a DC converter to control the variable

FIGURE 6.9 Level 1 and level 2 use onboard charger of EV.

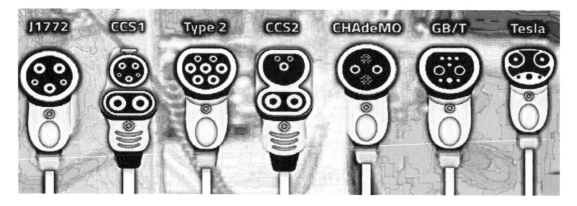

FIGURE 6.10 Different EV charging standards.

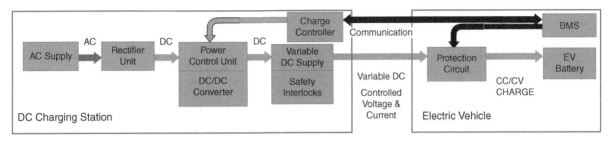

FIGURE 6.11 Block diagram of DC charger.

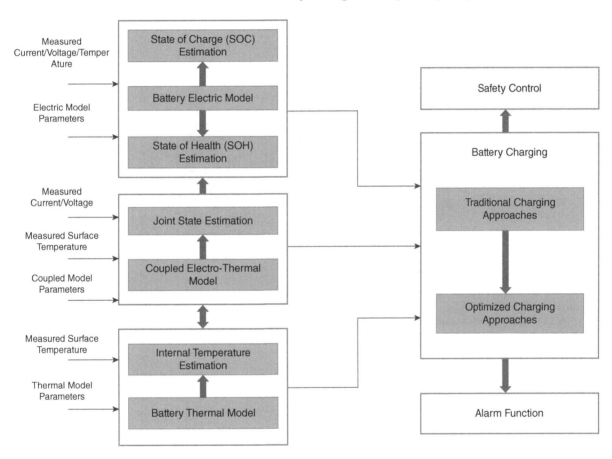

FIGURE 6.12 Battery management system (BMS).

DC power delivered to charge the battery. There are safety interlocks and protection circuits used to de-energize the EV connector and to stop the charging process whenever there is a fault condition or improper connection between the EV and the charger. The battery management system (BMS) plays the important role of communicating between the charging station and controlling the voltage and current delivered to the battery and operating the protection circuits in case of unsafe situation. The block diagram of BMS is given in the following along with its functions.

Control area network (CAN) and power line communication (PLC) are used for communication between the EV and the charger. There are five types of DC chargers used globally, which are shown here:

First is the CCS (combined charging system), called the combo 1 system, which is used mainly in the United States. Second is the CCS (combined charging system), called the combo 2 system, which is used mainly in Europe. Third is the CHAdeMO system, used globally for EVs built by Japanese manufacturers. Fourth is the Tesla DC connector,

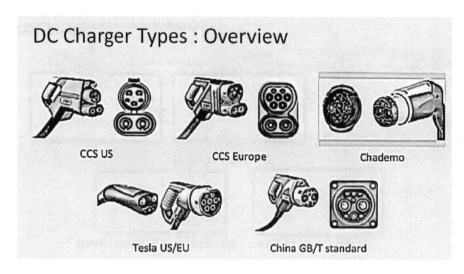

FIGURE 6.13 DC charger types.

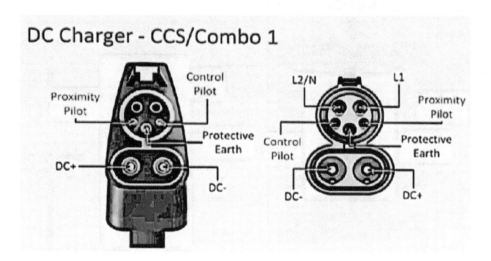

FIGURE 6.14 CCS (combined charging system) or combo 1 system with plug and inlet.

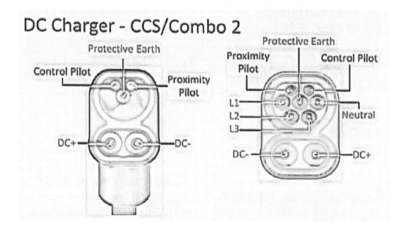

FIGURE 6.15 CCS (combined charging system) or combo 2 system with plug and inlet.

used by Tesla Motors, which can be used for AC charging as well and are used globally. Fifth is the Chinese GB/T standard, which is used by Chinese EV manufacturers. Let us now look at these chargers in detail.

The CCS (combined charging system) or the combo 1 system is an integrated system used for both DC and AC charging. They are derived from type 1 and type 2 connectors for AC charging by adding two extra pins at the bottom for high-current DC charging. The connectors derived from type 1 and type 2 are respectively called combo 1 and combo 2. In the previous figure, the connector is shown on the left side, while the vehicle inlet is shown on the right side. This charger pin is derived from the AC charging level 1 and retains the two signal pins, which are the proximity pilot and the control pilot. The proximity pilot is used by the EV to detect when the charging cable is connected, to prevent driving away until it is removed. In addition, communication (analog or digital, baseband or modulated) is carried over the control pilot for charging control. It has two DC power pins added for fast charging at the bottom of the connector.

The CCS (combined charging system) or the combo 2 system is derived from type 2 AC chargers and retains the earth pin and the two signal pins, namely, the proximity pilot and the control pilot. Two DC power pins are added on the bottom of the connector for high-power DC charging. On the vehicle inlet side, the upper side facilitates the AC charging from three-phase AC, and at the bottom part, we have the DC charging. Unlike type 1 and type 2 chargers, which use pulse width modulation (PWM) signaling on the control pilot, the power line communication of both the CCS (combined charging system) uses the power line communication (PLC) on the control pilot. PLC is a technology that carries data for communication on existing power lines and is used for simultaneous transfer of both signal and power. These chargers both can deliver up to 350A of maximum current with voltages between 200 and 1,000 V and maximum power of up to 350 kW.

The CHAdeMO chargers have three power pins and seven signal pins and use the control area network (CAN) protocol for communication between the charger and the battery pack of the EV. CAN is a robust vehicle communication standard which allows microcontrollers to communicate with each other in real time without a host computer. This charger supports voltages in the range between 50 and 500 V, with a maximum current up to 400 A and maximum power up to 200 kW.

Tesla uses their own propriety chargers with the unique aspect being that same connector can be used for both AC as well as DC charging. Tesla offers DC charging up to 120 Kw, and this is expected to increase in the future.

China uses its own charging standard, called the China GB/T (Guobiao), and uses control area network (CAN) for

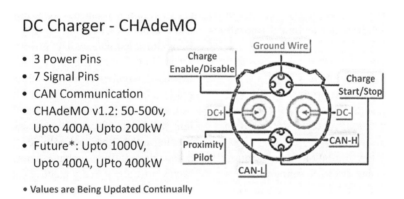

FIGURE 6.16 CHAdeMO DC charger plug.

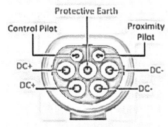
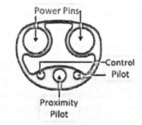

FIGURE 6.17 Tesla DC charging system with plug and inlet.

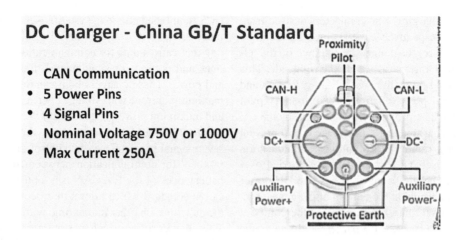

FIGURE 6.18 China GB/T DC charging plug.

communication. It has five power pins, in which two pins are for high-voltage DC, two pins are for auxiliary power, and one pin is for ground. It has four signal pins, where two pins are for the proximity pilot and two pins are for the control area network (CAN). The nominal voltage used here is 750 V or 1,000 V, with maximum current of 250 A.

DC fast charging looks to be very attractive because of the very fast charging supported by it, but these chargers have their own limitations. Fast charging power cannot be increased infinitely. The three major technical limitations of fast charging are:

1. Higher losses in charger and battery (I^2R).
2. Battery pack can be charged up to only 70–80% with fast charging, and lifetime of the battery reduces at a very fast rate if repeatedly charged on fast chargers.
3. The charging cable has limitation for maximum current it can carry, and if the charging power is further increased, the cables will become very thick; else, it will heat up due to the losses, and the thicker cable will make it difficult for people to carry the cable and connect it to their EVs.

6.5 ALTERNATIVES TO THE LITHIUM-ION BATTERY AND EMERGING BATTERY TECHNOLOGIES

Lithium has emerged as one of the most important elements driving the electric vehicle revolution ever since the development of rechargeable lithium-ion battery. These batteries are used extensively in a wide variety of electronic gadgets, ranging from smartphones to laptops, and even are used in the battery pack of almost all modern EVs, which we have already discussed in the previous sections. However, one major problem in the wide-range adoption of the lithium-ion battery apart from its capability of limited energy storage and long charging time is the availability of lithium. Lithium is a rare element, and around 39.5 million tons of lithium reserves have been found to date.

A vast majority of the lithium reserves are found in South America between the borders of Argentina, Bolivia, and Chile, forming a triangle known as lithium triangle. The lithium in this triangle is concentrated in various salt pans that exist along the Atacama Desert and neighboring arid areas. This area is thought to hold around 54% of the world's lithium reserves. Countries such as Australia, China, and the United States are the other major suppliers of lithium. India has very limited reserves of lithium, and recently, 14,100 tons were discovered in Mandya near Bengaluru, Karnataka. The government of India decided to set up India's first lithium refinery in Gujarat (Sanand or Dholera), which will process lithium ore to produce battery-grade material. However, even after all these efforts, it is estimated that by 2050, the currently known reserves of lithium will be insufficient to cater to the needs of the society. So alternate battery technologies are under research to find a feasible solution to what is known as the "lithium crisis." Today, just about every electric car uses lithium-ion batteries, and their performance is pretty good, but they are heavy and have long charging times for the amount of energy they can store. When we compare fossil fuels with lithium-ion battery, we find that 1 kg of gasoline contains about 48 MJ of energy, while 1 kg of lithium-ion battery pack contains only 0.3 MJ of energy, and these batteries degrade with each charging cycle, gradually losing capacity over the batteries' lifetime. Researchers often compare batteries by the number of full cycles until the battery has only 80% of its original charge capacity. The five new battery technologies that could change this present situation are:

1. Lithium air
2. Nanotechnology
3. Lithium sulfur
4. Solid-state
5. Dual-carbon

Metal air batteries, such as zinc air button cell, have been around for a while and have been extensively used in biomedical devices, such as hearing aids. Scaled-up aluminum and lithium air chemistries are also promising for the automotive and aerospace industries. The potential for lightweight batteries with high energy storage makes this battery technology promising. Lithium air batteries could have a maximum theoretical specific energy of 3,460 Wh/kg, which is ten times more than lithium-ion batteries. Realistic battery packs will be closer to 1,000 Wh/kg, which will still be three to five times more than Li-ion technology. As usual, this technology is not without its drawbacks. Current electrodes of lithium air batteries tend to clog with lithium salts after only a few tens of cycles. Most researchers are using porous forms of carbon to transmit air to the liquid electrolytes. Feeding pure oxygen to the batteries is a solution, but it is a potential safety hazard in the automotive environment. Researchers at the University of Illinoi, United States, have found that they can prevent this clogging by using molybdenum disulfide nanoflakes used to catalyze the formation of a thin coating of lithium peroxide on the electrodes. Their test batteries ran for 700 cycles compared to just 11 cycles of an equivalent with an uncoated electrode. While this isn't an enough lifetime for a car battery, it is a promising hint of things to come. NASA (National Aeronautics and Space Administration) is also working on lithium air battery development and expects to get around 800–900 Wh/kg from these batteries in the near future.

Nanotechnology has been a buzzword for several decades but is now finding applications in everything from nanoelectronics to biomedical engineering. Nanomaterials make use of particles between 1 and 100 nm in size, essentially one size up from the molecular scale. The magic is that they behave in unusual ways because their small size ridges the gap between that which operates under the rules of quantum physics and those of our familiar macroworld. As we have seen, one of our major challenges in the physical expansion of lithium electrodes as they charge. Researchers at Purdue University, United States, made use of antimony nanochain electrodes last year to enable this material to replace graphite or carbon metal composite electrodes.

By structuring the metalloid element in this nanochain net shape, extreme expansion can be accommodated within the electrode since it leaves a web of empty pores. The battery appears to charge rapidly and shows no detrition over the 100 charge cycles tested. Carbon nanostructures also show great promise, with graphene being one of the most exciting of these. Graphene is made up of a single atomic thickness sheet of graphite and has very interesting electrical properties, being a very thin semiconductor with high carrier mobility, meaning, that electrons are transmitted along it rapidly in the presence of electric field as inside an electric battery. It is also thermally conductive and has exceptional mechanical strength in the range of 50,000–60,000 MPa when compared with strengthened steel, which has mechanical strength of 2,500 MPa. GRABAT, a Spanish nanotechnology company, is perusing graphene polymer cathodes with metallic lithium anodes, which is a highly potent combination if their electrolyte can protect the metallic anode and prevent dendrite growth.

This battery promises to be capable of charging and discharging faster with greater energy capacity and is lighter and more robust than the current generation of batteries. Samsung Electronics has patented a technology which they call as "Graphene Balls." These are silicon oxide nanoparticles which are coated with graphene sheets which resemble popcorns and are used as cathodes as well as applied as a protective layer on the anode. Researchers found an increase in the volumetric density of a full cell of 27.6% compared to an uncoated equivalent, and the experimental cell retains almost 80% capacity after 500 charge cycles. Additionally, charging is accelerated and temperature control is improved. University of California–Irvine has even produced electrodes good for 200,000 charge cycles using gold nanowires and manganese dioxide polymer gel electrolyte, and many other research efforts are going on with other diverse materials.

Lithium sulfur batteries are one emerging technology which can offer greater energy density compared to lithium ion. The theoretical maximum specific energy capacity of this chemistry is 2,567 Wh/kg, compared to lithium ion's maximum specific energy capacity of 350 Wh/kg. This is a huge improvement, and a lithium sulfur battery could be up to 7.5 times lighter than a lithium-ion battery. Presently, lithium sulfur batteries are still under development, and ALISE, which is a pan-European collaboration, is working towards attaining a stable automotive battery of 500 Wh/kg based on this technology. In terms of economics, sulfur is much cheaper than the cobalt and manganese it would replace and can be extracted as by-product of fossil fuel refinement or mined from abundant natural deposits. Existing lithium-ion batteries are made up of an anode and a cathode, between which a liquid electrolyte allows dissolved lithium ions to travel. Lithium sulfur batteries are constructed similarly, except that the active element in the cathode is sulfur or the anode remains lithium-based. Researchers are facing a few challenges in bringing this technology to market, because sulfur is a poor conductor of electricity. Typically, sulfur atoms are embedded within the matrix of carbon atoms in graphite, an excellent electrical conductor. This arrangement is vulnerable to a process called as "shuttling," which causes the batteries to drain when not in use, while also corroding metallic lithium anodes, reducing capacity as the battery is cycled. Electrodes also physically swell up as the lithium ions bond to them in lithium sulfur batteries. A sulfur cathode can expand and contract by 78% or eight times more than typical lithium-ion batteries, which reduces capacity over time. One approach to solving this problem is to bind the cathodes with different polymers and to reduce their thickness so that the absolute change in dimension is not so extreme. Dr. Bimlesh Lochab and her team from Shivnadar University, based in Greater Noida, India, along with professors from IIT, Bombay, developed a sulfur battery recently based on green principle and green chemistry, because they used sulfur obtained as by-product from petroleum industry along with copolymer like cardanol from cashew nut processing and eugenol from clove

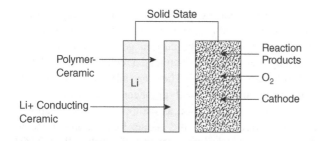

FIGURE 6.19 Solid-state batteries.

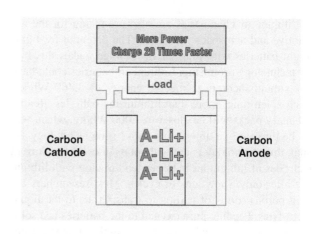

FIGURE 6.20 Dual-carbon battery.

oil. A new cathode material was also developed, and the battery developed had three times more energy storage capacity than lithium-ion battery. The battery developed is under further testing and evaluation and holds promise for the future. Such batteries are very important for countries like India which have very limited reserves of lithium and depend on exports to meet their lithium demands, which further increases the cost of lithium-ion batteries.

Solid-state electrolyte could also offer solution to the problems faced in lithium-ion batteries. They will replace flammable organic liquids with stable crystalline or glassy-state solids or a polymer base.

It is hoped that using these solid electrolytes would enable the use of metallic lithium electrodes to provide higher-output voltages and increase energy density. Research is going on in this field, and we are hopeful to see the use of such batteries in the near future, because companies such as Volkswagen, Toyota, and BMW have been investing in this technology. One of the difficulties in solid-state electrolyte design deals with the expansion of electrodes, which is more difficult to manage in solid materials. A solid electrolyte must be sufficiently flexible to permit this yet also tough enough to resist dendrite growth penetration. Panasonic has also been working in this field and is partnering with Tesla Motors. Toyota has also announced their collaboration with Panasonic for the development of solid-state batteries. Samsung Electronics is also working in this field and, in May 2020, described their technology based on silver and carbon anode, giving a generic electric car 500 mi range and survive over 1,000 carge cycles.

Two carbon electrodes and a nontoxic electrolyte are the basic things needed for dual-carbon batteries. They add the ability to extract more power from conventional lithium-ion battery, and they also charge 20 times faster than convential batteries. These batteries could be the future for EVs.

PJPEye, an offshoot of Japan Power Plus, and National Kyushu University, Fukuoka, Japan, have developed this technology and named it Cambrian batteries. They are supplying these batteries to an electric bicycle company, Maruishi Cycles. Currently, these are single-carbon electrode batteries, and their details are kept secret. However, research is going on for fully dual-carbon battery with two carbon electrodes, eventually to be manufactured from natural agricultural products such as cotton fibers. They are anticipated to achieve a performance similar to graphene-based batteries. These batteries boast higher specific power for the same mass as compared to lithium-ion batteries. These batteries can also be discharged fully, getting a 40% improvement in range over lithium-ion batteries of the same capacity. They also claim that the battery runs cool, doesn't require the heavy cooling systems found in current electric vehicles. They also claim that their battery degraded 10% after 8,000 charge cycles. This technology holds promise for the future and might power the EVs of the next generation.

6.6 CONCLUSION

So much diverse research is undergone in battery technology that it is just impossible to pick up five technologies. Lithium-ion batteries are found in almost any modern battery packs of products such as computers, phones, and vehicles. Even drones and quadcopters have been developed due to advancement in battery technology. Better batteries are also important for the advancement of stationary storage from renewable sources such as solar power. Recently, cost reduction is the main objective in most EV manufacturers' design strategy. Hence, new traction motor technologies such as synchronous reluctance machine, in which the rotor anisotropic geometry can eliminate expensive magnet, and aluminum or copper bar from the rotor structure became more attractive for automotive applications. Research is also going on for the development of alternate technologies, such as solar energy–powered vehicles and hydrogen-powered vehicles, but they are far from commercial implementation. As further research takes place, newer technologies might develop which will ultimately reduce our dependence on lithium and cobalt and propel a new revolution in electric vehicle industry. EVs hold promise for the future but need a few more years to reach their full potential, and companies like Tesla, Toyota, BMW, Hyundai, Mahindra, TATA, and Pravaig Dynamics are developing their EVs which have the capability to compete with their internal combustion counterparts. Pravaig Dynamics, which is a Bengaluru-based EV start-up, claims that its two-door, four-seater electric car Extinction MKI can run 504 km on a single charge powered by a 96 kWh lithium-ion battery pack using 200 bhp engine and a top speed of 100 kmph, which are all promising figures. By 2025, EVs will become a common sight on Indian roads due to the push given by government agencies to reduce pollution and our dependence on fossil fuels. A ban on petrol and diesel cars is likely to be adopted in Britain in 2030; Norway and various European countries are aiming for 2025, while India, Israel, France, and the Netherlands are aiming for 2030, with Germany, Taiwan, the United States, and China aiming for 2040 or earlier. The potential research areas in the field of e-mobility are:

- Motor design for EVs
- Control system for electrical drive system
- Development of better battery technology
- Other energy storage systems
- Better battery management system (BMS)
- Efficient energy management system (EMS)
- Integrated vehicular communication for driverless vehicles
- Better charging infrastructure
- DC fast charging converter
- Indigenous measuring and instruments development
- Data acquisition system
- Smart grid development
- Regulatory support, communications, and supply chain in EV adoption
- Electrified, shared, and autonomous vehicles

6.7 BIBLIOGRAPHY

1. Assessment of Electric Vehicle Impacts on Energy, Environment and Transportation Systems. International Energy Agency (1999), https://www.bing.com/search?q=Assessment+of+Electric+Vehicle+Impacts+on+Energy%2C+Environment+and+Transportation+Systems.+International+Energy+Agency+(1999)&cvid=1d9bfc24ecfc4a4f999b24304b6c8bb7&aqs=edge..69i57.329j0j9&FORM=ANAB01&PC=DCTS
2. Berndt, D. (1997), *Maintenance-Free Batteries: Lead-Acid, Nickel/Cadmium, Nickel/Hydride: A Handbook of Battery Technology.* Somerset, England: Research Studies Press and New York: John Wiley & Sons.
3. Blomen, L.J.M.J. and M.N. Mugerwa (1993), *Fuel Cell Systems.* New York: Plenum Press. [This book introduces an outlook of fuel cell systems and their feature]
4. Burke, A. (2000), "Ultracapacitors: Why, How and Where is the Technology", *Journal of Power Source*, Vol. 91, pp. 37–50 [This paper presents a comprehensive review on the performance of ultracapacitors]
5. Agenbroad, J., and B. Hollandl (2014), *Pulling Back the Veil on EV Charging Station Costs.* Rocky Mountain Institute, http://blog.rmi.org/blog_2014_04_29_pulling_back_the_veil_on_ev_charging_station_costs.
6. Ates, M.N. et al. (2016), "In Situ Formed Layered-Layered Metal Oxide as Bifunctional Catalyst for Li-Air Batteries", *Journal of the Electrochemical Society*, Vol. 163, No. 10, pp. A2464–A2474.
7. Sharma, S., A.K. Panwar, and M.M. Tripathi (2020, June), "Storage Technologies for EVs", *Journal of Traffic and Transportation Engineering*, Vol. 7, No. 3.
8. ResearchGate Chapter, January 2019. DOI: 10.1007/978-981-13-3290-6_7
9. 20th Anniversary – GM EV1 – Mentor Graphics, 2016.
10. Modern Electric Cars of Tesla Motors Company, October 2020.
11. Tesla Model 3's Motor- The Brilliant Engineering behind it, Learn Engineering EMWorks, December 2020.
12. Ehsani, M., Y. Gao, S.E. Gay, and A. Emadi (2005), *Modern Electric, Hybrid Electric & Fuel Cell Vehicles Vol. 1.* Boca Raton: CRC Press.
13. Delft University of Technology and D-INCERT on edx@ www.tiny.cc/ecarsx, 2019.
14. Electric Future – 5 New Battery Technologies That Could Change Everything, September 2020, https://www.gray.com/insights/5-new-battery-technologies-that-will-change-the-future/
15. Dallinger, D., G. Schubert, and M. Wietsche (2012), "Integration of Intermittent Renewable Power Supply Using Grid-connected Vehicles – a 2030 Case Study for California and Germany", No. S4, Fraunhofer Institute for Systems and Innovation Research ISI.
16. Fulton, L., G. Tal, and T. Turrentine (2016), "Can We Achieve 100 Million Plug-in Cars by 2030?", Institute of Transportation Studies, University of California, Davis.
17. Gerbaulet, C., and W. Schill (2015), "Power System Impacts of Electric Vehicles in Germany: Charging with Coal or Renewables?", *Applied Energy*, Vol. 156, pp. 185–196.
18. IRENA (2016a), *REmap: Roadmap for a Renewable Energy Future*, 2016 Edition. IRENA, Abu Dhabi, www.irena.org/DocumentDownloads/Publications/IRENA_REmap_2016_edition_report.pdf.
19. IRENA (2016b), *The Renewable Route to Sustainable Transport: A Working Paper Based on Remap*. IRENA, Abu Dhabi, www.irena.org/DocumentDownloads/Publications/IRENA_REmap_Transport_working_paper_2016.pdf.
20. IRENA (International Renewable Energy Agency) (2015), *Renewables and Electricity Storage: A Technology Roadmap for REmap 2030*. IRENA, Abu Dhabi, www.irena.org/DocumentDownloads/Publications/IRENA_REmap_Electricity_Storage_2015.pdf.
21. Kurani, K., N. Caperello, and J. Tyree-Hageman (2016), *New Car Buyers' Valuation of Zero-Emission Vehicles: California*, Institute of Transportation Studies. University of California, Davis, http://its.ucdavis.edu/research/publications/?frame=https%3A%2F%2Fitspubs.ucdavis.edu%2Findex.php%2Fresearch%2Fpublications%2Fpublicationdetail%2F%3Fpub_id%3D2682.
22. Mwasilu, F. et al. (2014), "Electric Vehicles and Smart Grid Interaction: A Review on Vehicle to Grid and Renewable Energy Sources Integration", *Renewable and Sustainable Energy Reviews*, Vol. 34, pp. 501–516.
23. Nykvist, B., and M. Nilsson (2015), "Rapidly Falling Costs of Battery Packs for Electric Vehicles", *Nature Climate Change*, Vol. 5, pp. 329–332, www.nature.com/nclimate/journal/v5/n4/full/nclimate2564.html.

7 Onboard Electric Vehicle Charger in G2V and V2G Modes Based on PI, PR, and SMC Controllers with Solar PV Charging Circuit

Premchand Mendem and Satish Kumar Gudey

CONTENTS

7.1 Introduction: Background and Driving Forces	81
7.2 Electric Vehicle Charging Circuit Configuration	82
7.2.1 Passive Components Design	82
7.2.1.1 DC Bus Capacitor Design (C_{dc})	82
7.2.1.2 Input-Side Filter (L_s and C_s)	83
7.2.1.3 Output Filter Design (L_o and C_o)	83
7.3 Operation of EV with a PI Controller	84
7.3.1 Charging Mode of Operation (G2V)	84
7.3.2 Discharging Mode of Operation (V2G)	87
7.4 Electric Vehicle Operation Using a PR Controller	89
7.4.1 Charging Mode of Operation (G2V)	89
7.4.2 Discharging Mode of Operation (V2G)	91
7.5 Operation of an Electric Vehicle Using Variable Structure Systems – SMC	91
7.5.1 Charging Mode of Operation (G2V)	91
7.5.2 Discharging Mode of Operation (V2G)	95
7.6 Solar-Based Electric Vehicle Charging	97
7.7 Conclusions and Future Scope	98
7.8 References	98

7.1 INTRODUCTION: BACKGROUND AND DRIVING FORCES

Research and development of electric vehicles (EV) and vehicle charging stations are increasing day by day due to the high efficiency, clean, and safe transportation offered by the EVs [1]. HEV (hybrid electric vehicles), BEV (battery electric vehicles), and PHEV (plug-in hybrid electric vehicles) are the types of electric vehicles available in the literature. In HEV, the internal combustion engine (ICE) is combined with battery-driven electric motor (BEM). The PHEV consists of ICE with BEM. The battery can be recharged by plugging a charging cable with an external source of power supply. BEV, also termed electric vehicle (EV), consists of a battery-driven electric motor. The battery plays an important role in an EV. There are various types of batteries available for an EV. Lead-acid, nickel-cadmium, and lithium-ion batteries are the prominent ones. The most commonly used batteries for EVs are Li-ion (lithium-ion) batteries; these batteries have good discharging capabilities, high specific energy, good life span with less maintenance, and low environmental impact.

Detailed information about the type and capacity of various EV batteries and chargers, along with the manufacturers, ˆs presented for readers' information in Table 9.1. EV chargers are classified as onboard and off-board, with unidirectional or bidirectional flow of power. The electric vehicle consists of three main parts, namely, bidirectional AC/DC converter, bidirectional DC/DC converter, and a battery [2]. *Unidirectional charger* refers to power flow from grid to battery, and in a bidirectional charger, the power flow is vice versa, that is, from grid to battery and battery to grid, following the demand of the convenience of the EV owner and the grid. Onboard chargers are referred as slow chargers, which are used in houses, parking places, and off-board chargers are fast chargers, which are utilized at charging stations. This chapter focuses on implementation of the controllers for EV to operate in charging and discharging modes using bidirectional converters. The bidirectional converter considered is an AC/DC H-bridge converter consisting of IGBT switches with their antiparallel diodes for bidirectional power flow. A bidirectional DC-to-DC converter is also used in the operation of the system. Onboard charging unit is considered for the EV.

TABLE 7.1
EV Chargers with Manufacturers

Model	Battery Capacity	Charger Time
Toyota Prius PHEV	4.4 kWh Li-ion, all-electric range 18 km	3 h at 115 V AC, 15 A
		1.5 h at 230 V AC, 15 A
Mitsubishi iMiEV	16 kWh Li-ion, range 128 km	13 h at 115 V AC, 15 A
		7 h at 230 V AC, 15 A
BMW i3	42 kWh Li-manganese, range 246 km	4 h with 11 kW onboard AC charger 30 min with 50 kW DC charger
Nissan Leaf	30 kWh Li-manganese, range 250 km	8 h at 230 V AC, 15 A
		4 h at 230 V AC, 30 A
Tesla S	70 kWh Li-ion, range 424 km	9 h with 10kW charger
		30 min with 120 kWh supercharger for 80% charge
Tesla 3	75 kWh Li-ion, range 496 km	30 min at 11.5 kW, 48 A
Chevy Bolt	60 kWh Li-ion, range 383 km	40 h at 115 V AC, 15 A
		10 h at 230 V AC, 30 A

Source: https://batteryuniversity.com/learn/article/bu-1003-electric-vehicle-ev.

A 2.5 kW to 5 kW battery charger with 100 Ah capacity is considered in this work. A total capacity of 20 kWh which requires 12–15 hours is used in this work. It is found in the literature; most of the researchers have concentrated on the controller for EV charging for vehicle to grid (V2G) or grid to vehicle (G2V) using the conventional PI controllers. In this work, advanced controllers like proportional plus resonant controller (PR) and variable structure systems like sliding mode controller (SMC) are designed for G2V and V2G operation, which provide operation in charging and discharging modes. During charging mode, AC-to-DC converter works as a rectifier, and bidirectional DC-to-DC converter acts as a buck converter to charge the battery to its full capacity. The reverse happens during discharging. The DC voltage and grid current are the control parameters, which are controlled effectively for smooth operation of the EV. The design of the passive components present in the EV circuit configuration is discussed in Section 7.2, EV with PI (proportional plus integral controller) in Section 7.3, EV with PR (proportional plus resonant controller) in Section 7.4, and EV with the sliding mode controller (SMC) in Section 7.5, respectively. A comparative analysis is performed among PI, PR, and SMC, and the results are tabulated. It is found that grid current waveform and its tracking performance and THD levels are as per IEEE 519 standards in SMC compared to PI and PR controllers for EVs in both charging and discharging modes of operation. A solar PV–based electric vehicle charging is presented in Section 7.6. Section 7.7 presents the conclusions and scope of the work for the readers.

7.2 ELECTRIC VEHICLE CHARGING CIRCUIT CONFIGURATION

Figure 7.1 shows the electric vehicle circuit configuration consisting of two stages. The AC-to-DC converter will act as either converter/inverter in the first stage of operation, and the second stage consists of a DC-to-DC converter, which operates as buck converter in charging mode and as boost converter in discharging mode of operation. The battery serves the purpose of storing the electrical energy during charging operation and supplying the energy to the load during discharging phenomenon [3, 4]. The input side is fed with an LC filter consisting of L_s and C_s, which improves the grid current waveform from harmonic pollution. On the output side, an additional L_o, C_o filter is used to remove the ripple content in the DC voltage [5, 6]. Hence, the life span of the battery increases. Alternatively, one can use an LCL filter on the input side.

7.2.1 Passive Components Design

An effective design of the passive filters is very much necessary for an efficient working of the EV [7].

7.2.1.1 DC Bus Capacitor Design (C_{dc})

In single-phase converters, the second-order harmonics are present in the output DC voltage; hence, a large value of DC capacitor on the DC side is very much essential. The DC capacitor acts as an energy buffer between the input and output. The capacitor value is selected based on energy storage requirement. Assuming input power factor to be unity, the input power is given by (7.1):

$$P_{in} = v_{in} \times i_{in} = \frac{VI}{2} - \frac{VI}{2}\cos 2\omega t \quad (7.1)$$

Power stored in the input inductor is given by (9.2):

$$P_L = \frac{\partial\left(\frac{1}{2}L(I\sin\omega t)^2\right)}{\partial t} = \omega LI^2 \sin\omega t \cos\omega t \quad (7.2)$$

The energy flows from the source-side inductor to the H-bridge converter, and it charges the DC capacitor. Assuming the device power losses are negligible, energy stored in the DC capacitor is equal to the comparison between the input

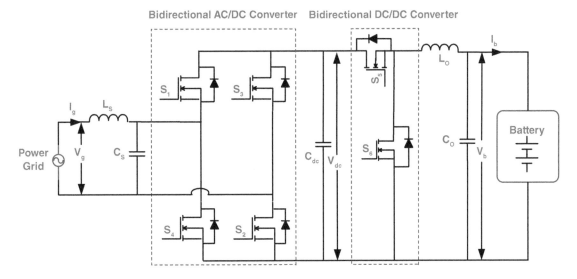

FIGURE 7.1 Circuit configuration of an electric vehicle.

energy and energy stored in inductor. Equation 7.3 gives the flow of power through the capacitor.

$$P_c = P_{in} - P_L = \left(\frac{VI}{2} - \frac{VI}{2}\cos 2\omega t\right) \\ -\left(\omega LI^2 \sin \omega t \cos \omega t\right) \quad (7.3)$$

In (7.3) the second-order components flow through the capacitor, which causes ripple in output DC voltage. By simplifying and taking the integration, the instantaneous power for a half cycle, the ripple energy is given by (7.4):

$$E_C = \int_0^{T/2} \sqrt{\frac{V^2 I^2}{4} + \frac{\omega^2 L^2 I^2}{4}} \sin 2\omega t\, dt \\ = \sqrt{\frac{V^2 I^2}{4} + \frac{\omega^2 L^2 I^2}{4}}\bigg/\omega \quad (7.4)$$

From (7.4), a correlation exists between the DC capacitor C_{dc}, input inductor current, and DC voltage V_{dc} as (9.5):

$$C_{dc} = \frac{\sqrt{\frac{V^2 I^2}{4} + \frac{\omega^2 L^2 I^4}{4}}}{2 \times V_{dc} \times \Delta V_{dc} \times \omega} \quad (7.5)$$

From the configuration variables shown in Table 7.2 and using (7.5), C_{dc} is derived as 2 mF.

7.2.1.2 Input-Side Filter (L_s and C_s)

Based on the amount of ripple current present in the current through the inductor (L_s), L_s is designed using (7.6).

$$L_s = \frac{V_{dc} \times \dfrac{V\sin\omega t}{Vdc} \times \left(1 - \dfrac{V\sin\omega t}{Vdc}\right)}{2\Delta I f_{sw}} \quad (7.6)$$

V_{dc} is the DC voltage with ripple, and $V\sin\omega t$ is the instantaneous value of AC input voltage. ΔI is the ripple current and is taken as 10% of the input current, and f_{sw} is the switching frequency. A filter capacitor C_s is used to reduce the harmonics and provide a well-regulated output voltage. It is obtained from (7.7) as 20 µF. The filter inductor is obtained as 0.75 µH.

$$Cs = \left(\frac{1}{2\pi f_{sw}}\right)^2 \times \frac{1}{L_s} \quad (7.7)$$

7.2.1.3 Output Filter Design (L_o and C_o)

Equations 7.8 and 7.9 are used to calculate output filter inductor L_o and capacitor C_o and are obtained as 41 µH and 600 µF.

$$L_o = \frac{\left(D(1-D)^2 r\right)}{2 f_{sw}} \quad (7.8)$$

$$C_o = \frac{(1-D)}{8 L_o \left(\dfrac{\Delta V_o}{V_o}\right) f_{sw}^2} \quad (7.9)$$

A very large value of output capacitor is required to obtain a well-stable regulated output voltage. D is the duty ratio, r is the internal resistance, and V_o is the output voltage; switching frequency of the buck/boost converters is f_{sw}, and ΔV_o is voltage ripple, which is assumed as 5% of the output voltage. All the obtained parameters are tabulated in Table 7.2. A 120 V, 60 Hz single-phase system is presented in this work.

TABLE 7.2
Circuit Configuration Parameters

Parameters	Values
Grid voltage V_g	120 V_{rms}
System frequency f_s	60 Hz
AC filter inductor L_s	0.75 mH
AC filter capacitor C_s	20 µF
DC capacitor C_{dc}	2 mF
Inductor L_o	41 µH
Capacitor C_o	600 µF
Battery capacity	100 Ah
Hysteresis band h	±0.5 A
PI parameters k_p, τ_i	50, 10 ms
PR parameters	100, 0.1 ms
SMC parameters (k_1, k_2)	10, 20,000

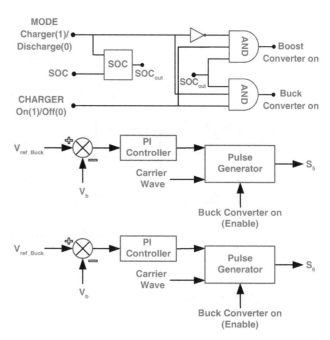

FIGURE 7.3 (a) Upper-level control and (b) lower-level control for a DC–DC converter.

7.3 OPERATION OF EV WITH A PI CONTROLLER

7.3.1 Charging Mode of Operation (G2V)

Figure 7.2 shows a PI controller used for a single-phase, bidirectional converter in discharging and charging modes. Voltage across the DC capacitor V_{dc} is sensed using a voltage sensor and taken as a feedback and compared with a reference voltage V_{dc}^*. Using a PI controller, the grid current is generated I_g^* for the current loop. Hysteresis logic is used to generate the switching pulses to the IGBT switches of the single-phase converter circuit. The upper-level and lower-level control mechanisms are shown in Figure 7.3(a) and Figure 7.3(b) for the bidirectional DC–DC converter circuit. SOC_{out} is derived using the upper-level control mechanism according to the battery state of charge (SOC) between 5% and 100% and mode of operation, that is, charging or discharging mode. Now, SOC_{out}, Charger ON/OFF, and mode of operation are applied to produce/enable signal to buck or boost converter. Similar operation is performed in lower-level control for obtaining gating pulses using the voltage error and buck/boost converter on-signal.

The switching pulses obtained during the EV charging mode are shown in Figure 7.4, and Figure 7.5 reveals the switching pulses produced to the buck converter circuit. Figure 7.6 shows the grid current and voltage waveforms in charging mode of operation using a PI controller. The grid current and voltage are said to be in-phase. Figure 7.7 shows the tracking characteristics of DC voltage. It is clear that there exists a DC voltage steady-state error of 2.5%. The time taken to track the reference DC voltage is around 0.5 sec. Figure 7.9 shows the harmonic spectrum of the DC voltage. It is found to be high during charging phenomenon. The battery voltage and current waveforms during battery charging mode are shown in Figure 7.10. It is observed that voltage waveform has a slight increment at 1.02 sec, with a magnitude of 220 V, whereas the current is said to be constant at 10 A. The battery SOC waveform when battery is in charging mode is shown in Figure 7.10.

Table 7.3 shows the percentage of odd-ordered harmonic components present in the grid current and voltage. The grid voltage THD and grid current THD are 0.054% and 4.447%, respectively, and are within the *IEEE*-519 standards. The third harmonic component is more due to magnetic saturation of the transformer.

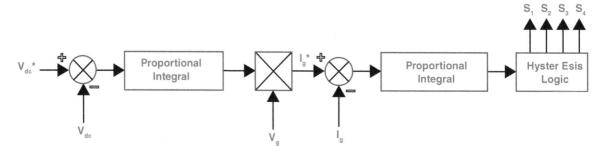

FIGURE 7.2 PI controller diagram for G2V and V2G operation.

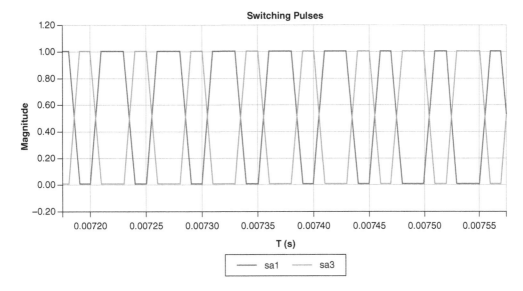

FIGURE 7.4 Switching pulses to the bidirectional AC–DC converter switches.

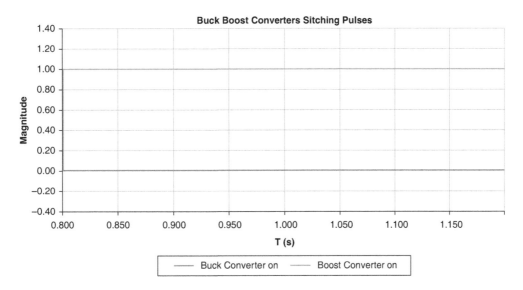

FIGURE 7.5 Switching pulses to the bidirectional DC–DC converter.

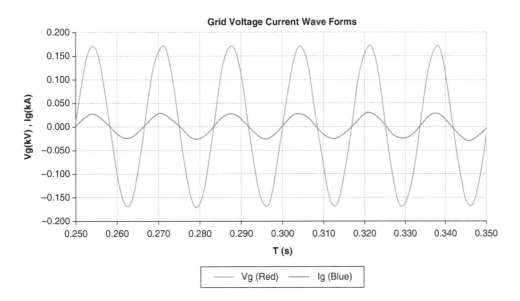

FIGURE 7.6 Grid voltage and current waveforms.

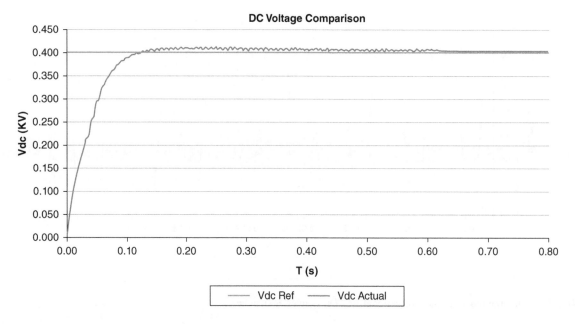

FIGURE 7.7 Tracking characteristics of DC voltage.

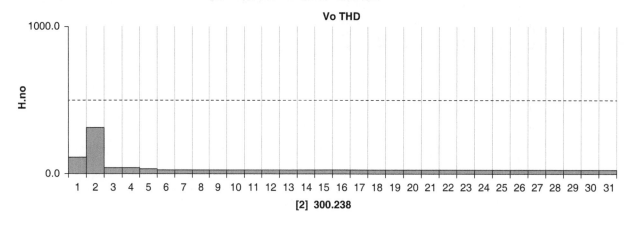

FIGURE 7.8 Harmonic spectrum of the DC voltage.

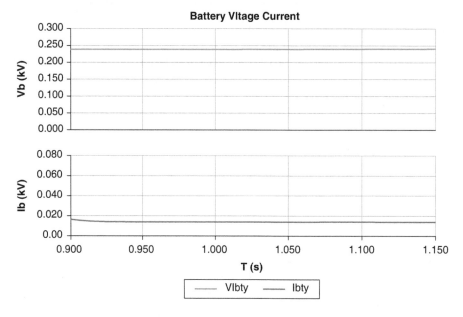

FIGURE 7.9 Battery voltage (V_b) and current (I_b).

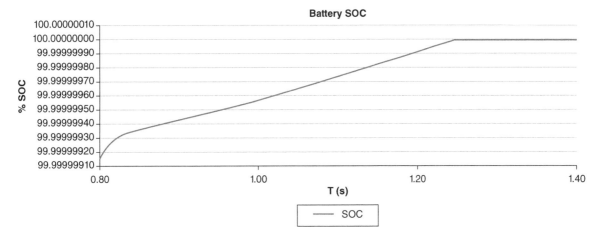

FIGURE 7.10 Battery state of charge (SOC).

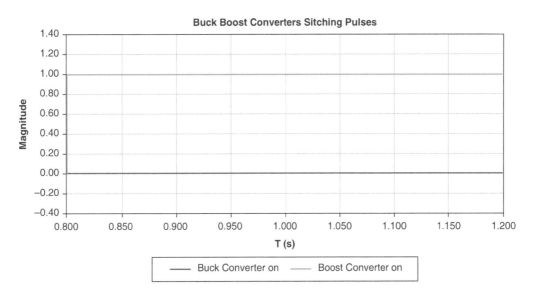

FIGURE 7.11 Switching pulses to the bidirectional DC–DC converter during discharging.

TABLE 7.3
Individual Harmonic Content Present in Grid Current and Voltage Using PI Controller

Harmonic Number	Individual Current THD (%)	Individual Voltage THD (%)
3	2.244	0.026
5	1.404	0.019
7	1.597	0.016
9	1.510	0.009
11	1.287	0.011
13	1.015	0.009

7.3.2 Discharging Mode of Operation (V2G)

The grid current is controlled during discharging mode of operation, and the bidirectional AC–DC converter operates as an inverter. In this mode, bidirectional DC–DC converter operates as a boost converter. In this mode, the battery supplies power to the grid. Figure 7.11 shows the switching pulse given to the DC–DC boost converter. Figure 7.12(a) shows the grid voltage and current waveforms during discharging operation. In charging mode, the grid current and voltage are said to be in-phase, but during discharging operation, the grid current is found to be in phase opposition with grid voltage. It indicates the flow of power from EV battery to the power grid. Figure 7.12(b) shows the grid current tracking, and the error during steady state is found to be 19.06%. Figure 7.12(c) shows the battery current and voltage plots during battery discharging. The voltage decreases with a constant current at 10 A. Hence, the EV during discharging mode of operation is said to perform well using a PI controller. Figure 7.12(d) shows the battery SOC during discharging.

It is found through simulations that the PI controller is not able to track sinusoidal grid current reference without

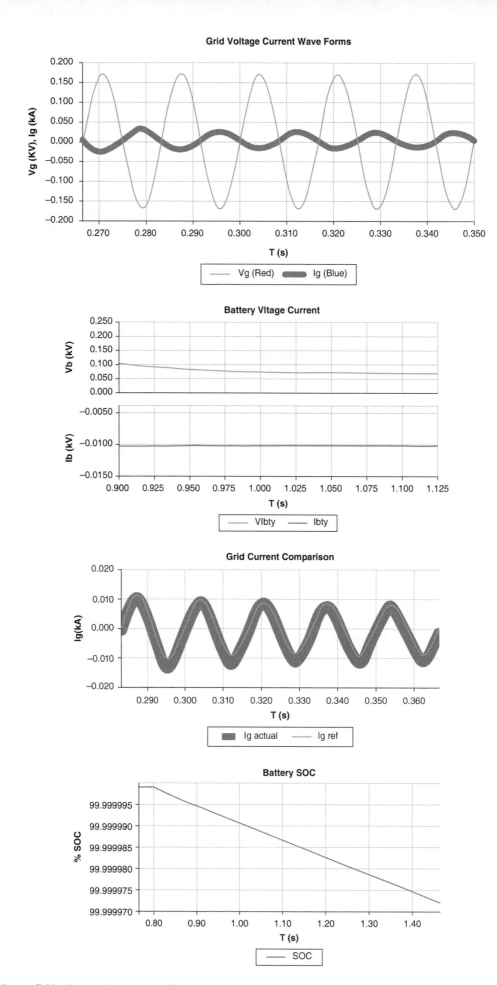

FIGURE 7.12 (a) Grid voltage and current waveforms, (b) battery voltage and current, (c) tracking of grid current, and (d) battery SOC.

steady-state error in both discharging and charging modes of operation. The settling time to track the DC voltage is found to be large. The grid current THD is more (4.447%), the second-order component in the DC voltage is more (300%), and hence, the life of the battery is said to degrade. Hence, to overcome the aforementioned disadvantages, a proportional plus integral (PR) controller–based EV is proposed by the authors and is discussed in Section 7.4.

7.4 ELECTRIC VEHICLE OPERATION USING A PR CONTROLLER

Using a PR controller, the shortcomings of a PI controller are overcome in an EV. Figure 7.14(a) shows the control block diagram of the PR controller. The PR controller introduces an infinite gain at the fundamental frequency, and it has the ability to track the sinusoidal current [8]. The transfer function $T_{PR}(s)$ for a PR controller is given by (7.10):

$$T_{PR}(s) = k_p + k_i \cdot \frac{2\omega_c s}{s^2 + 2\omega_c s + \omega_o^2} \quad (7.10)$$

In (7.10), the cutoff frequency is given by ω_c, which is far less than the resonant frequency ω_o. The value of the resonant frequency is 376.8 rad/s in this work. The dynamic response of the system is obtained by controlling k_p and the phase shift between the output, and the reference is controlled by k_i. The value of $k_i = k/\omega_c$ is taken as unity for understanding the effect of ω_c [9]. Figure 7.14(b) shows the frequency response characteristics obtained for various values of ω_c ranging from 0 to 25 rad/sec. k_p is taken as 1. For any value of ω_c, Bode plots converge to 20 db/dec. The peak magnitude of gain margin increases at lower values of ω_c. Hence, it is understood that the infinite gain occurs only at resonant frequency ω_o. Hence, the controller is very much sensitive between resonant frequency and fundamental frequency of the converter circuit. A small value of ω_c is sufficient for a stable operating system. Therefore, the value of ω_c is taken as 10 rad/sec. For proper selection, a trade-off should be made between k and ω_c.

7.4.1 Charging Mode of Operation (G2V)

Figure 7.15(a) shows the grid voltage and current waveforms. The grid voltage is in phase with the grid current. Figure 7.15(b) shows effective tracking of DC voltage with the reference voltage. The steady-state error is found to be 1%, and the time taken to track the reference voltage is around 0.16 sec. These are comparatively less than that of a

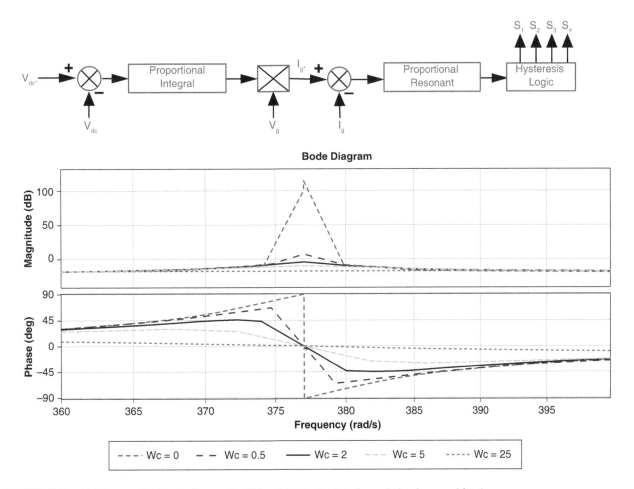

FIGURE 7.13 (a) PR controller block diagram for EV and (b) Bode plots for variation in ω_c and $k = 1$.

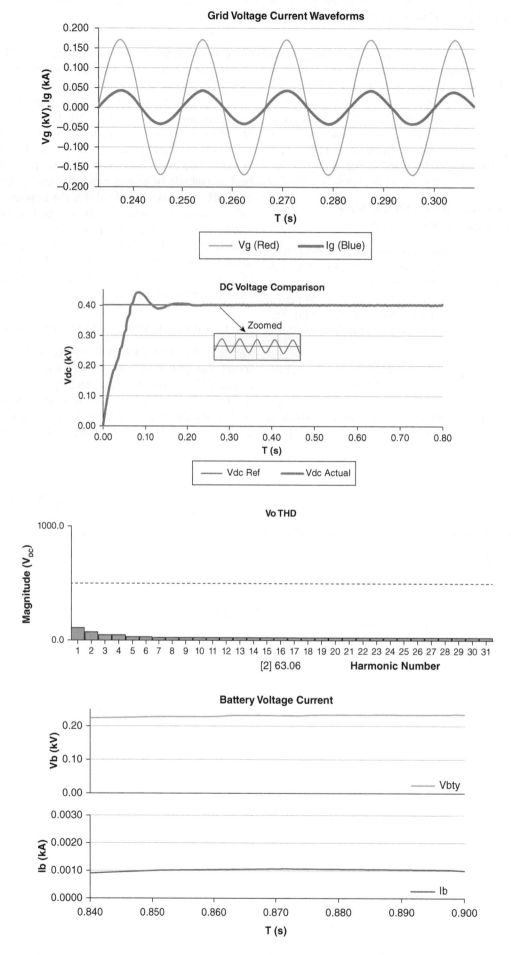

FIGURE 7.14 (a) Grid voltage and current waveforms, (b) DC voltage tracking, (c) harmonic spectrum of the DC voltage, (d) battery voltage and current, and (e) battery SOC during charging phenomenon.

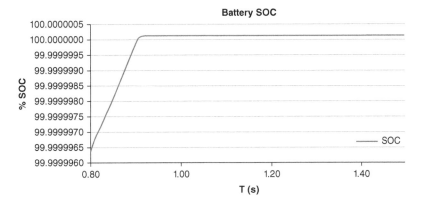

FIGURE 7.14 (Continued)

TABLE 7.4
Individual Harmonic Content of the Grid Voltage and Current Using PR Controller

Harmonic Number	Current THD (%)	Voltage THD (%)
3	1.804	0.0220
5	1.604	0.0072
7	0.708	0.0203
9	0.192	0.0052
11	0.590	0.0078
13	1.235	0.0133

PI controller. Figure 7.15(c) shows the harmonic spectrum of the DC voltage. It consists of 63% compared to 300% harmonic content in the DC voltage in case of PI controller. Figure 7.15(d) shows the battery current and voltage waveforms during battery charging. A small increase in voltage waveform is observed, with current waveform remaining constant. During charging of the battery, the SOC is as shown in Figure 7.15(e).

Table 7.4 shows the percentage of odd-ordered harmonic components present in the grid current and voltage. Grid voltage THD and THD in grid current are 0.039% and 3.66%, respectively, and are said to be within the *IEEE*-519 standards.

7.4.2 Discharging Mode of Operation (V2G)

Figure 7.15(a) shows the gird current and voltage waveforms during discharging operation. Figure 7.15(b) shows the grid current tracking, and the grid current steady-state error is found to be 3.1%.

Figure 7.15(c) shows the battery current and voltage waveforms during battery discharging. Figure 7.15(d) shows the battery SOC during discharging. Hence, it is very clear that the performance of the PR controller is better than that of the PI controller in both the modes of operation in terms of steady-state error, tracking characteristics, and settling time.

7.5 OPERATION OF AN ELECTRIC VEHICLE USING VARIABLE STRUCTURE SYSTEMS – SMC

The sliding mode controller (SMC) is widely used for stable tracking of voltage and current in a power electronic circuit. Its insensitivity to parametric variations and a power converter, owing to its variable structure systems, has influenced researchers to use this control mechanism. Its robustness is far better than those of conventional controllers presented in the literature so far [10]. Here, a linear sliding surface is considered, comprising of DC voltage error and the grid current error. The sliding coefficients are taken as k_1 and k_2. The time response of the system is obtained by taking $\tau = k_1/k_2$. For a desired time, response τ, and value of k_1, k_2 is derived [11]. The sliding surface s expression is given by (7.11):

$$s = k_1(v_{dc}^* - v_{dc}) + k_2(i_g^* - i_g) \quad (7.11)$$

The charging and discharging modes are discussed in this section using SMC. The control block diagram of the SMC controller is shown in Figure 7.16.

7.5.1 Charging Mode of Operation (G2V)

Figure 7.17(a) represents the variable switching frequency waveform with a maximum switching frequency of 11 kHz. Figure 7.17(b) shows the grid current and voltage waveforms during charging with the SMC controller; the grid current is in-phase with grid voltage [12, 13]. Figure 7.18(c) shows the DC voltage tracking; the steady-state error is 1%, and settling time is 0.1 sec. Figure 7.18(d) shows the THD spectrum of the DC voltage waveform.

The % THD obtained is 27% and is less compared to PI, PR controllers. Figure 7.18(a) shows the battery voltage and current in charging mode, and Figure 7.18(b) shows the battery SOC. Table 7.5 shows the percentage of odd-ordered harmonic components present in the grid current

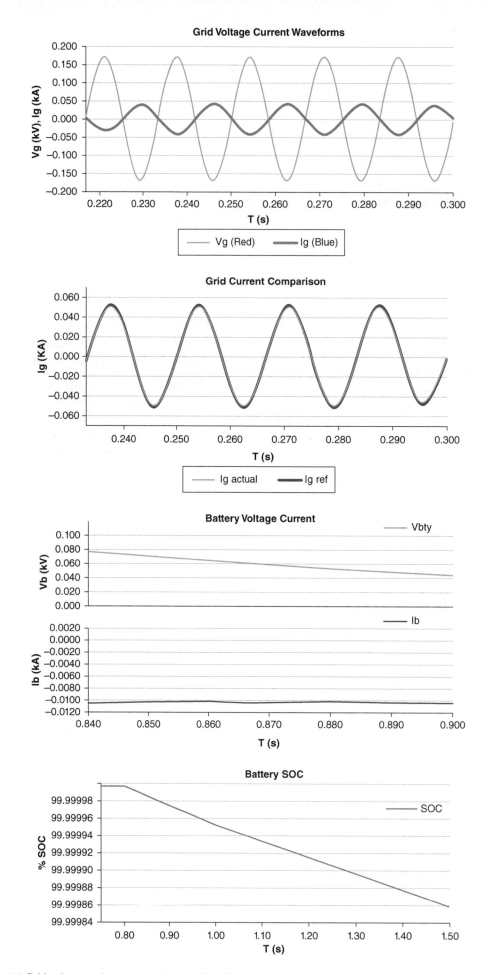

FIGURE 7.15 (a) Grid voltage and current waveforms, (b) grid current tracking, (c) battery voltage and current, and (d) battery SOC.

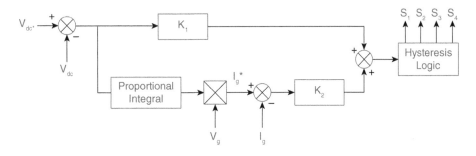

FIGURE 7.16 Simplified block diagram of an SMC controller.

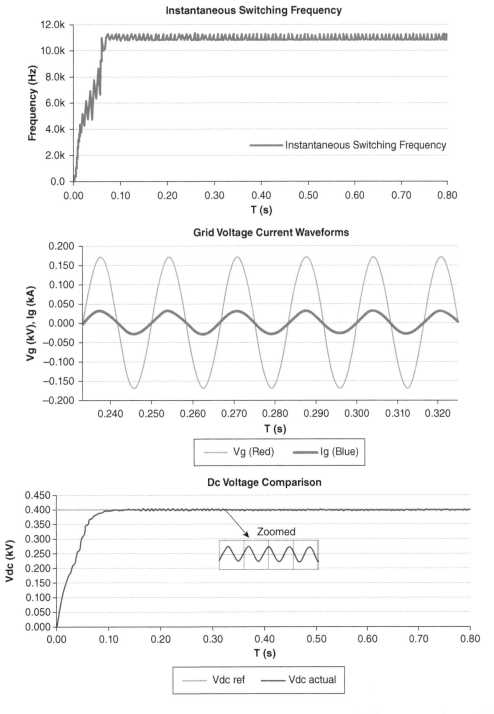

FIGURE 7.17 (a) Switching frequency waveform, (b) in-phase grid voltage and current, (c) DC voltage tracking, and (d) THD of the DC voltage.

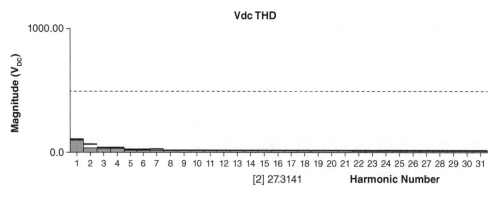

FIGURE 7.17 (Continued)

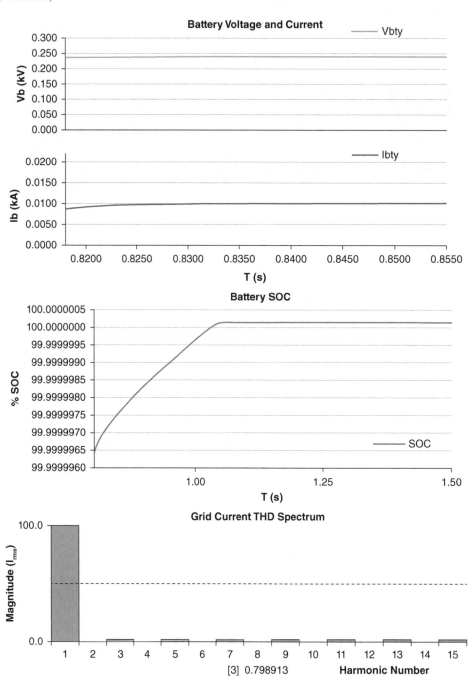

FIGURE 7.18 (a) Battery voltage and current, (b) battery SOC, and (c) grid current THD spectrum.

TABLE 7.5
Individual Harmonic Components of the Grid Current and Voltage Using SMC

Harmonic Number	Current THD (%)	Voltage THD (%)
3	0.798	0.0206
5	2.190	0.0138
7	1.399	0.0101
9	1.035	0.0228
11	0.950	0.0142
13	0.948	0.0086

and voltage. Grid voltage THD and grid current THD are 0.038% and 3.28%, respectively, and are within the *IEEE-519* standards. Gird current harmonic spectrum is shown in Figure 7.18(c).

7.5.2 Discharging Mode of Operation (V2G)

Figure 7.19(a) shows the grid current and voltage waveforms obtained during discharging. Figure 7.19(b) shows the tracking of grid current with steady-state error obtained as 1.52%.

Figure 7.19(c) shows the battery current and voltage waveforms during battery discharging. Figure 7.19(d) reflects the battery state of charge during discharging. Figure 7.19(e) shows the phase plot of SMC for convergence. It is seen that the system is stable, that is, reaching the origin within a small settling time.

Table 7.6 shows the comparison of the three controllers, namely, PI, PR, and SMC controllers, used for electric vehicle charging (G2V) and discharging modes (V2G). SMC is found to be better than PI and PR in both modes of operation. Hence, onboard electrical vehicle charging and discharging operations can be well obtained using an SMC.

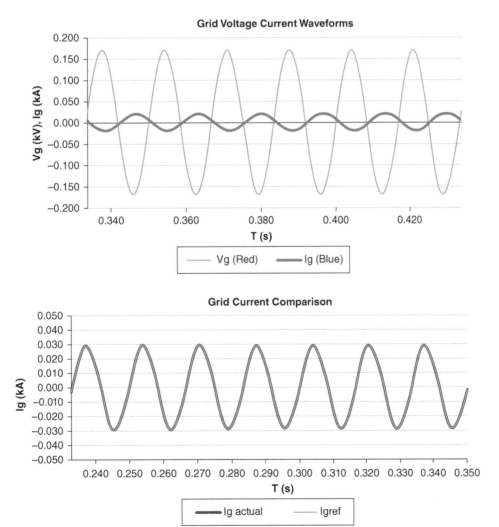

FIGURE 7.19 (a) Grid current and voltage in-phase opposition, (b) tracking of grid current waveform, (c) battery current and voltage during discharging, (d) battery SOC, and (e) phase plane plot of grid current.

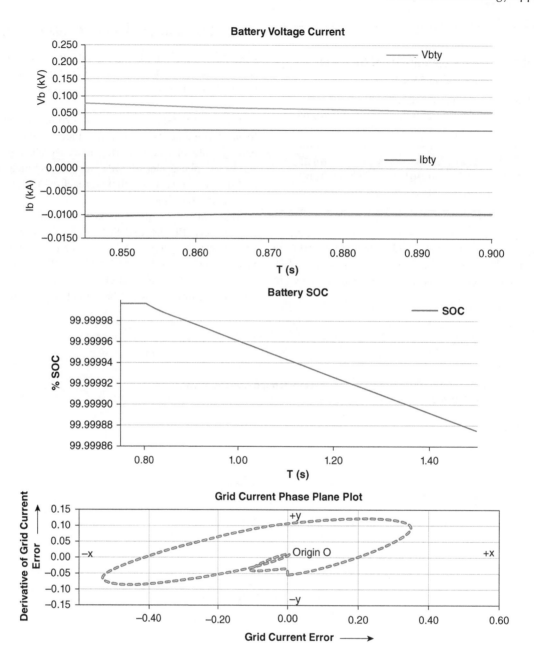

FIGURE 7.19 (Continued)

TABLE 7.6
Comparison of PI, PR, SMC Controllers for G2V and V2G Operation

Controller	Current THD (%)	Voltage THD (%)	Current Steady-State Error (%)	Voltage Steady-State Error (%)	Voltage Settling Time (s)
PI	4.447	0.054	19.06	2.5	0.50
PR	3.665	0.039	3.10	1.0	0.16
SMC	3.289	0.038	1.52	1.0	0.10

7.6 SOLAR-BASED ELECTRIC VEHICLE CHARGING

Due to the development of solar technology worldwide, utilizing solar energy for EV applications is the topical issue of interest. When EVs are directly integrated to the solar charging stations, there is very less burden on the main power grid. This work presents a solar PV array–based electric vehicle charging which can be developed in urban areas [14–19]. It is depicted in Figure 7.20(a). Simulation results are shown in Figure 7.20(b) of MPPT tracking, and Figure 7.20(c), the SOC of the battery. It shows that during normal operation, the power grid charges the battery and it takes a changeover period when supply is not available. Now the charging is done through the solar PV system connected in parallel.

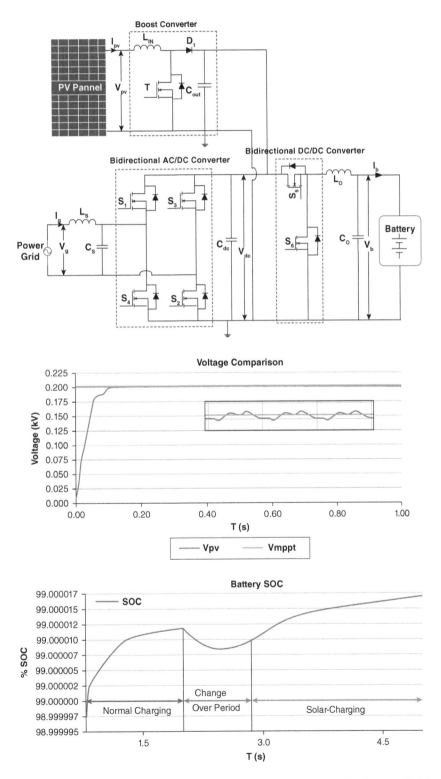

FIGURE 7.20 (a) Solar charger for EV, (b) solar MPPT voltage tracking characteristics, and (c) battery SOC for a solar charger.

7.7 CONCLUSIONS AND FUTURE SCOPE

A 2.5 kW–5 kW battery charger with 100 Ah capacity is used in a single-phase 120 V, 60 Hz EV configuration system. A bidirectional AC/DC and a bidirectional DC/DC converters are used as the power modulators between the source and the battery. Both V2G and G2V modes of operation are obtained in a single-stage conversion from the proposed EV circuit configuration. Onboard electric vehicle charging (G2V) and discharging (V2G) with three different controllers – conventional PI, PR, and the SMC – is presented in this work. Passive filter components used in the input side and output side are derived using the mathematical expressions and appropriately taken for simulation. Using a PI controller, it is found that the grid current sinusoid is unable to track properly. The THD present in DC voltage is more. This causes the battery to drain away and affects the life span of the battery. These drawbacks are overcome by using a PR controller. Due to its infinite gain at fundamental frequency, it is able to track the grid current effectively. Frequency response characteristics are presented to understand the effect of crossover frequency with the assumption of unity gain. Accordingly, the cutoff frequency is fixed at 10 rad/sec. The PR controller reduces the grid current THD by 18%, and the second-order harmonic component in DC voltage by 80%, with respect to a PI controller. It tracks the grid current with small steady-state error. The SMC controller has more tracking capabilities compared to PR and PI controllers. A linear sliding surface is used, with DC voltage error and grid current error as state variables. It is realized that the SMC is effective in generating pure sinusoidal grid current waveform with very less steady-state error of 1.52%. A phase plane plot of the grid current error and its derivative are presented to understand the convergence of the controller in reaching the origin. Settling time is also reduced. A reduction in the second harmonic components (27%) in the DC voltage in SMC as compared to PI and PR controller is realized. It is found to be better than PI and PR controllers during electric vehicle charging and discharging modes. The authors have presented also a solar-based EV charging circuit as an extension to the work. The simulation results obtained are found to be good in charging the battery with its rated voltage within a short period of time. Hence, future work lies in using a solar-based onboard charger at homes, which gives more flexibility to the user for charging and during time of power failures. The future scope lies in connecting a large number of EVs to grid during V2G mode for reactive power compensation. Application of machine learning concepts to an EV has a high potential in research work which the readers can carry out.

7.8 REFERENCES

[1] Mehrdad Ehsani, Yimin Gao, Sebastien E. Gay, and Ali Emadi: *Modern Electric, Hybrid Electric, and Fuel Cell Vehicles Fundamentals, Theory, and Deign*, 2nd edition, CRC Press, Boca Raton, London, New York, Washington, DC, 2005, pp. 1–10.

[2] M. Yilmaz and P. T. Krein: Review of battery charger topologies, charging power levels and infrastructure for plug-in electric and hybrid vehicles, *IEEE Trans. Power Electron*, vol. 28, no. 5, May 2013, pp. 2151–2169.

[3] V. Monteiro, J. G. Pinto, and J. L. Afonso: Operation modes for the electric vehicle in smart grids and smart homes: Present and proposed modes, *IEEE Trans. Veh. Technol.*, vol. 65, no. 3, March 2016, pp. 1007–1020.

[4] Siddhant Kumar and Adil Usman: A review of converter topologies for battery applications in plug-in hybrid electric vehicles, *IEEE Industry Applications Society Annual Meeting (IAS)*, November 2018, pp. 1–9.

[5] Vitor Monteiro, J. G. Pinto, Bruno Exposto, Henrique Goncalves, Joao C. Ferreira, Carlos Couto, and Joao L. Afonso: Assessment of a battery charger for Electric Vehicles with reactive power control, *IECON-38th Annual Conference on IEEE Industrial Electronics Society*, October 2012, pp. 5142–5147.

[6] M. C. Kisacikoglu, B. Ozpineci, and L. M. Tolbert: EV/PHEV bidirectional charger assessment for V2G reactive power operation, *IEEE Trans. Power Electron*, vol. 28, no. 12, December 2013, pp. 5717–5727.

[7] X. Zhou, S. Lukic, S. Bhattacharya, and A. Huang: Design and control of grid-connected converter in Bi-directional battery charger for plugin hybrid electric vehicle application, *Proc. IEEE Veh. Power and Propulsion Conf*, September 2009, pp. 1716–1721.

[8] D. N. Zmood and D. G. Holmes: Stationary frame current regulation of PWM inverters with zero steady-state error, *IEEE Trans. Power Electron*, vol. 18, no.3, May 2003, pp. 814–822.

[9] A. V. J. S. Praneeth, Najath A. Azeez, Lalit Patnaik, and Sheldon S. Williamson: Proportional resonant controllers in on-board battery chargers for electric transportation, *IEEE International Conference on Industrial Electronics for Sustainable Energy Systems (IESES)*, April 2018, pp. 237–242.

[10] H. Komurcugil, Saban Ozdemir, Ibrahim Sefa, Necmi Altin, and Osman Kukrer: Sliding-mode control for single-phase grid-connected LCL-filtered VSI with double-band hysteresis scheme, *IEEE Transactions on Industrial Electronics*, vol. 63, no. 2, February 2016, pp. 864–873.

[11] G. Satish Kumar and Rajesh Gupta: Sliding mode control in voltage Source inverter-based higher order circuits, *International Journal of Electronics*, vol. 102, no. 4, July 2014, pp. 668–689.

[12] N. Vinay Kumar and G. Satish Kumar: Design and analysis of a three phase transformerless hybrid series active power filter based on sliding mode control using PQ- theory and stationary reference frames, *Serbian Journal of Electrical Engineering*, vol. 16, no. 3, October 2019, pp. 289–310.

[13] V. Naik, S. K. Gudey, and A. R. Anuku: Design of robust feedback controller for a UPS system in industrial application, *IEEE 2018 Recent Advances on Engineering, Technology and Computational Sciences (RAETCS)*, 2018, pp. 1–6, doi: 10.1109/RAETCS.2018.8443809.

[14] Q. Yan, B. Zhang, and M. Kezunovic: Optimized operational cost reduction for an EV charging station integrated with battery energy storage and PV generation, *IEEE Trans. on Industrial Informatics,* vol. 14, no. 1, January 2018, pp. 2096–2106.

[15] K. Chaudhari, A. Ukil, K. N. Kumar, U. Manandhar, and S. K. Kollimalla: Hybrid optimization for economic deployment of ESS in PV-integrated EV charging stations, *IEEE Trans. Ind. Informat*, vol. 14, no. 1, January 2018, pp. 106–116.

[16] G. Satish Kumar and Bikram Sah: A comparative study of different MPPT techniques using different dc-dc converters in a standalone PV system, *IEEE TENCON*, Singapore, November 2016.

[17] Nian Liu, Qifang Chen, Xinyi Lu, Jie Liu, and Jianhua Zhang: A charging strategy for PV-based battery switch stations considering service availability and self-consumption of PV energy, *IEEE Trans. on Industrial Electronics*, vol. 62, no.8, August 2015, pp. 4878–4889.

[18] Tan Ma and Osama Mohammed: Optimal charging of plug-in electric vehicles for a car park infrastructure, *IEEE Trans. on Industrial Application*, vol. 50, no.4, August 2014, pp. 2323–2330.

[19] M. J. E. Alam, Kashem M. Muttaqi, and Danny Sutanto: Effective utilization of available PEV battery capacity for mitigation of solar PV impact and grid support with integrated V2G functionality, *IEEE Trans. On Smart Grid*, vol. 7, no. 3, May 2016, pp. 1562–1571.

8 Experimental Investigation on Hybrid Photovoltaic and Thermal Solar Collector System

P Narasimha Siva Teja, S. K. Gugulothu, and B. Bhasker

CONTENTS

8.1 Introduction 101
8.2 Solar Cell 102
8.3 Solar Thermal Collector 102
8.4 Theoretical Heat Transfer Assessment 103
    8.4.1 Heat Transfer by Conduction 103
    8.4.2 Heat Transfer by Convection 103
    8.4.3 Heat Transfer by Radiation 103
8.5 Design of Thermosyphon Cooling System 104
8.6 Design and Development of PVT System 104
    8.6.1 Hybrid Photovoltaic/Thermal Solar Systems 104
    8.6.2 Hybrid System Design Considerations 105
    8.6.3 Collector Design 105
    8.6.4 Experimental Setup and Analysis 105
8.7 Results and Discussions 106
8.8 Conclusions 107
8.9 References 107

8.1 INTRODUCTION

Energy plays a crucial role in the development of any country in the world. India is in the third position in the world both in production and consumption of electrical energy, with an installed capacity of 373 GW as of 2020 (*Source*: Ministry of Power, government of India). Indisputably, all kinds of energy on Earth are obtained directly or indirectly from the sun. However, conventional fossil fuels derive solar energy in a much-concentrated form. Current fossil fuels are environmentally hazardous and will be depleted in the near future. Unconventional solar energy is a possible solution to this problem. Solar power occupies nearly 8% of the total generation in the national grid, and it can be generated by using both light and heat. The farmer is called photovoltaic (PV) energy, and later is thermal energy. These solar cells can be manufactured as a single-crystalline and polycrystalline structure. Single-crystalline solar cells are more effective than polycrystalline cells. The parameters that affect the solar photovoltaic cell's efficiency are ambient temperature, dust on the solar panel, and other environmental conditions. Over the past decade, focus has been on reducing the proliferation of temperature effects [1], present errors, and order of accuracy of a hybrid PV/T system. Energy and exergy analysis of a photovoltaic–thermal collector with and without glass cover is discussed by [2]. An evaluation of a hybrid (PV/T) air collector's thermal performance is studied by [3]. In this performance, comparison of PV module with glass-to-Tedlar and glass-to-glass is studied. The experimental and theoretical evaluation of design, development, and performance prediction of a hybrid PV/T air collector by a simulation model was carried out by Niccolo Aste et al. [4]. Exergetic analysis and study of sustainable renewable energy resources were discussed by [5]. Electrical, thermal, and exergy studies on hybrid PV/T air collector for different Indian climatic conditions are evaluated [6]. Testing and analyzing a bifacial PV/T module with a set of reflecting planes for the water heating system were carried out [7]. In the research study "Integration of hybrid photovoltaic's/thermal collectors in buildings" [8], the maximization of solar energy for electricity and thermal generation by using the hybrid solar collectors is presented. Exergy studies of a parallel plate hybrid (PV/T) air collector at the Srinagar location are evaluated [9]. A dual heat extraction system of PV/T collector with water/air circulation to be used in residential buildings for thermal comfort is presented [10]. Performance of different PV panels fabricated with crystalline (c-) Si, α-Si, and $CuInSe_2$ for thermal contact analysis between the panel and collector was carried by [11]. The PV panel in the PV/T system installed is smaller than the solar collector. Cogeneration of preheated air, hot water, and electricity using an advanced PV/T collector was presented by

Y. B. Assoa et al. [12]. Industrial heat processing at 60°C and 80°C with polycrystalline and amorphous solar cells is studied, and the former shows more electricity production [13]. Performance studies on a finned double-pass PV/T solar collector using the mathematical model were carried out by [14]. It is reported that 65% of energy saving can be achieved using PV/T collector with cell covering factor of 0.63 and glazing transmissivity of 0.83 when the ratio of heating-load-to-collector-area exceeds 80 kg/m^2 [15]. T. T. Chow et al. [16, 17] conducted experimental investigations for different seasonal operating conditions on PV and hot water collectors mounted on the exterior face. The development and validation of a computational model of the PV/T collector system have been presented. It will be used to predict the installed system's energy output and the payback period. Parameters affecting a wall-mounted hybrid PV/T collector's performance are studied [18]. Simulation of hybrid PV/T solar systems for domestic hot water applications, both passive (thermosyphon) and active, with the help of TRANSYS, was carried out. Theoretical estimation of a PV/T system for residential heating and cooling applications can reduce the thermal load. Nevertheless, the PV/T collector's performance was estimated at different geographical locations and the system's total surface area [19]. However, even though several research studies have been carried out, developing energy-efficient cooling technique is still an advance. Therefore, there is a need for cooling the solar cell with the most energy-conservative techniques. One of the best possible cooling methods of PV modules is using photovoltaic–thermal (PV/T) collector. In this chapter, it is proposed to study the improvement of the PV/T system performance thermally and electrically by using additional glazing and a booster diffuse reflector individually, or both, by giving flexibility in system design.

8.2 SOLAR CELL

A solar photovoltaic cell is a device that generates electrical energy with the excitation of the photons by receiving sunlight. This cell is made of different semiconductors, like single-crystalline, polycrystalline, ribbon, and amorphous silicon materials. Among the mentioned, the single-crystalline cell material has more efficiency compared to others. Photovoltaic (PV) systems have different components. They are solar panels, inverters, charge controllers, batteries, panel mounts, and tracking system, as shown in Figure 8.1. A single solar cell can generate about half of the voltage. So a group of solar cells, termed as a module and its array, will increase the system's output. The arrays will be arranged in series, parallel connections to maximize the voltage or current.

PV arrays are installed depending on the metering system, either in off-grid, on-grid, or hybrid system. In the off-grid system, the energy generated with the PV system will not be supplied to the electrical grid, and it will be stored in the batteries. It is also called as stand-alone system. Whereas on-grid or grid-connected systems are coupled to the electrical grid and operated simultaneously with the utility grid. The solar cell's performance mainly varies depending on the ambient temperature. Though it is impossible to control the ambient air temperature, there is a possibility to monitor the cell's surface temperature by installing a hybrid solar thermal collector management technique. The PV cell specifications considered for the present study are shown in the Table 8.1.

8.3 SOLAR THERMAL COLLECTOR

It is a device to collect radiation energy from the sun to raise the working fluid's temperature for daily needs. The heat energy can be generated using highly reflective mirrors and absorber plates. There are various types of thermal collectors, like a flat plate, evacuated tube, point focus, and line focus. These collectors are generally used to generate steam for cooking or process or space heating or used in solar thermal power plants to generate electricity. Building an energy-efficient system is a big task in any solar devices.

As the thermal management of a solar PV module plays a crucial role in improving its performance, it has been attempted to build a hybrid solar photovoltaic/thermal collector system. The system utilizes the excess temperature

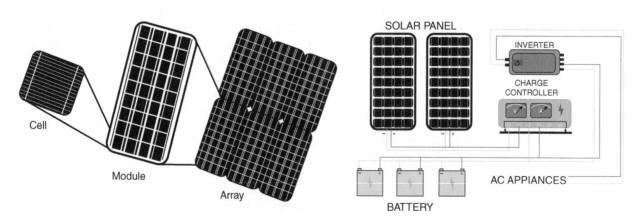

FIGURE 8.1 Solar PV system.

TABLE 8.1
Specifications of PV Module Components

Part in PV Module	Material	Thickness (δ) (mm)	Thermal Conductivity (k) (W/mk)	Area (A) (sq.m)
Glazing	Ceria-doped glass	3	1.4	0.083
Adhesive	Ethyl vinyl acetate	0.15	0.585	0.083
Solar cell	Silicon	0.1	50	0.0014
Solder	Tinned lead	0.1	50	0.0014
Substrate	Polyvinyl	0.2	0.711	0.083

on the PV module with thermal collectors using the thermosyphon cooling technique.

8.4 THEORETICAL HEAT TRANSFER ASSESSMENT

Heat is transferred from the module by conduction, convection, and radiation modes of heat transfer. One has to calculate the amount of heat transferred through three modes, converted into power, and retained in the solar cells. The mass flow rate of water circulated to remove the heat from the solar cell can be calculated by knowing the amount of heat retained in the solar cell.

In this study, the solar photovoltaic cell can be considered as a composite plane wall, and heat transfer analysis has been carried out on this wall. The PV cell specifications for the calculations are taken from Table 8.1.

Measured values of temperature (at 1:00 p.m. on a typical day) are:

$T_{cell} = 80^0 C$, $T_{glazing} = 50^0 C$, $T_{substrate} = 45^0 C$, Solar flux $= 880 \ W/m^2$

8.4.1 Heat Transfer by Conduction

Considering steady-state condition,

$$Q_{Glazing} = Q_{Adhesive} = Q_{Cell} = Q_{Solder} = Q_{Substrate} \quad (8.1)$$

$$\frac{Q}{A} \propto \frac{\partial T}{\partial x} \quad (8.2)$$

$$Q = -kA\frac{\partial T}{\partial x} \quad (8.3)$$

$$Q_{cond} = \frac{\Delta T}{\left(\frac{\delta}{kA}\right)} \quad (8.4)$$

$R = \left(\frac{\delta}{kA}\right)$ is the resistance offered by the material to the conduction of heat.

Final heat conduction equation for the solar cell is:

$$Q_{cond} = \frac{2T_{cell} - T_{glazing} - T_{substrate}}{R_{Glazing} + R_{Adhesive} + R_{Cell} + R_{Substrate}} \quad (8.5)$$

Taking all the resistances into account, conduction heat transfer is calculated as 1.892 W (2.59% of input radiation falling on solar cell).

8.4.2 Heat Transfer by Convection

According to Newton's law of cooling:

$$Q_{conv} = hA(T_w - T_a) \quad (8.6)$$

Here, the heat transfer coefficient is calculated by natural convection correlations by calculating the Grashof number.

$$\text{heat transfer coefficient } h = 0.58\sqrt[4]{\frac{\Delta T}{L}} \text{ for} \quad (8.7)$$

$10^4 < Gr.\Pr < 10^9$

where $\Delta T = T_{surface} - T_a = 50 - 30 = 20^0 C$
L is the horizontal dimension of solar panel = 0.4 m.

$$\text{Grashof number is given by } Gr = \frac{l^3 g \beta \Delta T}{v^2} \quad (8.8)$$

The fluid properties are taken for the mean temperature of 40°C.

The heat lost due to convection is 3.54% of input radiation falling on solar cell.

8.4.3 Heat Transfer by Radiation

According to Stefan–Boltzmann law, the heat transferred by radiation is directly proportional to the difference between the fourth power of the absolute temperature and is given by:

$$Q_{radiation} = \varepsilon\sigma\left(T_{glazing}^4 - T_a^4\right) \quad (8.9)$$

$Q_{radiation}$ is 18.41% of input radiation falling on solar cell.

$$I \times A = Q_{cond} + Q_{conv} + Q_{radiation} + P_{el} + Q_{cooling} \quad (8.10)$$

Where I is the solar heat flux, which is measured to be 880 W/m².

Energy required to cool the solar panel is calculated as $Q_{Cooling} = 42.108\ W$.

From previous calculations, it is noted that, from the received solar radiation falling on the solar PV panel,

- Only 15% can be converted to power;
- 26% is transferred by conduction, convection, and radiation; and
- 59% is retained in the solar PV cell so that its temperature rises to 80°C.

From the preveding conclusion, it is required to design a thermosyphon cooling system. This cools the photovoltaic cell and retains the efficiency of the solar cells

8.5 DESIGN OF THERMOSYPHON COOLING SYSTEM

Thermosyphon is a heat exchange method in which the fluid circulates in the system without using any external device. It works on a gravity-based principle. In the thermosyphon cooling system, the fluid circulates in the system without any pump, therefore natural convection. In this context, it is to calculate the mass of water required to be flown on the photovoltaic cell in a day by assuming the heat removal capacity and the heat gained by water should be dissipated to the atmosphere by choosing a suitable material for the chamber.

Mass of water to be circulated in a day can be calculated as given by:

$$\dot{m} = \frac{P \times H \times A}{C_p \times \Delta T} \quad (8.11)$$

Where $\dot{m}$ is the mass of water to be circulated in a day in kg/day, P is the percentage of heat removal capacity of the system, H is the solar radiation falling on solar cell in a day (measured value on a typical day 11.56 MJ/m2), A is the area of solar cell (0.76 m2), C_p is the specific heat of water (4.2 kJ/ kg K), and ΔT is the temperature difference of water $= T_f - T_i$.

The heat gained by the water is 580.860 kJ/day.

The material used for thermosyphon should have high thermal conductivity; here, copper and aluminum are used. Design of the thermosyphon consists of determining the following:

- Length of the tube and surface area of the sheet required
- Heat transfer coefficient for both the materials
- Heat transfer coefficient of aluminum can be calculated as:

$$h = \frac{Nu \times k}{l} \quad (8.12)$$

Nusselt number is given by the correlation

$$Nu = 0.68 + \frac{0.67(Gr + Pr)^{0.25}}{\left[1 + (0.492/Pr)^{0.5625}\right]^{0.45}} \quad (8.13)$$

Pr and other properties are taken for the mean temperature of 310.5 K.

Grashof number is given by $Gr = \dfrac{g\beta(T_w - T_a)l^3}{v^2}$.

From the properties, Nu is calculated as 65.58, and heat transfer coefficient as 22406.5 W/m²K.

Heat transferred by aluminum per square meter area is calculated as:

$$\frac{q}{A} = h\Delta T = 336.09 \times 10^3 \quad (8.14)$$

Heat transfer coefficient of copper tube is determined by using:

$$Nu = 0.53 \times (Gr \times Pr)^{0.25} \quad (8.15)$$

The Nu number is calculated as 3.15, and the heat transfer coefficient of copper tube is:

$$h_{cu} = 121590\ W/m^2 K$$

The heat transferred by copper tube per meter length is given by:

$$\frac{q}{l} = h \times \pi \times d \times \Delta T = 57.29 \times 10^3\ \frac{W}{m} \quad (8.16)$$

By taking the length of copper tubes as 5 m, the surface area required for a copper is calculated as $A_{cu} = 2m^2$.

Hence, the surface area required is obtained, and thermosyphon cooling system is fabricated.

8.6 DESIGN AND DEVELOPMENT OF PVT SYSTEM

8.6.1 Hybrid Photovoltaic/Thermal Solar Systems

Solar energy comprises both light and radiation. PV panels convert some of the light energy falling on them into electrical energy. The radiation energy and remaining light energy increase the PV panel's surface temperature and reduce electrical efficiency. Reducing the panel's temperature by heat extraction mechanism with a proper natural or forced convection can improve the PV panel's efficiency. Solar hybrid PV/T system is a technology that integrates PV cells and thermal collector to improve electrical efficiency by simultaneously cooling it. The combined electrical and thermal energy output of the PV/T system depends on parameters like solar energy input, the surrounding temperature, the air velocity, the system parts' operating temperature, and the heat extraction mode. The thermal system unit is operated to maximize the electrical output. Past research studies reported that employing air gap with glazing and reducing thermal losses in PV/T collector raise the thermal efficiencies up to 65%. The theoretical models estimated the thermal efficiencies up to 55% for maximum airflow rate, minimum air duct depth, and lengthy PVT system.

8.6.2 Hybrid System Design Considerations

The photovoltaic system should be operated at optimum lower temperatures to maximize the efficiency for the input radiation, surrounding air temperature, and wind velocities. This can be achieved by reducing the temperature of the system using fluid flow convection. The rise in temperature of the fluid will be used for air and water heating system and space heating applications. The heat removal from the bottom surface of the PV system by fluid circulation is an efficient cooling method.

8.6.3 Collector Design

The principal requirement of the photovoltaic/thermal collector is the incorporation of high thermal conductive and better electrical insulating material betwixt the PV cells and the thermal absorber plate. The substrate of this collector is manufactured with copper alloy battens which consist of male and female connecting members on either side, as shown in Figure 8.2. Enhanced heat transfer has been ensured with the large heat transfer area between the absorbing surface and circulating fluid. The absorber's leveled top surface facilitates the lamination between PV cells and the absorber. Depending on the requirement of the collector size, numerous amounts of battens can be assembled easily.

Black silicon cells with single-crystalline are made to cover the absorber surface owing to their improved electrical conversion efficiency along with thermal absorption despite having higher thermal emissivity than blue silicon cells with polycrystalline structure. The arrangement involves transparent TPT, EVA, single-crystalline silicon cells, and silica gel on the absorber plate, with an insulation thickness of 30 mm at a depth of 25 mm from the front glazing. The transparent TPT, with a thickness not exceeding 0.2 mm, has a better performance in electrical insulation and thermal conduction. EVA with a thickness of 0.4–0.6 mm acts as an adhesive. As EVA and TPT are high-transmissive materials, these two are laminated to the PV module and absorber.

8.6.4 Experimental Setup and Analysis

The collector with an aperture area of 2 m² is connected to a well-insulated 100 L water tank. Single-crystalline silicon-based solar cells with a conversion efficiency of 14.5% are used. The PV module is attached to the top surface of the copper absorber. The laminated peak power is 75 W. The closed enclosure absorber is fabricated with the assembly of multiple extruded round tube copper modules with their ends connected with headers. The incoming and outgoing circulation of the working fluid is carried through the upper transverse header with an inserted stopper placed after the first longitudinal module. By this, the inlet fluid flows down from the first module to the lower traverse header, then afterwards flows upwards to all the other modules, employing the thermosyphon effect. The tank can be emptied with the help of a drain cork and refilled with a working fluid to a fixed quantity. The overall experimental setup contains the data-logging device, the equipment to monitor and record solar conditions, and the collector operating temperatures for an interval of 15 min. There is an insulating

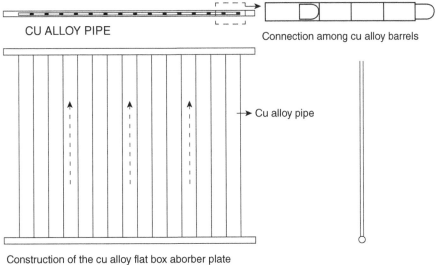

FIGURE 8.2 Copper alloy absorber plate.

air layer between the front glazing and the PV encapsulation. The enclosure is equipped with encapsulated materials; the transparent TPT (Tedlar polyester Tedlar) and the EVA (ethylene-vinyl acetate) layers on the upper surface, the EVA and opaque TPT layers at the bottom surface, and even below that, the thermal insulation layer have been created. The previously mentioned arrangement is assembled in an aluminum-frame back cover.

The experimental analysis of the hybrid PV/T system has been carried out using the setup as shown in Figure 8.3(a). The present setup was fabricated using nine battens with a heat collection area (A_c) of 1.86 m². The system comprises one thermal collector, one storage tank for working fluid with pipelines and valves. The storage tank was fixed with 30 mm thick polyurethane foam to an Al-alloy bracket. Thermocouples have been placed at the upper surface of the tank. The PV/T module consists of black silicon single-crystalline cells arranged in series, batteries of four number. The single solar cell has a conversion efficiency of 14.5% at a standard irradiance condition of 1,000 W/m² and a temperature of 25°C. PV/T collector is facing south, with a tilt angle of 350 all over the testing period. The operating temperature, weather conditions like solar radiation, and mass glow rate of the fluid are the parameters considered for the analysis.

8.7 RESULTS AND DISCUSSIONS

Experimental analysis of a hybrid solar PV panel incorporating a thermal collector was performed. The tests are conducted from morning, at 8:00 a.m., to evening, at 5:00 p.m. The irradiance received from the sun, the temperature at

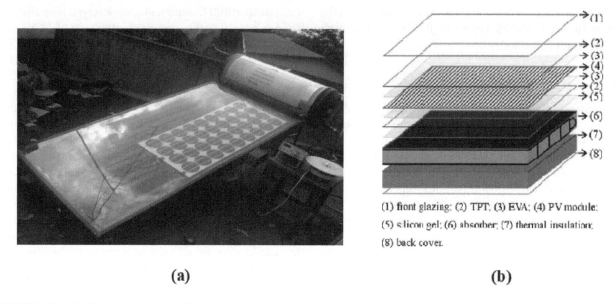

FIGURE 8.3 (a) Experimental setup of hybrid photovoltaic system; (b) different layers of PVT system.

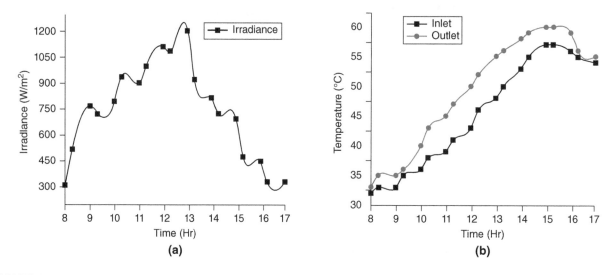

FIGURE 8.4 (a) Irradiance received by the solar collector in a day; (b) the circulated water inlet and outlet temperatures.

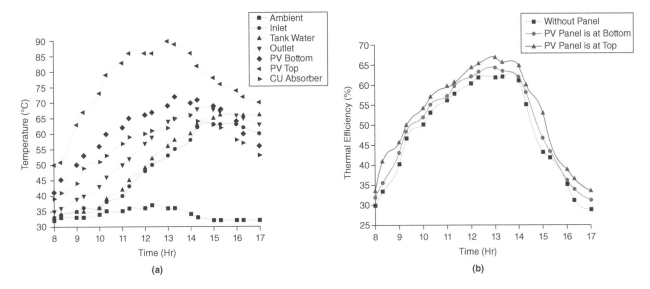

FIGURE 8.5 (a) Temperatures recorded at various locations; (b) thermal efficiency of thermal collector with and without the panel.

various locations of the panel, inlet and outlet cooling fluid, and performance parameters are calculated.

Irradiance received by the solar collector on a typical day is shown in Figure 8.4(a). The maximum irradiance of 1.204 kW/m² was observed at 1300 hr. Figure 8.4(b) shows the inlet and outlet temperature of water flowing inside the tube placed in the thermal collector. The temperature of the fluid rises linearly with time to reduce the panel temperatures.

Figure 8.5 illustrates the temperatures recorded at various locations of the hybrid PVT collector. In the total system, the least temperatures are accorded at the water inlet and the highest temperatures at top of the PV panel. In each case, the maximum temperatures were recorded between 1200 and 1300 hrs. The tests are carried out to estimate the thermal efficiency for the thermal collector with and without the panel. It has been observed that the thermal collector embodied with a PV panel on the top side has a maximum efficiency of 66.97% at 1300 hrs. An improvement of 4% electrical efficiency and 6% thermal efficiency is observed compared to the collector without panel.

8.8 CONCLUSIONS

The experimental design and analysis of closed-box copper alloy solar PV/T collector are carried out and compared with the normal flat-box collectors.

The PV/T water heater has been organized for the free circulation of a working fluid. The PV/T collector employed with the cell cowling factor of 0.63 and front-glazing transmissivity of 0.83 can work for a heating load of 80 kg/m² with an electrical efficiency of 10.15% and the thermal efficiency of 45% – in turn, the daily total efficiency of 52%. An energy saving of about 65% has been recorded. An improvement of 4% electrical and 6% thermal efficiency is observed while using a hybrid PV/T system than considering a normal collector.

With the incorporation of shared front glazing, back cover, and fixed frame, the PV/T collector can achieve high energy yield with less investment than the typical series-operated solar PV and thermal systems per unit surface area. PV/T systems will be a possible solution when more PV panels are to be installed in small facade areas.

8.9 REFERENCES

[1] Erzat Erdil, Mustafa Ilkan, Fuat Egelioglu, "An experimental study on energy generation with a photovoltaic (PV) solar thermal hybrid system", Article history: Received 26 December 2007 www.elsevier.com/locate/energy.

[2] K.F. Fong, Z. Lin, A.L.S. Chan J. Ji, T.T. Chow, G. Pei, "Energy and exergy analysis of photovoltaic – thermal collector with and without glass cover" www.elsevier.com/locate/energy.

[3] A.S. Joshi, A. Tiwari, G.N. Tiwari, I. Dincer, B.V. Reddy, "Performance evaluation of a hybrid photovoltaic thermal (PV/T) (glass-to-glass) system", Received 23 January 2008 www.elsevier.com/locate/ijts

[4] Niccolo Aste, Giancarlo Chiesa, Francesco Verri, "Design, development and performance monitoring of aphotovoltaic-thermal (PVT) air collector", *Renewable Energy* 33 (2008) 914–927.

[5] Arif Hepbasli, "A key review on exergetic analysis and assessment of renewable energy resources for a sustainable future", *Renewable and Sustainable Energy Reviews* 12 (2008) 593–661.

[6] Vivek Raman, G.N. Tiwari, "Life cycle cost analysis of HPVT air collector under different Indian climatic conditions", *Energy Policy* 36 (2008) 603–611.

[7] B. Robles-Ocampoa, E. Ruı z-Vasquezb, H. Canseco-Sanchezb, R.C. Cornejo-Mezac, G. Tra paga-Martı´ nezd, F.J. Garcı -Rodrigueza, J. Gonza lez- Herna ndeze, Yu.V. Vorobievd, "Photovoltaic/thermal solar hybrid system with bifacial PV module and transparent plane collector", *Solar Energy Materials & Solar Cells* 91 (2007) 1966–1971.

[8] G. Fraisse, C. Me ne zo, K. Johannes, "Energy performance of water hybrid PV/T collectors applied to combisystems of Direct Solar Floor type", *Solar Energy* 81 (2007) 1426–1438.

[9] Anand S. Joshi, Arvind Tiwari, "Energy and exergy efficiencies of a hybrid photovoltaic- thermal (PV/T) air collector", *Renewable Energy* 32 (2007) 2223–2241.

[10] Y. Tripanagnostopoulo, "Aspects and improvements of hybrid photovoltaic/thermal solar energy systems", *Solar Energy* 81 (2007) 1117–1131.

[11] R. Zakharchenkoa, L. Licea-Jiméneza, S.A. P erez-García, P. Vorobievb, U. Dehesa-Carrascoc, J.F. Perez-Roblesa, J. González-Hernándeza, Yu. Vorobieva, "Photovoltaic solar panel for a hybrid PV/thermal system", *Solar Energy Materials & Solar Cells* 82 (2004) 253–261

[12] Y.B. Assoa, C. Menezo, G. Fraisse, R. Yezou, J. Brau, "Study of a new concept of photovoltaic – thermal hybrid collector", *Solar Energy* 81 (2007) 1132–1143

[13] S.A. Kalogirou, Y. Tripanagnostopoulos, "Industrial application of PV/T solar energy systems", *Applied Thermal Engineering* 27 (2007) 1259–1270

[14] Mohd. Yusof Othmana, Baharudin Yatima, Kamaruzzaman Sopianb, Mohd. Nazari Abu Bakara, "Performance studies on a finned double-pass Photovoltaic-thermal (PV/T) solar collector", *Desalination* 209 (2007)

[15] Jie Ji, Jian-Ping Lu, Tin-Tai Chow, Wei He, Gang Pei, "A sensitivity study of a hybrid photovoltaic/thermal water-heating system with natural circulation", *Applied Energy* 84 (2007) 222–237

[16] T.T. Chow, W. He, J. Ji, "An experimental study of façade-integrated Photovoltaic/water-heating system", *Applied Thermal Engineering* 27 (2007) 37–45

[17] T.T. Chow, W. He, J. Ji, A.L.S. Chan, "Performance evaluation of Photovoltaic – thermosyphon system for subtropical climate application", *Solar Energy* 81 (2007) 123–130

[18] S.A. Kalogirou, Y. Tripanagnostopoulos, "Hybrid PV/T solar systems for domestic hot water and electricity production", *Energy Conversion and Management* 47 (2006) 3368–3382

[19] G. Vokas, N. Christandonis, F. Skittides, "Hybrid photovoltaic – thermal systems for domestic heating and cooling – A theoretical approach", *Solar Energy* 80 (2006) 607–615

9 Concentrated Solar Integrated Hydrothermal Liquefaction of Wastes and Algal Feedstock
Recent Advances and Challenges

Namrata Sengar, Matthew Pearce, Christopher Sansom, Xavier Tonnellier, and Heather Almond

CONTENTS

9.1 Introduction: Background and Driving Forces 109
9.2 Biomass Conversion to Biofuels 110
9.3 Algae Growth and Potential 111
9.4 Thermochemical Routes for Biomass Conversion to Fuels 113
9.5 Hydrothermal Liquefaction 113
9.6 Solar Collectors – Flat and Concentrating 114
9.7 Studies on Solar Integrated Hydrothermal Liquefaction 114
9.8 Challenges and Opportunities 118
9.9 References 119

9.1 INTRODUCTION: BACKGROUND AND DRIVING FORCES

Energy demands are continually increasing, with increased concerns for climate change. Fuels used commonly to provide power and aid in transportation and daily requirements are gasoline, diesel, coal, natural gas, liquid petroleum gas, uranium, and biomass. Out of these fuels, major consumption is of fossil fuels in the form of gasoline, diesel, coal, natural gas, and LPG, which results in fast depletion and poses environmental threats in the form of greenhouse gas (GHG) emissions and climate change. This scenario is leading to a search for pathways to shift the energy consumption towards cleaner options; one such field is biomass-based fuels through solar energy processing.

Biomass is a complex mixture of carbohydrates, lignin, proteins, and lipids formed with the presence of elements carbon, hydrogen, nitrogen, sulfur, and oxygen [1]. Biomass has been used as fuel since time immemorial; still, biomass is the major fuel for developing countries, with common use in cooking, water heating, or space heating applications. The traditional use of biomass with conventional cookstoves is inefficient and polluting, leading to increased disease burden and around 4 million deaths per year [2]. In order to convert biomass to more efficient biofuels which can supplement and take the place of fossil fuels, various technologies have been developed. Biofuels obtained through biomass feedstock are considered carbon-neutral because, upon use, they release almost as much CO_2 as these consume when biomass is created through photosynthesis.

Biomass feedstock can be classified as dry and wet feedstock, as shown in Figure 9.1. Dry feedstock is lignocellulose, which consists of cellulose, hemicellulose, and lignin. The wet feedstock is mainly algae, which has fractions in form of lipids/fats and proteins/amino acids. Biomass wastes and algae have attracted attention due to their potential for conversion to biofuels. Management of waste and wastewater is a serious issue worldwide. Especially, developing countries face a two-way problem, with lack of infrastructure on one hand and the increased disease burden on the other hand. Among various methods for biofuel production, studies on hydrothermal liquefaction (HTL) are gaining momentum and can prove to be a promising option for future energy needs. HTL has major limitation of being an energy-intensive technique, so efforts are going on to integrate the process with solar thermal energy through solar concentrating collectors.

The chapter discusses the concept of integration of solar energy with the hydrothermal liquefaction process and presents an overview of recent advances and studies conducted in this area. This field of research is considerably new and promising; it presents ample opportunities for developing the technologies for clean fuel in near future. Research in this area may offer solutions for efficient energy conversion, waste management, and tackling climate change. This chapter provides a brief introduction to biofuel generations and biomass conversion processes, provides details on hydrothermal liquefaction, presents an overview of studies on solar integrated HTL, and discusses the opportunities and challenges in this field.

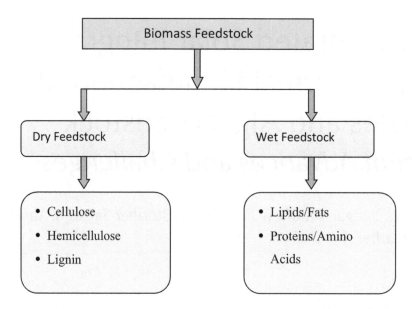

FIGURE 9.1 Components of biomass feedstock.

9.2 BIOMASS CONVERSION TO BIOFUELS

Biofuels can be classified in four generations on the basis of feedstock source, as shown in Figure 9.2 [3]. First-generation biofuels are the conventional biofuels, such as ethanol and biodiesel, produced mainly from sugar, starch, or vegetable oil obtained from food crops. The processes involved in the production of first-generation biofuels are fermentation, distillation, and transesterification, which have been in use in centuries. The technologies for first-generation biofuels are well developed and mature, but there are some disadvantages associated with these biofuels, as these require food crops, are expensive, require large land and water, and need power for processing.

The second-generation biofuels are sourced from wood, agricultural waste, food waste, nonedible plants like Jatropha, silver grass, switchgrass. The processing is similar to the first-generation biofuels, but pretreatment is required in the form of biochemical or thermochemical reactions. Through biochemical route, ethanol is obtained, whereas through thermochemical route, syngas is produced. The advantage with second-generation biofuels is that these do not require food crops, but still, limitations are there in the form of large land requirement, water requirement, high cost, and technology still under development for improved efficiency.

Third-generation biofuels make use of algae biomass, such as macroalgae and microalgae for production of bio-oil. Algae have high carbon dioxide sequestering capabilities and high photosynthesis levels. Microalgae can be easily grown on wastewater, sewage, salt water; hence, it does not require as much land and water as in the case of the first two generations of biofuels. Using microalgae as feedstock has advantages of high growth rate, lesser time duration of cultivation (5–6 days), low cultivation cost, and high oil content. Algal biomass can be processed through fermentation, transesterification, thermochemical methods, and microbial fuel cells to produce bioethanol, biogas, biohydrogen, biodiesel, jet fuel, syngas, gasoline, jet fuel, and bioelectricity [4]. Biofuels obtained from genetically modified microalgae are known as fourth-generation biofuels. Genetic modifications are introduced in microorganisms to increase their uptake of carbon dioxide and improve biofuel production. In some studies, it is reported that the lipid content and triglyceride content are increased by 20–30% due to genetic modifications in microalgae, and enhanced greenhouse gases (GHG) fixation of around 40% is achieved [3].

Biofuels can be obtained through three major routes of biochemical, chemical (transesterification), and thermochemical conversion, as shown in Figure 9.3. Biochemical methods commonly used are anaerobic digestion and fermentation, which make use of microorganisms, bacteria, and enzymes to convert biomass to liquid or gaseous fuels such as ethanol and biogas. Transesterification is a chemical method which results in the formation of biodiesel from the biomass through chemical reactions. The chemical binding agents of plant matter are lignin, cellulose, and hemicellulose. These nonfermentable polysaccharide sugars cannot readily be broken down by microorganisms for the production of alcohol-based biofuels, such as sugarcane. The nonfermentable solid residue left after pressing out water and dissolvable sugars or oil from oilseed crops has been widely studied for both biological and chemical methods to release this embedded plant-based biochemical waste product energy. Thermochemical conversion of biomass involves the use of heat to bring about chemical transformations to produce energy, fuels, and value-added products [5]. Thermochemical processes can further be subdivided into combustion, pyrolysis, gasification, and hydrothermal processing, as shown in Figure 9.3.

Concentrated Solar Integrated Hydrothermal Liquefaction

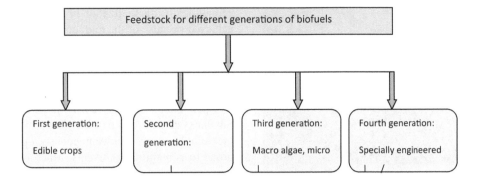

FIGURE 9.2 Feedstock for the four generations of biofuels.

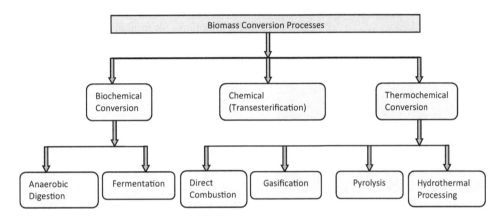

FIGURE 9.3 Different processes for conversion of biomass to biofuels.

9.3 ALGAE GROWTH AND POTENTIAL

Microalgae can be grown and produced from solar-based photosynthetic radiation, artificial illumination, or heterotrophic carbon-based assimilation growth without an obligatory requirement for light energy inputs and photosynthetic conversion pathways. The commercial focus for microalgae fuel production has been open-pond photosynthetic growth systems because of higher overall productivity and volumetric yields.

Much endeavor has been made into the pursuit of microalgae-based fuels over the last couple of decades. Table 9.1 shows the potential of microalgae in delivering superior oil content as compared to the other plant sources, such as corn, soybean, canola, Jatropha, coconut, and palm.

From Table 9.1 it can be seen that the oil yield in liters per sq. m corresponding to microalgae is almost ten to hundred times more than the other plant sources. Thus, use of microalgae will require less land as compared to their terrestrial counterparts, if technological microalgae production and processing innovation issues might be able to be overcome.

The quest for algae-based fuel production is a not new phenomenon. In the early 1970s, following a political global crude oil supply and distribution problem, the United States began its Aquatic Species Program for hydrogen and biodiesel from microalgae [7]. The program identified 3,000 species of microalgae containing high levels of oils, of which 300 candidates were highlighted with potential for conversion into fuels. Open-pond algal production systems as the lowest-cost algae production system were considered and experimentally tested for algae production and conversion into fuels. Some visual examples of algae are shown in Figure 9.4.

Algae grow quickly, and under correct light, nutrient, and temperature conditions, some species can double their biomass within four to six hours. Algae also store lipids and oils as energy storage compounds that, upon conversion into

TABLE 9.1
Comparison of Biodiesel Feedstock (Data Adopted from Chisti, Y. [6])

Source	Oil Yield (L/sq.m.)
Corn	0.0172
Soybean	0.0446
Canola	0.119
Jatropha	0.1892
Coconut	0.2689
Palm Oil	0.595
Microalgae (30% oil by weight)	5.87
Microalgae (70% oil by weight)	13.69

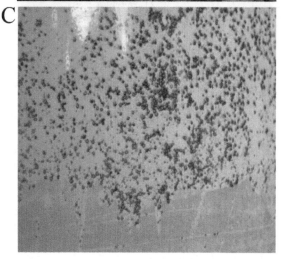

FIGURE 9.4 Visual examples of (a) macroalgae drying on a desert beach (*Laminaria spp.*) from Southern Pacific Ocean, (b) microalgae *Spirulina maxima sp.*, and (c) microalgae *Chlorella vulgaris sp.*, showing cell clustering and adhesion in close proximity to (1) the internal wall of the photobioreactor facing towards the illuminated source and (2) internal aeration bubbles

Source: Dr. Matthew Pearce.

biodiesel fuel, have similar combustion properties to conventional petroleum-based fuels; hence, these have attracted attention as biological candidates for fuel generation. There has been much-claimed potential for algal biofuels, not just in the scientific community, but also in the financial and investment community. However, energetically and economically, the realization of this desire has not yet been achieved because of the high costs of fertilization, production, harvesting, drying, and conversion into fuels. Before algal biofuels can contribute to energy sustainability and security, costs per liter of algal fuels need to come down significantly [8–10].

Open ponds are less costly than the photobioreactor tubular systems because of a higher volume of algal growth medium; however, disadvantageously, open ponds are more prone to contamination events. Much of the algal growth mass-production technology is borrowed from the open-pond microalgae species that are known as *extremophile* species because they are tolerant to high temperatures, salinities, or pH. Extremophile species include those used for the production of antioxidants, whole-cell microalgae dried biomass for human consumption, and highly unsaturated long-chain omega-3 oils as vegetarian alternatives to fish oils containing DHA and EPA lipids.

Increasing the contribution of biofuels to the energy mix can be achieved by one or a combination of four methods, namely:

i. Increase in the overall production of biomass
ii. Increase in the system efficiency or resilience
iii. Increase in the yield per unit area or per unit volume
iv. Exploitation of the waste products

Microalgal nutrient production inputs and harvesting are known to be the most expensive components of the production system. Phototrophic microalgae photosynthesize carbon from dissolved carbon dioxide rather than obtaining energy from dissolved sugars. If microalgae are going to provide enough biodiesel or bio-oil, they will need to be grown in huge volumes and to be grown phototrophically. Conversely, heterotrophic growth of microalgae uses sugar-based crops, requiring fertilizer production, which is energetically fossil fuel expensive. Fermentation facilities are not as scalable as microalgae photobioreactors or open ponds due to the high capital expense and the requirement for sterility to maintain axenic monocultures of microalgae.

Algae have greater productivity than terrestrial crops and can be a nonfood resource [9, 11]. Algae may use seawater or fresh water, waste carbon dioxide, and are able to combine with wastewater treatment processes. Microalgae can use wastewater, including NH_4^+, NO_3^-, PO_4^{3-} from the aqueous waste streams, as nutrient sources; can be grown on areas unsuitable for agriculture, therefore do not compete with food production; can be used for coproduction of proteins; and can remediate gases from industrial processes [12, 13].

Microalgal production costs are the main limiting factor for development of markets for microalgae biofuels and food. Costs relate to the complexity of the cultivation phase and the downstream processes required for extracting the high-value products. Energy consumption required for harvesting alone is estimated at 90% of the energy input of the life cycle analysis and 25% of the total production costs [14, 15].

9.4 THERMOCHEMICAL ROUTES FOR BIOMASS CONVERSION TO FUELS

Thermochemical conversion processes can be further classified mainly into direct combustion, pyrolysis, gasification, and hydrothermal processing, as shown in Figure 9.3.

Direct combustion is burning of biomass in open air or excess of air to convert the chemical energy of biomass to heat and gases. The pyrolysis process occurs in the temperature range of 400–700°C in the absence of oxygen, which results in thermal decomposition of biomass. Gasification process consists of controlled supply of heat, steam, and oxygen for conversion of biomass to gaseous fuels. All the three processes, namely, direct combustion, pyrolysis, and gasification, work with dry biomass; biomass with water content less than 10% is preferred. With wet biomass, these techniques require a preliminary drying step, which is energy-consuming. Another aspect of a key differentiation between combustion and other thermochemical technique is the presence or absence of oxygen in the thermochemical conversion process. Fire or combustion requires heat, fuel, and oxygen. Pyrolytic technologies have an absolute requirement for an absence of oxygen. In other words, although both combustion and pyrolytic processes employ heat, combustion uses exothermic heat to expend fuel, whereas pyrolysis uses endothermic heat to create fuel.

A key feature of biomass fuels is their high water content, as plants have high moisture content postharvest. The latent heat or specific heat capacity of water is about 2,250 KJ/kg^{-1}. This means that pretreatment for drying of fuels designated for combustion is of paramount importance in assessing an optimal energy return on energy investment. Species choice, therefore, might depend upon climatic conditions, suitability for drying (biochemical structure), atmospheric humidity and temperature, particle form and size, and chemical composition (carbon, nitrogen, and water ratios). Nonorganic chemical composition affects the pyrolytic or combustion conversion from biomass-based fertilizer remnants of potassium and phosphate into ash, known as sintering, tar cracking, or slagging, which has implications for the viscous and crystalline deposits within the biothermal conversion system. Pyrolytic technologies encompassing charcoal flue bed carbonization and gasification are widely advocated as potential entrants into the biomass-to-fuel sector. Due to the water content of many choice species for biomass crops, these require pretreatment to increase the energy density and/or reduce the activation energy for energy return on energy invested. Low-cost drying technologies require high surface area exposure to the environment for both solar, heat, and air movement, or there is the detrimental, compromised decomposition of biomass by microbiological contamination in the form of mold, fungus, and bacteria. High-cost drying technologies as a pretreatment methodology themselves expend energy; hence, overall efficiency is compromised.

Thermochemical conversion processes are relatively faster in operation than biological conversion processes. For example, biogas generation from anaerobic digestion is a cumulative and continuous industrial process that, if achieved as a start-to-finish batch process, would endure for a few weeks for the entire volumetric yield of biogas. By contrast, combustion processes endure for minutes or hours. The vast majority of thermochemical conversion processes employed within the commercial sector are dependent upon combustion technologies, capture of heat, or the conversion of heat into steam to subsequently produce electricity. Unfavorably, drying to between 5% and 10% moisture is an essential pretreatment process to combustion.

Hydrothermal process can make use of wet biomass with water as reaction agent for conversion to fuels in the form of char, oil, and gas. Hydrothermal processes can also be further subdivided as hydrothermal carbonization, hydrothermal liquefaction, and hydrothermal gasification, depending on the parameters. Hydrothermal carbonization takes place at low temperature (180–250°C) and pressures in the range 20–100 bar to produce a solid carbonaceous product known as hydrochar. Hydrothermal gasification takes place either in subcritical or supercritical water conditions to produce gaseous fuel consisting of methane, hydrogen, and carbon monoxide, and other products [16].

During hydrothermal liquefaction (HTL), wet biomass is depolymerized at moderate temperatures in the range around 200–370°C and high pressures around 50–200 bar. The water acts as a reactant and catalyst in the HTL process. The macromolecules in the presence of water are hydrolyzed to produce highly reactive molecules which repolymerize to form bio-oil. The advantage of the HTL process is that biomass need not be prior dried to produce bio-oil, but still, HTL is an energy-intensive process. It requires energy for maintaining temperatures and pressures for the duration of the reaction, usually for 30–60 min.

9.5 HYDROTHERMAL LIQUEFACTION

Hydrothermal liquefaction is a biomass-to-bioliquid conversion route carried out in water at moderate temperature of 280–370°C and high pressure (10–25 MPa). It has a liquid biocrude as the main product, along with gaseous, aqueous, and solid-phase by-products. Unique physiochemical properties of subcritical and supercritical, water produce water and bio-oils as the main products. The dielectric constant decreases from 78.5 Fm^{-1} at 25°C and 0.1 MPa to 14.07 Fm^{-1} at 350°C and 25 MPa [17]. There are many sources of waste which can be valorized, including plastic, vehicle tires, agricultural waste from sugarcane, rice straw, and the deep-rooted Indian invasive plant species, such as *Prosopis juliflora*.

Energetics of HTL are dominated by the energy required to heat the reactor, 6.51 MJ (1 kg microalgae) [17]. Multiple-carbon-containing feedstock materials can be used in HTL; lower feedstock costs, together with solar heat, could decrease the minimum fuel sales price more, potentially closer to cost parity with conventional fossil fuels. Heating rate and optimal processing temperature and pressure are essential for product formation.

TABLE 9.2
Comparison of Hydrothermal Liquefaction and Pyrolysis Processes for Biomass Conversion [18–22]

	Pyrolysis	Hydrothermal Liquefaction
Drying	Necessary	Unnecessary
Pressure (MPa)	0.1–0.5	5–20
Temperature (°C)	370–526	200–400
Catalyst	No	Sometimes
Heating Value	Low (~ 17 MJ/kg)	High (~ 30 MJ/Kg)
Viscosity	Low	High
Upgrade	Hard	Easy

TABLE 9.3
Comparison of Biocrude by HTL and Pyrolysis

Elemental Composition	Hydrothermal Liquefaction	Pyrolysis
C (wt%)	73	58
H (wt%)	8	6
O (wt%)	16	36
S (ppm)	<45	29
Moisture	5.1	24.8
HHV (MJ/kg)	35.7	22.6
Viscosity (cPs)	15,000	59

Source: From Gollakota et al. [1] under Creative Commons License, adopted from Elliot D.C. [23].

A comparison between pyrolysis and hydrothermal liquefaction compares the parameters of interest in Table 9.2.

Table 9.3 compares the fuel composition qualities of two thermochemically derived biocrude fuels: pyrolysis biocrude without pressure in the absence of water and HTL biocrude with pressure [1, 23].

The higher heating value of HTL biocrude in Table 9.3 of 35.7 MJ/kg is about eight units more than the HHV of pyrolysis-based crude oil. Biocrude, like conventional crude oil, has the potential for upgrading and with subsequent downstream processing into a variety of products.

There are multiple benefits in HTL technology, such as that it aids in:

1. *Reducing emissions.* Reducing reliance on fossil fuels by developing an integrated approach to produce biofuels from waste or biomass, causing lower emissions and carbon mitigation. An added benefit is the use of wastewater and plastic waste as feedstock to produce fuel.
2. *Improving security of energy supply.* Use of aqueous by-products for secondary additional nutrient supplies, reducing fertilizer consumption in agricultural chain feed and food production processes, as well as development of non-agriculturally productive land for fuel production.
3. *Reducing the costs of energy production.* Direct solar radiation could provide energy for the heat required for the thermoconversion of biomass and waste into liquid bio-oil and nutrient-rich aqueous waste outputs, resulting in significant cost savings in the conversion process. The integration of waste streams with biomass production processes increases the potential to be more cost-effective.

9.6 SOLAR COLLECTORS – FLAT AND CONCENTRATING

Major energy sources on Earth can be directly or indirectly linked to solar energy, such as biomass, wind, and water. Conventionally, solar energy has been used for various day-to-day applications. Presently, the use of solar energy can be broadly divided into two main areas – solar thermal and solar photovoltaics. Solar thermal applications consist of the conversion of solar radiation to heat through solar collectors for cooking, water heating, space heating, or any other heating purposes. Solar photovoltaics involve the conversion of solar radiation to electrical power through the use of photovoltaic cells.

Here in this chapter we will be focusing on the solar thermal part. For conversion of solar energy to heat, selection of appropriate solar collector is essential. Solar collectors can be broadly classified as nonconcentrating collectors and concentrating collectors, as shown in Figure 9.5. Nonconcentrating collectors consisting of flat plate collectors and evacuated tube collectors are suitable for the temperature range 80–150°C, whereas concentrating collectors are used for higher temperatures. There are different types of concentrating collectors which can be used for a wide variety of temperature applications, depending on the design and concentration ratio.

Concentrating solar collectors are more suited for hydrothermal processes as these achieve high temperature. Studies have been reported on solar reactors employing parabolic trough, multidish concentrator, and toroidal solar concentrator for hydrothermal processing of biomass [16].

9.7 STUDIES ON SOLAR INTEGRATED HYDROTHERMAL LIQUEFACTION

In order to reduce the energy consumption involved in the hydrothermal liquefaction process, there have been recent advances in the field of integration of solar thermal systems for hydrothermal liquefaction. Solar integrated hydrothermal liquefaction offers a new possibility of producing biofuels in a cleaner manner and may even provide promising solutions for treatment of wastes to produce biofuels or value-added products. Solar concentrating systems such as parabolic troughs are suitable for producing temperature and pressure in the required range. Technoeconomic analysis of integration of concentrating solar collectors with hydrothermal liquefaction plant has been studied by Pearce

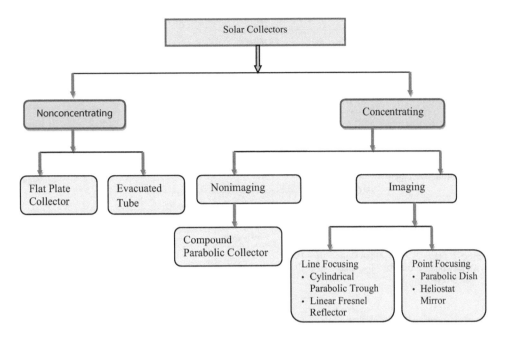

FIGURE 9.5 Classification of solar collectors.

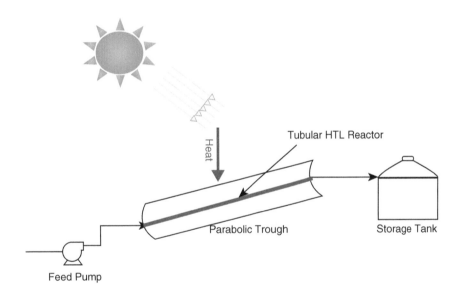

FIGURE 9.6 Solar parabolic trough integrated hydrothermal liquefaction plant with storage tank for post-HTL products.

Source: Reprinted from Pearce M., Shemfe M., Sansom C., "Techno-economic analysis of solar integrated hydrothermal liquefaction of microalgae," *Applied Energy*, 2016, 166, 19–26, with permission from Elsevier.

et al. (2016) [24]. It reported that the nutrient recycling and CSP/HTL integration can produce microalgae bio-oil in a sustainable manner. Figure 9.6 shows the schematic diagram of solar parabolic trough collector with tubular HTL reactor at its focal line considered in the previous study. Microalgae biomass feedstock is semicontinually pumped into the reaction vessel core of the concentrating solar parabolic trough, and the post-HTL products are stored in the storage tank. The concentrating solar system size has been considered for 1,000 L HTL feedstock. The parabolic troughs are of 100 m length, with aperture area of each trough as 2.26 sq.m.; thus, total aperture area has been taken as 226 sq.m. It has been proposed that the system can run for three batch processes per day at 20% (w/v) microalgae – 160 L HTL reaction core volume. Minimum fuel sale price (MFSP) of $1.23/kg and a positive cash flow have been demonstrated in the study on the basis of the calculations.

Giaconia et.al. [25] reported a conceptual study on combining a concentrating solar power plant to biorefinary process for hydrothermal liquefaction of microalgae to meet

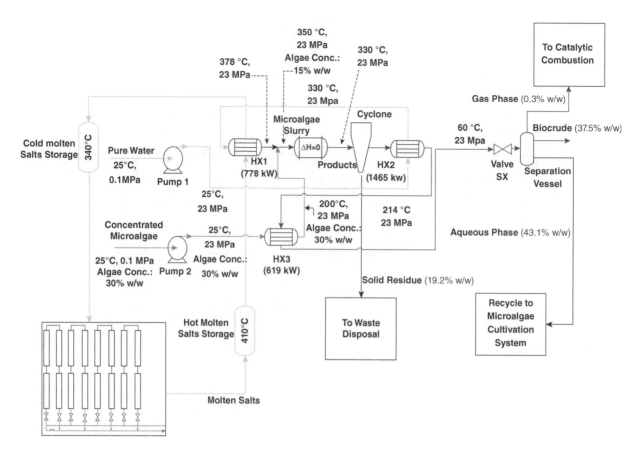

FIGURE 9.7 Concentrated solar integrated hydrothermal liquefaction plant conceptual design. Taken from Giaconia et al. [25]

Source: Reprinted from "Biorefinery process for hydrothermal liquefaction of microalgae powered by a concentrating solar plant: A conceptual study," Giaconia A. et.al., *Applied Energy*, 2017, 208, 1139–1149, with permission from Elsevier.

the high heat demand. They investigated the coupling of parabolic troughs to a hydrothermal liquefaction plant. They proposed a system configuration which utilizes an indirect solar reactor – that is, feed is not directly added to the solar collectors; instead, the chemical reactor is provided heat by the heat transfer fluids heated by the solar concentrating collectors. The schematic diagram is shown in Figure 9.7. It has been designed to process 10 kT of microalgae per year. From the figure it can be seen that cold molten salts are stored at 340°C, which are heated by solar collectors to 410°C. Two input streams are proposed; one is of pure water at 25°C, 1 bar pressure, and flow rate 1.07 kg/s, and the other is concentrated microalgae at 30% w/w at 25°C, 1 bar pressure, and mass flow rate of 1.07 kg /s. The two streamlines are pressurized at 230 bar by pumps 1 and 2 and connected to heat exchangers to raise the temperature. There are three heat exchangers in the proposed system: HX1, HX2, and HX3. HX1 uses molten salts heated by solar collectors, and HX2 uses hot effluent from cyclone. Stream 1 of pure water is heated by the heat exchangers HX1 and HX2. The residual effluent from HX2 is sent to HX3, where stream 2 of concentrated algae is preheated. Streams 1 and 2, after being compressed and heated, are allowed to mix rapidly, resulting in 15% w/w microalgae slurry at 350°C. This microalgae slurry is sent to the reactor, which has been modeled as a single-tube adiabatic reactor. The microalgae slurry inside the reactor gets converted to a liquid biocrude, an aqueous phase, a gas phase, and a solid residue. This hot stream from the reactor is taken to a cyclone, where the solid part is removed and the purified stream is cooled in HX2 and HX3 to 145°C and 23 MPa. The dielectric constant of water increases to 46 in these conditions, resulting in completion of biocrude liquid phase separation, and a two-phase compressed stream is then sent to a separation vessel through a flash valve, where the three phases get separated by gravity.

Experimental studies on the integration of solar collectors with HTL process have been conducted by collaborative works of Cranfield University, UK; Phycofeeds Ltd. UK; and Kota University, India [26–28]. Initially, before the development of the prototype system, experiments were conducted on a small-size reactor pressure vessel to evaluate HTL process for bio-oil and associated by-products [26]. A small-size Swagelok reactor pressure vessel of 10 ml was placed in a heated sand bath with pressure tube of 1/2-inch Hoke Gyrolok able to withstand temperatures in the range -235°C to +426°C. The dimensions of the pressure tube were 170 mm length and 10.28 mm inner diameter. The system was heated through an insulated ceramic plate with base heat source. *Chlorella vulgaris, Chlorella*

Concentrated Solar Integrated Hydrothermal Liquefaction

salina, *Nannochloropsis salina*, and *Tetraselmis chui* were the microalgae samples used in the experiments. For the lab-scale HTL experiments, the reactor was loaded with 1.8 g of dried biomass and 9 ml of deionized water (20% w/v). When the reactor attained 280°C temperature, it was allowed to undergo HTL for 20 min. After that, it was quenched in a water bath to arrive at room temperature. The gases were vented, and the liquid was tested through gas chromatography (GC) and total organic carbon (TOC) analysis. During bio-oil evaluation, as shown in Figure 9.8, it was found that lipids corresponding to C14, C15, C16–17, C20 were present. Lipid fractions and other chemical compounds were also formed, which were grouped in unclassified lipids.

As a next step for the development of a prototype "proof of concept" demonstrator for CSP-HTL integration, solar parabolic trough collectors, as shown in Figure 9.9, were installed at the University of Kota, India [27]. The parabolic trough collectors were supplied by Global CSP Ltd. UK and modified by Cranfield University, UK, for the purpose of a bioreactor. The dimensions of the parabolic trough are 690 mm × 1960 mm, with focal length of 120 mm. It has a

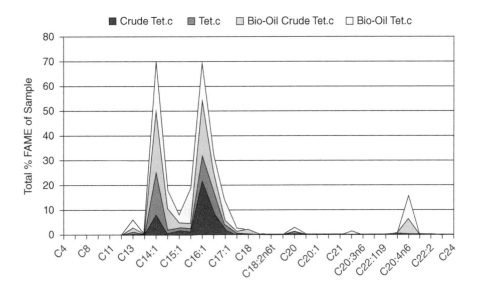

FIGURE 9.8 Bio-oil evaluation.

Source: From Pearce et al., AIP Proceedings 2018 [26].

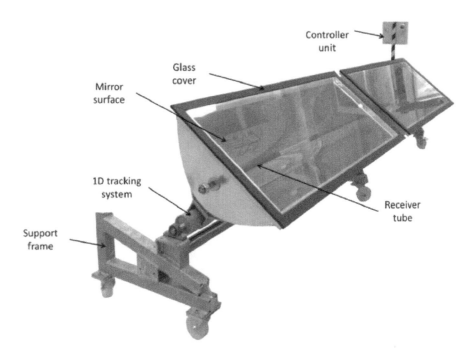

FIGURE 9.9 Solar parabolic trough collectors – Global CSP Ltd.

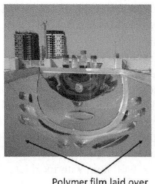
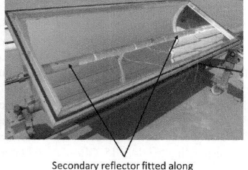

Polymer film laid over original metal parabolic reflector surface

Secondary reflector fitted along the focal line

FIGURE 9.10 Modified solar parabolic trough strips of polymer film (left) and the secondary reflector in place (right).

Source: From Almond et al., AIP Conference Proceedings 2019 [27].

glass cover on top to prevent heat loss and soiling/scratching of reflecting surface. It has a manually controlled single-axis tracking system.

The receiver tube was redesigned and modified with Swagelok pipes and fittings to withstand high pressures in the range 15–200 bar and temperatures 250–350°C for the purpose of the bioreactor. The receiver tube is equipped with pressure valves and gauges. Four longitudinally orientated, matte-black-painted parallel pressure tubes with an internal diameter of 6 mm and external diameter of 8 mm were equidistantly spaced and placed within a borosilicate glass sheath of internal diameter of 30 mm for the solar reactor. Original Global CSP parabolic trough with stainless steel reflecting surface achieves temperatures around 250°C; to achieve higher temperatures, Skyfuel® ReflecTech Plus polymer film was added to improve the reflectivity, and a secondary aluminum reflector was affixed, as shown in Figure 9.10.

Experimental testing at Kota University revealed that the temperatures in the HTL range 280°C–350°C were achieved with the modified solar concentrating collectors. With successful testing of the solar concentrating system for the HTL temperature range, experiments were performed on the system for processing of waste and microalgae samples at Kota University, Rajasthan, India, with different raw material mixtures comprising algae, PET plastic, sugarcane bagasse, and animal manure [28]. The materials were placed in the reactor in the ratio 20% solids and 80% water. Tests were performed on clear days, with measurement of solar radiation through pyranometer and recording of temperatures and pressure through sensors and gauges. After the temperatures reached around 320°C and stayed in the range for 30 min., the solar collectors were defocused for 10 min. to cool down, and the mixture was extracted for gas chromatography mass spectroscopy (GC-MS) testing. Samples obtained after experiments at Kota University were tested by the Indian Institute of Petroleum through GC-MS for identifying biocrude composition. On analysis, a prominence of C6 to C12 aliphatic and aromatic hydrocarbons and derivative of carbonitrile and cyclohexane compounds was found. These experiments, performed as initial, early-phase demonstration, seem promising, though still there are many technical challenges which need to be addressed.

9.8 CHALLENGES AND OPPORTUNITIES

There exist multiple opportunities and challenges in the field of integration of solar thermal systems with hydrothermal liquefaction processes. Firstly, as these technologies are still in the development phase for commercial exploitation, there is much scope for multiple potential manifestations and innovations. Secondly, these technologies are closely linked to the issues related to energy and waste management; their successful development may present promising solutions environmental protection and remediation. Thirdly, biofuel technologies provide a drop-in fuel solution to incumbent transport systems that many electrical, gas, and hydrogen alternative transport solutions cannot make a smooth transition to fulfill, due to the engineering and infrastructure differences that exist between these transport propulsion technologies. Thus, successful, cost-effective development of the solar-HTL integrated technologies may pave a way towards a sustainable world in the future, addressing the issues of waste, water, and energy.

There are several other challenges and opportunities for future work in this area of solar-HTL integration, such as [16, 27, 28]:

- For this technology to attain commercial status, it is important to produce biofuels in cost comparable to those of fossil fuels.
- Integration of solar concentrating collectors with HTL requires special technical and engineering skills, as solar systems are not specifically designed for this application.
- These technologies require high initial investment.

- Material research is required to develop materials suitable for high-temperature, high-pressure applications without corrosion, scaling, and UV impacts.
- In case of direct reactors where load is put directly in receiver, problems related to corrosion, scaling, plugging of tubes need to be addressed.
- In case of indirect reactors with thermal storage system, careful designing and handling of molten salts is required to avoid temperature variations, corrosion, and blockages.
- Optimal use of heat exchange and heat recovery systems is needed to enhance the efficiency of energy conversion for the process.

Before it becomes possible to ramp up the production of HTL-solar biofuels, it will be necessary to demonstrate incremental scalable production at cost-effective price relative to the conventional fuels and second-generation biofuels. This depends upon factors, including:

- Production throughput optimization of unit volume per time in terms of liters per hour or tons per day.
- Justification and validation of fuel composition and capability for upgrading fuels into suitable fuels for direct application to transport vehicles.
- Valuation of end output products and quantification of gases, liquids, and solids and separation efficacy for aqueous and organic liquids.
- Whole life cycle analysis incorporating capital expenditure, equipment infrastructure, operational running costs, and longevity.
- Resilience to cyclical temperature fluctuations and systems engineering factors, including wear on gaskets, flanges, and valves.

This growing field of solar-HTL integration presents ample opportunities for scientists, engineers, and researchers to develop innovative solutions to take this technology to the commercial level.

9.9 REFERENCES

[1] Gollakota, A.R.K., Kishore, N. and Gu, S., A review on hydrothermal liquefaction of biomass, *Renewable and Sustainable Energy Reviews*, 2018, 81, 1378–1392.

[2] World Health Organisation, www.who.int/news-room/factsheets/detail/household-air-pollution-and-health, accessed on 5–8–2021.

[3] Aron, N.S.M. et al. Sustainability of four generations of biofuels-A review, *International Journal of Energy Research*, 2020, 44:12, 9266–9282, https://doi.org/10.1002/er.5557

[4] Saad, M.G., Dosoky, N.S., Zoromba, M.S. and Shafik, H.M., Algal biofuels: Current status and key challenges, *Energies*, 2019, 12:10, 1920, 1–22, https://doi.org/10.3390/en12101920

[5] Verma, M., Godbout, S., Brar, S.K., Solomatnikova, O., Lemay, S.P. and Larouche, J.P., Biofuels production from biomass by thermochemical conversion technologies, *International Journal of Chemical Engineering, Recent Advances in Valorization Methods of Inorganic/Organic Solid, Liquid, and Gas Wastes*, 2012, Article ID 542426, 18 pages, https://doi.org/10.1155/2012/542426

[6] Chisti, Y., Biodiesel from microalgae, *Biotechnology Advances*, 2007, 25, 294–306.

[7] Sheehan, J., Dunahay, T., Benemann, J. and Roessler, P., A look back at the U.S. department of energy's aquatic species program: Biodiesel from algae; National Renewable Energy Laboratory, 1998, www.nrel.gov/docs/legosti/fy98/24190.pdf

[8] Greenwell, H.C., Laurens, L.M., Shields, R.J., Lovitt, R.W. and Flynn, K.J., Placing microalgae on the biofuels priority list: A review of the technological challenges, *Journal of the Royal Society Interface*, 2010, 7:46, 703–726, https://doi.org/10.1098/rsif.2009.0322

[9] Williams, PJle-B. and Laurens, L.M.L., Microalgae as biodiesel & biomass feedstocks: Review & analysis of the biochemistry, energetics & economics, *Energy & Environmental Science*, 2010, 3, 554–590, https://doi.org/10.1039/B924978H

[10] Clarens, A.F., Resurreccion, E.P., White, M.A. and Colosi, L.M., Environmental life cycle comparison of algae to other bioenergy feedstock, *Environmental Science & Technology*, 2010, 44:5, 1813–1819, https://doi.org/10.1021/es902838n

[11] Pienkos, P.T., The potential for biofuels from algae, Algae Biomass Summit, San Francisco, CA. 15th November 2007, National Renewable Energy Laboratory.

[12] Mata, T.M., Martins, A.A. and Caetano, N.S., Microalgae for biodiesel production and other applications: A review, *Renewable Sustainable Energy Reviews*, 2010, 14:1, 217–232, http://dx.doi.org/10.1016/j.rser.2009.07.020

[13] Sawayama, S., Inoue, S., Dote, Y., Yokoyama, S.Y., CO_2 fixation and oil production through microalgae, *Energy Conversion and Management*, 1995, 36, 729–731, https://doi.org/10.1016/0196-8904(95)00108-P

[14] Sander, K.B., Downstream processing of microalgal biomass for biofuels, M.Sc. thesis, 2010, Oregon State University.

[15] Molina Grima, E., Belarbi, E-H., Acién Fernández, F.G., Robles Medina, A. and Chisti, Y., Recovery of microalgal biomass and metabolites: Process options and economics, *Biotechnology Advances*, 2003, 20, 491–515, https://doi.org/10.1016/s0734-9750(02)00050-2

[16] Ayala Cortes, A. et al., Solar integrated hydrothermal processes: A review, *Renewable and Sustainable Energy Reviews*, 2021, 139, 110575, https://doi.org/10.1016/j.rser.2020.110575

[17] Toor, S.S., Rosendahl, L. and Rudolf, A., Hydrothermal liquefaction of biomass a review of subcritical water technologies, *Energy*, 2011, 36, 2328–2342, https://doi.org/10.1016/j.energy.2011.03.013

[18] Centi, G. and Santen, R.A., editors, 2007, *Catalysis for renewables: From feedstock to energy production*. Weinheim: Wiley/VCH

[19] Demirbas, A., Mechanisms of liquefaction and pyrolysis reactions of biomass, *Energy Conversion Management*, 2000, 41, 633–646, https://doi.org/10.1016/S0196-8904(99)00130-2

[20] Elliott, D.C., Hart, T.R., Neuenschwander, G.G., Rotness, L.J. and Zacher, A.H., Thermochemical biomass liquefaction and fuels production. Presented at the Frontiers in Bioenergy Symposium, May 25, 2010.

[21] Xu, C. and Lancaster, J., Conversion of secondary pulp/paper sludge powder to liquid oil products for energy

recovery by direct liquefaction in hot-compressed water, *Water Research*, 2008, 42:6–7, 1571–1582, https://doi.org/10.1016/j.watres.2007.11.007

[22] Zhang, B., Keitz, M. and Valentas, K., Thermal effects on hydrothermal biomass liquefaction, *Applied Biochemistry and Biotechnology*, 2008, 147:1–3, 143–150, https://doi.org/10.1007/s12010-008-8131-5

[23] Elliot, D.C., Sealock, L.J., Phelps, M.R. and Neuenschwander, G.G., 1st Biomass Conference of the Americas, Burlington, Vermont, August 30 - September 2, 1993.

[24] Pearce, M., Shemfe, M. and Sansom, C., Techno-economic analysis of solar integrated hydrothermal liquefaction of microalgae, *Applied Energy*, 2016, 166, 19–26, http://dx.doi.org/10.1016/j.apenergy.2016.01.005

[25] Giaconia, A. et al., Biorefinery process for hydrothermal liquefaction of microalgae powered by a concentrating solar plant: A conceptual study, *Applied Energy*, 2017, 208, 1139–1149, https://doi.org/10.1016/j.apenergy.2017.09.038

[26] Pearce, M., Tonnellier, X., Sengar, N. and Sansom, C., Commercial development of bio-combustible fuels from hydrothermal liquefaction of waste using solar collectors, *AIP Conference Proceedings 2033*, 2018, 130011, https://doi.org/10.1063/1.5067145

[27] Almond, H., Tonnellier, X., Sansom, C., Pearce, M. and Sengar, N., The design and modification of a parabolic trough system for the hydrothermal liquefaction of waste, *AIP Conference Proceedings*, 2019, 2126, 120001, https://doi.org/10.1063/1.5117619.

[28] Pearce, M., Tonnellier, X., Sengar, N. and Sansom, C., Fuel from hydrothermal liquefaction of waste in solar parabolic troughs, *AIP Conference Proceedings 2126*, 2019, 180015, https://doi.org/10.1063/1.5117695

10 Integrated PV-Wind-Battery-Based Single-Phase System

B. Mangu, P. Satish Kumar, and A. Jayaprakash

CONTENTS

- 10.1 Introduction 121
- 10.2 Existing Stand-Alone Systems 122
  - 10.2.1 Four-Stage Configuration..122
  - 10.2.2 Three-Stage Configuration...122
  - 10.2.3 Two-Stage Configuration 122
  - 10.2.4 Hybrid Stand-Alone Systems...123
- 10.3 Proposed Converter Configuration 124
  - 10.3.1 Limitations and Design Issues 126
  - 10.3.2 Control Scheme for Power Flow Management 126
  - 10.3.3 Simulation Results and Discussion 127
  - 10.3.4 Experimental Validation 128
  - 10.3.5 Summary...129
- 10.4 Integrated PV-Wind-Battery-Based Single-Phase Grid-Connected System 130
  - 10.4.1 Proposed Converter Configuration..130
  - 10.4.2 Proposed Control Scheme for Power Flow Management 132
  - 10.4.3 Simulation Results and Discussion 133
  - 10.4.4 Summary...133
- 10.5 References..134

10.1 INTRODUCTION

The rapid depletion of fossil fuel reserves, an ever-increasing energy demand, and concerns over climate changes motivate power generation from renewable energy sources. Out of these, solar photovoltaic and wind have emerged as popular energy sources due to their eco-friendly nature and cost-effectiveness. However, these sources are intermittent in nature. Hence, it is a challenge to supply stable and continuous power using these sources. This can be addressed by efficiently combining with energy storage technologies. The interesting complementary behavior of solar insolation and wind velocity pattern coupled with the aforementioned advantages has led to research on their integration, resulting in the hybrid photovoltaic (PV) wind systems. For achieving the integration of multiple renewable sources, the traditional approach involves using dedicated single-input converters, one for each source, which are connected to a common DC bus. However, these converters are not effectively utilized due to the intermittent nature of the renewable sources. In addition, there are multiple power conversion stages which reduce the efficiency of the system. Not many attempts are made to optimize the circuit configuration of these systems that could reduce cost and increase efficiency and reliability.

The use of multi-input converter for hybrid power systems is attracting increasing attention because of reduced component count, enhanced power density, compactness, and centralized control. The work presented in this chapter pertains to the development of multi-input converter topologies and their control for effective utilization of integrated PV-wind-battery-based systems. It focuses on the hybrid PV-wind-battery-based stand-alone and grid-connected system. A novel multi-input transformer–coupled DC–DC converter, followed by a full-bridge inverter, is proposed. The power circuit configuration of the system is shown in Figure 10.1. A control strategy is proposed to operate the proposed converter in all the possible modes. The controller of a stand-alone system has to perform the following tasks: (a) maximum power extraction from both the photovoltaic and wind power sources; (b) protection of battery from overcharging and discharging; (c) DC–AC conversion and load voltage regulation. To achieve these functionalities, a stand-alone system operates in four modes, namely, maximum power point tracking (MPPT) mode, non-MPPT mode, battery-only mode, and shutdown mode.

Hybrid PV and wind-based generation of electricity and its interface with the power grid are the important research areas. In continuation of the work presented previously, a control strategy for power flow management of a grid-connected PV-wind-battery-based system with an efficient multi-input transformer–coupled bidirectional DC–DC converter is presented. The proposed system aims to satisfy

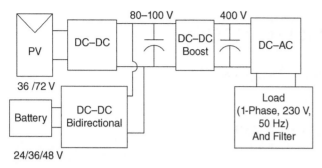

FIGURE 10.1 Block diagram of a conventional stand-alone system with four-stage configuration and two controlled DC links [4].

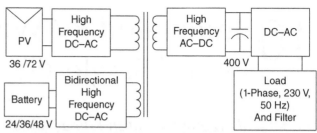

FIGURE 10.2 Three-port high-frequency transformer with four converters [5].

the load demand, manage the power flow from different sources, inject surplus power into the grid, and charge the battery from grid as and when required. The control philosophy for power flow management of the multisource system is developed based on the power balance principle.

10.2 EXISTING STAND-ALONE SYSTEMS

A large number of households in rural areas do not have access to electricity, as extension of grid is not economically viable. To improve the scenario of rural electrification, several alternatives are being considered. Majority of these are targeted towards deployment of renewable energy source–based stand-alone systems. Among various renewable energy sources, photovoltaic arrays and wind turbines are widely explored, owing to their sustainable and environmental features. However, these sources are influenced by the operating environment, which significantly decreases stable and reliable energy to the loads. Hence, backup storage elements are incorporated in the system.

Several stand-alone schemes have been reported in the literature, which vary in complexity, effectiveness, component requirement, etc. In [1–3], stand-alone schemes with low voltage requirement at the load terminals are specifically designed for household DC appliances. Hence, the voltage gain requirement from the intermediate converter(s) in these schemes is less. Schemes reported in literature for feeding AC loads have voltage rating of 230 V rms and thus require high-voltage gain from the intermediate converters.

Based on the number of intermediate converters between the input and output, they can be categorized as follows:

10.2.1 Four-Stage Configuration

Schematic block diagram of a four-stage configuration is shown in Figure 10.1. Here, three DC–DC converters are generally used in addition to a DC–AC inverter [4]. Out of these converters, one DC–DC converter is used for maximum power point tracking (MPPT), second for charge.

Various four-stage configurations are described in [6–11]. Generally, nonisolated DC–DC converters are used for MPP tracking and battery charge control, while a transformer-coupled DC–DC converter is used to realize voltage boosting. An alternative approach utilizing a multiport transformer–based scheme is also reported in the literature [5, 12], and its schematic block diagram is shown in Figure 10.2. Though this configuration is suitable for achieving high gain, it requires a complex control scheme. The phase angles of terminal voltages of the transformer windings are varied for controlling the power flow.

The higher number of power conversion stages used in the previous approaches reduce efficiency and reliability of the system as the component count is high. Further, these configurations also suffer from low battery charging efficiency as two converter stages exist in the battery charging path. Furthermore, the converter used for extracting maximum power remains idle when there is no PV power, thereby reducing the utilization of the system.

10.2.2 Three-Stage Configuration

In the literature, schemes having battery placed in the cascaded path are proposed. This type of configuration [7, 8] has the following drawbacks: (a) battery has to negotiate frequent unwanted charging/discharging with sudden change in load or PV power, thereby increasing stress, and (b) the battery voltage has to be high. Otherwise, the step-up DC–DC converter following the battery has to be designed for high voltage gain.

In a three-stage configuration, a battery is placed in parallel to the converters through a bidirectional converter [9–11]. Figure 10.3 shows three-stage configuration topology proposed by R. Chattopadhyay et al. in [13]. In this topology, a transformer-coupled boost half-bridge converter is used to harness power from PV, and energy storage element balances the power needs through a buck/boost converter.

10.2.3 Two-Stage Configuration

In order to achieve further reduction in the number of power conversion stages, a two-stage configuration is reported in [14, 15]. In [14], it is realized by eliminating "high step-up DC–DC" converter from the configuration of Figure 10.5. This scheme requires high voltage level for the PV array and the battery.

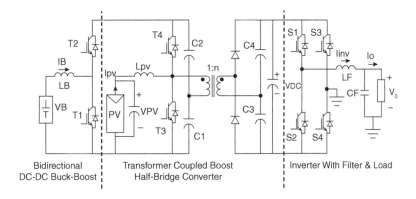

FIGURE 10.3 Transformer-coupled boost half-bridge converter with bidirectional buck/boost converter [13].

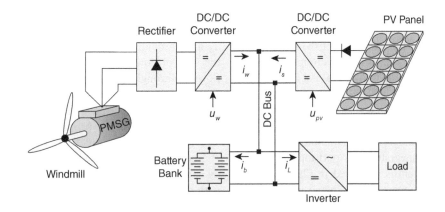

FIGURE 10.4 Block diagram of a typical low-voltage DC bus stand-alone hybrid PV-wind energy system [16].

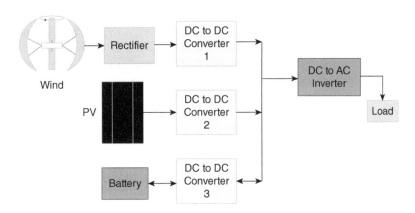

FIGURE 10.5 Block diagram of a typical high-voltage DC bus stand-alone hybrid PV-wind energy system [17].

10.2.4 Hybrid Stand-Alone Systems

PV and wind have emerged as popular energy sources as they are considered to be clean, cost-effective, with no extra fuel consumption. An interesting complementary behavior of solar insolation and wind velocity pattern has led to their integration, resulting in the hybrid PV-wind systems. They provide reliable energy and high-quality power to consumers. Typically, multiple renewable sources combined through a DC bus are more common due to the ease of integrated monitoring and control compared to AC coupling.

A hybrid PV-wind-battery-based system connected to a low-voltage (LV) DC bus is shown in Figure 10.4, and that to a high-voltage (HV) DC bus system is shown in Figure 10.5. In case of a typical LV bus system, there are three power conversion stages from the source to the AC load, and two stages in case of an HV bus system. However, the battery is connected directly to the DC bus in the LV system, while it is connected through a DC–DC converter in case of an HV system. Significant amount of literature exists on the integration of solar and wind energy as a hybrid energy generation system, with focus mainly on its sizing and optimization [18, 19]. In [18], the sizing of generators in a hybrid system is investigated. In this system, the sources and storage are interfaced at the DC link through their dedicated converters. Other contributions are made on their modeling aspects and control techniques for a stand-alone hybrid energy system in [16, 17, 20–27]. Dynamic performance of a stand-alone hybrid PV-wind system with battery storage is analyzed in [20]. In [16], a passivity/sliding mode control is presented which controls the operation of wind energy system to complement the solar energy generating system. In [18–20, 23–27], both PV and wind sources have their own power converters. They are not utilized effectively because of the highly intermittent nature of these sources. In addition, there are multiple power conversion stages which reduce the efficiency of the system. Not many attempts are made to optimize the circuit configuration of these systems that could reduce the cost and increase efficiency and reliability. In [28–30], integrated DC–DC converters for PV and wind energy systems are presented. The PV-wind hybrid system, proposed by Daniel et al. [28], has a simple power topology, but it is suitable for stand-alone applications. An integrated four-port topology based on hybrid PV-wind system is proposed in [29]. However, despite simple topology, the control scheme used is complex. In [30], to feed the DC loads, a low-capacity multiport converter for a hybrid system is presented.

10.3 PROPOSED CONVERTER CONFIGURATION

The proposed converter consists of a transformer-coupled boost half-bridge converter fused with bidirectional buck-boost converter and a full-bridge DC–AC inverter. The schematic diagram of the converter is depicted in Figure 10.6(a). It has two sources, one storage element, and a total of eight control switches. The boost half-bridge converter has two DC links on both sides of the high-frequency transformer. Controlling the voltage of one of the DC links ensures controlling the voltage of the other. This makes the control strategy simple. Moreover, additional converters can be integrated with any one of the two DC links. A bidirectional buck-boost DC–DC converter is integrated with the primary-side DC link, and a full-bridge inverter is connected to the DC link of the secondary side. The input of the half-bridge converter is formed by connecting the PV array and battery in series, thereby incorporating an inherent boosting stage for the scheme. In addition to this, the high-frequency step-up transformer further increases the boosting capability. The transformer also ensures galvanic isolation of the load from the sources and the battery. Bidirectional buck-boost converter is used for harnessing power from PV along with battery charging/discharging control. The unique feature of this converter is that MPP tracking, battery charge control, and voltage boosting are accomplished through a single converter. Transformer-coupled boost half-bridge converter is used for harnessing power from wind, and a full-bridge inverter is used for feeding AC loads. High boost of PV and wind voltages, battery current control, and required DC link voltage of 400 V are realized with four controllable switches. Thus, the proposed converter has reduced number of power conversion stages, reduced component count, and high efficiency compared to the existing stand-alone schemes.

The power flow from wind source is controlled through a boost half-bridge converter. When switch T_3 is turned on, the current flowing through the source inductor increases. The capacitor C_1 discharges through the transformer primary and T_3, as shown in Figure 10.1(b). Capacitor C_3 charges through transformer secondary and diode D_1. When T_3 is turned off and T_4 is turned on, initially the inductor current flows through antiparallel diode of switch T_4 and through the capacitor bank. During this interval, the current flowing through the diode decreases, and that flowing through the transformer primary increases. The path of current is shown in Figure 10.1(c). When current flowing through the inductor becomes equal to that flowing through transformer primary, the diode turns off. Since T_4 is gated on during this time, the capacitor C_2 now discharges through it and the transformer primary. During the on time of T_4, D_2 conducts to charge the capacitor C_4. The path of current flow is shown in Figure 10.6(d). During the on time of T_3, the primary voltage $V_P = -V_{C1}$. The secondary voltage $V_S = nV_P = -nV_{C1} = -V_{C3}$, or $V_{C3} = nV_{C1}$, and voltage across primary inductor L_w is V_w. When T_4 is turned on, the primary voltage $V_P = V_{C2}$. The secondary voltage $V_S = nV_P = nVC_2 = VC_4$, and voltage across primary inductor L_w is $V_w - (V_{C1} + V_{C2})$. It can be proved that $(V_{C1} + V_{C2}) = V_w/(1 - D_w)$, where D_w is the duty ratio of switch T_3. The capacitor voltages are considered constant in steady state, and they settle at $V_{C3} = nV_{C1}$, $V_{C4} = nV_{C2}$. Hence, the output voltage is given by:

$$V_{dc} = n(V_{c1} + V_{c2}) = n(V_b + V_{pv}) = \frac{nV_w}{(1 - D_w)}$$

$$I_b = I_{pv}\left(\frac{1 - D_{pv}}{D_{pv}}\right) + I_w\left(\frac{1 - D_{pv}}{D_{pv}}\right) - I_g\left(\frac{m_a n}{\sqrt{2D_{pv}}}\right) \quad (10.1)$$

Therefore, the output voltage of the secondary-side DC link is a function of the duty cycle of the primary-side converter and turns ratio of transformer. The transformer-coupled bidirectional buck-boost converter charges/discharges the capacitor bank C_1–C_2 of boost half-bridge converter based on the load demand. During battery charging mode, when switch T_1 is on, the energy is stored in the inductor L. When T_1 is turned off and T_2 is turned on, energy stored in L is transferred to the battery. During battery discharging mode, inductor current becomes negative. Here, the stored energy in the inductor increases when T_2 is turned on and decreases when T_1 is turned on.

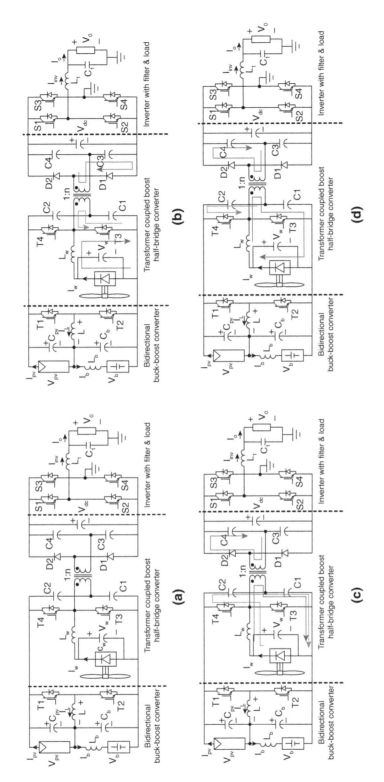

FIGURE 10.6 Operating modes of the proposed multi-input transformer–coupled DC–DC converter for PV-wind-battery-based stand-alone system. (a) Proposed converter configuration. (b) Operation when switch T3 is turned on. (c) When switch T4 is on, charging the capacitor bank. (d) When switch T4 is on, capacitor C2 discharging.

It can be proved that $V_b = D/(1-D) V_{pv}$, where D is the duty ratio of switch T_1. The output voltage of the transformer-coupled boost half-bridge converter is given by equation (10.2). This voltage is n times the voltage at the primary side, which can be controlled by the boost half-bridge converter or bidirectional buck-boost converter. The average value of inductor current over a switching cycle is $I_L = I_b + I_{pv}$. It can be seen that I_b and I_{pv} can be controlled by controlling I_L. Therefore, the MPP operation is assured by controlling I_L while maintaining proper battery charge

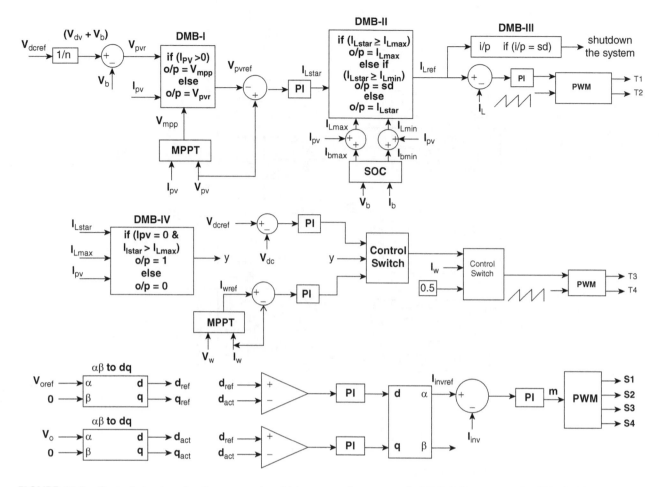

FIGURE 10.7 Control structure for the proposed multi-input transformer–coupled DC–DC converter for PV-wind-battery-based stand-alone system.

level. Both the converters are operated in continuous conduction mode.

$$V_{dc} = n(V_{c1} + V_{c2}) = n(V_b + V_{pv}) = \frac{nV_w}{(1-D_w)} \quad (10.2)$$

10.3.1 Limitations and Design Issues

The output voltage V_{dc} of a half-bridge converter depends on MPP voltage of PV array V_{PV} mpp, the battery voltage V_b, and the transformer turns ratio n. Since the environmental conditions influence PV array voltage and the battery voltage depends on its charge level, the output DC link voltage V_{dc} is also influenced by these factors. However, the PV array voltage exhibits narrow variation in voltage range with wide variation in environmental conditions, and the battery voltage is generally stiff and remains within a limited range over its entire charge–discharge cycle. Further, the SOC limits the operating range of the batteries used in a stand-alone mode (e.g., 30–80%) to avoid overcharge or discharge. Therefore, with proper selection of n, PV, and battery voltage, the output DC link voltage V_{dc} can be kept within an allowable range, though not fully controllable.

In addition, when there is no PV power, by controlling the PV capacitor voltage, the output DC link voltage V_{dc} can be controlled. At the output stage, the load voltage of the full-bridge DC–AC inverter is controlled at 230 V by employing a closed-loop control. Therefore, despite the variation in the DC link voltage V_{dc}, the desired voltage across the load can be maintained at 230 V_{ac}.

10.3.2 Control Scheme for Power Flow Management

The controller of a stand-alone system is required to perform the following tasks: (a) maximum power extraction from both the photovoltaic and wind power sources; (b) protection of battery from overcharging and discharging; and (c) DC–AC conversion and load voltage regulation. To achieve these functionalities, stand-alone systems operate in four modes, namely, MPPT mode, non-MPPT mode, battery-only mode, and shutdown mode.

MPPT mode. In this mode, maximum power is tracked from both the sources. For operating in MPPT mode, the following conditions must be satisfied: (1) available maximum power is more than the load demand ($P_{mpp} > P_{load}$), and the battery should absorb the surplus power without being overcharged; (2) when $P_{mpp} < P_{load}$ and battery has the ability to provide the deficient power $P_{load} - P_{mpp}$ without exceeding its minimum discharge limit. The total power in MPPT mode is given by $P_{total} = P_{mpp} = P_b + P_l$. Here, P_b is the battery power, defined as positive while charging and negative while discharging.

Non-MPPT mode. The charging current of the battery is required to be limited to a maximum permissible limit, I_{bmax}, based on its SOC level. I_{bmax} restricts the maximum power that can be absorbed by the battery to $P_{bmax} = I_{bmax} V_b$. When $P_{mpp} > P_l$ and the surplus power is more than P_{bmax}, the system cannot be operated in MPPT mode, as it would overcharge the battery. During this condition, the system operates in non-MPPT mode and power extraction from PV is reduced.

Battery-only mode. When the power from both the sources is not available and the battery has the ability to provide the entire load demand without exceeding the minimum discharge limit, the system then operates in battery-only mode.

Shutdown mode. When $P_{mpp} < P_{load}$ and battery reaches below its safe discharge limit b_{lim}, the system is shut down to prevent the battery from being completely discharged. The control structure of the whole system is shown in Figure 10.7. When both PV and wind power are available, to operate the system in MPPT mode, the references V_{mpp} or I_{mpp} are chosen from the output of the respective MPPT algorithm blocks. The DC link voltage $V_{dc} = n(V_{C1} + V_{C2}) = n(V_b + V_{pvmpp})$ is controlled by the PV source. In the absence of PV power, the DC link voltage can be controlled either by the battery or wind source. In the absence of both PV and wind power, the DC link voltage is controlled by the battery (battery-only mode). In order to maintain the DC link voltage at 400 V, the voltage across the PV capacitor is controlled with the help of a battery. The generated voltage reference for PV capacitor is compared with its actual value, and error is fed to the PI controller, which generates a reference for the inductor current, i_{Lstar}. In order to protect against overcharging or discharging for the battery, an appropriate upper limit and lower limit, I_{Lmax} and I_{Lmin}, are set in the reference for the inductor current. These limits are derived based on the following relationship: $I_{Lmax} = I_{bmax} + I_{pv}$ and $I_{Lmin} = I_{bmin} + I_{pv}$, where I_{bmax} and I_{bmin} are the maximum and minimum current limits for the battery, respectively. These limits are obtained from the SOC control block of the battery and are the inputs to the decision-making block, DMB-II, which generates final reference for the inductor current i_{Lref}. When the battery charging current exceeds I_{Lmax}, DMB-II limits i_{Lref} to I_{Lmax}. This limits the battery charging current to its maximum value I_{bmax}. Thus, the power flow to the battery gets reduced to P_{blim}. Similarly, after reaching the minimum discharge limit of the battery, i_{Lstar} reaches I_{Lmin}, and the DMB-II limits i_{Lref} to I_{Lmin}. When this occurs, DMB-III shuts down the system by withdrawing the switching pulses. When PV source is not available and wind source is available, the DC link voltage is controlled by the battery until the battery reaches its maximum charging limit I_{bmax}. If battery reaches I_{bmax}, it loses control over the regulation of the DC link voltage. Therefore, to ensure power balance, the wind source has to deviate from MPPT and regulate the DC link voltage. The decision-making block DMB-IV decides whether the wind source should operate in MPPT mode or in DC link regulation mode (non-MPPT mode). IF Y = 1, $I_{pv} = 0$, and $I_{Lstar} > I_{Lmax}$ and wind source regulates the DC link voltage, otherwise, wind source operates in MPPT mode. Stand-alone systems for AC loads require fast output voltage control. A d-q reference frame approach [31] is used for controlling the voltage.

10.3.3 Simulation Results and Discussion

Detailed simulation studies are carried out on MATLAB/Simulink platform, and the results for various operating modes are presented in this section. Values of parameters

TABLE 10.1
Simulation Parameters

Parameter	Value
Load power rating	1,000 VA
Solar PV power	525 W (I_{mpp} = 14.8 A, V_{mpp} = 35.4 V)
Wind power	560 W (I_{mpp} = 16 A, V_{mpp} = 35 V)
Switching frequency	15 kHz
Transformer rating and turns ratio	1 kVA and 5.5
Inductor half-bridge boost converter, L_w	500 µH
Inductor bidirectional converter L	3000 µH
Primary-side capacitors C_1–C_2	500 µF
Secondary-side capacitors C_3–C_4	500 µF
Secondary-side capacitor for the entire DC link	2000 µF
Battery capacity and voltage	400 Ah, 36 V

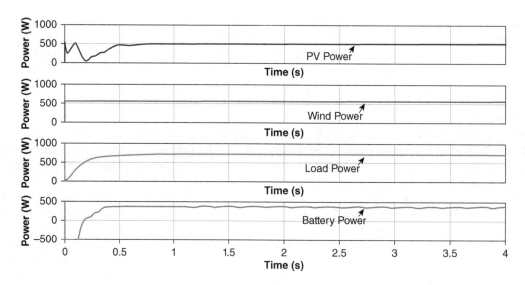

FIGURE 10.8 Plot of PV, wind, load, and battery power vs. time for steady-state operation in MPPT mode.

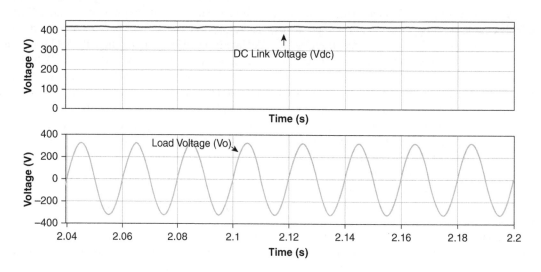

FIGURE 10.9 Plot of DC link voltage and load voltage vs. time for steady-state operation in MPPT mode.

used in the simulation model are listed in Table 10.1. The primary-side DC link voltage $(V_b + V_{pv}) = 72$ V, and the transformer turns ratio n is 1:5.5, to obtain 400 V DC link voltage.

The steady-state response of the system in the MPPT mode of operation is shown in Figures 10.8 and 10.9. The load demand is kept at 700 W, which is less than the total maximum power P_{mpp} that is generated by both the sources. It can be seen from Figure 10.8 that PV power and wind power are at their MPP values, while battery power is positive, implying it is charging to consume the surplus power. Figure 10.9 shows the voltage profile of DC link and load voltage, which is maintained at 230 V rms.

Simulated results of the system for step changes in load demand while operating in MPPT mode and transition from MPPT to non-MPPT modes follow.

10.3.4 Experimental Validation

In order to verify the simulation results, experimental tests are carried out on a laboratory prototype, shown in Figure 10.10. The specifications of experimental setup are given in Table 10.2. Two solar array simulator modules each of 0–95 V, 6.3 A, 600 W, Agilent-E4360 are used to emulate two sources, namely, solar PV and wind. The control strategy of the system is implemented using Texas Instruments floating-point DSP, TMS320F28335. The

Integrated PV-Wind-Battery-Based Single-Phase System

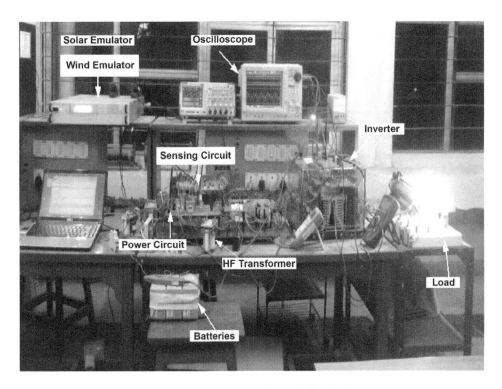

FIGURE 10.10 Experimental setup of the proposed hybrid PV-wind system.

TABLE 10.2
Specifications of Experimental Setup

Parameter	Value	Part Number
Solar PV power	250 W	
Wind power	250 W	
Total load power	500 W	
MOSFET, T1–T4	200 V, 90 A	IRFP4668PbF
IGBT, S1–S4	1200 V, 20 A	IRG7PH35UD1PbF
Diode, D1–D2	1000 V, 60 A	STTH6010W
Capacitor, Cb	1,000 µF, 100 V	SLPX102M100A3P3
Capacitor, C1–C2	560 µF, 100 V	100ZLJ560M
Capacitor, C3–C4	560 µF, 400 V	MCLPR400V567M
Capacitor, Cw	1,000 µF, 63 V	ECA1JHG102
Capacitor, Cpv	2,000 µF, 200 V	CGS202T200V4C
Inductor, L	3,000 µH, 40 A	
Inductor, Lw	500 µH, 50 A	
Inductor, Lb	1,000 µH, 30 A	
Battery	12 × 3 V, 7.2 Ah	

steady-state response of the system in the MPPT mode of operation is shown in Figure 10.11. The load demand is set at 400 W. The parameters for source 1 (PV source) and source 2 (wind source) are set at 40 V (Vmpp) and 5 A (Impp), respectively. It can be seen that V_{pv} and I_{pv} of source 1, and V_w and I_w of source 2, attain set values required for MPP operation. The load voltage V_o is maintained at 230 V rms. The battery current I_b is negative, indicating that it is getting discharged to supply the deficit power.

10.3.5 Summary

A hybrid PV-wind-based stand-alone power evacuation scheme for rural household application is proposed. It is realized by a novel multi-input transformer–coupled DC–DC converter and full-bridge DC–AC inverter. Detailed simulation studies are carried out to ascertain the viability of the scheme, and these results are validated through detailed studies.

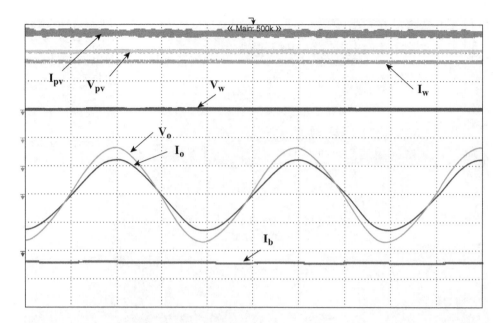

FIGURE 10.11 Steady-state operation in MPPT mode (Vpv = 10V/div; Ipv = 2A/div; Vw = 20V/div; Iw = 2A/div; Vo = 200V/div; Io = 2A/div; Ib = 5A/div).

10.4 INTEGRATED PV-WIND-BATTERY-BASED SINGLE-PHASE GRID-CONNECTED SYSTEM

The integration of renewable sources to the grid is growing rapidly due to the enhanced power electronics technology. The inverter should be able to operate in grid feeding mode as well as stand-alone mode to supply uninterrupted power to the critical loads during power outages. These systems involve one or more renewable power sources, load, grid, and battery. Hence, a power flow management system is essential to balance the power flow among all these sources. The main objectives of these systems are as follows:

- To explore a multiobjective control scheme for optimal charging of the battery using multiple sources.
- To supply uninterruptible power to loads.
- To ensure evacuation of surplus power from renewable sources to the grid and to charge the battery from grid as and when required.

A grid-connected hybrid PV-wind-battery-based system for household application based on the multi-input structure is shown in Figure 10.12. This interface works either in stand-alone or grid-connected mode. This system is more suited for household applications, where a low-cost, simple, and compact topology capable of autonomous operation is desirable. The core of the proposed system is the multi-input transformer–coupled bidirectional DC–DC converter that interconnects various power sources and the storage element. It has an effective power flow management scheme for providing uninterrupted power supply to the AC loads from a grid-connected hybrid PV-wind-battery-based system while facilitating the evacuation of excess power to the grid. Thus, the proposed configuration and control scheme provide an elegant integration of multiple sources.

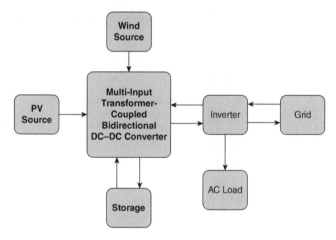

FIGURE 10.12 Grid-connected hybrid PV-wind-battery-based system for household applications.

10.4.1 Proposed Converter Configuration

The topology proposed in Chapter 3 is suitably modified to interface with the power grid and charge the battery from grid as and when required. The modified topology consists of a transformer-coupled boost dual-half-bridge bidirectional converter fused with bidirectional buck-boost converter and a single-phase full-bridge inverter. The proposed converter has reduced number of power conversion stages with less component count and high efficiency compared to the existing grid-connected schemes. The topology is simple and needs only six power switches. The schematic diagram of the converter is depicted in Figure 10.13(a). The power flow from wind source is controlled through a boost half-bridge converter. When switch T_3 is on, the current flowing through the source inductor increases. The capacitor C_1 discharges through the transformer

Integrated PV-Wind-Battery-Based Single-Phase System

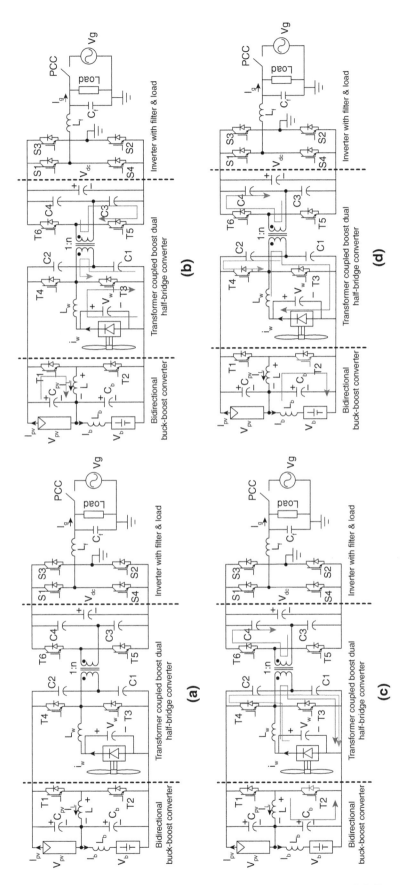

FIGURE 10.13 Operating modes of proposed multi-input transformer–coupled bidirectional DC–DC converter. (a) Proposed converter configuration. (b) Operation when switch T3 is turned on. (c) Operation when switch T4 is on, charging the capacitor bank. (d) Operation when switch T4 is on, capacitor C_2 discharging.

primary and switch T_3, as shown in Figure 10.13(b). On the secondary-side capacitor, C_3 charges through transformer secondary and antiparallel diode of switch T_5. When T_3 is off and T_4 is turned on, initially the inductor current flows through the antiparallel diode of switch T_4 and through the capacitor bank. The path of current is shown in Figure 10.13(c). During this interval, the current flowing through the diode decreases and that flowing through the transformer primary increases. When current flowing through the inductor becomes equal to that flowing through transformer primary, the diode turns off. Since, T4 is gated on during this time, the capacitor C2 now discharges through switch T_4 and transformer primary. During the on time of T_4, antiparallel diode of switch T6 conducts to charge the capacitor C4. The path of current flow is shown in Figure 10.13(d). During the on time of T_3, the primary voltage $V_P = -V_{C1}$. The secondary voltage $V_S = nV_P = -nV_{C1} = -V_{C3}$, or $V_{C3} = nV_{C1}$, and voltage across primary inductor L_w is V_w. When T_4 is turned on, the primary voltage $V_P = V_{C2}$. The secondary voltage $V_S = nV_P = nV_{C2} = V_{C4}$, and voltage across primary inductor L_w is $V_w - (V_{C1} + V_{C2})$. It can be proved that $(V_{C1} + V_{C2}) = V_w$. The capacitor voltages are considered constant in steady state, and they settle at $V_{C3} = nV_{C1}$, $V_{C4} = nV_{C2}$. Hence, the output voltage is given by:

$$V_{dc} = V_{c3} + V_{c4} = n\frac{V_w}{(1-D_w)} \quad 10.3$$

Therefore, the output voltage of the secondary-side DC link is a function of the duty cycle of the primary-side converter and turns ratio of transformer.

The bidirectional buck-boost converter charges/discharges the capacitor bank C_1–C_2 of boost half-bridge converter based on the load demand. During battery charging mode, when switch T1 is on, the energy is stored in the inductor L. When switch T1 is turned off and T2 is turned on, energy stored in L is transferred to the battery. During battery discharging mode, inductor current becomes negative. Here, the stored energy in the inductor increases when T2 is turned on and decreases when T_1 is turned on. It can be proved that $V_b = D/(1 - D) V_{pv}$. The output voltage of the transformer-coupled boost half-bridge converter is given by:

$$V_{dc} = n(V_{c1} + V_{c2}) = n(V_b + V_{pv}) = \frac{nV_w}{(1-D_w)} \quad 10.4$$

To charge the battery from the grid, a single-phase full-bridge converter formed by the devices S1–S4 works in the rectifying mode, while switches T5 and T6 (which constitute the half-bridge converter) transfer the grid power to charge the battery through the multi-input transformer.

10.4.2 Proposed Control Scheme for Power Flow Management

A grid-connected hybrid PV-wind-battery-based system consisting of four power sources (grid, PV, wind source, and battery) and three power sinks (grid, battery, and load) requires a control scheme for power flow management.

The control philosophy for power flow management of the multisource system is developed based on the power balance principle. In the stand-alone case, PV and wind source generate their corresponding MPP power and load takes the required power. In this case, the power balance is achieved by using a battery until it reaches its maximum charging current limit I_{bmax}. Upon reaching this limit, the power absorbed by the battery is restricted. Under this condition, to ensure power balance, one of the sources or both have to deviate from their MPP power based on the load demand. In the grid-connected system, both the sources always operate at their MPP power irrespective of the power restriction imposed by the battery. In the absence of both the sources, the power is drawn from the grid to charge the battery as and when required. The equation for the power balance of the system is given by:

$$V_{pv}I_{pv} + V_w I_w = V_b I_b + V_g I_g \quad 10.5$$

Where V_g is the grid voltage and m_a is modulation index.

The peak value of the output voltage for a single-phase full-bridge inverter is:

$$\hat{v} = m_a V_{dc} \quad 10.6$$

And from 4.2:

$$V_{dc} = n(V_{pv} + V_b) \quad 10.7$$

Hence, by substituting for V_{dc} in (4.4), it gives:

$$V_g = \frac{1}{\sqrt{2}} m_a n(V_{pv} + V_b) \quad 10.8$$

In the boost half-bridge converter (4.2):

$$V_w = (1-D_w)(V_{pv} + V_b) \quad 10.9$$

Now, substituting V_w and V_g in (4.3):

$$V_{pv}I_{pv} + (V_{pv} + V_b)(1-D_w)I_w$$
$$= V_b I_b + \frac{1}{\sqrt{2}} m_a n(V_{pv} + V_b)I_g \quad 10.10$$

After simplification:

$$I_b = I_{pv}\left(\frac{1-D_{pv}}{D_{pv}}\right) + I_w\left(\frac{1-D_{pv}}{D_{pv}}\right) - I_g\left(\frac{m_a n}{\sqrt{2D_{pv}}}\right) \quad 10.11$$

From the previous equation it is evident that, if there is a change in power extracted from either PV or wind source, the battery current can be regulated by controlling the grid current Ig. Hence, the control of a single-phase full-bridge bidirectional converter depends on the availability of grid, power from PV and wind sources, and battery charge status. Its control strategy is illustrated in Figure 10.14. The actual battery voltage is compared with the reference voltage, and the error is fed to a PI controller, which generates a reference for the battery current I_{bref}. To protect against overcharging or discharging of the battery, an appropriate upper limit and lower limit are set in the reference for the battery current. This is compared with the actual battery current, and the error is fed to a PI controller, which generates the reference grid current. To ensure the supply of uninterrupted power to critical loads, priority is given to charge the batteries. After reaching the maximum battery charging current limit I_{bmax}, the surplus power from renewable sources is fed to the grid. In the absence of these sources, battery is charged from the grid.

10.4.3 Simulation Results and Discussion

Detailed simulation studies are carried out on MATLAB/Simulink platform, and the results obtained for various operating conditions are presented in this section. Values of parameters used in the model for simulation are listed in Table 10.3. The steady-state response of the system during the MPPT mode of operation is shown in Figure 10.15. The values for source 1 (PV source) is set at 35.4 V (V_{mpp}) and 14.8 A (I_{mppp}), and for source 2 (wind source) is set at 37.5 V (V_{mpp}) and 8 A (I_{mppp}). It can be seen that Vpv and Ipv of source 1, and V_w and I_w of source 2, attain set values required for MPP operation. The battery is charged with constant magnitude of current, and remaining power is fed to grid.

10.4.4 Summary

A grid-connected hybrid PV-wind-battery-based power evacuation scheme for household application is proposed. The proposed hybrid system provides an elegant integration of PV and wind source to extract maximum energy from the two sources. It is realized by a novel multi-input transformer–coupled bidirectional DC–DC converter, followed by a conventional full-bridge inverter. A versatile control strategy which achieves better utilization

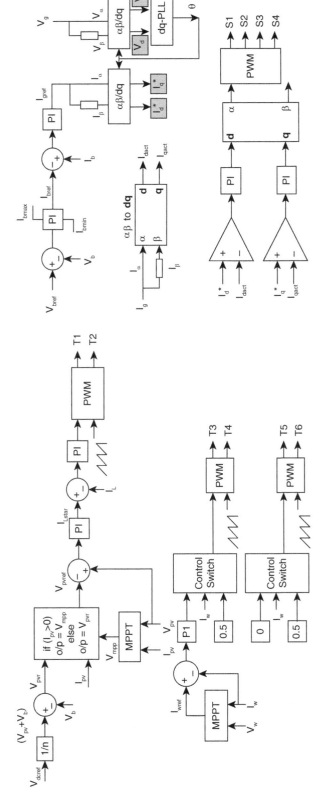

FIGURE 10.14 Proposed control scheme for power flow management of a grid-connected hybrid PV-wind-battery-based system.

TABLE 10.3
Simulation Parameters

Parameter	Value
Solar PV power	525 W (Impp = 14.8 A, Vmpp = 35.4 V)
Wind power	300 W (Impp = 8 A, Vmpp = 37.5 V)
Switching frequency	15 kHz
Transformer turns ratio	5.5
Inductor half-bridge boost converter, Lw	500 μH
Inductor bidirectional converter, L	3,000 μH
Primary-side capacitors, C1–C2	500 μF
Secondary-side capacitors, C3–C4	500 μF
Secondary-side capacitor for the entire DC link	2,000 μF
Battery capacity and voltage	400 Ah, 36 V

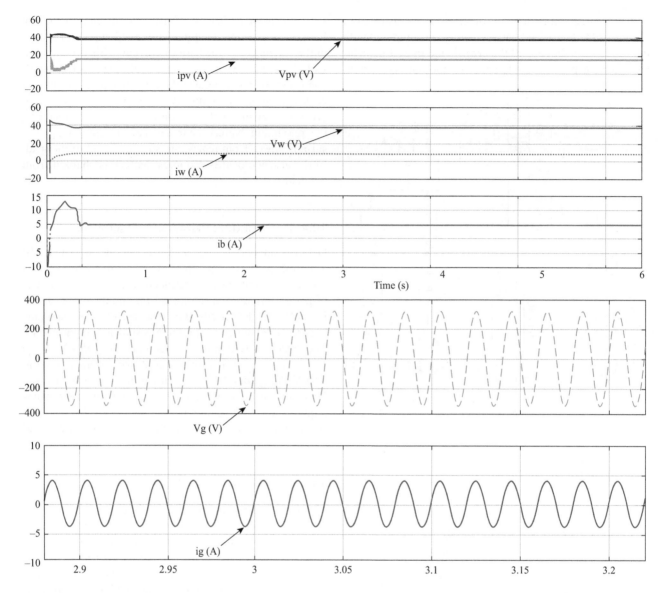

FIGURE 10.15 Steady-state operation in MPPT mode.

of PV, wind power, battery capacities without affecting life of battery and power flow management in a grid-connected hybrid PV-wind-battery-based system feeding AC loads is presented. Detailed simulation studies are carried out to ascertain the viability of the scheme. The experimental results obtained are in close agreement with simulations and are supportive in demonstrating the capability of the system to operate either in grid-feeding or stand-alone mode. The proposed configuration is capable of supplying uninterruptible power to AC loads and ensures evacuation of surplus PV and wind power into the grid.

10.5 REFERENCES

[1] F. Locment, M. Sechilariu, and I. Houssamo, "Batteries and DC charge control of stand-alone photovoltaic system. Experimental validation," *in Proc. 14th Int. Power Electronron. Motion Control Conf. EPE*, pp. T12–43-T12–48, 2010.

[2] V. Mummadi, "Buck-Integrated SEPIC Converter For Photovoltaic Power Conver- sion," *in Proc. 31st Int. Telecommunications Energy Conf.*, pp. 1–5, 2009.

[3] H. Nakayama, E. Hiraki, T. Tanaka, N. Koda, N. Takahashi, and S. Noda, "Stand- alone Photovoltaic Generation system with combined storage using lead battery and EDLC," *in Proc. 13th Power Electronron. and Motion Control Conf. (EPE-PEMC)*, pp. 1877–1883, 2008.

[4] H. Matsuo and F. Kurokawa, "New solar cell power supply system using a boost type bidirectinal dc-dc converter," *IEEE Trans. Ind. Electron.*, vol. Ie-31, no. 1, pp. 1118–1126, 1984.

[5] L. Shengyong, Z. Xing, G. Haibin, and X. Jun, "Multiport dc/dc converter for stand- alone photovoltaic lighting system with battery storage," *in Proc. Int. Conf. on Electron. and Control Eng. (ICECE)*, pp. 3894–3897, 2010.

[6] H. Wang and B. Li, "The cooperated MPPT control of stand-alone PV power generation system," *in Proc. of the World Congress on Intelligent Control and Automation (WCICA)*, pp. 2228–2231, 2010.

[7] H. Aghazadeh, H. M. Kojabadi, and A. S. Yazdankhah, "Stand-alone PV generation system with maximum power

point tracking," *in proc. 9th Int. Conf. on Environment and Electron. Eng. (EEEIC)*, pp. 549–552, 2010.

[8] L. F. Costa and R. P. Torrico-Bascope, "Stand-Alone Photovoltaic System with Three Energy Processing Stages," *in Proc. Brazilian Power Electron. Conf. (COBEP)*, pp. 651–656, 2011.

[9] X. Liu, P. Wang, and P. C. Loh, "Coordinated control scheme for stand-alone PV system with nonlinear load," *in Proc. of 2010 IEEE PES Transmission and Distribution Conf. and Exposition*, pp. 1–8, April 2010.

[10] Y. M. Chen, A. Q. Huang, and Y. Xunwei, "A high step-up three-port dc-dc converter for stand-alone PV/battery power systems," *IEEE Trans. Power Electron.*, vol. 28, no. 11, pp. 5049–5062, Nov. 2013.

[11] F. Ishengoma, F. Schimpf, and L. Norum, "DSP-controlled photovoltaic inverter for universal application in research and education," *in Proc. IEEE Trondheim PowerTech Conf.*, pp. 1–6, 2011.

[12] Z. Qian, O. Abdel-Rahman, H. Hu, and I. Batarseh, "An integrated three-port in- verter for stand-alone PV applications," *in Proc. IEEE Energy Convers. Congr. Expo.*, pp. 1471–1478, Sep. 2010.

[13] R. Chattopadhyay and K. Chatterjee, "PV based stand alone single phase power generating unit," *in Proc. IEEE Ind. Electron. Soc. Conf., IECON'* 2012, pp. 1138–1144, Oct. 2012.

[14] N. Adhikari, B. Singh, and A. L. Vyas, "Design and performance of low power solar- PV energy generating system with zeta converter," *in Proc. IEEE Int. Conf. on Ind. and Information Systems (ICIIS)*, pp. 1–6, 2012.

[15] D. Debnath and K. Chatterjee, "Two-stage solar photovoltaic-based stand-alone scheme having battery as energy storage element for rural deployment," *IEEE Trans. Ind. Electron.*, vol. 62, no. 7, pp. 4148–4157, Jul. 2015.

[16] F. Valenciaga, P. F. Puleston, and P. E. Battaiotto, "Power control of a solar/wind generation system without wind measurement: A passivity/sliding mode approach," *IEEE Trans. Energy Convers.*, vol. 18, no. 4, pp. 501–507, Dec. 2003.

[17] W. Qi, J. Liu, X. Chen, and P. D. Christofides, "Supervisory predictive control of standalone wind/solar energy generation systems," *IEEE Trans. Control Sys. Tech.*, vol. 19, no. 1, pp. 199–207, Jan. 2011.

[18] W. Kellogg, M. Nehrir, G. Venkataramanan, and V. Gerez, "Generation unit sizing and cost analysis for stand-alone wind, photovoltaic and hybrid wind/PV systems," *IEEE Trans. Ind. Electron.*, vol. 13, no. 1, pp. 70–75, Mar. 1998.

[19] L. Xu, X. Ruan, C. Mao, B. Zhang, and Y. Luo, "An improved optimal sizing method for wind-solar-battery hybrid power system," *IEEE Trans. Sustainable Enery.*, vol. 4, no. 3, pp. 774785, Jul. 2013.

[20] B. S. Borowy and Z. M. Salameh, "Dynamic response of a stand-alone wind energy conversion system with battery energy storage to a wind gust," *IEEE Trans. Energy Convers.*, vol. 12, no. 1, pp. 73–78, Mar. 1997.

[21] S. Bae and A. Kwasinski, "Dynamic modeling and operation strategy for a microgrid with wind and photovoltaic resources," *IEEE Trans. Smart Grid*, vol. 3, no. 4, pp. 1867–1876, Dec. 2012.

[22] C. W. Chen, C. Y. Liao, K. H. Chen and Y. M. Chen, "Modeling and controller design of a semi isolated multi input converter for a hybrid PV/wind power charger system," *IEEE Trans. Power Electron.*, vol. 30, no. 9, pp. 4843–4853, Sept. 2015.

[23] F. Valenciaga and P. F. Puleston, "Supervisor control for a stand-alone hybrid gen- eration system using wind and photovoltaic energy," *IEEE Trans. Energy Convers.*, vol. 20, no. 2, pp. 398–405, Jun. 2005.

[24] M. H. Nehrir, B. J. LaMeres, G. Venkataramanan, V. Gerez, and L. A. Alvarado, "An approach to evaluate the general performance of stand-alone wind/photovoltaic generating systems," *IEEE Trans. Energy Convers.*, vol. 15, no. 4, pp. 433–439, Dec. 2000.

[25] C. Liu, K. T. Chau and X. Zhang, "An efficient wind-photovoltaic hybrid generation system using doubly excited permanent-magnet brushless machine," *IEEE Trans. Ind. Electron.*, vol. 57, no. 3, pp. 831–839, Mar. 2010.

[26] W. M. Lin, C. M. Hong, and C. H. Chen, "Neural network-based MPPT control of a stand-alone hybrid power generation system," *IEEE Trans. Power Electron.*, vol. 26, no. 12, pp. 3571–3581, Dec. 2011.

[27] T. Hirose and H. Matsuo, "Standalone hybrid wind-solar power generation system applying dump power control without dump load," *IEEE Trans. Ind. Electron.*, vol. 59, no. 2, pp. 988–997, Feb. 2012.

[28] S. A. Daniel and N. A. Gounden, "A novel hybrid isolated generating system based on PV fed inverter-assisted wind-driven induction generators," *IEEE Trans. Energy Convers.*, vol. 19, no. 2, pp. 416–422, Jun. 2004.

[29] Z. Qian, O. A. Rahman, and I. Batarseh, "An integrated four-Port DC/DC converter for renewable energy applications," *IEEE Trans. Power Electron.*, vol. 25, no. 7, pp. 1877–1887, July. 2010.

[30] F. Nejabatkhah, S. Danyali, S. Hosseini, M. Sabahi, and S. Niapour, "Modeling and control of a new three-input DCDC boost converter for hybrid PV/FC/battery power system," *IEEE Trans. Power Electron.*, vol. 27, no. 5, pp. 2309–905, Feb. 2014.

[31] D. Dong, "Modeling and control design of a bidirectional PWM converter for single-phase energy systems," *Master Thesis, Virginia Tech*, 2009.

11 Modeling of Power Management Strategy Using Hybrid Energy Generating Sources

Hinal Surati

CONTENTS

11.1 Introduction 137
11.2 World Energy Consumption Growth Chart 137
11.3 Power Management Strategies 138
11.4 Photovoltaic System 138
11.5 Fuel Cell Energy Storage System 139
      11.5.1 Introduction 139
11.6 Wind Energy Generation System 140
      11.6.1 Modeling of Wind Turbine 140
11.7 For Obtaining the Efficient Power Management Strategy, the Individual Components Have Been Sectioned and Mathematically Modeled Using MATLAB 140
      11.7.1 P-V and I-V Characteristics of Single PV Cell 140
      11.7.2 PV Module Simulation Model 140
      11.7.3 PV Module Simulations Using MPPT Technique (P&O Algorithm) 141
11.8 Modeling of SOFC Cell 141
      11.8.1 Simulation Results and Discussions 145
11.9 Integration of PV Module with MPPT Technique and Fuel Cell 145
11.10 Wind Generation System 145
11.11 Final Modules: "Integration of PV/Wind/Fuel Cell Energy Generation System [11] Shown in Figure 14 145
11.12 References 149

11.1 INTRODUCTION

A hybrid power generating resource such as nonconventional (renewable) is combining wind energy with solar photovoltaic or any other nonrenewable energy generating sources when integrated with inverter, energy storage medium as fuel cell, and other auxiliary components. In the literature, several examples are presented showing the effectiveness of PV-wind as a hybrid power generating system, which is utilized to overcome the system load demand. Once the power generating resources (solar and wind energy) are sufficient, then the excess generated power is given to the fuel cell storage system. Thus, the fuel cell will come into action only when the energy generating renewable sources (PV-wind) power is not able to fulfill the load demand until the storage is depleted. The operation of hybrid structure of PV-wind system relies on the performance of all individual elements. Firstly, the individual system and its single components are modeled, and thereafter, their hybrid combinations are integrated and examined to meet the desired dependability and to achieve the maximum output from each component. So the electric power production through this element will offer electrical power at the very least charge.

11.2 WORLD ENERGY CONSUMPTION GROWTH CHART

It is being reflected through Figure 11.1, "World Energy Consumption Growth Chart, 1970–2025" [1–3], that the amount of energy to be consumed across the world is projected to increase about 57% in the coming era [1]. To date, the major section of generation of energy is about 86% in 2002 by the conventional resources. In the near future, the perspective of the growth of the world will be totally dependent on how this elevating energy demand can be met. Therefore, it provides a big opportunity and challenge to the whole world to broaden up the study on applications to fulfill the energy demand using the available energy generating sources.

It has been assumed that the consumption of electrical energy is to approach above 50% by 2025 [1–3]. Governments are becoming more flexible in establishing the targets for solar and any other nonhydro renewable energy technology. It is being proposed that in the near future, China, India, and Germany are required to meet their desired solar targets by building up further 70 GW, 68 GW, and 48 GW, respectively, by 2030 or earlier.

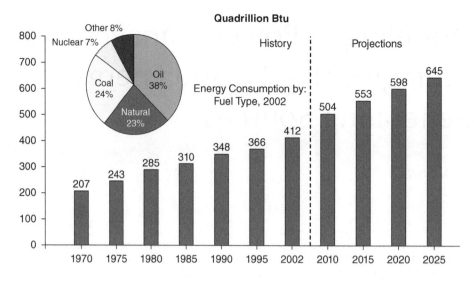

FIGURE 11.1 World Energy Consumption Growth Chart, 1970–2025 [1–3].

11.3 POWER MANAGEMENT STRATEGIES

Power management strategies (PMS) [2] in Figure 11.2 show the coordination between different power generating sources, especially solar and wind. The PMSs is adopted to satisfy the desired load requirements utilizing the benefits of renewable energy sources (RES) and with energy storage unit. Thus, the PMS in hybrid structure is too important for balancing between power and efficiency of hybrid generation systems. However, varying power demands and the unpredictable nature of load are the uncertainties which can't be avoided fully [2–3]. Power management strategies are defined in Figure 11.2, and to achieve the power management requirements, fuel cell storage device and the sizing of system elements are done using MATLAB/Simulink.

(i) If P load exceeds the available generated power by PV (Ppv) and wind (Pwind), the fuel cell (Pfc) will come into the action.

Pload = Pwind + Ppv + PFC, Psys < 0

(ii) If wind energy as well as solar energy generation surpass the actual load demand, the surplus energy is again redirected towards the energy storage device (fuel cell).

PFC = Pwind + Ppv − Pload, Psys > 0

(iii) If the wind energy and solar energy generation matches the load demand, then the whole generated power by the renewable sources is injected to the load.

Pload = Pwind + Ppv, Psys = 0

The efficiency of the proposed system using the proposed block diagram of PV/wind/fuel cell is shown in Figure 11.3, demonstrated using the simulation tool. The simulations of only PV array, only wind generator, and only PV array with wind generator, all with energy storing element fuel cell, are modeled and developed using MATLAB/Simulink.

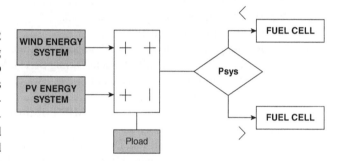

FIGURE 11.2 Power management strategies (PMS) [2].

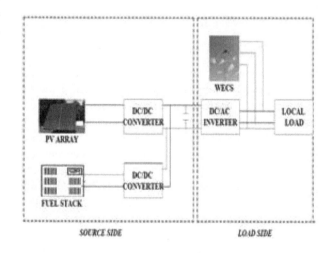

FIGURE 11.3 Block diagram of PV/wind/fuel cell.

11.4 PHOTOVOLTAIC SYSTEM

The solar photovoltaics (PV) industry is a rapidly growing sector that has created an enlarged job market demanding highly skilled engineers. The growth of the PV industry is foreseen to reach TW-scale generation by 2050 [4]. Simulink-based simulation includes building a single-diode

Power Management Using Hybrid Energy Generating Sources

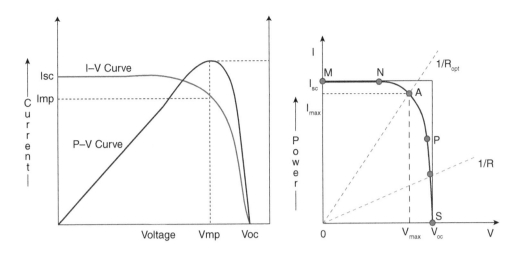

FIGURE 11.4 Theoretical characteristics of a PV cell (VI and PV).

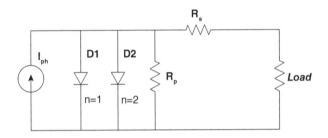

FIGURE 11.5 More accurate model.

model of a solar cell and analyzing its simulated current–voltage (I–V) and power–voltage (P–V) curves and comparing it with the theoretical characteristics of a PV cell (VI and PV), as shown in Figure 11.4, so that obtained results can be analyzed in a precise manner [5].

To achieve better power management strategy for solar PV system, a more accurate model is used [5–7], as shown in Figure 11.5, for simulating PV database, including minor variation in series resistance, parallel resistance, and recombination.

The performance of the PV model is obtained from an equivalent circuit, which includes a current source (I_{ph}), diodes (D1 and D2), and two resistors (Rp and Rs), as shown in Figure 11.5. I_{ph} acts as a current source that represents the generation of charge carriers caused by insolations. D1 and D2 are the shunt diode, which represents the process of recombination of charge carriers at a forward-bias condition. Rp is the shunt resistor that gives the path for the high current through the semiconductor through the mechanical defects and dislocations, whereas the series resistor _Rs is outside the semiconductor regions.

For a module under even irradiance that has a common unique peak point on the P–V characteristic curve, a DC–DC boost converter is used to boost the output-side voltage at the source site. And the proper source-side-to-load-side impedance can be matched by precisely varying the value of duty cycle ratio of the DC–DC boost converter. Perturb and observe (P&O) is widely used in MPPT for the reason that it has a simple structure and a few measured parameters which are required.

11.5 FUEL CELL ENERGY STORAGE SYSTEM

11.5.1 Introduction

Recent stringent energy-related problems have raised an indefinite interest in renewable (nonconventional) energy generating resources. Depending upon applications (both for stationary and mobile), the selection of the energy storage device for energy generating renewable resources is still under debate [8]. Therefore, integration of energy storage devices with nonconventional energy sources undergoing an electrochemical reaction process, converting directly chemical energy into electrical energy through an electrolyzes, is acquired as a distinctly sustainable process for the utilization of energy [8, 9]. As compared to a battery, fuel cell requires fuel for its continuous storage and operation. Among the different categories of fuel cell technologies, as specified in the literature, proton exchange membrane fuel cell (PEMFC) and solid oxide (SOFC) type of fuel cell–based energy storage system have shown wide applicability. Looking to the literature, that they have lower efficiency and are dependent on hydrogen as only fuel, they are not in much use for static power applications. Overcoming the drawbacks of PEMFCs, SOFCs, which operate at higher temperature (1,800°F), are regarded as ideal for DG applications [9].

$$Vfc = No\left[Eo + \frac{RT}{2F}ln\left[\frac{pH_2 pO_2^5}{pH_2O}\right]\right] - rI_{fc} \quad (11.1)$$

Where *Vfc* is the operating voltage _DC (v); *No* the number of cells in the stack; *Eo* the standard reversible cell potential (v); Rs the universal gas constant (Jol/mol K); *T* the stack temperature (k); and *F* the Faraday constant (C/mol).

Equations representing the characteristics of an SOFC are as follows:

$$P_{ref} = V_{fc} * I_{ref} \quad (11.2)$$

$$\frac{dI_{fc}}{dt} = \frac{1}{T_a}\left[-I_{fc} + I_{ref}\right] \quad (11.3)$$

$$\frac{dq^{in}_{H2}}{dt} = \frac{1}{T_f}\left[-q^{in}_{H2} + \frac{2Kr}{Uopt}I_{fc}\right] \quad (11.4)$$

$$\frac{dP_{H2}}{dt} = \frac{1}{\tau H2}\left[-P_{H2} + \frac{1}{K_{H2}}\left[q^{in}_{H2} - 2K_r I_{fc}\right]\right] \quad (11.5)$$

11.6 WIND ENERGY GENERATION SYSTEM

The turbine is a source of converting the wind's motivity into mechanical power, and a generator converts that mechanical power into useful electrical energy [10]. For the utilization of this wind velocity, there are two sectors, one by adopting a fixed-speed method or, two, by variable-speed method. Both of them have their own importance in terms of configuration, robustness, and efficiency. In this technique, the asynchronous (SCIG) is interconnected to the grid [11]. Since the introduction of SCIG will absorb the reactive power, thus a compensator, such as a bank of a capacitor, is needed to reduce the excess power demand. It can be achieved by switching the capacitor banks continuously, looking into the amount of generated active power.

11.6.1 MODELING OF WIND TURBINE

One can compute the mechanical output power of this turbine (Pm) using the following expression [9, 11]:

$$P_m = C_p(\lambda, \beta)\frac{\rho A}{2}v^3 \quad (11.6)$$

Where cp is the performance coefficient of the turbine; λ the tip speed ratio of the rotor blade tip speed to wind speed; β the pitch angle of the wind generator's blades; ρ the air density [kg/m3]; A the area swept by the turbine's blades; and V the wind velocity.

The electrical system dynamics are represented as follows [6, 9]:

$$V_{qs} = R_s I_{qs} + \frac{d\varphi_{qs}}{dt} + w\varphi_{ds} \quad (11.7)$$

$$V_{ds} = R_s I_{ds} + \frac{d\varphi_{ds}}{dt} - w\varphi_{qs} \quad (11.8)$$

$$V'_{qr} = R'_r I'_{qr} + \frac{d\varphi'_{qr}}{dt} + (w - w_r)\varphi'_{dr} \quad (11.9)$$

$$V'_{dr} = R'_r I'_{dr} + \frac{d\varphi'_{dr}}{dt} + (w - w_r)\varphi'_{qr} \quad (11.10)$$

$$T_e = 1.5.p.(\varphi_{ds}.i_{qs} - \varphi_{qs}.i_{ds}) \quad (11.11)$$

11.7 FOR OBTAINING THE EFFICIENT POWER MANAGEMENT STRATEGY, THE INDIVIDUAL COMPONENTS HAVE BEEN SECTIONED AND MATHEMATICALLY MODELED USING MATLAB

i. PV module simulations without MPPT
ii. PV module simulations with MPPT

11.7.1 P-V AND I-V CHARACTERISTICS OF SINGLE PV CELL

Firstly, the obtained simulation result for a single PV cell is shown in Figure 11.6, which determines the PV and IV characteristics of single PV cell, and then further the connection of such cells in parallel/series forms a PV array to obtain the desired output voltages/currents [2, 5].

11.7.2 PV MODULE SIMULATION MODEL

PV module simulation is shown Figure 11.7, where at different varying insolations, the value of the obtained output at a desired power as Vpv and Ipv is measured and

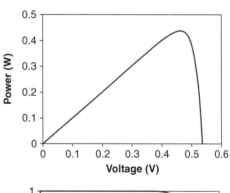

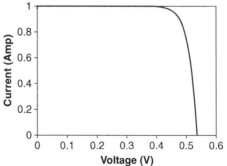

FIGURE 11.6 The P–V and I–V characteristics of single PV cell.

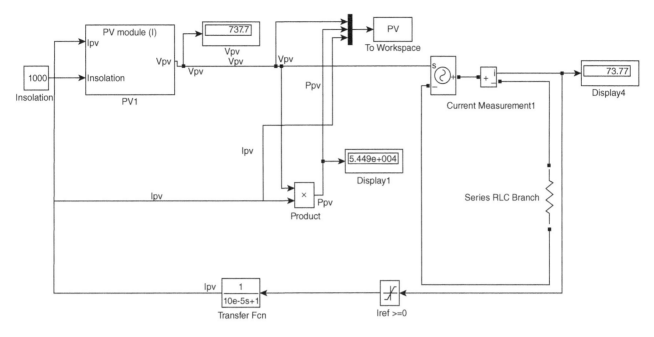

FIGURE 11.7 PV module simulation model.

TABLE 11.1
Voltage, Current, and Power at Different–Different Insolations for a PV Module

Insolations (W/m²)	Resistance (OHM)	Voltage (V)	Power (W)	Current (A)
1,000	10	737.7	54420.9	73.77
800	10	655.7	42994.24	65.57
600	10	500.6	25060.10	50.06
200	10	166.1	2674.0	16.61

analyzed through the obtained results of voltage, current, and power at different–different insolations for a PV module, which is shown in Table 11.1, where it can be seen that with the increase in the insolations, the power increases, and vice versa. So as to overcome this, MPPT algorithms have to be adopted to achieve maximum and constant output.

11.7.3 PV Module Simulations Using MPPT Technique (P&O Algorithm)

As shown Figure 11.8, PV array with MPPT (P&O algorithm) shows the control algorithm for PV array [6]. If the value of voltage and current is incrementing, the perturbation process will carry on in the same direction; otherwise, the direction is altered, as shown in Table 11.2, "Representation of the Implemented MPPT Algorithm," where the power management strategy, with its obtained results, shows the adopted and implemented P&O algorithm is efficient enough to track the maximum power from the PV array to obtain better-managed power with the proposed methodology to obtain Vab rms (V) as 415 V.

From Figure 11.9, the obtained results for V_{dc}, V_{ab}-inv, V_{ab} load, m, and Vout(pu) when a PV module is simulated using MPPT technique (P&O algorithm) show that with the change in insolations, there is a variable change in the voltage, current, and power across the PV array. But with the application and implementation of P&O algorithm, the system is able to observe and perturb to track the maximum power from the PV array as a efficient power management strategy.

11.8 MODELING OF SOFC CELL

For the modeling (Figure 11.10(a)) of SOFC, assuming the cell temperature to be constant, the fuel cell gases are ideal, with the applicability of the Nernst equation [4, 10].

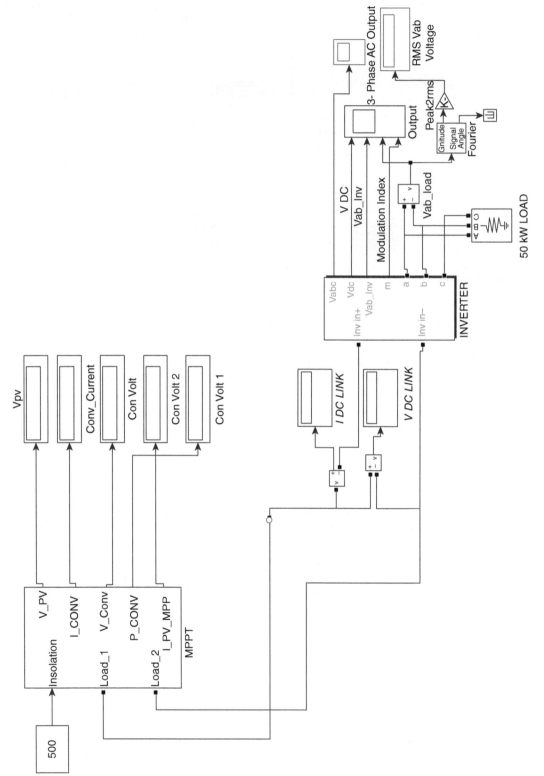

FIGURE 11.8 PV module simulation model using MPPT technique (P&O algorithm).

TABLE 11.2
Representation of the Implemented MPPT Algorithm

Insolation (W/m²)	Vpv (V)	Ipv (A)	D	Vcon (V)	Icon (I)	Power (W)	Vabrms (V)
1,000	642.3	78	0.529	1281	36.71	50100	415.2
900	636.6	71	0.553	1346	31.68	45200	415.2
800	637.1	63	0.566	1395	27.31	40100	415
700	641	54.6	0.5711	1431	23.39	35000	415.3
600	722.9	30	0.2	857.9	24.73	21600	415.3
500	635.1	38.6	0.5791	1463	16.22	24500	415.1
400	636	30.2	0.4617	1153	16.23	19200	410.8

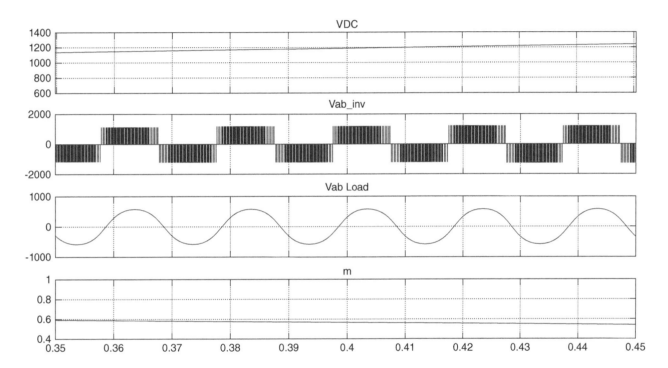

FIGURE 11.9(A) Waveform of V_{dc}, $V_{ab\text{-}inv}$, V_{abload}.

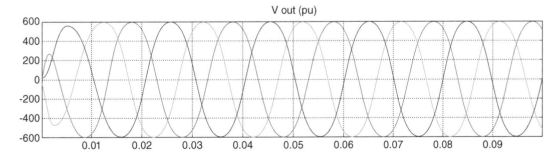

FIGURE 11.9(B) Waveform of V_{out}(pu).

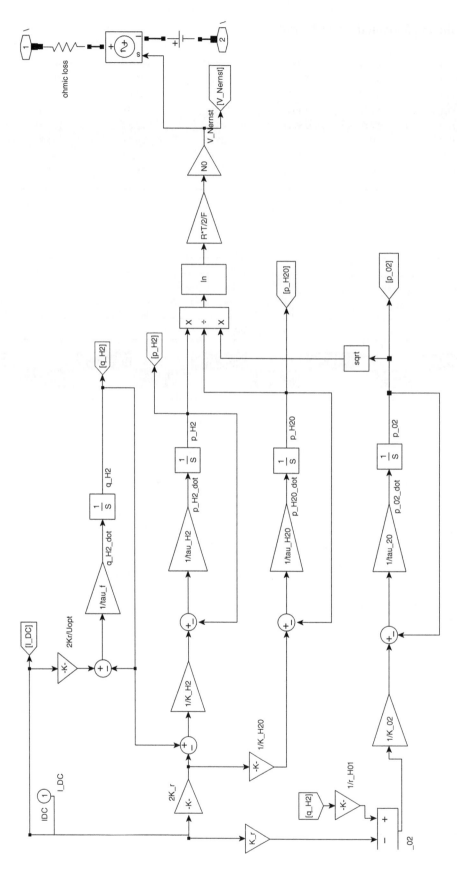

FIGURE 11.10(A) Modeling of SOFC Cell.

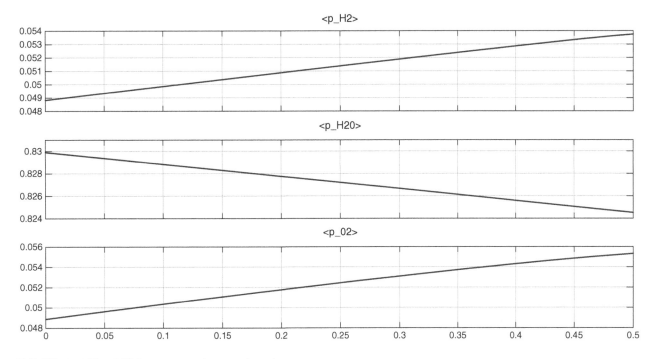

FIGURE 11.10(B) SOFC output waveforms and results.

11.8.1 Simulation Results and Discussions

The fuel utilization (Figure 11.10(b)) under the standard fuel condition, if the value of Uf > 0.9, can cause permanent damage to the cell, and if the value of Uf < 0.7, it can lead to higher cell voltages. So the value of Uf should range between 0.8 and 0.9 to achieve a better performance and satisfy the health of fuel (preventing overuse and underuse) [10].

11.9 INTEGRATION OF PV MODULE WITH MPPT TECHNIQUE AND FUEL CELL

Figure 11.11 portrays the simulated model of hybrid PV fuel cell performance at different loading and sun irradiance level. The integration of PV module with MPPT with fuel cell as energy storage system is shown (Figures 11.12(a) and (b)), with necessary output waveforms at varying isolations [3, 5, 7], that is, under load of 45 kw and maximum insolation of 1,000 w/m^2, as well as under the load of 45 kw and minimum insolation of 400 w/m^2. Initially, the fuel cell contributes the power with zero PV power. After certain duration of time, as the solar insolation level is increased to 1,000 W/m^2, the PV generated power is sufficient enough to run the load. If the solar insolations is reduced to 400 W/m^2, in that case, fuel cell is responsible to deliver the required power.

11.10 WIND GENERATION SYSTEM

In this section, the conventional directly grid-coupled SCIG has been designed. The model is being implemented on MATLAB/Simulink.

Figures 11.13(a) and (b) represent the mechanical power (in pu) for several wind speeds. As the amount of power generated varies, the speed of rotor as well as the slip of the squirrel cage induction generator changes [6].

11.11 FINAL MODULES: "INTEGRATION OF PV/WIND/FUEL CELL ENERGY GENERATION SYSTEM [11] SHOWN IN FIGURE 14

NOTE: The integration of all the three sources of energy generation is more reliable to feed the load so as to not require the energy storage system (fuel cell) to work full day, which will give the effective regulated output [9, 12]. Here in Table 11.6, two different cases have been formed for the efficient operation during the integration of PV/fuel cell/wind energy generation system under constant wind speed and varying insolation as well as varying wind speed and varying insolation. It has been noted through the obtained observations that as the wind speed is kept constant at 12 m/s and the PV insolation is varied, the variation in voltage and current is incrementing and perturbing

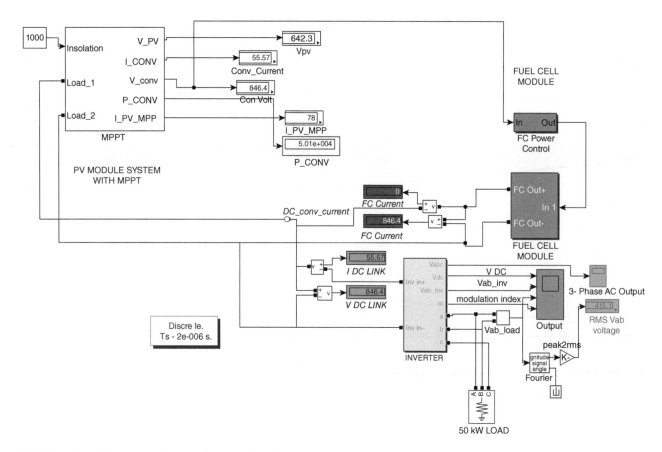

FIGURE 11.11 Integration of PV module with MPPT technique and fuel cell.

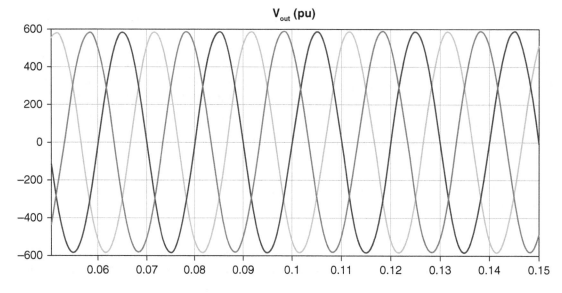

FIGURE 11.12(A) Waveforms of V_{out} (pu).

Power Management Using Hybrid Energy Generating Sources

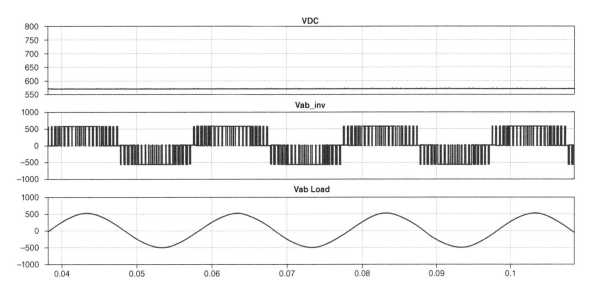

FIGURE 11.12(B) Waveforms under load of 45 KW and insolation 400 w/m^2.

TABLE 11.3
Values of Active and Reactive Power at Different Wind Velocities

Wind Speed (m/sec)	P (Active Power)	Q (Reactive Power)
12	345.2	8.289
11	269.2	26.89
10	201.9	39.5
9	144.7	47.44
8	92.83	52.48
7	50.06	55.1
6	15.41	56.22
5	−0.7394	56.44
4	−0.7394	56.44
3	−0.7394	56.44
2	−0.7394	56.44

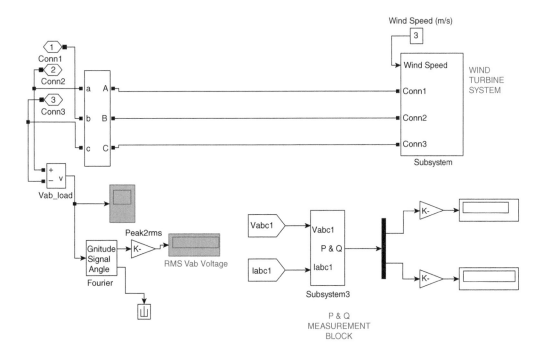

FIGURE 11.13(A) Simulated SCIG model.

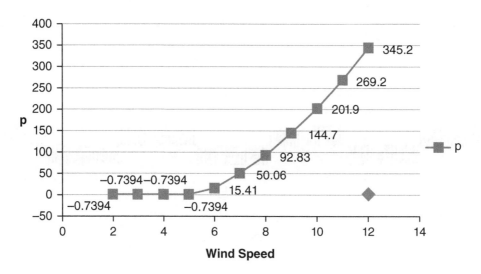

FIGURE 11.13(B) Speed vs. power.

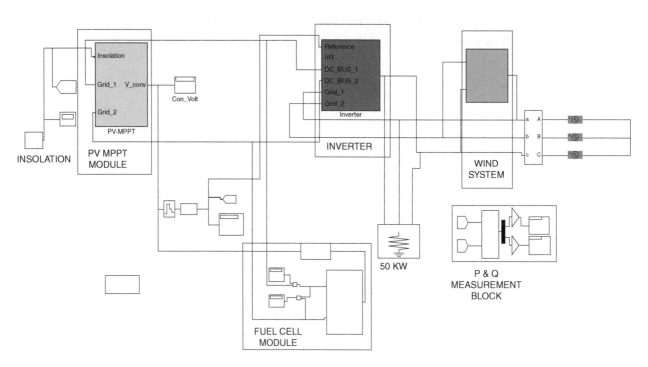

FIGURE 11.14 Simulink model for integration of PV/wind/fuel cell energy generation system.

at near duty cycle, leading to their interconnection with fuel cell to achieve the final output power of approximately 420 V constant near to the load.

Case 1: *Variable Wind Speed, Varying Insolation*

Through Table 11.6, variations of wind speed (max/min) and insolation (max/min) are obtained, which represent voltage and current at different–different values of wind speed and insolation before and after grid integration and at inverter side with their fft analysis. (It is assumed that the wave forms under the varying wind speed of 12m/s and varying insolations at 1,000 w/m²). This showed to b eefficient enough to achieve the stable output at the load with the proposed power management strategy [13]. In conclusion, the system operates and moves between various modes, dependent on the power requirement and needs of the load and the amount of power needed to be supplied by each

TABLE 11.4
Constant Wind Speed and Varying Insolation for PV/Wind/Fuel Cell System

Wind m/s	Insolations w/m²	MPPT PV				DUTY cycle	Fuel Cell				
		Vpv (V)	Vcon (V)	Icon (A)	Power		Vfc (V)	Ifc (A)	Vdc (V)	P	Q
12m/s	1000	630.5	675.3	78	4.92E+04	0.875	675.3	40	419.9	−420.4	−10.16
	800	594.3	674.6	69.4	4.12E+04	0.829	674.5	45	419.1	−410.2	−10.8
	600	600.2	675.5	51	3.06E+04	0.85	675.4	40	419.8	−396.6	−10.1
	400	600.2	675.5	51	1.97E+04	0.849	675.4	40	419.8	−384.4	−10.9
10m/s	1000	630.5	675.3	78	4.92E+04	0.875	675.3	40	419.9	−277.1	−41.37
	800	594.3	674.6	69.4	4.12E+04	0.829	674.5	45	419.1	−266.8	−41.31
	600	600.2	675.5	51	3.06E+04	0.85	675.4	40	419.8	−253.3	−41.31
	400	590.9	675.6	33.4	1.97E+04	0.849	675.6	40	419.8	−241.4	−41.31
8m/s	1000	630.5	675.3	78	4.92E+04	0.875	675.3	40	419.9	−168	−54.35
	800	594.3	674.5	69.4	4.12E+04	0.829	674.5	45	419.1	−157.7	−54.29
	600	600.3	675.5	51	3.06E+04	0.85	675.4	40	419.8	−144.2	−54.29
	400	590.9	675.6	33.4	1.97E+04	0.849	675.6	40	419.8	−132	−54.27
6m/s	1000	630.5	675.3	78	4.92E+04	0.875	675.3	40	419.9	−90.58	−58.09
	800	594.3	674.6	69.4	4.12E+04	0.829	674.5	45	419.1	−80.35	−58.03
	600	600.2	675.5	51	3.06E+04	0.85	675.4	40	419.8	−66.35	−58.03
	400	590.9	675.6	33.4	1.97E+04	0.849	675.6	40	419.8	−54.63	−553.02
4m/s	1000	630.5	675.3	78	4.92E+04	0.875	675.3	40	419.9	−74.4	−58.31
	800	594.3	674.6	69.4	4.12E+04	0.829	674.5	45	419.1	−64.18	−58.25
	600	600.2	675.5	51	3.06E+04	0.85	675.4	40	419.8	−50.63	−58.25
	400	590.9	675.6	33.4	1.97E+04	0.849	675.6	40	419.8	−38.45	−58.23

of the sources. Accordingly, the whole system operates in three stages:

Stage 1

When it totally relies on the energy generated by the PV to satisfy the load demand. Though the wind turbines are connected to the system, it is not utilized to satisfy the load demand in this mode. So the produced energy of the wind turbine is given to the fuel cell, which is then utilized for producing hydrogen fuel.

Stage 2

In this mode, if the energy that is available from the PV and the wind turbine integration is in excess of what is required by the load, in that case, the sufficient power which is available from the PV is used to satisfy the load simultaneously; the excess energy from the wind turbine can be utilized for producing hydrogen fuel in stage 1 [12]. It is a case when the power that has been generated by PV is not sufficient to satisfy the load.

Decision Mode

In the decision stage, if the integrated power of the PV/wind is not sufficient to supply the load and also not able to satisfy the scarcity of power required [12], in that case, the fuel cell is used to meet the load requirements.

11.12 REFERENCES

[1] International Energy Outlook, "Energy Information Administration (EIA)," 2012, www.eia.doe.gov/iea.
[2] M. Mahalakshmi, Dr. S. Latha, "Modeling, Simulation and Sizing of Photovoltaic/Wind/ Fuel Cell Hybrid Generation System," *International Journal of Engineering Science and Tech (IJEST)*, Volume 4, Issue 5, May 2012.
[3] M. Harini, R. Ramaprabha and B. L. Mathur, "Power-Management Strategies for a Grid- Connected PV-FC Hybrid Systems", *ARPN Journal of Engineering and Applied Sciences*, Volume 7, Issue 4, pp. 1157–1161, September 2012.
[4] B. Vishnu Priya, G. Vijay Kumar, "Power-Management Strategies For A Grid-Connected PV-FC Hybrid Systems," *International Journal of Engineering Research & Technology (IJERT)*, Volume 1, Issue 8, pp. 1–7, October 2012.
[5] S. Gomathy, S. Saravanan, Dr. S. Thangavel, "Design and Implementation of Maximum Power Point Tracking (MPPT) Algorithm for a Standalone PV System," *International Journal of Scientific & Engineering Research*, Volume 3, Issue 3, March 2012.
[6] Michał Knapczyk, Krzysztof Pieńkowskif, "Analysis of Pulse Width Modulation Techniques for AC/DC Line-side Converters," *Prace Naukowe Instytutu Maszyn, Napędów i Pomiarów Elektrycznych, Studia I materialy*, Nr 26, 2009.
[7] Saeid Esmaeili, Mehdi S. Hafiee, "Simulation of the Dynamic Response of Small Wind – Photovoltaic-Fuelcell Hybrid Energy System," *Smart Grid and Renewable Energy*, 2012 (www.SciRP.Org/journel/sgree)z.
[8] Fuel Cell Basics, "Fuel Cell Today," *The Leading Authority on Fuelcell*, May 2012.

[9] N. Prema Kumar, K. Nirmala, K.M. Rosalina, "Modeling Design of Solid Oxide Fuelcell Power System for Distributed Generation Applications", *IJARCET*, November 2012

[10] Suman Nath, Somnath Rana, "The Modeling and Simulation of Wind Energy Based Power System using MATLAB," *International Journal of Power System Operation and Energy Management*, Volume 1, Issue 2, 2011.

[11] Mohamayee Mohapatra, B. Chitti Babu, "Fixed and Sinusoidal-Band Hysteresis Current Controller for PWM Voltage Source Inverter with LC Filter", *IEEE Student Technology Symposium (TechSym)*, pp. 1–11, April 2010.

[12] Rosana Melendez, Dr. Ali Zilouchian, Dr. H. Abtahi, "Power Management System Applied to Solar/Fuel Cell Hybrid Energy Systems", *8th Latin American and Caribbean Conference for Engineering and Technology*, June 1–4, 2010.

[13] Nitin Adhav, Pg Student Shilpa Agarwal, "Comparison and Implementation of Different PWM Schemes of Inverter in Wind Turbine", *IJITEE*, Volume 2, Issue 2, pp. 2522–2531, January 2013.

12 Photovoltaic Transformerless Inverter Topologies for Grid-Integrated High-Efficiency Applications

Ahmad Syed, S. Tara Kalyani, Xiaoqiang Guo, and Freddy Tan Kheng Suan

CONTENTS

12.1 Introduction 151
12.2 Study of Common-Mode Current..153
12.3 Study of Junction Capacitances of Power Devices 153
12.4 Proposed Low-Switching-Count Configurations...154
12.5 Operating Modes of Proposed LSC Topologies 155
12.6 Simulation Results 158
12.7 Output Performance and Common-Mode Behavior..158
12.8 Loss Distributions and Comparisons..160
12.9 Experimental Results...161
12.10 Conclusion 167
12.11 References...167

12.1 INTRODUCTION

Currently, photovoltaic (PV) energy is one of the most preferred resources due its enormous characteristics, such as being free of pollution and having low module price in the present market [1, 2]. Nonetheless, inverters play a massive role in converting direct current (DC) into alternating current (AC) between the PV system and the grid. Based on the operating principles, it can be classified into two types, namely, isolated inverters and nonisolated inverters/transformerless inverters (TLI). Isolated inverters are adopted with transformers, which results in increased size/weight, high cost, and lower efficiency [3]. Therefore, nonisolated inverters are introduced to overcome the aforementioned issues, except isolation problems [4], such as it leads to dangerous ground current through stray capacitors to the ground. As per the DIN VDE 0126-1-1 standards, it needs to limit the ground current below 300 mA [5]. On the other hand, to improve the efficiency of the PV system, many attractive MOSFET solutions have been made in terms of clamping circuits, PWM schemes, and control strategies, but their design is not straightforward, as compared with their IGBT counterpart [6–9]. For the understanding of the readers, here several conventional H-bridge MOSFET inverter (CHB-MI) topologies adopted with unclamping (Un-CL) and clamping (CL) techniques have been reviewed to address the issues related with the ground current in [9–12].

An Un-CL-based H5 inverter [13] has been proposed by SMA Technologies, such as one additional switch (S5) installed at the DC side of the conventional H-bridge MOSFET inverter (CH-BMI), as shown in Figure 12.1(a). Similarly, two additional switches (S5, S6) are installed at the AC side of the CH-BMI, namely, the highly efficient reliable inverter concept (HERIC) [14], as shown in Figure 12.1(b). However, due to lack of clamping circuit, common-mode voltage (CMV) is floating, and hence higher ground current.

Later CL-based topologies have been installed with an additional circuit at midpoint of the DC link capacitor to keep constant CMV during the freewheeling period as shown in Figure 12.1(a). Nonetheless, the CMV is time-varying, which depends on the junction capacitor of the power devices. In [15], an improved HERIC topology (as shown in Figure 12.1(b)) has been proposed by installing two additional switches (S7, S8) at the midpoint of the DC link capacitor, named as OHERIC, as shown in Figure 12.1(c). It can maintain the CMV as constant during the freewheeling periods and hence eliminate ground current. Similarly, another topology in [16] is with rectifier bridge at midpoint of the DC link capacitor, named as H-bridge zero voltage rectifier (HBZVR), as shown in Figure 12.1(d). Here the clamping branch is made with an additional installed diode (D5) to achieve constant CMV in the whole grid cycle.

Unfortunately, HBZVR failed to maintain constant CMV during freewheeling periods, which can be overcome via HBZVR-D topology in [17], such as one additional diode (D6) installed at the midpoint of the DC link capacitor to the existing HBZVR topology, as shown in Figure 12.1(e), but unclear about the CMV performance during dead time.

Recently, in [18] an un-CL hybrid-bridge (HB) topology has been proposed to reduce the ground current, as shown in

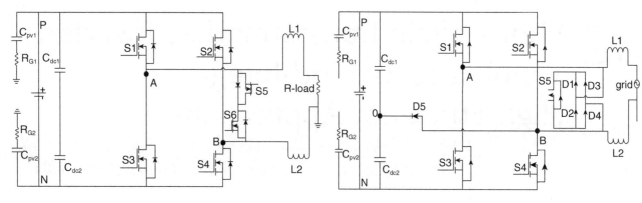

FIGURE 12.1(a) CL-based topology with constant CMV.

FIGURE 12.1(e) HBZVR with DC link capacitor.

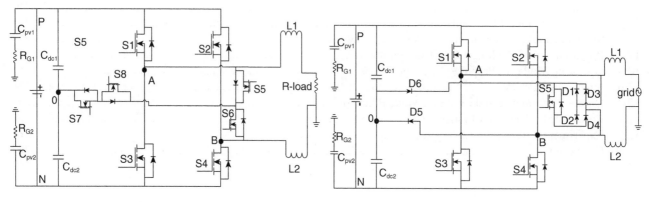

FIGURE 12.1(b) CL-based topology with additional switches (S7, S8).

FIGURE 12.1(f) CL Hybrid bridge topology.

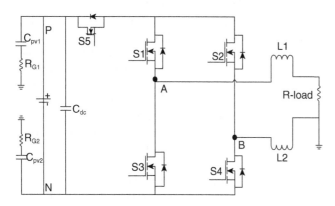

FIGURE 12.1(c) OHERIC topology.

FIGURE 12.1(g) Improved H-bridge topologu.

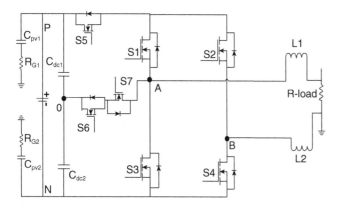

FIGURE 12.1(d) HBZVR topology.

Figure 12.1(f), but lack of clamping branch CMV is floating throughout the cycle. It can be overcome to install two additional switches at the midpoint of the DC link capacitor, named improved H-bridge (I-HB) [19], as shown in Figure 12.1(g). But losses are more due to higher switching count in the freewheeling periods. Based on the previous discussions, it's confirmed that CL-based TLITs are more attractive and effectureive than un-CL topologies in tackling ground current elimination [17–21].

However, major gaps have been found recently based on the switching count in the circuit, which is an active role to increase overall efficiency [13–21]. In order to consider the aforementioned point, here low-switching-count H-bridge MOSFET inverter (LSC-HBMI) topologies and corresponding modulation schemes have been proposed without compromising the overall system performance. The proposed clamping branch is installed with a rectifier bridge at the midpoint of the DC link capacitor to limit the ground current with constant CMV in the whole grid cycle. A detailed mode of operation and theoretical findings of the LSC-HBMI topology is validated via MATLAB/Simulink. And further, it can be tested with a 1 kW prototype in the laboratory. The results confirm that the LSC-HBMI has excellent performance characteristics in terms of CMV, ground current, losses, total harmonic distortion (% THD), and efficiency respectively.

12.2 STUDY OF COMMON-MODE CURRENT

In transformerless inverter topologies (TLITs), the common-mode current is more dependent on the variation of the magnitude, such as charging and discharging of the stray capacitors. Here V_{dc} refers to input DC voltage; C_{PV} is the stray capacitance of the PV module. Positive and negative terminals of the DC source and phase, neutral, are represented as P, N and A, B, respectively. And the importance of the CMC is exposed based on few relations, namely, CMV and differential mode voltage [3–5], as shown next.

The CMV can be computed as the average of two-phase leg voltages (V_{AN}, V_{BN}):

$$V_{cm} = \frac{V_{AN} + V_{BN}}{2} \quad (12.1)$$

$$V_{dm} = V_{AN} - V_{BN} \quad (12.2)$$

And expressions for V_{CM-DM} and total CMV (V_{TCMV}) are shown in (3) and (4) due to asymmetries in the filter inductors.

$$V_{cm-dm} = V_{dm}\frac{L_2 - L_1}{2(L_2 + L_1)} \quad (12.3)$$

$$V_{tcm} = V_{cm} + V_{CM-dm} \quad (12.4)$$

$$V_{CMC} = V_{PV}\frac{dV_{tCMV}}{dt} \quad (12.5)$$

From equation 12.3, if L1 = L2 (i.e., for a well-designed magnetic circuit), the voltage V_{tcmv} = Vcm [12]. From equation 12.5 it is clear that magnitude of C_{PV} and oscillations in V_{tcmv} are controlling the CMC. Finally, the magnitude of C_{PV} is not controllable because it completely depends on the environmental conditions. Hence, to eliminate the CMC is to reduce the oscillations in V_{tcmv}, which depends on the structure and PWM scheme of an inverter.

12.3 STUDY OF JUNCTION CAPACITANCES OF POWER DEVICES

In real time, common-mode current (CMC) analysis due to junction capacitances of power devices is not ignored, such as

in positive half cycle, $V_{AN} = V_{dc}$, $V_{BN} = 0$, and $V_{CM} = 0.5V_{dc}$, and in negative half cycle, $V_{AN} = 0$, $V_{BN} = V_{dc}$, and $V_{CM} = 0.5V_{dc}$, as shown in Figure 12.2(a) to Figure 12.2(c). And hence, CMV is constant in both active modes of operations. But in freewheeling modes, inductor currents continuously freewheel through the grid and power is not transferred from the input DC source, which is shown in Figures 12.2(c)–12.2(d). However, transferring from active mode to freewheeling modes, the power devices S4 and S5 are turned off simultaneously. But in transient period, junction capacitances of power devices S4, S5 are charging and discharging through S2, S3, as shown in Figure 12.2(e).

And one complete transition, only the body diode of S3 is conducted to provide the freewheeling path. Due to charging and discharging, the V_{AN} is increased and V_{BN} decreased accordingly [8]. Figure 12.2(f) shows the simplified equivalent circuit of junction capacitances, and its terminal voltages are evaluated by equation 12.6 via Kirchhoff's current law (KCL). Similarly, steady-state voltages during freewheeling periods are evaluated by equation 12.6. Hence, CMV is not constant in whole grid periods and results in more changes in CMC. Similarly, equation 12.7 and equation 12.8 show the

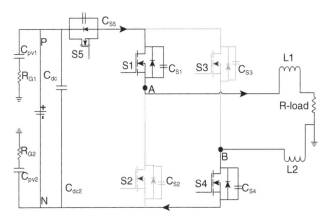

FIGURE 12.2(a)

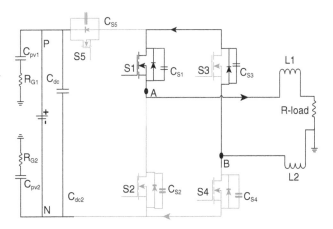

FIGURE 12.2(b)

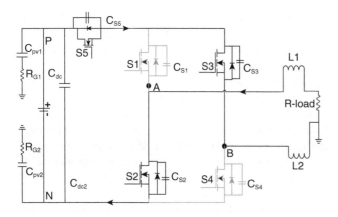

FIGURE 12.2(c)

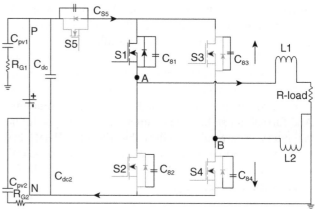

FIGURE 12.2(e)

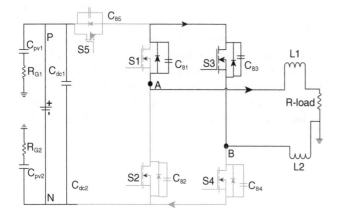

FIGURE 12.2(d)

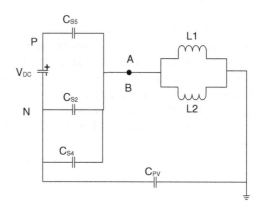

FIGURE 12.2(f)

terminal voltages of the HERIC and H6 TPVI topologies [22].

$$V_{AN} = V_{BN} = \frac{(C_{S2} + C_{S5})}{C_{S2} + C_{S5} + C_{S4}} V_{DC} \quad (12.6)$$

$$V_{AN} = V_{BN} = \frac{(C_{S2} + C_{S3})}{C_{S1} + C_{S2} + C_{S3} + C_{S4}} V_{DC} \quad (12.7)$$

$$V_{AN} = V_{BN} = \frac{(C_{S1} + C_{S2})}{C_{S1} + C_{S2} + \frac{C_{S3} \cdot C_{S5}}{C_{S3} + C_{S5}} + C_{S4}} V_{DC} \quad (12.8)$$

From the preceding discussions, it confirms that during the freewheeling periods, phase leg voltages (V_{AN} and V_{BN}) of H5, HERIC, and H6 are not constant at 0.5 Vdc. Hence, common-mode voltage is oscillating in all the modes of inverter operation, which results in the magnitude of CMC being further increased; here the effect of parasitic capacitances is ignored [17].

12.4 PROPOSED LOW-SWITCHING-COUNT CONFIGURATIONS

In this section, various low-count TLITs and corresponding modulation schemes are proposed based on the existing

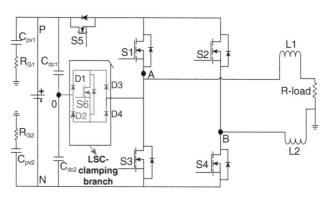

FIGURE 12.3(a)

configurations, which are shown in Figure 12.3. The proposed clamping branch is made with low switching count during the freewheeling period, such as a rectifier bridge which consists of four diodes and one switch at the midpoint of the DC link to keep half-input DC voltage in the whole grid cycle. Nonetheless, the proposed modified switching scheme is able to control reactive power during the freewheeling period.

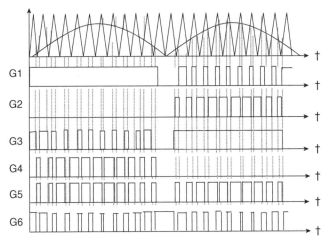

FIGURE 12.3(b)

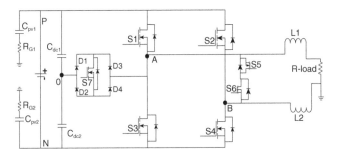

FIGURE 12.3(c)

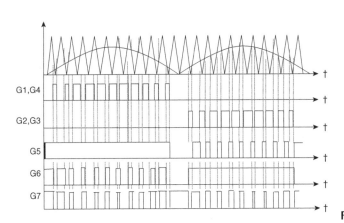

FIGURE 12.3(d)

12.5 OPERATING MODES OF PROPOSED LSC TOPOLOGIES

The PWM waveforms of proposed LSC-TLITs with UPF are illustrated in Figure 12.3(a) to Figure 12.3(c), where all switches, S1–S7, are operating at switching frequency. The operating modes of the LSC-H5, LSC-HERIC, and LSC-H6 TLI topologies are illustrated with three-level output voltages (+V_{dc}, 0, and -V_{dc}), which are presented in Figure 12.4 to Figure 12.6. And the operating modes of both conventional (H5, HERIC, H6) and proposed (LSC-H5, HERIC, H6) in positive half cycle are the same, and CMV, V_{dm}, and switching functions are presented in detail in Table 12.1.

Mode 1. During positive half cycle (PHC) in LSC-H5, S1, S4, S5; in LSC-HERIC, S1, S4; and in LSC-H6, S1, S4, S6 are turned on and remains are turned off. And the current flows via source to load or load source, as shown in Figures 12.4(a), 12.5(a), and 12.6(a). Therefore, CMV and differential mode voltage become:

$$V_{CM} = \frac{V_{AN} + V_{BN}}{2} = \frac{1}{2}(V_{dc} + 0) = \frac{V_{dc}}{2} \quad (12.9)$$

$$V_{AB} = V_{AN} - V_{BN} = V_{dc} - 0 = V_{dc} \quad (12.10)$$

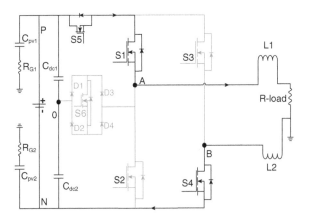

FIGURE 12.4(a)

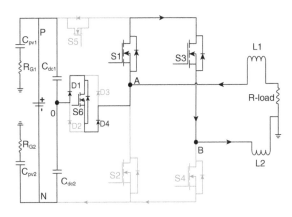

FIGURE 12.4(b)

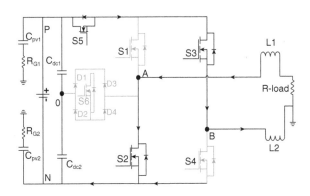

FIGURE 12.4(c)

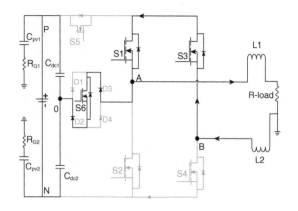

FIGURE 12.4(d)

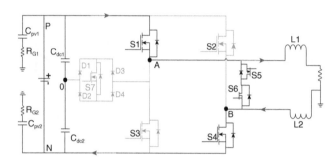

FIGURE 12.5(a)

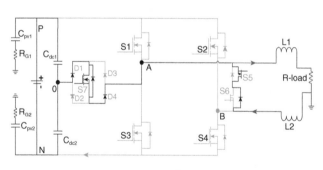

FIGURE 12.5(b)

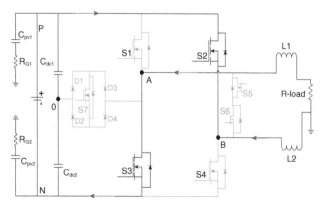

FIGURE 12.5(c)

Mode 2. During the positive freewheeling period, the DC source is completely isolated from the grid. In LSC-H5 S1, body diode of S3; in LSC-HERIC S5, body diode of S6; and in LSC-H6, S4, D5 are turned on and remains are turned off. And the current freewheels via the inductors and the load, as shown in Figures 12.4(b), 12.5(b), and 12.6(b). The voltage V_{AN} decreases and V_{BN} increases until their values reach the common point, $V_{dc}/2$, and thus, V_{cm} and V_{dm} become:

$$V_{CM} = \frac{V_{AN} + V_{BN}}{2} = \frac{1}{2}\left(\frac{V_{dc}}{2} + \frac{V_{dc}}{2}\right) = \frac{V_{dc}}{2} \quad (12.11)$$

$$V_{DM} = \frac{V_{dc}}{2} - \frac{V_{dc}}{2} = 0 \quad (12.12)$$

Mode 3. During positive half cycle (PHC) in LC-H5, S2, S3, S5; in LSC-HERIC, S2, S3; and in LSC-H6, S2, S3, S5 are turned on and remains are turned off. And the current flows via source to load or load source, as shown in Figures 12.4(c), 12.5(c), and 12.6(c). Thus, CMV and differential mode voltage become:

$$V_{CM} = \frac{V_{AN} + V_{BN}}{2} = \frac{1}{2}\left(0 + V_{dc}\right) = \frac{V_{dc}}{2} \quad (12.13)$$

$$V_{DM} = 0 - V_{dc} = -V_{dc} \quad (12.14)$$

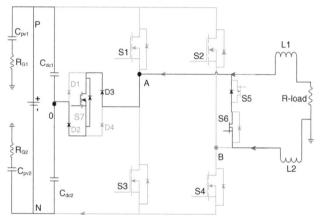

FIGURE 12.5(d)

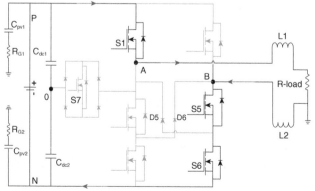

FIGURE 12.6(a)

Photovoltaic Transformerless Inverter Topologies

Mode 4. During the positive freewheeling period, the DC source is completely isolated from the grid. In LSC-H5 S3, body diode of S1; in LSC-HERIC S6, body diode of S5; and in LSC-H6 S3, D6 are turned on and remains are turned off. And the current freewheels via the inductors and the load, as shown in Figures 12.4(d), 12.5(d), and 12.6(d). The voltage V_{AN} increases and V_{BN} decreases until their values reach the common point, $V_{dc}/2$, and hence V_{cm} and V_{dm} become:

$$V_{CM} = \frac{V_{AN} + V_{BN}}{2} = \frac{1}{2}\left(\frac{V_{dc}}{2} + \frac{V_{dc}}{2}\right) = \frac{V_{dc}}{2} \quad (12.15)$$

$$V_{DM} = \frac{V_{dc}}{2} - \frac{V_{dc}}{2} = 0 \quad (12.16)$$

From equations 12.9–12.16, it confirms that CMV is constant in four modes (mode 1–mode 4) because of improved clamping branch during freewheeling periods, and hence, magnitude of common-mode current is eliminated, including reduced size of the common mode filter. Nonetheless, proposed LSC-based TLI configurations for four MOSFETs are operating at lower voltage rating and also simple design clamping branches with low switching count during the freewheeling periods; hence, proposed LSC-based clamping topologies are extremely suitable for high-efficiency PV applications.

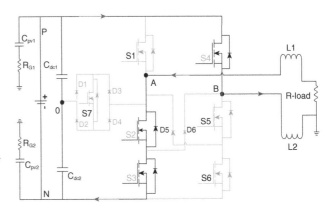

FIGURE 12.6(c)

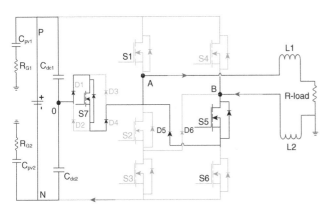

FIGURE 12.6(b)

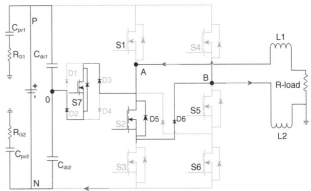

FIGURE 12.6(d)

TABLE 12.1

Switching Modes and CMV Evaluations on Proposed LSC-Based H5, HERIC, and H6 Topologies

Proposed LSC TLITs		S1	DS1	S2	DS2	S3	DS3	S4	DS4	S5	DS5	S6	DS6	S7	DS7	D1	D2	D3	D4	D5	D6	V_{AN}	V_{BN}	V_{cm}
H5	Mode1	1	0	0	0	0	0	1	0	1	0	0	0	0	0	0	0	0	0	0	0	Vdc	0	0.5Vdc
	Mode2	0	0	1	0	1	0	0	0	1	0	0	0	0	0	0	0	0	0	0	0	0	Vdc	0.5Vdc
	Mode3	1	0	0	0		1	0	0	0	0	0	0	1	0	1	0	0	1	0	0	0.5Vdc	0.5Vdc	0.5Vdc
	Mode4	0	1	0	0	1	0	0	0	0	0	0	0	1	0	0	1	1	0	0	0	0.5Vdc	0.5Vdc	0.5Vdc
HERIC	Mode1	1	0	0	0	0	0	1	0	1	0	0	0	0	0	0	0	0	0	0	0	0.5Vdc	0	0.5Vdc
	Mode2	0	0	1	0	1	0	0	0	0	0	1	0	0	0	0	0	0	0	0	0	0	0.5V_{dc}	0.5Vdc
	Mode3	0	0	0	0	0	0	0	0	1	0	1	1	0	1	0	0	1	0	0	0	0.5Vdc	0.5Vdc	0.5Vdc
	Mode4	0	0	0	0	0	0	0	0	1	1	0	1	0	1	1	0	0	0	0	0	0.5Vdc	0.5Vdc	0.5Vdc
H6	Mode1	1	0	0	0	0	0	1	0	0	0	1	0	0	0	0	0	0	0	0	0	0.5Vdc	0	0.5Vdc
	Mode2	0	0	1	0	1	0	0	0	1	0	0	0	0	0	0	0	0	0	0	0	0.5Vdc	0.5Vdc	0.5Vdc
	Mode3	0	0	0	0	0	0	1	0	0	0	0	1	0	0	0	0	0	0	1	0	0.5Vdc	0.5Vdc	0.5Vdc
	Mode4	0	0	1	0	0	0	0	0	0	0	0	1	0	0	0	0	0	0	0	1	0.5Vdc	0.5Vdc	0.5Vdc

12.6 SIMULATION RESULTS

In order to validate and compare the performance of the proposed clamping topologies with conventional unclamping structures using a simulation result, the following parameters are used: 400 V input DC voltage, stary capacitors (C_{PV1}, C_{PV2}) of 100 nF, ground resistance (R_{G1}, R_{G2}) 11Ω, two symmetrical filter inductors (L1, L2) with value of 3 mH, switching frequency 10 kHz, and 230 V, 50 Hz grid voltage (rms) system [4].

12.7 OUTPUT PERFORMANCE AND COMMON-MODE BEHAVIOR

The output and common-mode characteristics in terms of V_{out}, I_{out}, i_{leak}, and V_{AN}, V_{CM}, V_{BN} of clamping, unclamping topologies are explored in Figure 12.7 and Figure 12.8. It confirms that discussed topologies are three-level (+V_{dc}, 0, -V_{dc}) output voltage and sinusoidal output current, which is shown in Figure 12.7(a) to Figure 12.7(i).

On the other hand, conventional H5, HERIC, HBZVR topologies are unable to control the oscillations during the freewheeling periods, which results in V_{AN}, V_{BN} floating and hence oscillating at ~215V, as shown in Figure 12.8(a) to Figure 12.8(b).

As a result, high common-mode current in the whole grid cycle, which is illustrated in Figure 12.7(a) to Figure 12.7(b).

Therefore, the proposed LC-based clamping branch is for overcoming the CMV issues and clamps at $0.5V_{dc}$ in all operating modes, and hence CMC is completely eliminated, as shown in Figure 12.8. The total harmonic distortions (% THD) of the previously discussed topologies are as follows: with magnitude of 0.88%, 0.88%, 2.41%, 1.84%, 0.90%, 1.41%, 0.92%, 1.39%, 0.90%, respectively, and which are followed as per IEEE1547 standards [23], as shown in Figure 12.9(a) to Figure 12.9(i).

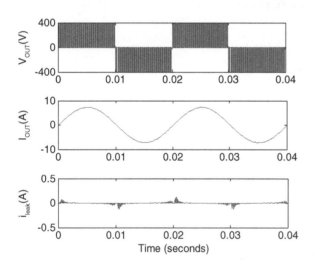

FIGURE 12.7(A)

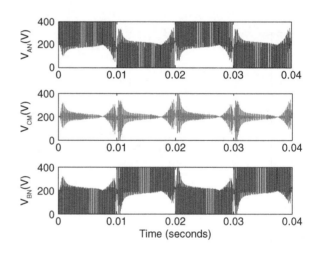

FIGURE 12.8(A)

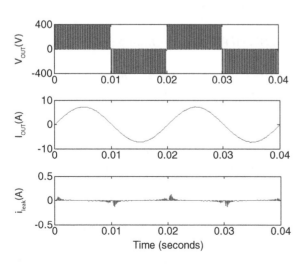

FIGURE 12.7(B)

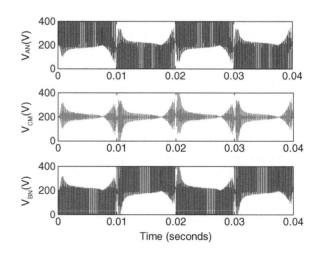

FIGURE 12.8(B)

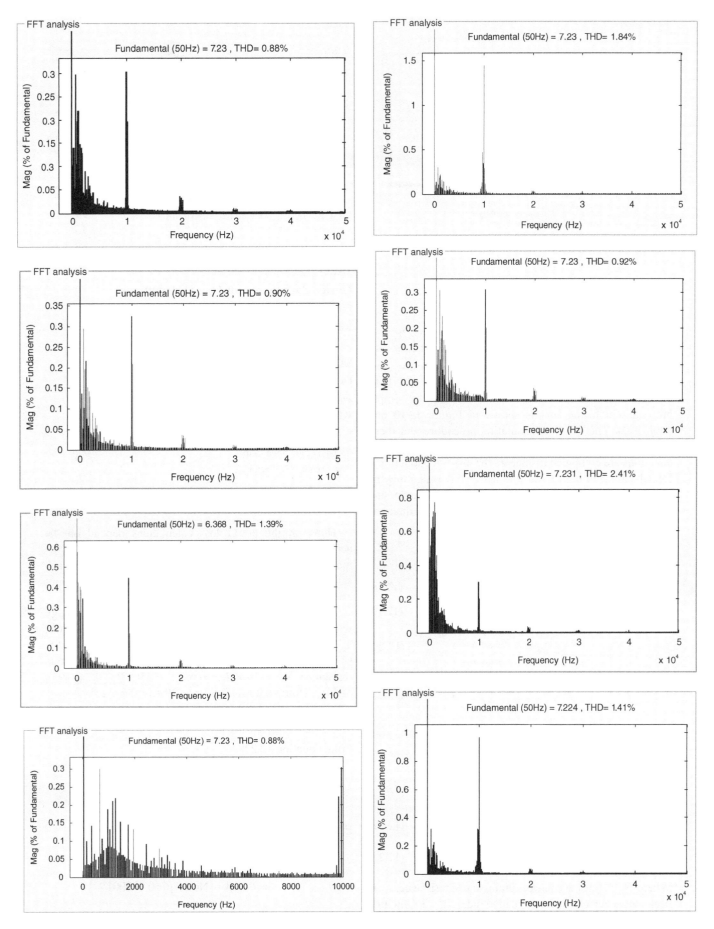

FIGURE 12.9 Total harmonic distortion for (a) H5, (b) HERIC, (c) H6, (d) BDC-H5, (e) BDC-HERIC, (f) BDC-H6, (g) proposed LSC-H5, (h) LSC-HERIC, and (i) LSC-H6 TLTIs.

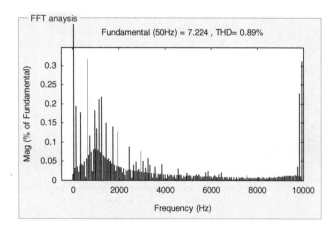

FIGURE 12.9 (Continued)

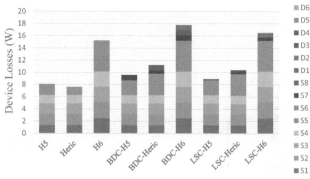

FIGURE 12.10 Device loss distributions.

12.8 LOSS DISTRIBUTIONS AND COMPARISONS

In this section, the loss distributions of various un-CL- and CL-based TLITs rated at 1 kW are analyzed in detail. And losses are measured via thermal modules in PSIM software, and device parameters are listed in Table 12.2 [24]. For the clear understanding of readers, losses are simply presented in histogram, as shown in Figure 12.10 and Table 12.4. The detailed calculation procedure and theoretical studies have been discussed in [14–18]. And few equations are computed to analyze the losses (conduction + freewheeling) in both IGBT and MOSFET, which may be useful to the readers. For IGBTs, conductions, diode, and switching losses are calculated by the following equations (12.17–12.23) [17].

$$P_{conduction} = V_{CE(SAT)} I_C \quad (12.17)$$

$$P_{conduction} = V_F I_F \quad (12.18)$$

$$P_{SW} = P_{SW\_ON} + P_{SW\_OFF} \quad (12.19)$$

$$P_{SW-ON} = \frac{E_{on} f V_{dc}}{V_{dc-datasheet}} \quad (12.20)$$

$$P_{SW-ON} = \frac{E_{off} f V_{dc}}{V_{dc-datasheet}} \quad (12.21)$$

Where I_C defines the on-state current, V_F is forward voltage drop, I_F defines freewheeling current, and $V_{dc-datasheet}$ is actual DC bus voltage.

And MOSFET losses are calculated by [14–16]:

$$P_{Conduction} = \frac{I_m^2 R_{ds} 2M}{3\pi} \quad (12.22)$$

$$P_{switching} = f_{sw} E_{oss} \quad (12.23)$$

Where R_{ds} defines the on-state drain-source resistance, I_m is peak output current, M is modulation index, f_{sw} is switching frequency, and E_{oss} is energy loss from the device data sheet.

TABLE 12.2
Device Parameters for Loss Distributions

Parameters	Value
IGBT (600V)	HGTG20N60A4D
MOSFET (600V)	SPN04N60C2
SiC diode	IDD08SG60C
Frequency	50 Hz
Junction temperature, $T_{j(max)}$	150°C
Pcond_Q calibration factor	1
Psw_Q calibration factor	1
Pcond_D calibration factor	1
Psw_D calibration factor	1

However, the power loss calculations depend on the accuracy of the device data sheet, which is provided by the manufacturer. For the understanding of the readers, a few sample loss calculations of H5, HERIC, and H6 topologies are given next.

The operating conditions of the HERIC topology with MOSFET-SPN04N60C2 is V_{in} = 600V, V_{out} = 540V, T_J = 25°C, 100°C, 150°C, R_{ds} = 0.8Ω, f_{s1} = 10kHz, M = 0.95, COSφ = 0.994, I_m = 0.8–4A, and the drain-source on-state resistance (V_{GS} = 10 V, I_D = 0.65, T_J = 25C, $R_{DS(ON)}$ = 0.95), and E_{oss} from cool MOSFET at 400V is 28μJ [25].

For H5:

From equation 12.22:

$$P_{Conduction-H5} = \frac{2^2 * 0.95 * 2 * 0.95}{3\pi} = 7.55W$$

From equation 12.23:

$$P_{switching} = 10000 * 28\mu j = 10000 * 0.000028$$
$$= 0.28W \quad [\text{Note}: 1\mu j = 0.000001W]$$

$$P_{Total = P_{Condu} + P_{switch}} = 7.55 + 0 + 28 = 7.83W$$

TABLE 12.3
Measured Device Losses for the Unclamping and Clamping TLITs

Topology		Losses (W)													Total losses (W)
		Conduction Losses								External Diode Losses					
	H5	S1	S2	S3	S4	S5	S6	S7	S8	D1	D2	D3	D4	D5	D6	
Unclamping		1.37	1.78	1.78	1.37	1.78	0	0	0	0	0	0	0	0	0	8.08
	HERIC	1.37	1.78	1.78	1.37	0.66	0.64	0	0	0	0	0	0	0	0	7.6
	H6	2.52	2.56	2.56	2.52	2.56	2.51	0	0	0	0	0	0	0	0	12.67
Clamping	BDC-H5	1.37	1.78	1.78	1.37	1.78	0.63	0.89	0	0	0	0	0	0	0	9.6
	BDC-HERIC	1.37	1.78	1.78	1.37	1.78	1.74	0.53	0.87	0	0	0	0	0	0	11.22
	BDC-H6	2.52	2.56	2.56	2.52	2.56	2.51	0.83	0.34	0.22	0.77	0	0	0	0	17.39
	LSC-H5	1.37	1.78	1.78	1.37	1.78	0.63	0	N.C	0.02	0.04	0.05	0.08	0	0	8.89
	LSC-HERIC	1.37	1.78	1.68	1.37	1.78	1.74	0.53	N.C	0.02	0.03	0.04	0.07	0	0	10.51
	LSC-H6	2.52	2.56	2.56	2.52	2.56	2.51	0.53	N.C	0.21	0.28	0.08	0.06	0.07	0.09	16.55

From equation 12.22:

$$P_{Conduction-Heric} = \frac{2^2 * 0.8 * 2 * 0.95}{3\pi} = 6.36W$$

From equation 12.23:

$$P_{switching} = 10000 * 28\mu j = 10000 * 0.000028$$
$$= 0.28W \ [\text{Note: 1}\mu j = 0.000001W]$$

$$P_{Total = P_{Condu} + P_{switch} = 6.36 + 0 + 28 = 6.64W}$$

$$P_{Conduction-Heric} = \frac{2^2 * 0.8 * 2 * 0.95}{3\pi} = 6.36W$$

For H6:
From equation 12.22:

$$P_{Conduction-H5} = \frac{2.5^2 * 0.95 * 2 * 0.95}{3\pi} = 11.80W$$

From equation 12.23:

$$P_{switching} = 10000 * 46\mu j = 10000 * 0.000054$$
$$= 0.54W \ [\text{Note: 1}\mu j = 0.000001W]$$

$$P_{Total = P_{Condu} + P_{switch} = 7.55 + 0.54 = 12.26W}$$

However, the theoretical losses are lower than the simulation results because all values are running at real-time environment and forward voltage of the device diode. But theoretical values are closer to the simulation results.

It is noted that, here, losses are included with the conduction loss, switching loss, and freewheeling loss, respectively. Table 9.6 gives the device operation of the discussed un-CL and CL topologies to justify the performance among the LSC-based TLITs. As expected, unclamping H5, HERIC TLITs have the lowest losses due to reduced switching count (only two MOSFETs) in the active modes (mode 1, mode 3), and only one switch and one diode are active (mode 2, mode 4) during the freewheeling periods. On the other hand, BDC-based TLITs (H5, HERIC, H6 family) have the highest device losses as compared with LSC-based TLITs due to excessive switching count during freewheeling periods. However, the device losses of proposed LSC topologies are higher as compared with unclamping topologies (H5, HERIC, H6), but poor performance in terms of CMV and leakage current, which is discussed in the earlier section.

12.9 EXPERIMENTAL RESULTS

To verify the competency of proposed LC-based TLITs, a universal prototype has been built in the laboratory, as shown in Figure 12.11, and its parameters are listed in Table 9.4.

FIGURE 12.11 The control algorithms are implemented in FPGA SPARTAN-6. As per the theoretical validation, the clamping and unclamping TLITs give the sinusoidal output current (I_{out}), which is presented in Figure 12. And the common-mode performances of both unclamping and clamping topologies are shown in Figure 13. As expected, the unclamping H5, HERIC TLITs have poor CM performance, such as oscillating CMV, and hence higher CMC, with magnitudes of 29.20 mA, 29.10mA, and 29.08mA.

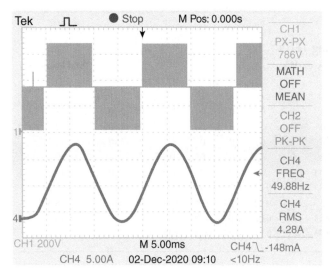
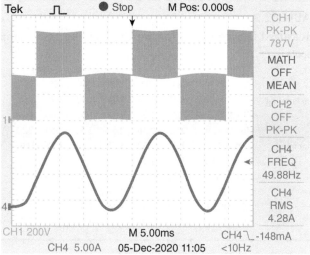
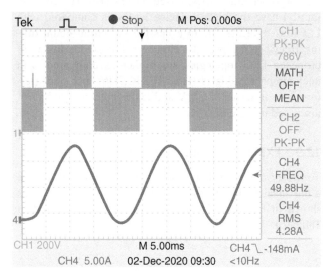
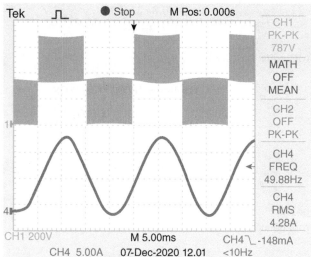
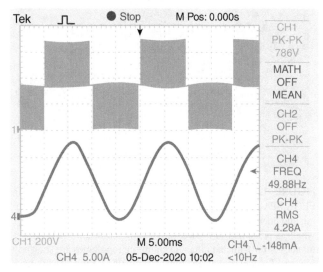
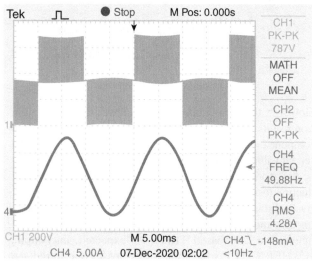

FIGURE 12.12 Clamping and unclamping TLITs sinusoidal output current (I_{out}).

Photovoltaic Transformerless Inverter Topologies

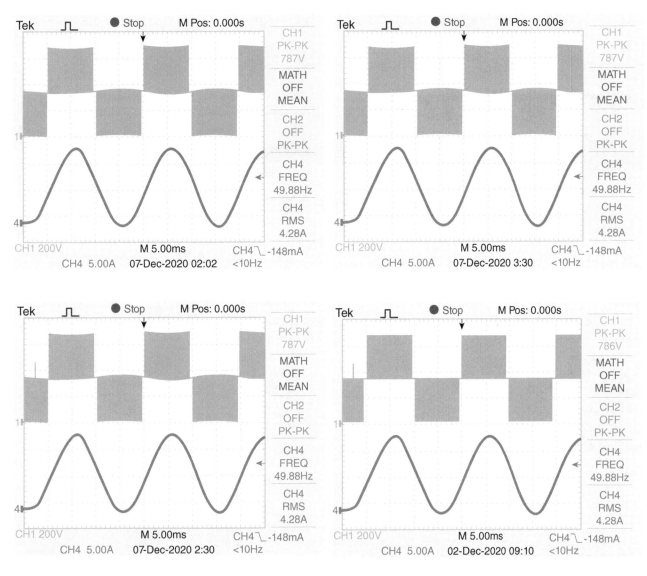

FIGURE 12.12 (Continued)

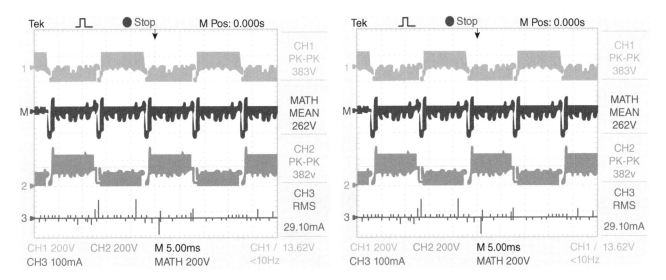

FIGURE 12.13 Experimental waveforms of V_{AN}, $2V_{cm}$, V_{BN}, and GLC (i_{leak}) for (a) H5, (b) HERIC, (c) H6, (d) BDC-H5, (e) BDC-HERIC, (f) BDC-H6, (g) proposed LSC-H5, (h) LSC-HERIC, and (i) LSC-H6 TLTIs.

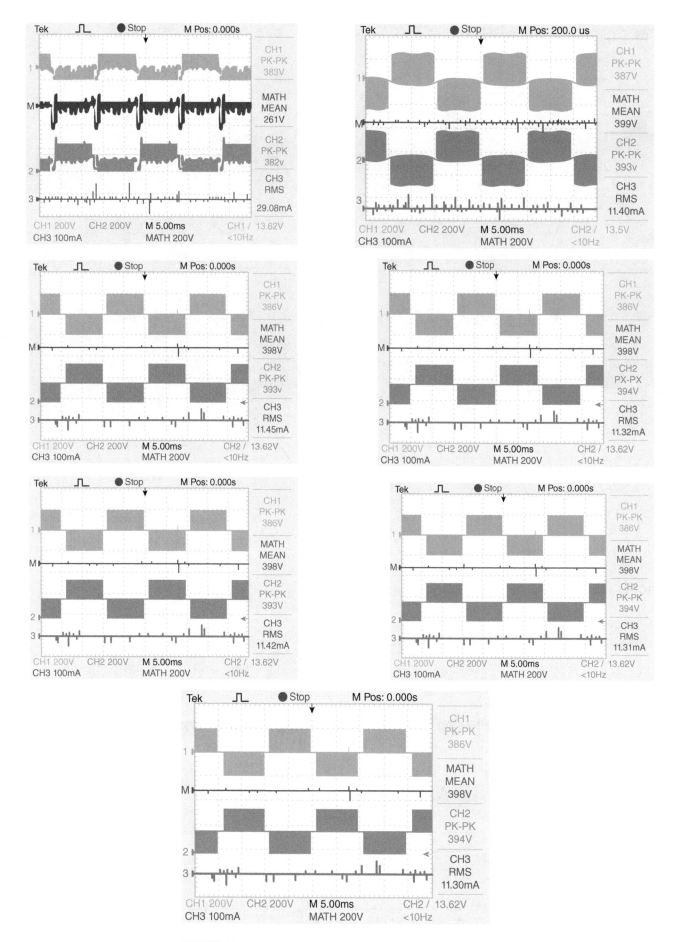

FIGURE 12.13 (Continued)

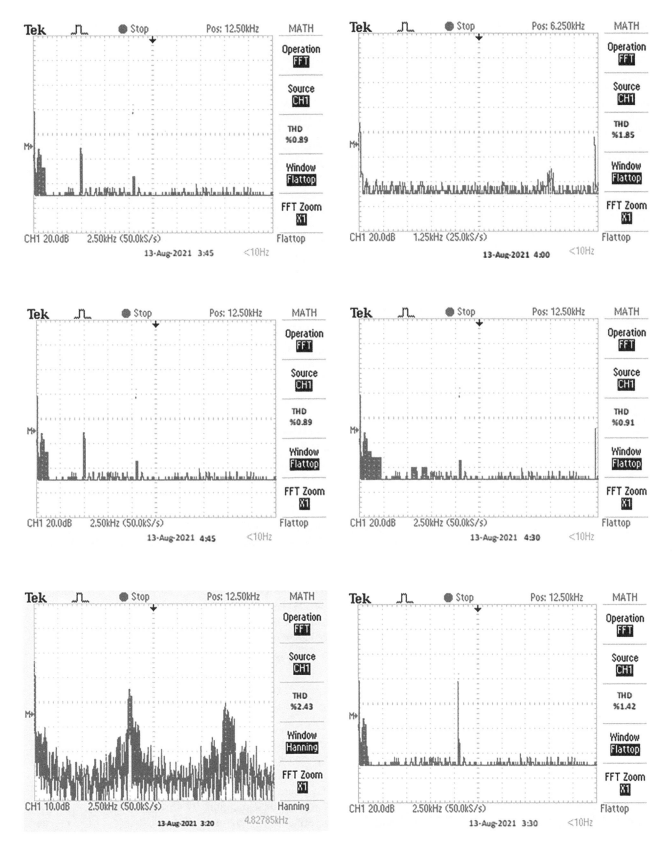

FIGURE 12.14 (a) to (i) % THD of various topologies.

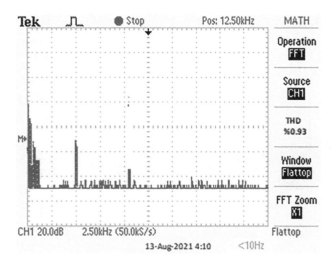

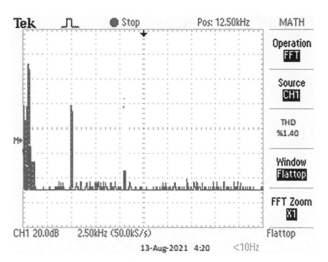

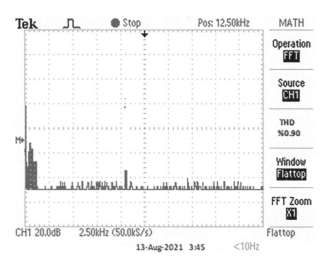

FIGURE 12.14 (Continued)

TABLE 12.4
Universal TLITs Specifications

Parameter		Value
Input voltage		400 V
Grid voltage/frequency		230 V/50 Hz
Rated power		1,000 W
Switching frequency		10 kHz
DC bus capacitor (C_{DC1}, C_{DC2})		1 mF
IGBT	600 V	HGTG20N60A4D
MOSFET	600 V	SPN04N60C2
SiC diodes		IDD08SG60C
Filter inductors (L_1, L_2)		3 mH
Filter inductors (C_f)		4 µf
Parasitic capacitor (C_{PV1}, C_{PV2})		100 nFure
Resistive load		50 Ω
Controller		FPGA SPARTAN-6

This effect is eliminated via BDC- and LC-based TLITs, namely, H5, HERIC, and H6 family, such as constant CMV, in all operating periods. As a result, CMC is reduced significantly, with a magnitude of 11.45 mA, 11.42 mA, 11.40 mA, 11.32 mA, 11.31 mA, and 11.30 mA, respectively. The total harmonic distortion (THD) in terms of output load current is measured by the FLUKE 434 Series II power analyzer [24]. The % THD of various topologies are as follows: with magnitude of 0.89%, 0.89%, 2.43%, 1.85%, 0.91%, 1.42%, 0.93%, 1.40%, 0.90%, which are depicted in Figure 12.14(a) to Figure 12.14(i). And compared to simulation, experimental % THD is relatively higher due to grid affects and it parameters modeling in real-time environment [21].

Table 12.4 lists the experimental results and their comparisons in terms of CMC, CMV, % THD, and efficiency, respectively. It confirms that clamping TLITs (BDC-based and LC-based H5, HERIC, H6 family) have relatively lower % THD than unclamping (H5, HERIC) TLITs due to reduced CMC. The HIOKI 3197 power analyzer is used for efficiency measurements. The European efficiency is computed using the following equation (12.21) [15–20]:

$$\eta_{EU} = 0.03\eta_{5\%} + 0.06\eta_{10\%} + 0.13\eta_{20\%} \\ + 0.10\eta_{30\%} + 0.48\eta_{50\%} + 0.2\eta_{100\%} \quad (12.21)$$

Figure 12.15 shows the efficiency analysis of unclamping- and clamping-based TLITs. And the calculated European efficiency of the unclamping- and clamping-based TLITs are 96.30%, 98.89%, 96.87%, 98.18%, 97.12%, 97.675%, 98.21%, and 97.45%, respectively. However, the efficiency of the proposed LSC-based TLITs registers lower than the conventional TLITs topologies due to more active elements in the grid cycle. Nonetheless, the maximum efficiency of the proposed LSC-based topologies is

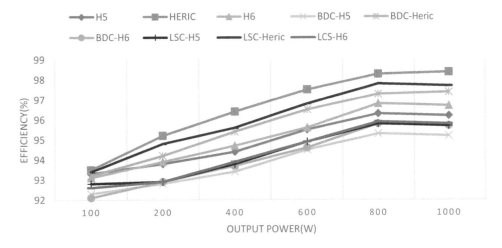

FIGURE 12.15 Efficiency analysis.

TABLE 12.5
Performance Analysis of Unclamping and BDC- and LSC-Based Clamping Topologies

| | Unclamping | | | Clamping | | | | | |
| | | | | BDC-Based | | | LSC-Based | | |
Topology	H5	HERIC	H6	H5	HERIC	H6	H5	HERIC	H6
Output voltage	Unipolar	Unipolar	Unipolar	Unipolar	Unipolar	Unipolar	Unipolar	Unipolar	Unipolar
No. of switching count	5	6	6	7	8	8	6	7	7
No. of diodes	0	0	2	0	0	2	4	4	6
CMV	Floating	Floating	Floating	Constant	Constant	Constant	Constant	Constant	Constant
CMC (mA$_{rms}$)	29.20	29.10	29.08	11.45	11.42	11.40	11.32	11.30	11.31
THD$_i$ (%)	0.88	0.88	2.41	1.84	0.90	1.41	0.92	1.39	0.89
European efficiency (%)	96.82	98.89	96.78	96.23	98.18	96.12	97.67	98.21	97.45

relatively higher than that of the BDC-type topologies because of higher switching count during freewheeling periods. And overall performance comparisons are summarized in Table 12.5. Therefore, it is experimentally confirmed that the LC-based TLITs combine the features of both low-loss AC-clamping method (HERIC) and leakage-current-elimination (BDC-H5, HERIC, H6 family) of the clamping method. However, proposed LSC-TLITs are more attractive for high-efficiency PV applications.

12.10 CONCLUSION

In this chapter, several low-switching-count TLITs have been reviewed first. It is evident that unclamping (HERIC) TLITs have excellent efficiency but poor CM performance. On the other hand, BDC-based clamping TLITs (H5, HERIC, H6 family) have reduced CMC with higher losses. On the other hand, proposed LSC-based TLITs (H5, HERIC, H6) combine the features of low-loss unclamping (HERIC) method and leakage-current-elimination of the clamping (H5, HERIC, H6 family) method. Hence, proposed LSC-based TLITs are well suited for high-efficiency PV applications.

12.11 REFERENCES

[1] F. T. K. Suan, N. A. Rahim, and H. W. Ping, "Modeling, analysis and control of various types of transformerless grid connected PV inverters," in *Proc. IEEE Clean Energy Technol.*, Jun. 2011, pp. 51–56.

[2] M. Islam, S. Mekhilef, and M. Hasan, "Single-phase transformerless topologies for grid-tied photovoltaic system: A review," *Renewable and Sustainable Energy Reviews*, vol. 45, pp. 69–86, 2015.

[3] Wuhua Li, Yunjie Gu, Haoze Luo, Wenfeng Cui, Xiangning He, and Changliang Xia, "Topology review and derivation methodology of single-phase transformerless photovoltaic inverters for ground leakage current suppression," *Industrial Electronics, IEEE Transactions On*, vol. 62, pp. 4537–4551, 2015.

[4] E. Gubia, P. Sanchis, A. Ursúa, J. Lopez, and L. Marroyo, "Ground currents in single-phase transformerless photovoltaic systems," *Prog. Photovolt., Res. Appl.*, vol. 15, no. 7, pp. 629–650, Nov. 2007.

[5] I. Patrao, E. Figueres, F. Gonzalez-Espın, and G. Garcera, "Transformerless topologies for grid-connected single-phase photovoltaic inverters," *Renew. Sustainable Energy Rev.*, vol. 15, pp. 3423–3431, 2011.

[6] Automatic Disconnection Device Between a Generator and the Public Low-Voltage Grid, DIN VDE V 0126-1-1, 2006.

[7] H. Xiao, S. Xi e, Y. Chen, and R. Hu ang, "An optimized transformerless photovoltaic grid-connected inverter," *IEEE Trans Ind Electron.*, vol. 58, no. 5, pp. 1887–1895, May. 2011.

[8] W. Yu, J.-S. Lai, H. Qian, and C. Hutchens, "High-efficiency MOSFET inverter with H6-type configuration for photovoltaic non-isolated AC-module applications," *IEEE Trans. Power Electron.*, vol. 26, no. 4, pp. 1253–1260, Apr. 2011.

[9] B. Ji, J. Wang, and J. Zhao, "High-efficiency single-phase transformer less PV H6 inverter with hybrid modulation method," *IEEE Trans. Ind. Electron.*, vol. 60, no. 5, pp. 2104–2115, May 2013.

[10] L. Zhang, K. Sun, Y. Xing, and M. Xing, "H6 transformerless full-bridge PV grid-tied inverters," *IEEE Trans. Power Electron.*, vol. 29, no. 3, pp. 1229–1238, Mar. 2014.

[11] M. Islam and S. Mekhilef, "An improved transformerless grid connected photovoltaic inverter with reduced ground leakage current," *Energy Convers. Manage.*, vol. 88, pp. 854–862, 2014.

[12] B. Chen, B. Gu, L. Zhang, Z. U. Zahid, J.-S. Lai, Z. Liao, and R. Hao, "A high efficiency MOSFET transformerless inverter for non-isolated micro-inverter applications," *IEEE Trans. Power Electron.*, vol. 30, no. 7, pp. 3610–3622, Jul. 2015.

[13] M. Islam and S. Mekhilef, "Efficient transformerless MOSFET inverter for a grid-tied photovoltaic system," *IEEE Trans. Power Electron.*, vol. 31, no. 9, pp. 6310–6316, Sept. 2016.

[14] M. Victor, K. Greizer, and A. Bremicker, "Method of converting a direct current voltage from a source of direct current voltage, more specifically from a photovoltaic source of direct current voltage, into a alternating current voltage," U.S. Patent 2005 028 6281 A1, Apr. 23, 1998.

[15] D. Schmidt, D. Siedle, and J. Ketterer, "Inverter for transforming a DC voltage into an AC current or an AC voltage," EP Paten 1 369 985, Dec. 10, 2003.

[16] Hua F. Xiao, Xipu Liu, Ke Lan, "Optimised full-bridge transformerless photovoltaic grid-connected inverter with low conduction loss and low leakage current," *IET Power Electronics*, vol. 7, no. 4, pp. 1008–1015, Apr. 2014.

[17] T. Kerekes, R. Teodorescu, P. Rodriguez et al., "A new high-efficiency single-phase transformerless PV inverter topology," *IEEE Transactions on Industrial Electronics*, vol. 58, pp. 184–191, 2011.

[18] H. Li, Y. Zeng, T. Q. Zheng, and B. Zhang, "A novel H5-D topology for transformerless photovoltaic g rid-connected inverter application," *IEEE I nt. Power Electron. Motion Control Conf. and ECCE Asia*, Hefei, May 2016, pp. 731–735.

[19] Li Zhang, Kai, Feng, Hongfei and Yana "A family of neutral point clamped full bridge topologies for transformerless photovoltaic grid tied inverter," *IEEE Tran on Power Electronics*, vol. 28, pp. 730–739, Feb. 2013.

[20] T. K. S. Freddy, N. A. Rahim, W. P. Hew, and H. S. Che, "Comparison and analysis of single-phase transformerless grid- connected PV inverters," *IEEE Trans. Power Electron.*, vol. 29, no. 10, pp. 5358–5369, Oct. 2014.

[21] Z. Ahmad and S. N. Singh," An improved single phase transformerless inverter topology for grid connected PV system with reduce GLC and reactive power capability," *Solar Energy*, 157 pp. 133–146, 2017.

[22] K. Sateesh kumar, A. Kirubakaran, and N. Subrahmanyam, "Bi-directional clamping based H5, HERIC and H6 transformerless inverter topologies with improved modulation technique," *Pros IEEE PESGRE Conference Cochin*, India, pp. 1–6, Jan. 2020.

[23] PSIM User's Guide. (2007) [Online]. www.psim europe. com.

[24] Md Noman H. Khan, Yam P. Siwakoti, Li Li, Frede Blaabjerg, "H-bridge zero-voltage switch controlled rectifier (HB-ZVSCR) transformerless mid-point-clamped inverter for photovoltaic applications," *IEEE Journal of Emerging and selected Topics in Power Electronics*, early to access.

[25] "IEEE Standard conformance test procedures for equipment interconnecting distributed resources with electric power systems," IEEE Std 1547.1–2005, pp. 0_1–54, 2005 and Exposition (ECCE), Sept. 2015, pp. 442–449.

13 Performance Analysis of Rooftop Grid-Connected Solar PV System Under Net Metering System
A Case Study

T. Bramhananda Reddy, G. Sreenivasa Reddy, and Y. V. Siva Reddy

CONTENTS

13.1 Introduction 169
13.2 Energy Demand of India Due to Population 169
13.3 Necessity of Installation of Solar Power Plant by GPREC 170
13.4 Net Metering System 171
13.5 Method of Billing with Net Metering 171
13.6 Main Objective of This Chapter Is Carried Out Following Case Studies at Selected Location 171
13.7 Estimating Payback Period of a 400 kWp Solar PV System at GPREC: For Calculating the Payback Period of Any Solar Photovoltaic System Requires the Following 172
13.8 Key Conclusions of the Case Study 173
13.9 The Important Characteristics Recommended for the Grid-Connected PV Systems 173
13.10 Acknowledgments 174
13.11 References 174

13.1 INTRODUCTION

Electrical energy demand has expanded faster than any other energy generating source because of the consistently rising fuel cost, a dangerous atmospheric condition deviation, and depletion of fossil fuels. Consequently, the portion of electrical energy utilization worldwide has reached a level beyond the expected. Renewable energy (RE) innovative techniques empower us to produce clean energy that has a lower environmental effect than conventional technologies. They differ from fossil fuels principally in their diversity, abundance, and potential for use anywhere on the planet. Increased generation of green energy will, in turn, reduce greenhouse emissions, thereby helping to mitigate global warming. Now, renewable energy sources have contributed a major portion in electricity generation [1–3]. It is predicted by the World Energy Forum that there will be exhaustion of fossil fuel reserves in less than ten decades, which accounts for about 79% of primary energy consumption in the world. This has resulted in forcing planners and policy makers to look for alternate energy sources [4].

13.2 ENERGY DEMAND OF INDIA DUE TO POPULATION

India is not only the country with the second-largest population but is also a very fast-developing country in the present world. It is assessed that the population of India will overwhelm China by the year 2025. India is able to meet its energy demand as per the increase in population, but it still has less per capita power utilization among developed and developing nations.

Energy from thermal stations provides about 61% of the capacity of total installation. India's commitment towards renewable energy generation, such as the Jawaharlal Nehru National Solar Mission (JNSS) and Faster Adoption and Manufacturing of Hybrid and & Electric Vehicles (FAME) scheme, has resulted in the decrease of involvement of thermal energy, burden of India's foreign reserves spending on fossil fuels, etc. In addition to this, the creation of sustainable jobs for a growing youth population is also a result. Geographically, India is a blessed for the utilization of renewable energy as it has almost 300 sunny days per year [5].

The sun is a significant source of unlimited free energy (i.e., solar energy) for the planet Earth. At present, new innovations are being utilized to generate power from solar energy. These methodologies have just been proven and are broadly drilled all throughout the world as sustainable options in contrast to ordinary nonhydro technologies. The improvement of novel solar power technologies is viewed as one of the solutions towards narrowing the demand for energy [6].

Due to huge technological improvements, affordable price, and good performance, the rooftop building models' development has been increased. Thereby, the installed capacity of the solar PV system in recent years has rapidly increased as well. In addition to this, there is a requirement for wide expansion in the solar energy contribution at socially acceptable cost, resulting in cost-competitiveness with fossil fuels.

"Solar energy may be called upon to play a much larger role in the global energy system by midcentury." About two-thirds of CO_2 emissions from fossil fuels are associated with electricity generation, heating, and transportation, and to reduce these emissions, new kinds of energy sources, such as wind and solar PV energy, solar thermal, and battery-operated vehicles are gaining more attention. Of these sources, solar PV–based energy generation is being given more importance by researchers [7]. Solar energy generation is the more economical and eco-friendly among all other renewable energy generating systems. With the advancement in power electronics conversion devices and controllers, the cost of solar energy generation is getting cheaper.

The two types of solar energy systems are photovoltaic solar technology, which directly converts sunlight into electricity using solar panels made of semiconductor cells, and, second, solar thermal technology, which captures the sun's heat. Solar PV system can be operated in three modes of operation based on the requirement (i.e., grid-tied, off-grid, and grid with battery backup, or hybrid). After simulation study, Karthik et.al proved that crystalline solar cells give better performance under the Indian weather conditions [8]. Whenever both grid supply fails at unfavorable weather conditions, GPREC load runs with diesel generators. The motivation of using PV in place of diesel generators and battery comes from three aspects: they are a mutual relationship between environment, economics, and technical system issues. These are strongly interrelated to power network performance [9].

13.3 NECESSITY OF INSTALLATION OF SOLAR POWER PLANT BY GPREC

Geographically, GPREC is located at 15°49'59.88" N 78°02'60.00" E, which is shown in Figure 13.1.

GPREC is connected with the local power distribution company (APSPDCL), having 320 kVA contract maximum demand (CMD), HT connection. Though it is connected with a 11 kV feeder with 500 kVA distributed transformer, it is facing continuous power outages, causing disturbance to academics, because of its locality under rural area. During power outages, the institution runs two diesel generators (DG) rated at 250 kVA and 400 kVA, respectively, based on the load demand. Since the use of a DG set causes environmental pollution as well as are expensive in operation, the management has opted for solar PV generation to minimize power interruptions as well as the operating cost of the DG. Apart from this, it provides an attractive incentive from the government of India and is also an affordable investment that, as well, has negligible operating cost – these are the reasons for stepping into the solar generating system.

Solar power plant installation has been done in three stages: 100 kWp in the year 2014, 100 kWp in 2015, 200 kWp in 2016. All these power plants are established at various rooftops of the department blocks. With an effective coordination between the 400 kWp solar plant and grid supply, the institution load is being operated satisfactorily.

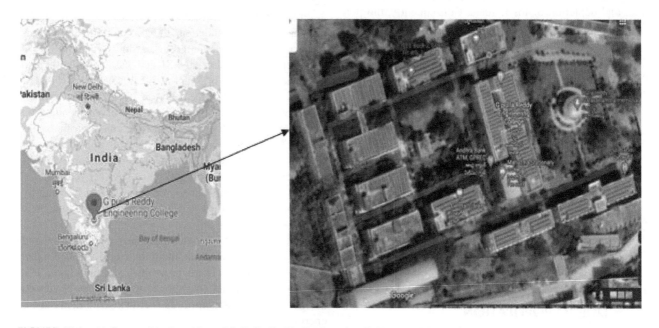

FIGURE 13.1 (a) Geographical position of G. Pulla Reddy Engineering College, and (b) rooftop solar PV plant.

Source: Google Maps

13.4 NET METERING SYSTEM

An off-grid system is generally a stand-alone system. The independent functioning of generation and load at this system is technically not a feasible method as the load may increase or generation may decrease. Whereas an on-grid system is connected to the main utility grid and exchange of power can be done for consumer satisfaction. This on-grid system incorporates the policy of net metering.

The mechanism in which the surplus energy is fed back to the grid by the domestic, commercial, and any other user who generates electricity using solar PV is known as net metering system. This system gives opportunity to the owners to gain extra revenue by selling their excess power to the grid and also making up for shortfalls with the utility grid. The owner gets compensation for the excess amount of energy generated. However, energy will be imported for the grid if the amount of energy consumed is more than the amount of energy generated, and the owner pays only the net amount. In India, net metering was introduced as an initiative to make renewable energy more economical and accessible.

13.5 METHOD OF BILLING WITH NET METERING

In net metering, billing is only for the "net" energy that is used, as mentioned in the preceding text. Hence, surplus energy earns revenue, while the shortage of energy is covered by the grid. In India, every state has their own regulations for the type of net meter used to record the power. Few of them require only one meter to measure the net amount of power consumed by the owner, while the others require two meters to record the total amount of units consumed from the grid along with the total amount of power generated. As per the terms and conditions of the Andhra Pradesh State electricity power distribution company (Southern Power Distribution Company Limited of Andhra Pradesh), GPREC has set up a complete solar PV system with net metering system. This system is interconnected with the grid as per the grid standards. The rewards of a net metering system are an extra monetary credit, elimination of the need for battery storage, less expense and low maintenance, and ease in extending the generating plant capacity. The only disadvantage in a net metering policy is, a battery storage system is not allowed.

13.6 MAIN OBJECTIVE OF THIS CHAPTER IS CARRIED OUT FOLLOWING CASE STUDIES AT SELECTED LOCATION

GPREC

Case 1: During normal running conditions of the grid and institute, majority of the power requirements are met by the PV system and less power will be drawn from the grid.

(i) At normal operating condition, considering fair weather conditions, the complete institution load will run smoothly with the generation of own solar power, and only excessive load, if any, will be met by importing power from the utility grid. In case the institutional load is less than the solar power generation, then extra power will be exported to the grid.

(ii) Similarly, at unfair weather conditions, power generation from the solar PV will not be up to the mark (generation capacity of plant is not equal to the installed capacity due to the variation of temperature and/or irradiation). Maximum possible generation from the solar plant along with grid supply will run the load, avoiding academic disturbances.

(iii) At very poor (unfavorable) climatic conditions, the total load will be run by importing power from the utility grid only. Here, net metering will play a crucial role for selecting the priority of power drawn and running the load with the solar PV plant.

Case 2: During normal running condition of the grid and light load conditions of the institute (usually on a holiday), majority power will be exported to the grid:

In this case, with fair weather conditions, the solar power will generate equal to the installed capacity, but due to light load or no load conditions (may be in general holidays like Sunday, festival, midterms, and end of the academic year), the institution requires less power consumption. Hence, the excess power generation will be sent to the grid and no power is imported from the grid. This is very useful for owners without losing their power generation.

Case 3: Abnormal conditions.

(a) **Connection to the grid is lost and institute is running at normal load condition with a diesel generator (DG) set.**

One of the disadvantages of the net metering system is failure of inverters when the grid supply fails. Even at good weather condition, complete solar PV generation will come to a standstill because of the shutdown of inverters.

Under normal conditions, the institute draws power from the grid in order to meet the load demand, and this is in addition to the generated solar power. When the grid is not operating, the system has to operate in an island mode, wherein the demand is met locally instead of by the grid-connected mode. In order to meet the existing demand, a generator is operated in parallel to solar power generation. At present, only 100 kWp is available in this mode. The DG solar combination will operate satisfactorily as long as generation meets the demand.

(b) Connection to the grid is lost and institute is running at light load condition with a DG set.

If connection to the grid is lost, the phase II plant (100 kWp) will generate the power under favorable climatic conditions. The system will operate satisfactorily until the load is nearer to the one generated from solar. At light load conditions, there is a flow of current from inverter to the DG (reversal of current in the armature of the alternator), leading to large fluctuations in the operation of the DG set. This issue is not present in case of moderate or heavy loads, since both the generating sources will share the load.

(c) Connection to the grid is lost and institute is running at normal condition under unfavorable weather conditions with a DG set.

Because of the failure of PV inverter due to the loss of grid, the total load demand (light load, medium load, or heavy load) will be met by the DG set. Due to the unfavorable weather condition, only a small amount of power is generated, and based on the load demand, the DGs will operate. If the load demand is small, only 250 kVA DG will be committed, and at full load conditions, 400 kVA generator will be committed.

If the load is high, then both the generators will be operated in parallel with the separate autosynchronization device. The problem of load following will be difficult during unfavorable weather conditions, leading to fluctuations in the generator due to cloud movement when the load is very small.

(d) Connection to the grid is lost and institute is running at light load condition under unfavorable weather conditions with a DG set.

When connection to the grid is lost, the phase I and phase III solar PV plants will not be able to generate power due to shutdown of PV inverters (as per the net metering policy). Phase II plant will provide the entire load, and the generator (to provide a reference voltage) floats on bus bars, resulting in less efficiency. This plant has to shut down when the irradiation falls below a level required for generation. At this situation, load will be run with the 250 kVA DG.

Case 4: The economic aspects (payback period analysis).

One of the important reasons to go for solar PV is the continuous run of load from power outages and significant reduction of electricity bill. An average monthly electric power consumption bill of GPREC is around INR 800,000 before the setting up of solar PV plant. After the installation of solar PV system in three stages, the average monthly electrical bill came down to about INR 400,000 only.

The electricity bill gets reduced when solar PV–based energy generation is opted, and electricity costs are no longer a significant factor in budget. Savings with solar depends on different factors, like how much energy the PV system will produce and how much energy is consumed by own, but the biggest factor is the capital cost and available weather conditions (like irradiation and temperature). Because the saving of power will reflect more in electricity bill, demand has been increasing continuously, since electricity prices will continue to rise, so savings will continue to grow every year over the 25-year life span of a solar PV system. The benefits of solar are not just economic, because of the ever-rising electricity prices and power interruptions. Thus, installation of solar PV system is essential. In this analysis, the total capital cost is cumulatively added and made a single value (since GPREC has installed a 400 kWp solar power plant in three phases).

13.7 ESTIMATING PAYBACK PERIOD OF A 400 KWP SOLAR PV SYSTEM AT GPREC: FOR CALCULATING THE PAYBACK PERIOD OF ANY SOLAR PHOTOVOLTAIC SYSTEM REQUIRES THE FOLLOWING

The installed capacity of the PV, the daily/monthly/yearly generation in kWh (PV system can generate power based on the availability of solar radiation, temperature), and the installation of a grid-tied system that exports its surplus production to the electric grid, cost per unit, and capital cost of the plant.

Note: For the calculations, the level of availability of irradiation (G) and temperature (T) are considered as very important factors. But practically, it is not possible to get constant irradiation and temperature as it varies from time to time and place to place, so it is not possible to predict and find accurate calculation of solar PV power generation.

Solar systems work best when the sun is shining. The strength of the sun's radiation affects the amount of solar power generated by the solar PV plant. In India, usually solar generation is high during 11:00 a.m. to 4:00 p.m., and this country is endowed with abundant solar energy and is also blessed with around 300 sunny days in a year and solar irradiation of 4–7 kWh per sq. m per day. As per the IRDEA, the effective generation of a solar PV panel is 4.5 hours.

Total installed capacity of the PV system = **400 kWp** (100 kWp + 100 kWp + 200 kWp)

Total cost of the SPV plant = **Rs 1,95,88,099.00 (Rupees one crore ninety-five lakh eighty-eight thousand and ninety-nine only)**

Daily generation of 400 kWp plant = **Installed capacity of the plant × 4.5 hrs**

= 400 kW × 4.5
= 1,800 kWh = 1,800 units
The cost per unit when it is exported to grid = **Rs 6.50 paise**
Cost of daily generated unit exported to grid = **1,800 × 6.5 = 11,700.00**
Generation cost in a year = **11,700 × 365 = 42,70,500.00**
Payback period = **Capital cost/revenue generated per year**
= 1,95,88,099/42,70,500
= 4.59 years
= 4 years and 7 months

Therefore, the payback is obtained after **four years and seven months**. This payback is calculated without considering the operation and maintenance (O&M) costs and the degradation. The technical persons will rectify the issues, if any, in a solar plant as the study location is a technological institute and PV modules require only cleaning, so O&M cost is negligible. Degradation, which is expressed as a percentage reduction per year, of PV modules occur because of their exposition to sunlight. The net effect is reduction in the energy generated by the PV modules. Typically, it is assumed to be 0.5% per year. Hence, the expected payback may be increased by three more months. Therefore, GPREC will get total investment cost of the PV plant back after five years. After this time period, the institution will earn profits.

Case 5: Analysis of actual power exported to the grid per month.

In a period of one year, the excess power generation sent to the grid is tabulated in Table 13.1. By observing Table 13.1, GPREC solar plant is sending electrical power to the utility grid after meeting the load demand.

Because of net metering system installation in the institute, excess power generation will be sent to the grid on holidays and at light load conditions.

13.8 KEY CONCLUSIONS OF THE CASE STUDY

- Most of the problems studied which are mentioned previously are not yet that much serious at present, and also in the future, if the systems are dealt properly with planning, design of controllers with existing technologies.
- Of the problems selected in this case study, when dealing with failure of utility grid, supply to run load satisfactorily is by enhancing the existing PV generation.
- Furthermore, advancements in power conditioning units are required to minimize the impact of harmonics due to DC injection from the transformerless PV inverters.
- The possibility of unconstrained islanding issues is less, though the risk involved due to this is more; hence, there is a need to protect the system.
- Many constraints, including overvoltage, can be eliminated while upgrading the system ratings, proper designing of distribution capacities, and grid configurations may be made to meet future capacity growth.

13.9 THE IMPORTANT CHARACTERISTICS RECOMMENDED FOR THE GRID-CONNECTED PV SYSTEMS

- Enhancing of solar PV plant generation capacity
- Online power system management using ICT (information and communications technology)
- Expansion of distribution capacity

TABLE 13.1
Energy Export Units to Grid

S. No.	Month and Year	Present Reading	Previous Reading	Difference (or) Exporting Units to Grid
1	Jan. 2020	88,791	79,601	9,190
2	Feb. 2020	105,043	88,791	16,252
3	Mar. 2020	117,247	105,043	12,204
4	Apr. 2020	138,029	117,247	20,782
5	May 2020	173,665	138,029	35,636
6	Jun. 2020	206,592	173,665	32,927
7	Jul. 2020	231,031	206,592	24,439
8	Aug. 2020	258,120	231,031	27,089
9	Sep. 2020	279,367	258,120	21,247
10	Oct. 2020	291,243	279,367	11,876
11	Nov. 2020	315,495	291,243	24,252
12	Dec. 2020	331,646	315,495	16,151

- Encouraging extra tariff to prosumers who exports to the grid during peak demand
- Improvement and widespread use of storage technology in net metering system or integration of either grid load management or building load control with PV generation output
- Proper selection of power converters for maintaining the quality of power

13.10 ACKNOWLEDGMENTS

The authors would like to acknowledge the Management and Principal of G. Pulla Reddy Engineering College (Autonomous), Kurnool, Andhra Pradesh, India for granting permission to carry out the research study on 400 kWp solar PV plant in the premises of the college.

13.11 REFERENCES

[1] www.mnre.gov.in/
[2] "The role of renewable energy in the global energy transformation", Dolf Gielen, Francisco Boshell, Deger Saygin, Morgan D. Bazilian, Nicholas Wagner, Ricardo Gorini, *Energy Strategy Reviews*, Elsevier 24, pp:38–50, 2019
[3] "Renewable energy for sustainable development in India: Current status, future prospects, challenges, employment, and investment opportunities", Charles Rajesh Kumar J., M. A. Majid, *Energy, Sustainability and Society*, pp:2–36, 2020.
[4] "Renewable energy in India: Current status and future potentials", Ashwani Kumar, Kapil Kumar, Naresh Kaushik, Satyawati Sharma, Saroj Mishra, *Renewable and Sustainable Energy Reviews*, vol:14. Issue:8, October 2010, Elsevier, Science Direct.
[5] "Role of renewable energy in Indian economy", Aaditya Ranjan Srivastava, Manavvar Khan, Faraz Yusuf Khan, Shrish Bajpai, *IOP Conf. Series: Materials Science and Engineering*, vol:404, p. 012046, 2018. doi:10.1088/1757-899X/404/1/012046.
[6] "Solar energy: Potential and future prospects", Ehsanul Kabira, Pawan Kumarb, Sandeep Kumarc, Adedeji A. Adelodund, Ki-Hyun Kime, *Renewable and Sustainable Energy Reviews*, pp: 894–900, 82, Elsevier, 2018.
[7] "Feasibility study on the provision of solar energy in rural area using solar panel", Emeka C. Okafor, *Journal of Pure and Applied Sciences*, pp. 1–19, March 2008.
[8] "Performance of rooftop solar PV system with crystalline solar cells", Karthik Atluri, Sunny M. Hananya, Bhogula Navothna, IEEE National Power Engineering Conference (NPEC), 2018.
[9] "Comprehensive overview of optimizing PV-DG allocation in power system and solar energy resource potential assessments", Raimon O. Bawazir, Numan S. Cetin, pp:173–208, https://doi.org/10.1016/j.egyr.2019.12.0102352-4847/© 2019 Energy Reports – 6, Elsevier Ltd, 2020.

14 Isolated Bidirectional Dual Active Bridge (DAB) Converter for Photovoltaic System
An Overview

Nishit Tiwary, Venkataramana Naik N, Anup Kumar Panda, Rajesh Kumar Lenka, and Ankireddy Narendra

CONTENTS

14.1 Introduction 175
    14.1.1 Power Electronics Converter in PV System 175
14.2 Topologies and Control Strategies of DAB 176
    14.2.1 Voltage-Fed DAB Converter 176
    14.2.2 Current-Fed DAB Converter 177
    14.2.3 Control Strategies of DAB Converter 179
14.3 DAB Converter: Analysis and Design for PV Application 181
    14.3.1 Analysis of DAB Converter Circuit 181
    14.3.2 Equivalent Inductor (L_k) Design 183
    14.3.3 Input Capacitor Design 184
    14.3.4 PV System Design 184
    14.3.5 High-Frequency Transformer Design 185
14.4 Conclusion 185
14.5 References 185

14.1 INTRODUCTION

The advantages of renewable energy sources over fossil fuels are familiar to the world now. The major issues of global warming and ocean acidification are now recognized and given attention across the globe, which originates with the use of conventional energy resources. Renewable energy sources are widely explored to replace fossil fuels and fulfill the energy demand. Solar photovoltaic (PV) is one of the most sustainable, clean, and promising energy sources and is being utilized with large-, medium-, and small-scale generation. The extraction of energy from PV requires the use of power electronics converters, and power generation is dependent on these converters. PV installations can vary in size and capacity as it can be of few kW in individual houses to MW range in large solar PV farms. The efficient handling of power and its extraction plays a major role in any PV installations.

14.1.1 POWER ELECTRONICS CONVERTER IN PV SYSTEM

The power generation from a PV array depends on irradiation and temperature. Therefore, for the extraction of maximum power, a power electronics converter plays an essential role. These converters also improve the power quality and act as an interface between the PV panel and the load.

The output voltage and current are regulated with a power electronics converter to achieve the maximum power point (MPP); hence, efficiency and control are important aspects of these converters. As the voltage and current from a single PV panel can be lower than the load requirement, multiple panels are connected in series and parallel to meet the voltage requirement. Subsequently, the power electronics converter also provides the voltage gain suitable to feed the load. To summarize, the following are the functions of various power electronics converter in the PV system:

- Provide voltage gain
- Extract maximum power from PV panel (MPP)
- Voltage regulation
- Improve efficiency
- Power conversion from DC to AC for grid integration

Various converter topologies have been proposed and utilized for the PV application, which incorporates the aforementioned functionalities [1–4]. High efficiency and high gain are major functions of the interfacing converters, irrespective of the installation's size and power rating. The boost converters are the most common and widely used converter for PV applications. The boost converter's efficiency is a major drawback when used for high voltage conversion [5–7]. Moreover, it provides gain limited to two to three times for stable operations. Other converters proposed in the literature for PV application include three-phase EDR boost converter [8], dual boost converter [9], ćuk-derived transformerless converter [10], and coupled inductor-based converter [11]. The limitations reported for these converters are the use of several inductor and capacitor, less power density, less efficiency, and less voltage gain. The disadvantages of the converters mentioned here are somewhat removed using a galvanically isolated converter [12]. These converters provide high voltage gain, high power density, although the voltage and current stress increases with the increase in gain. The soft switching capability becomes a major challenge to reduce voltage and current stress across switches, as hard switching can decrease efficiency and dilute these converters' advantages.

With the emergence of dual active bridge converter (DAB) [13–19], the converter has proved to be one of the most promising, with various advantages, namely, high efficiency, galvanic isolation, soft switching, high power density, bidirectional power flow, and smaller size and weight. The isolation provided by the DAB converter is the essential feature for the safety of photovoltaic (PV) installation, because the source and load are decoupled. Moreover, the transformer turns ratio is utilized for high voltage conversion. The high power density is achievable with a high-frequency transformer, making the converter suitable for high-power PV systems. The soft switching capability ensures the reduction in voltage and current stress on the switching devices, thereby increasing the power conversion efficiency. Additionally, the bidirectional power flow control is easier with the DAB converter due to the symmetric structure. Thus, it is preferable to interface the renewable energy sources with energy storage components such as battery bank, ultracapacitors, etc. [20–23]. Over the past few years, significant research has been carried out to address the technical challenges associated with the DAB converter's modulations and controls for PV systems.

14.2 TOPOLOGIES AND CONTROL STRATEGIES OF DAB

The DAB converter topologies are broadly classified into two categories: voltage-fed DAB converter and current-fed DAB converter. The classification is based on the input source for the converter. In voltage-fed converters, the input and output terminals contain voltage sources and the source is connected directly to the semiconductor switches. However, the source can be regarded as a current source if the inductors are placed in between the voltage source and switch nodes of the converter. The voltage-fed DAB uses input capacitors for energy storage and provides a constant voltage to the DAB input terminal. Whereas in a current-fed converter, a large inductor is placed in series with the voltage source, which provides constant input current to the DAB converter. The current-fed converters are mainly used in applications where the current ripple needs to be minimized drastically. The voltage-fed converters are applied where constant voltage is required with strict regulation. Thus, it can be classified for DAB converter application, as the current-fed converter for the PV system can be applied mainly for a battery storage system. The PV power is used to charge large battery capacities and storage and can be used at nighttime or whenever required by the grid. However, the voltage-fed converters can be applied for grid-connected solar installation. The PV system provides continuous power to the grid with the inverter tied to the DAB converter's output terminal. Some of the topologies reported for the voltage-fed and current-fed DAB, which is suitable for PV application, are discussed next.

14.2.1 Voltage-Fed DAB Converter

The basic DAB reported in earlier stages of research was a voltage-fed converter. Similar topologies were reported with reduced components and other modification. These variants can be considered as the voltage-fed category, as the input terminal contains the voltage source, which can be replaced by the PV panel to obtain a constant voltage. The input capacitors play an important role in these topologies and storages of energy in capacitive form. The double half-bridge topology is proposed in [24], as shown in Figure 14.1.

The topology uses four semiconductor switches for two half-bridges on the primary and secondary side. The

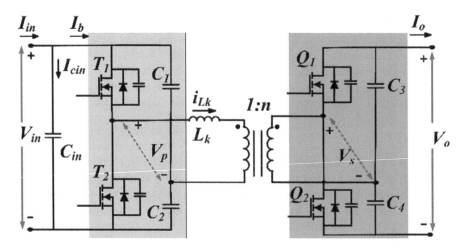

FIGURE 14.1 Double half-bridge topology.

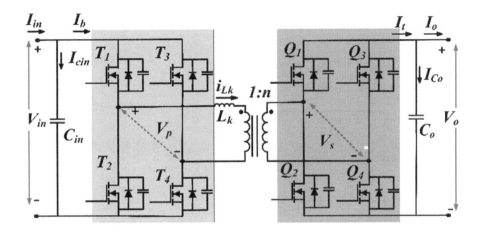

FIGURE 14.2 Double full-bridge topology.

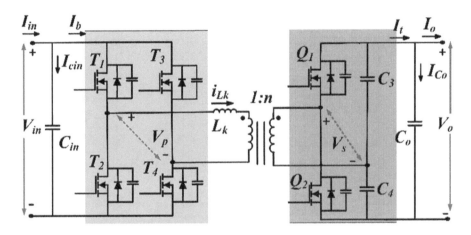

FIGURE 14.3 One full-bridge and one half-bridge topology.

ZVS is realized for all four switches. However, the ZVS range is small, as it depends on the leakage inductor and load situation. The power transfer capability is less than the double bridge topology. For light load conditions, the leakage inductor stores less energy and ZVS realization is difficult.

The topology has the advantage of eliminating the DC voltage component across the high-frequency transformer as the large capacitors on both bridges provide a constant voltage. The double full-bridge topology is proposed in [25], as shown in Figure 14.2. The two H-bridge on the primary and secondary side provide high power density and enhanced control for DAB operation. The soft switching is natural for this topology, which provides better efficiency for high-power and high-voltage application. Advance phase shift modulations such as double-phase shift, triple-phase shift, and extended phase shift can be implemented for increasing the ZVS range and efficiency. These phase shift controls are discussed later in Section 1.2.3. The other topology, as proposed in [26], is shown in Figure 14.3, which contains one full-bridge and one half-bridge.

The topology provides the capability of high power transfer as the current rating for the bridge is increased with the addition of two switches in the primary bridge. The capacitors in the secondary eliminate the output voltage ripple.

However, the current components are not affected, ensuring the ZVS transition of switches on the primary side. The modified topology based on a voltage-doubler rectifier for high voltage gain is proposed in [27]. The topology is depicted in Figure 14.4 with a full-bridge on the LV side, and the three-level voltage doubler rectifier is used on the HV side. The switches Q1 and Q2 on the secondary side are operated for the flow of current inductor L_k, which ensures the ZVS on the secondary side. Consequently, the three-level voltage is generated in the secondary bridge. The PWM and phase shift modulation can be used simultaneously on the bridges.

14.2.2 Current-Fed DAB Converter

For this type of topology, the input voltage source is connected to a large series inductor before the bridge. The double full-bridge current-fed converter is shown in Figure 14.5, which is obtained when an inductor is inserted between the source and the bridge [28].

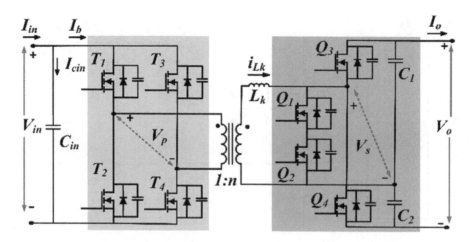

FIGURE 14.4 Full-bridge on the LV side, and the three-level voltage doubler rectifier is used on the HV side.

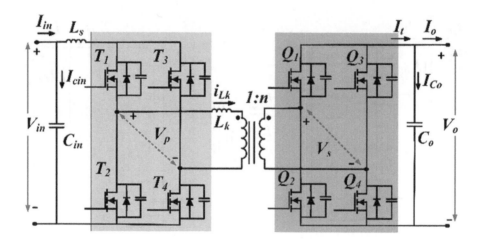

FIGURE 14.5 The double full-bridge current-fed converter.

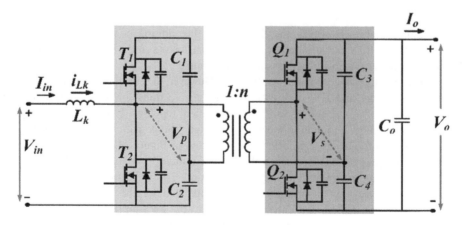

FIGURE 14.6 Topology of a current-fed DAB converter.

The overlapping PWM signals are given to the switches of one bridge to create the path for circulating current. The zero-current switching (ZCS) for the primary-side switches can be achieved easily, as discussed in [29]. The ZCS is effective in high-current, low-voltage application. However, for high-voltage and low-current applications, ZVS is more effective in reducing switching losses and enhancing the efficiency of the converter.

Another topology of a current-fed DAB converter is proposed in [30] to obtain a wide ZVS range and is shown in Figure 14.6. The low-voltage side consists of the boost converter–based half-bridge, which steps up the

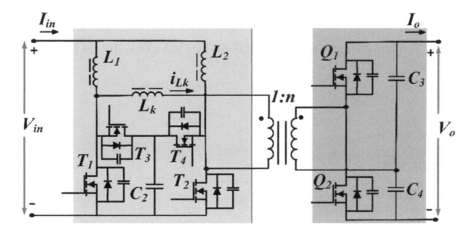

FIGURE 14.7 Topology of interleaved current-fed DAB converter.

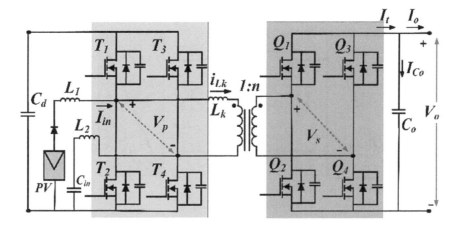

FIGURE 14.8 The two-level current-fed DAB.

voltage and then converts it to high-frequency AC voltage to feed the high-frequency transformer. The interleaved current-fed DAB topology [31] can be used as shown in Figure 14.7. The topology consists of two boost converters with input parallel and output parallel connection. This topology can achieve a low current ripple at the low-voltage side and wide ZVS ranges for all the power switches, which make it popular in high-voltage-gain applications.

The two-level current-fed DAB [32] is shown in Figure 14.8. The two interleaved inductor reduces the current ripple in the LV side. The phase shift control is employed for power transfer. The topology is employed for high voltage conversion with enhanced ZVS range and power flow.

The comparison between the double full-bridge voltage-fed and current-fed DAB is shown in Table 14.1. The circulating current is more in voltage-fed topology, due to which the nonactive power exists in the converter, thereby increasing the VA rating of the high-frequency transformer.

14.2.3 Control Strategies of DAB Converter

The phase-shift modulation techniques are the most effective and common control techniques for DAB converter. The section describes various control strategies considering the basic two-level DAB with double full-bridge, as shown in Figure 14.2. The equivalent circuit diagram for the DAB converter is shown in Figure 14.9 with the square wave bridge voltages and an equivalent inductance between them. The transformer turns ratio, on the requirement, can be applied for step-up and step-down of the primary- to secondary-side voltage. The power transfer from the primary to secondary side occurs when a phase shift is introduced in these voltage waveforms, similar to the power transfer between two voltage buses in a power system.

The phase shift makes one of the bridges as a leading and the other as a lagging bridge with power flow from the leading to the lagging bridge. Thus, the power direction can be easily controlled and reversed with the phase-shift value as the bridges are controlled simultaneously. The phase shift between the primary- and the secondary-side bridge voltage controls the power flow across the high-frequency transformer, which directly controls the output bus voltage of the DAB converter. The single phase-shift (SPS) control [33–35] is the most common technique with one control variable, that is, the phase-shift angle (δ). The voltages and

TABLE 14.1
Comparison of Voltage_Fed and Current-Fed Double Full-Bridge DAB Converter for the Same Voltage and Power Ratings

Parameters	Voltage-Fed DAB	Current-Fed DAB
Peak inductor current (i_{lk})	High	Low
Current stress in primary-bridge switches	High	Low
Current stress in secondary-bridge switches	Low	High
Voltage stress in primary-bridge switches	Low	High
Voltage stress in secondary-bridge switches	High	Low
Transformer VA rating	High	Low
Conduction losses in primary-bridge switches	High	Low
Conduction losses in secondary-bridge switches	Low	High
Turn-on loss	0 (ZVS)	Low
Turn-off loss	High	Very low (ZCS)

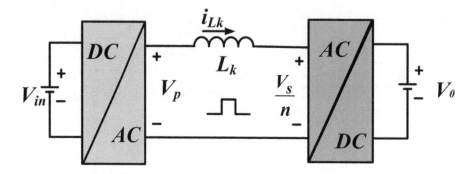

FIGURE 14.9 The equivalent circuit diagram for the DAB converter.

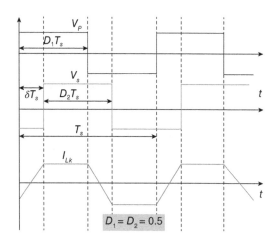

FIGURE 14.10 The voltages and inductor current waveforms under SPS control.

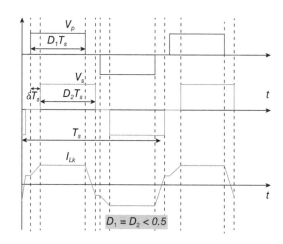

FIGURE 14.11 The voltages and inductor current waveforms under SPS control.

inductor current waveforms under SPS control are shown in Figure 14.10. The method is easy to implement and provides good dynamic results for the converter. However, the SPS method is suitable for fixed voltage conversion, and the conversion efficiency decreases when the voltage conversion ratio is high. The ZVS range is low and highly depends on the load condition as the circulating current and nonactive power increase when the conversion ratio changes.

The double phase-shift (DPS) [36–39] eliminates the limitations of SPS control. The voltages and inductor current waveforms under DPS control are shown in Figure 14.11. The duty ratio (D_1 and D_2) of the bridges are adjusted along with the phase shift (δ) in this control. This

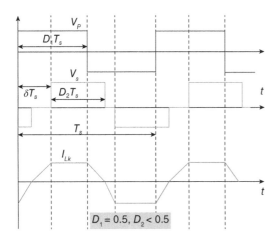

FIGURE 14.12 The voltages and inductor current waveforms under EPS control.

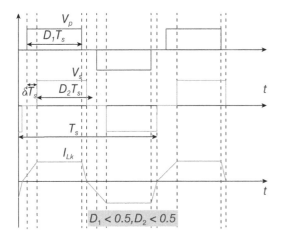

FIGURE 14.13 The voltages and inductor current waveforms under TPS control.

gives another control variable when the voltage conversion ratio is high. The wider ZVS range, increased power transfer capability, and improved voltage regulation are obtained with DPS control. However, the duty ratio is kept the same for both the bridges to simplify the control. The duty ratio control is further utilized in extended phase-shift (EPS) control, where the bridges are operated with different duty ratios [40–41]. The primary-bridge duty ratio is kept fixed at 50%, while the secondary-bridge duty ratio is altered along with the phase shift. The control provides more flexibility for power transfer and ZVS range optimization when loading and voltage level changes. However, the transition of power flow and voltage level is not smooth with EPS control. The voltages and inductor current waveforms under EPS control are shown in Figure 14.12.

The control technique with three control variables was introduced and named triple phase-shift control [42–45]. The power transfer is controlled with the phase shift between the bridges along with the two duty ratios for both bridges. The optimized control is possible to enhance the ZVS range and reduce the circulating currents in the converter. For large voltage gain, the mismatch in voltage across the high-frequency transformer is adjusted with the respective bridge duty ratio. However, the control needs complex calculation and optimization for conduction loss and maximum efficiency. The voltages and inductor current waveforms under TPS control are shown in Figure 14.13. The comprehensive review for the modulation strategies of DAB is presented in [46] with a detailed analysis of all possible phase-shift controls. Various advanced control schemes are incorporated with these modulation techniques for the converter to obtain efficient transient and steady-state performance of the DAB converter [47–53].

14.3 DAB CONVERTER: ANALYSIS AND DESIGN FOR PV APPLICATION

The voltage-fed double full-bridge DAB converter is the most common topology selected for PV application, as it includes various advantages among other topologies of DAB and provides flexible and efficient control over power and voltage regulation. The design and analysis of the DAB converter for PV application are presented in [54–55]. A novel MPPT control of DAB for PV integration with microgrid is presented in [56]. The circuit topology of a DAB converter with PV is shown in Figure 14.14, where V_{in} and V_o are the voltages in input and output DC link, and L_k is the equivalent inductance, which is actually representing the total sum of leakage inductance of the transformer and an auxiliary inductance in series, required for the power transfer. The semiconductor switches T_1–T_4 and Q_1–Q_4 make the two symmetric H-bridges on either side of the high-frequency transformer. The DC voltage is converted to high-frequency square wave AC voltage in the first stage of conversion implemented with the help of the first H-bridge, denoted as the primary bridge. Thus, the stage can be denoted as the inverter stage, and V_p denoting the primary bridge voltage. The transformer transfers the power from its primary winding to secondary winding. The secondary bridge then transforms the square wave AC voltage to DC voltage, which is the output of the DAB. The stage can be denoted as the rectifier stage, and V_s denoting the secondary bridge voltage.

The duty ratio of all switches is 50%, and the diagonal switches are turned on and off together to obtain a square wave. These voltages, as well as the equivalent inductor current waveforms, operated with SPS modulation are depicted in Figure 14.10, where T_s is the switching frequency and δ denotes the phase shift between the voltages of two bridges and n denotes the transformer turns ratio.

14.3.1 Analysis of DAB Converter Circuit

For the analysis of the DAB converter, a simplified circuit diagram is used as shown in Figure 14.10, and the steady-state waveform is depicted in Figure 14.10 with the phase

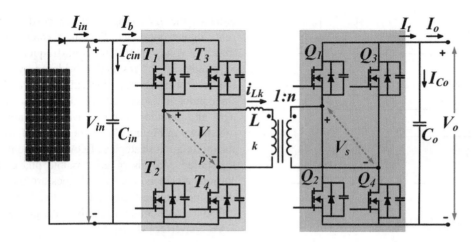

FIGURE 14.14 The circuit topology of a DAB converter with PV.

shift of δ in bridge voltage waveforms, inductor voltage V_{Lk}, and inductor current I_{Lk}. The inductor voltage and current are symmetrical about the time axis, dependent on the bridge voltages and phase shift. Therefore, the maximum value of the inductor current in the positive half cycle is equal to the maximum value of the negative half cycle. The power flow on the DAB converter is presented in [57] using a Fourier series analysis of bridge voltage and inductor current. The SPS modulation is utilized for the analysis of power transfer, and the expression for power flow is derived by averaging the power in one switching period, as expressed in (14.1) and (14.2).

$$P_t = \sum_{h=1,3,5...}^{\infty} P_h \quad (14.1)$$

$$P_h = \frac{\left(8.V_{in}.\dfrac{V_o}{n}\right)}{\pi^2 \omega_s L_k} \cdot \frac{\sin(h.\delta.\pi)}{h^3} \quad (14.2)$$

Thus, expression (2) shows that the power transfer from the input to the output terminal in the DAB converter depends upon the phase shift between the voltage waveform, equivalent inductance, and switching frequency. However, for the fixed value of the equivalent inductor and switching frequency, the phase shift value is the only parameter available to control the power flow. Also, the power flow capacity is inversely proportional to the equivalent inductance L_k, which is actually the sum of transformer leakage inductance and an auxiliary inductance connected in series. The auxiliary inductance is designed for the DAB operation to handle the specified power after the proper design of the transformer and considering its leakage inductance. The input capacitor plays an important role in the DAB interface with a renewable energy source to mitigate the voltage ripple and reverse current flow. Therefore, the analysis is focused on the design of the equivalent inductor L_k and input capacitor C_{in}. The output capacitor C_o is designed using the standard design process of any DC link capacitor to handle the voltage ripple. So the transformer design and output capacitor design are not discussed due to the limited scope of the chapter.

The power extraction from the renewable energy source depends on the DAB operation and is controlled by the phase shift value δ. The current analysis in the input node is essential to design the converter parameters L_k and C_{in}. The current waveforms I_{in}, I_{Cin}, and I_b at the input node along with the inductor current I_{Lk} are shown in Figure 14.16 for a single switching period. The Kirchhoff current law at input node is denoted as:

$$i_{in}(t) = i_{C_{in}}(t) + i_b(t) \quad (14.3)$$

And the current in the primary bridge (4) is formulated with the switching function S_1 in (5):

$$i_b(t) = i_{Lk}(t).S_1(t) \quad (14.4)$$

$$S_1(t) = \begin{cases} 1 & ; \ 0 < t < \dfrac{T_s}{2} \\ -1 & ; \ \dfrac{T_s}{2} < t < T_s \end{cases} \quad (14.5)$$

With the consideration that in the steady-state situation, the average capacitor current is zero, the average value of input current (3) can be obtained as:

$$\langle i_{in} \rangle = I_{in} = \langle i_{Lk}(t) S_1(t) \rangle \quad (14.6)$$

Therefore, expression (6) shows that the input current from renewable energy source depends upon the equivalent inductor current. Consequently, the power extraction from source and its delivery to DC bus thus depends on the value of inductor L_k and phase shift value δ. The inductor current analysis is required for the design of the equivalent inductor and input capacitor. The inductor current is shown in Figure 1, with the maximum value denoted as I_{max} and minimum value as I_{min}. The steady-state inductor current is symmetric around the time axis; thus, the magnitude of the maximum and minimum current can be considered as same in the positive and negative half cycle, that is, $|I_{max}| = |I_{min}|$. So in one half cycle, inductor current can be described by two linear equations as:

$$I_{Lk}(t) = I_{min} + \left(V_{in} + \frac{V_o}{n}\right)\cdot\frac{t}{L_k} \quad ; 0 < t < \frac{\delta T_s}{2} \quad (14.7)$$

$$I_{Lk}(t) = A + \left(V_{in} - \frac{V_o}{n}\right)\cdot\frac{t}{L_k} \quad ; \frac{\delta T_s}{2} < t < \frac{T_s}{2} \quad (14.8)$$

The constant A can be evaluated from (8) at $t = \frac{\delta T_s}{2}$ as:

$$A = I_{Lk\left(\frac{\delta T_s}{2}\right)} - \left(V_{in} - \frac{V_o}{n}\right)\cdot\frac{\delta T_s}{2L_k} \quad (14.9)$$

Therefore, from (14.7) and (14.8), the value of inductor current at $t = \frac{\delta T_s}{2}$ is:

$$I_{Lk\left(\frac{\delta T_s}{2}\right)} = I_{min} + \left(V_{in} + \frac{V_o}{n}\right)\cdot\frac{\delta T_s}{2L_k} \quad (14.10)$$

Replacing the value of $I_{Lk\left(\frac{\delta T_s}{2}\right)}$ from (14.10) in (14.9), the value of constant A is obtained as:

$$A = I_{min} + \frac{\delta T_s V_o}{n L_k} \quad (14.11)$$

Further, the maximum inductor current value can be derived with (14.8) and (14.11) at $t = \frac{T_s}{2}$ and using the condition $|I_{max}| = |I_{min}|$:

$$I_{max} = \frac{T_s}{4L_k}\left[V_{in} + (2\delta - 1)\frac{V_o}{n}\right] \quad (14.12)$$

Substituting (14.12) in (14.10), the value of inductor current at $t = \frac{\delta T_s}{2}$ results in equation (14.13):

$$I_{Lk\left(\frac{\delta T_s}{2}\right)} = \frac{T_s}{4L_k}\left[(2\delta - 1)V_{in} + \frac{V_o}{n}\right] \quad (14.13)$$

Finally, the expression for the inductor current is formed for the half cycle as:

$$I_{Lk}(t) = \begin{cases} I_{Lk(1)} & ; 0 < t \leq \frac{\delta T_s}{2} \\ \\ I_{Lk(2)} & ; \frac{\delta T_s}{2} < t \leq \frac{T_s}{2} \end{cases} \quad (14.14)$$

Where:

$$\begin{cases} I_{Lk(1)} = \left(V_{in} + \frac{V_o}{n}\right)\cdot\frac{t}{L_k} - \left(V_{in} + (2\delta - 1)\frac{V_o}{n}\right)\cdot\frac{T_s}{4L_k} \\ \\ I_{Lk(2)} = \left(V_{in} - \frac{V_o}{n}\right)\cdot\frac{t}{L_k} - \left(V_{in} - (2\delta + 1)\frac{V_o}{n}\right)\cdot\frac{T_s}{4L_k} \end{cases} \quad (14.15)$$

14.3.2 Equivalent Inductor (L_k) Design

The power transferred by the DAB converter from input to output through the transformer is denoted by expression (14.1). This is further represented in the form as:

$$P_t = \frac{8V_{in}\left(\frac{V_o}{n}\right)}{\pi^2 \cdot \omega_s \cdot L_k} \cdot \sum_{h=1,3,5...}^{\infty} \frac{\sin(h\cdot\delta\cdot\pi)}{h^3} \quad (14.16)$$

It clearly indicates that the power flow depends upon the input and output voltage, that is, (V_{in}) and (V_o), switching frequency ($\omega_s = 2\pi f_s$), transformer turns ratio (n), phase shift (δ), and equivalent inductance (L_k). The inductor value is denoted from (16) as:

$$L_k = \frac{8V_{in}\left(\frac{V_o}{n}\right)}{\pi^2 \cdot \omega_s \cdot P_t} \cdot \sum_{h=1,3,5...}^{\infty} \frac{\sin(h\cdot\delta\cdot\pi)}{h^3} \quad (14.17)$$

As the power flow equation is inversely proportional to the equivalent inductor value, the high value of the inductor will make the maximum power extraction from renewable source impossible. Thus, with a defined maximum power of PV denoted as $P_{t(mpp)}$, a critical inductance value is defined to extract and transfer the maximum possible power. The phase-shift condition for the maximum power transfer is also considered from (14.16). As the range of phase shift is defined for one direction power flow as $0 < \delta < 0.5$, the maximum value of all harmonic components occurs for the phase shift value $\delta = 0.5$. Thus, putting this value of phase shift in (14.17), the maximum or critical inductance value is obtained as:

$$L_{k(mpp)} = \frac{V_{in(mpp)}\cdot V_o \cdot \pi}{4\cdot n\cdot \omega_s \cdot P_{t(mpp)}} \quad (14.18)$$

This value of equivalent inductance is, in fact, the total inductance provided by the transformer leakage inductance with an additional inductance in series. The additional inductance, described as an auxiliary inductance, is required for the power transfer. The auxiliary inductor can be avoided if the transformer is properly designed and has the leakage inductance less than or equal to the value mentioned in (14.18).

14.3.3 Input Capacitor Design

The input voltage fluctuation is a common problem for the converters interfaced with any renewable energy sources. The input capacitor is an important component as it is directly interfacing the renewable energy source and minimizes the voltage ripple at the input terminal.

The capacitor also absorbs the reverse current from DAB in each cycle and maintains a constant voltage in the input terminal. The design of the input capacitor is important for DAB converters to provide a stable output voltage containing less ripple. The design process initiates with the input terminal current equation (14.3) and considering the current waveform shown in Figure 14.15. Thus, the capacitor current is denoted as:

$$i_{C_{in}}(t) = i_{in}(t) - i_b(t) \tag{14.19}$$

With the input voltage ripple ΔV_{in}, the charge accumulated in the capacitor during the charging process in the interval Δt is:

$$Q = 2.\Delta V_{in}.C_{in} = \frac{1}{2}\Delta t(I_{max} + I_{in}) \tag{14.20}$$

The input current is obtained from (14.6) and (14.15) as:

$$I_{in} = \frac{2}{T_s}\left[\int_0^{\frac{\delta T_s}{2}} I_{Lk(1)} + \int_0^{\frac{T_s}{2}} I_{Lk(2)}\right] \tag{14.21}$$

$$I_{in} = \frac{T_s V_o \delta(1-\delta)}{2nL_k} \tag{14.22}$$

Now, using (14.4) and (14.22) and considering the maximum power flow situation, the capacitor current is simplified as:

$$I_{C_{in}}(t) = \frac{T_s}{4L_k}\left[V_{in} - (2.\delta^2 - 4\delta + 1)\frac{V_o}{n} - \left(V_{in} + \frac{V_o}{n}\right)\frac{4t}{T_s}\right] \tag{14.23}$$

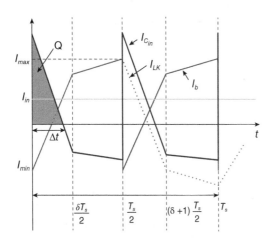

FIGURE 14.15 Input capacitor current waveform.

With the initial condition of $i_{C_{in}} = 0$, the time interval is expressed as:

$$\Delta t = \frac{T_s}{4}\left[\frac{V_{in} - (2.\delta^2 - 4.\delta + 1)\frac{V_o}{n}}{V_{in} + \frac{V_o}{n}}\right] \tag{14.24}$$

Substituting (14.12), (144.22), and (14.24) in (14.20), the input capacitor value is obtained with desired input ripple as:

$$C_{in} = \frac{T_s^2}{64.\Delta V_{in}.L_k}\left[\frac{\left[\frac{V_o}{n}(2.\delta^2 - 4\delta + 1) - V_{in}\right]^2}{\frac{V_o}{n} + V_{in}}\right] \tag{14.25}$$

Expression (14.25) shows that the derived capacitor value depends on the phase-shift value, and thereby the capacitor should be selected to mitigate the maximum input voltage ripple. The relation of capacitance value with respect to phase shift can be analyzed as:

$$\frac{\partial C_{in}}{\partial \delta} > 0 \tag{14.26}$$

As the maximum ripple occurs at the highest value of δ, the value of the capacitor is taken as:

$$C_{in} = \frac{T_s^2}{64.\Delta V_{in}.L_k}\left[\frac{\left(\frac{V_o}{2n} + V_{in}\right)^2}{\frac{V_o}{n} + V_{in}}\right] \tag{14.27}$$

14.3.4 PV System Design

Considering the single-diode model, the basic equivalent circuit of a PV cell array is as depicted in Figure 14.16. The equivalent circuit is designed to achieve the practical I–V characteristics of a PV cell array, with consideration of isolation and temperature conditions.

The corresponding equation defining the characteristics of the PV cell array can be derived as follows:

$$I_{PV} = N_p I_G - N_p I_{sat}\left(\exp\left(\frac{V_{PV} + I_{PV}R_S}{a}\right) - 1\right) - \frac{V_{PV} + I_{PV}R_S}{R_P} \tag{14.28}$$

$$I_G = \left(I_{SC} + K_I(T_c - T_r)\right)\frac{\lambda}{\lambda_r} \tag{14.29}$$

$$I_{sat} = I_{rs}\left(\frac{T}{T_r}\right)^3 \exp\left(\frac{qE_{g0}\left(\frac{1}{T_r} - \frac{1}{T_c}\right)}{AK}\right) \tag{14.30}$$

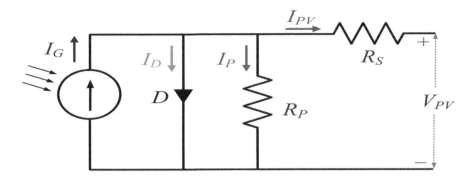

FIGURE 14.16 The basic equivalent circuit of a PV cell array.

$$I_{rs} = \frac{I_{sc}}{\exp\left(\frac{qV_{oc}}{AN_s KT}\right) - 1} \quad (14.31)$$

$$a = AN_s V_T = \frac{N_s A k T_c}{q} \quad (14.32)$$

Where I_G is the photo generated current; I_{sat} is saturation currentl N_p is the number of parallel strings; N_s is the number of series strings; R_P is equivalent parallel resistance; R_S is the equivalent series resistance; I_{rs} is the reverse saturation current; I_{sc} is short circuit current; E_{g0} is the bandgap energy; V_{oc} is open-circuit voltage; V_{PV} and I_{PV} are the output voltage and current of PV array, respectively; K_I is the short-circuit current temperature coefficient; K is the Boltzmann constant (1.38×10^{-23}); q is the charge of electron (1.602×10^{-19}); A is the ideality factor (1.3); λ_r and λ_c are rated and actual irradiance, respectively; and T_r and T_c are rated and actual temperature in kelvin, respectively.

The power from the PV array is thereby obtained as:

$$P_{PV} = V_{PV} \times I_{PV} \quad (14.33)$$

14.3.5 HIGH-FREQUENCY TRANSFORMER DESIGN

The high-frequency transformer design is an important aspect for DAB converter design and operation. The transformer design includes selection of core type, magnetic material, and type of winding. Further, the skin and proximity effects need to be considered and calculated properly to achieve minimum losses. The detail about the transformer design can be found in [58, 59], describing the major design procedure and calculations. These procedures for transformer design are not discussed here, as it require an in-depth discussion and is out of scope for this chapter.

14.4 CONCLUSION

This chapter presents a brief review of the DAB converter for a PV system. The DAB offers various advantages suitable for interfacing low-voltage PV modules to high-voltage DC bus or for a battery storage system. This chapter presents various topological variants and control schemes of DAB converter, developed in recent time, and are suitable for PV systems. The chapter includes a detailed analysis of the DAB converter to design the leakage inductor and input capacitor. The leakage inductor is an essential component for the extraction of maximum power from PV installation at variable solar irradiance. The input capacitor design is vital, as it provides a stable voltage input with reduced ripple and maximizes the MPPT efficiency. With the exploration of wide-bandgap semiconductor devices, high-frequency switches along with high-voltage capacity are possible. Thus, the DAB converters will be one of the choices for PV systems and other renewable energy systems.

14.5 REFERENCES

[1] A. K. Bhattacharjee, N. Kutkut and I. Batarseh, "Review of Multiport Converters for Solar and Energy Storage Integration," in *IEEE Transactions on Power Electronics*, vol. 34, no. 2, pp. 1431–1445, Feb. 2019.

[2] K. V. G. Raghavendra, K. Zeb, A. Muthusamy, T. N. V. Krishna, S. V. S. V. P. Kumar, D.-H. Kim, M.-S. Kim, H.-G. Cho, and H.-J. Kim, "A Comprehensive Review of DC – DC Converter Topologies and Modulation Strategies with Recent Advances in Solar Photovoltaic Systems," *Electronics*, vol. 9, no. 1, p. 31, Dec. 2019.

[3] M. Uno and K. Sugiyama, "Switched Capacitor Converter Based Multiport Converter Integrating Bidirectional PWM and Series-Resonant Converters for Standalone Photovoltaic Systems," in *IEEE Transactions on Power Electronics*, vol. 34, no. 2, pp. 1394–1406, Feb. 2019.

[4] H. Choi, M. Ciobotaru, M. Jang and V. G. Agelidis, "Performance of Medium-Voltage DC-Bus PV System Architecture Utilizing High-Gain DC – DC Converter," in *IEEE Transactions on Sustainable Energy*, vol. 6, no. 2, pp. 464–473, April 2015.

[5] A. M. S. S. Andrade, L. Schuch and M. L. da Silva Martins, "Analysis and Design of High-Efficiency Hybrid High Step-Up DC – DC Converter for Distributed PV Generation Systems," in *IEEE Transactions on Industrial Electronics*, vol. 66, no. 5, pp. 3860–3868, May 2019.

[6] H. Lee and J. Yun, "Quasi-Resonant Voltage Doubler With Snubber Capacitor for Boost Half-Bridge DC – DC Converter in Photovoltaic Micro-Inverter," in *IEEE Transactions on Power Electronics*, vol. 34, no. 9, pp. 8377–8388, Sept. 2019.

[7] K. Li, Y. Hu and A. Ioinovici, "Generation of the Large DC Gain Step-Up Nonisolated Converters in Conjunction

with Renewable Energy Sources Starting from a Proposed Geometric Structure," in *IEEE Transactions on Power Electronics*, vol. 32, no. 7, pp. 5323–5340, July 2017.

[8] J. Roy, Y. Xia and R. Ayyanar, "High Step-Up Transformerless Inverter for AC Module Applications with Active Power Decoupling," in *IEEE Transactions on Industrial Electronics*, vol. 66, no. 5, pp. 3891–3901, May 2019.

[9] D. Binu Ben Jose, N. Ammasai Gounden and J Ravishankar, "Simple Power Electronic Controller for Photovoltaic Fed Grid-tied Systems Using Line Commutated Inverter with Fixed Firing Angle," in *IET Power Electronics*, vol. 7, pp. 1424–1434.2014.

[10] V. Gautam and P. Sensarma, "Design of Ćuk-Derived Transformerless Common-Grounded PV Microinverter in CCM," in *IEEE Transactions on Industrial Electronics*, vol. 64, no. 8, pp. 6245–6254, Aug. 2017.

[11] R. K. Surapaneni and P. Das, "A Z-Source-Derived Coupled-Inductor-Based High Voltage Gain Microinverter," in *IEEE Transactions on Industrial Electronics*, vol. 65, no. 6, pp. 5114–5124, June 2018.

[12] D. Vinnikov, A. Chub, E. Liivik, R. Kosenko and O. Korkh, "Solar Optiverter – A Novel Hybrid Approach to the Photovoltaic Module Level Power Electronics," in *IEEE Transactions on Industrial Electronics*, vol. 66, no. 5, pp. 3869–3880, May 2019.

[13] H. Lee and J. Yun, "Quasi-Resonant Voltage Doubler With Snubber Capacitor for Boost Half-Bridge DC – DC Converter in Photovoltaic Micro-Inverter," in *IEEE Transactions on Power Electronics*, vol. 34, no. 9, pp. 8377–8388, Sept. 2019.

[14] N. Sukesh, M. Pahlevaninezhad and P. K. Jain, "Analysis and Implementation of a Single-Stage Flyback PV Microinverter With Soft Switching," in *IEEE Transactions on Industrial Electronics*, vol. 61, no. 4, pp. 1819–1833, April 2014.

[15] R. K. Surapaneni and A. K. Rathore, "A Single-Stage CCM Zeta Microinverter for Solar Photovoltaic AC Module," in *IEEE Journal of Emerging and Selected Topics in Power Electronics*, vol. 3, no. 4, pp. 892–900, Dec. 2015.

[16] S. Lee, W. Cha, J. Kwon and B. Kwon, "Control Strategy of Flyback Microinverter With Hybrid Mode for PV AC Modules," in *IEEE Transactions on Industrial Electronics*, vol. 63, no. 2, pp. 995–1002, Feb. 2016.

[17] S. Lee, W. Cha, B. Kwon and M. Kim, "Discrete-Time Repetitive Control of Flyback CCM Inverter for PV Power Applications," in *IEEE Transactions on Industrial Electronics*, vol. 63, no. 2, pp. 976–984, Feb. 2016.

[18] M. N. Kheraluwala, R. W. Gascoigne, D. M. Divan and E. D. Baumann, "Performance Characterization of a High-power Dual Active Bridge DC-to-DC Converter," in *IEEE Transactions on Industry Applications*, vol. 28, no. 6, pp. 1294–1301, Nov.-Dec. 1992.

[19] B. Zhao, Q. Song, W. Liu and Y. Sun, "Overview of Dual-Active-Bridge Isolated Bidirectional DC – DC Converter for High-Frequency-Link Power-Conversion System," in *IEEE Transactions on Power Electronics*, vol. 29, no. 8, pp. 4091–4106, Aug. 2014.

[20] Y. Shi, R. Li, Y. Xue and H. Li, "Optimized Operation of Current-Fed Dual Active Bridge DC – DC Converter for PV Applications," in *IEEE Transactions on Industrial Electronics*, vol. 62, no. 11, pp. 6986–6995, Nov. 2015.

[21] H. Zhou, Tran Duong, Siew Tuck Sing and A. M. Khambadkone, "Interleaved bi-directional Dual Active Bridge DC-DC converter for interfacing ultracapacitor in micro-grid application," *2010 IEEE International Symposium on Industrial Electronics*, Bari, 2010, pp. 2229–2234.

[22] S. Talbi, A. M. Mabwe and A. E. Hajjaji, "Control of a Bidirectional Dual Active Bridge Converter for Charge and Discharge of a Li-Ion Battery," *IECON 2015–41st Annual Conference of the IEEE Industrial Electronics Society*, Yokohama, 2015, pp. 000849–000856.

[23] H. Wu, K. Sun, L. Chen, L. Zhu and Y. Xing, "High Step-Up/Step-Down Soft-Switching Bidirectional DC – DC Converter With Coupled-Inductor and Voltage Matching Control for Energy Storage Systems," in *IEEE Transactions on Industrial Electronics*, vol. 63, no. 5, pp. 2892–2903, May 2016.

[24] X. Xu, A. M. Khambadkone and R. Oruganti, "A Soft-Switched Back-to-Back Bi-directional DC/DC Converter with a FPGA based Digital Control for Automotive applications," *IECON 2007–33rd Annual Conference of the IEEE Industrial Electronics Society*, Taipei, Taiwan, 2007, pp. 262–267.

[25] Y. Yan, H. Bai, A. Foote and W. Wang, "Securing Full-Power-Range Zero-Voltage Switching in Both Steady-State and Transient Operations for a Dual-Active-Bridge-Based Bidirectional Electric Vehicle Charger," in *IEEE Transactions on Power Electronics*, vol. 35, no. 7, pp. 7506–7519, July 2020.

[26] R. Morrison and M. Egan, "A New Single Transformer, Power Factor Corrected UPS Design," *APEC'98 Thirteenth Annual Applied Power Electronics Conference and Exposition*, Anaheim, CA, USA, 1998, pp. 237–243 vol.1.

[27] X. Zhan, H. Wu, Y. Xing, H. Ge and X. Xiao, "A High Step-up Bidirectional Isolated Dual-active-bridge Converter with Three-level Voltage-Doubler Rectifier for Energy Storage Applications," *2016 IEEE Applied Power Electronics Conference and Exposition (APEC)*, Long Beach, CA, USA, 2016, pp. 1424–1429.

[28] S. Bal, D. B. Yelaverthi, A. K. Rathore and D. Srinivasan, "Improved Modulation Strategy Using Dual Phase Shift Modulation for Active Commutated Current-Fed Dual Active Bridge," in *IEEE Transactions on Power Electronics*, vol. 33, no. 9, pp. 7359–7375, Sept. 2018.

[29] P. Xuewei and A. K. Rathore, "Novel Bidirectional Snubberless Soft-switching Naturally Clamped Zero Current Commutated Current-fed Dual Active Bridge (CFDAB) Converter for Fuel Cell Vehicles," *2013 IEEE Energy Conversion Congress and Exposition*, Denver, CO, USA, 2013, pp. 1894–1901.

[30] F. Z. Peng, Hui Li, Gui-Jia Su and J. S. Lawler, "A New ZVS Bidirectional DC-DC Converter for Fuel Cell and Battery Application," in *IEEE Transactions on Power Electronics*, vol. 19, no. 1, pp. 54–65, Jan. 2004.

[31] H. Xiao and S. Xie, "A ZVS Bidirectional DC – DC Converter With Phase-Shift Plus PWM Control Scheme," in *IEEE Transactions on Power Electronics*, vol. 23, no. 2, pp. 813–823, March 2008.

[32] Y. Shi, R. Li, Y. Xue and H. Li, "High-Frequency-Link-Based Grid-Tied PV System With Small DC-Link Capacitor and Low-Frequency Ripple-Free Maximum Power Point Tracking," in *IEEE Transactions on Power Electronics*, vol. 31, no. 1, pp. 328–339, Jan. 2016.

[33] Z. Zhang, H. Zhao, S. Fu, J. Shi and X. He, "Voltage and Power Balance Control Strategy for Three-phase Modular Cascaded Solid Stated Transformer," *2016 IEEE Applied Power Electronics Conference and Exposition (APEC)*, Long Beach, CA, USA, 2016, pp. 1475–1480.

[34] B. Zhao, Q. Song, W. Liu, G. Liu and Y. Zhao, "Universal High-Frequency-Link Characterization and Practical Fundamental-Optimal Strategy for Dual-Active-Bridge DC-DC Converter Under PWM Plus Phase-Shift Control,"

in *IEEE Transactions on Power Electronics*, vol. 30, no. 12, pp. 6488–6494, Dec. 2015.

[35] Z. Shan, J. Jatskevich, H. H. Iu and T. Fernando, "Simplified Load-Feedforward Control Design for Dual-Active-Bridge Converters With Current-Mode Modulation," in *IEEE Journal of Emerging and Selected Topics in Power Electronics*, vol. 6, no. 4, pp. 2073–2085, Dec. 2018.

[36] B. Zhao, Q. Song and W. Liu, "Power Characterization of Isolated Bidirectional Dual-Active-Bridge DC – DC Converter with Dual-Phase-Shift Control," in *IEEE Transactions on Power Electronics*, vol. 27, no. 9, pp. 4172–4176, Sept. 2012.

[37] M. Tsai, C. Chu and C. Chin, "Design a Dual Active Bridge Converter with Symmetrical Dual Phase-shift Strategy," *2018 IEEE International Conference on Applied System Invention (ICASI)*, Chiba, Japan, 2018, pp. 1002–1005.

[38] B. Feng, Y. Wang and J. Man, "A Novel Dual-phase-shift Control Strategy for Dual-active-bridge DC-DC Converter," *IECON 2014–40th Annual Conference of the IEEE Industrial Electronics Society*, Dallas, TX, USA, 2014, pp. 4140–4145.

[39] H. Bai and C. Mi, "Eliminate Reactive Power and Increase System Efficiency of Isolated Bidirectional Dual-Active-Bridge DC – DC Converters Using Novel Dual-Phase-Shift Control," in *IEEE Transactions on Power Electronics*, vol. 23, no. 6, pp. 2905–2914, Nov. 2008.

[40] B. Zhao, Q. Yu and W. Sun, "Extended-Phase-Shift Control of Isolated Bidirectional DC – DC Converter for Power Distribution in Microgrid," in *IEEE Transactions on Power Electronics*, vol. 27, no. 11, pp. 4667–4680, Nov. 2012.

[41] Z. Zhang, J. Sun, P. Wang, Z. Cai, J. Kong, X. Bai, et al., "An Improved DC Bias Elimination Strategy with Extended Phase Shift Control for Dual-active-bridge DCDC", *2019 Chinese Automation Congress (CAC)*, pp. 4274–4279, Nov 2019.

[42] H. Gu, D. Jiang, R. Yin, S. Huang, Y. Liang and Y. Wang, "Power Characteristics Analysis of Bidirectional Full-bridge DC-DC Converter with Triple-phase-shift control," *2015 IEEE 10th Conference on Industrial Electronics and Applications (ICIEA)*, Auckland, New Zealand, 2015, pp. 363–368.

[43] X. Ma, A. Tong, B. Li, L. Hang, G. Li and P. Shen, "ZVS Operation of DAB Converter Based on Triple-phase-shift Modulation Scheme with Optimized Inductor Current," *IECON 2017–43rd Annual Conference of the IEEE Industrial Electronics Society*, Beijing, China, 2017, pp. 4702–4707.

[44] A. K. Bhattacharjee and I. Batarseh, "Optimum Hybrid Modulation for Improvement of Efficiency Over Wide Operating Range for Triple-Phase-Shift Dual-Active-Bridge Converter," in *IEEE Transactions on Power Electronics*, vol. 35, no. 5, pp. 4804–4818, May 2020.

[45] Q. Bu and H. Wen, "Triple-Phase-Shifted Bidirectional Full-Bridge Converter with Wide range ZVS," *2018 IEEE International Conference on Power Electronics, Drives and Energy Systems (PEDES)*, Chennai, India, 2018, pp. 1–6.

[46] N. Hou and Y. W. Li, "Overview and Comparison of Modulation and Control Strategies for a Nonresonant Single-Phase Dual-Active-Bridge DC – DC Converter," in *IEEE Transactions on Power Electronics*, vol. 35, no. 3, pp. 3148–3172, March 2020.

[47] K. Takagi and H. Fujita, "Dynamic Control and Performance of a Dual-Active-Bridge DC – DC Converter," in *IEEE Transactions on Power Electronics*, vol. 33, no. 9, pp. 7858–7866, Sept. 2018.

[48] X. Li and Y. Li, "An Optimized Phase-Shift Modulation For Fast Transient Response in a Dual-Active-Bridge Converter," in *IEEE Transactions on Power Electronics*, vol. 29, no. 6, pp. 2661–2665, June 2014.

[49] S. Dutta, S. Hazra and S. Bhattacharya, "A Digital Predictive Current-Mode Controller for a Single-Phase High-Frequency Transformer-Isolated Dual-Active Bridge DC-to-DC Converter," in *IEEE Transactions on Industrial Electronics*, vol. 63, no. 9, pp. 5943–5952, Sept. 2016.

[50] W. Song, N. Hou and M. Wu, "Virtual Direct Power Control Scheme of Dual Active Bridge DC – DC Converters for Fast Dynamic Response," in *IEEE Transactions on Power Electronics*, vol. 33, no. 2, pp. 1750–1759, Feb. 2018.

[51] D. Nguyen, D. Nguyen, T. Funabashi, and G. Fujita, "Sensorless Control of Dual-active-Bridge Converter with Reduced-order Proportional-integral Observer," *Energies*, vol. 11, pp. 931–948, Apr. 2018.

[52] F. Xiong, J. Wu, Z. Liu and L. Hao, "Current Sensorless Control for Dual Active Bridge DC – DC Converter with Estimated Load-Current Feedforward," in *IEEE Transactions on Power Electronics*, vol. 33, no. 4, pp. 3552–3566, April 2018.

[53] N. Tiwary, N. N. Venkataramana, A. k. Panda and A. Narendra, "Direct Power Control of Dual Active Bridge Bidirectional DC-DC Converter," *2019 International Conference on Power Electronics, Control and Automation (ICPECA)*, New Delhi, India, 2019, pp. 1–4.

[54] P. Joebges, J. Hu and R. W. De Doncker, "Design Method and Efficiency Analysis of a DAB Converter for PV Integration in DC Grids," *2016 IEEE 2nd Annual Southern Power Electronics Conference (SPEC)*, Auckland, 2016, pp. 1–6.

[55] Y. Shi, R. Li, Y. Xue and H. Li, "Optimized Operation of Current-Fed Dual Active Bridge DC – DC Converter for PV Applications," in *IEEE Transactions on Industrial Electronics*, vol. 62, no. 11, pp. 6986–6995, Nov. 2015.

[56] J. Hu, P. Joebges, G. C. Pasupuleti, N. R. Averous and R. W. De Doncker, "A Maximum-Output-Power-Point-Tracking-Controlled Dual-Active Bridge Converter for Photovoltaic Energy Integration Into MVDC Grids," in *IEEE Transactions on Energy Conversion*, vol. 34, no. 1, pp. 170–180, March 2019.

[57] D. Segaran, D. G. Holmes and B. P. McGrath, "Enhanced Load Step Response for a Bidirectional DC – DC Converter," in *IEEE Transactions on Power Electronics*, vol. 28, no. 1, pp. 371–379, Jan. 2013.

[58] Iuravin, Egor. "Transformer Design for Dual Active Bridge Converter." PhD diss., Miami University, 2018.

[59] S. Thakur, G. Gohil and P. T. Balsara, "Design of Planar Transformer for Dual Active Bridge for Renewable Energy Sources and Grid Integration Applications," *2021 IEEE Applied Power Electronics Conference and Exposition (APEC)*, 2021, pp. 2857–2864.

15 Sustainable Energy Management in Lighting Urban Public Places

Melita Rozman Cafuta and Peter Virtič

CONTENTS

15.1 Introduction 189
15.2 Lighting of Public Urban Places 189
    15.2.1 Suitable Lighting for Everyone 189
    15.2.2 Environmental Acceptance of the Light 190
    15.2.3 Economic Efficiency Lighting 190
    15.2.4 User-Oriented Approach of Urban Artificial Night Lighting Management 190
15.3 The Experiment 191
15.4 Statistical Results and Data Evaluation 191
    15.4.1 Users' Priorities Regarding Illumination Necessity 192
    15.4.2 Users' Priorities Regarding Illumination Intensity 193
    15.4.3 Users' Priorities Regarding Illumination Costs 195
15.5 Conclusion and Discussion 195
15.6 References 196

15.1 INTRODUCTION

Nowadays, sustainable city lighting is a product of environmental responsibility, economic efficiency, and social cohesion like suitability for the user. It takes into account the interests of the individual and of society as a whole (Rozman Cafuta, 2015). Sustainability is based on assessing the possibility of restricting human activities and aims to find solutions to achieve such a minimum light intensity that the level of nature conservation is compatible with human activities, their subjective perception of illuminated areas, and the technological development of lighting.

Lighting design is one of the levels of urban development. It means achieving such quality and quantity of light sources that the level of illumination is in line with the principles of sustainable development in terms of preserving human health, preserving ecosystems, reducing energy consumption, and contributing to human well-being, sociality, and economic vitality. As lighting is highly subject to technological advances, it is important for its successful management to comply with the sustainability principles. A balance is sought between the positive effects of lighting and the remediation of the most negative effects.

The parameters of urban space are constantly changing. Sustainability is not achievable as a permanent state, but only as a set of desirable properties of a complex system. High-quality lighting ambiences have the ability to adapt to all spatial circumstances. In practice, circumstances change due to varying traffic densities and a smaller number of users in the early-morning or late-evening hours. In order to ensure good living conditions for everyone involved, it is necessary to adapt the lighting ambiences. Strong illumination alone is not enough; the lighting must be functional.

Modern city concepts, such as smart and sustainable city, all claim to be human-centered and inclusive (Rebernik et al., 2020). By improving the understanding of citizens, their needs, perceptions, and behaviors, it is possible to promote government measures towards inclusive urban design (Rebernik et al, 2019). Sustainable lighting management means that not only the technological properties of the luminaires and the light intensity are important but also the subjective effect of light on the users. The decisive factor here is the planning of the distribution and the intensity of light sources.

15.2 LIGHTING OF PUBLIC URBAN PLACES

15.2.1 SUITABLE LIGHTING FOR EVERYONE

Today, it is accepted in various disciplines that artificial night-light is a feature that enables normal activities at a time when daylight is not available and supplemental lighting is urgently needed. Personal perception indicates whether an illuminated area is attractive or not. Urban areas are always used by many users: drivers, motorists, bicyclists, pedestrians, residents, and animals that use areas at the same time. Each group has its own lighting needs. To satisfy them all, too many lighting objects are placed, which can cause negative side effects, such as lighting glow.

The intensity of light has a significant impact on the ability to perceive. According to Dabbagh (2019), color and light intensity play an important role in the beautification and readability of the city. If the lighting does not support the spatial requirements, the environment is frustrating. What kind of light do people appreciate, and what repels them? Mizon (2002) points out the difficulties in implementing appropriate lighting. Often, the criteria of luminance are too much in the foreground and not the target environment. An illuminated object that stands out from the dark surroundings is perceived as the most visually effective. Reducing the contrast between the illuminated object and the dark environment increases the difficulty of recognizing the spatial order.

Pellegrino (2006) looks for solutions in the direction of technological and aesthetic improvements. He believes that lighting technology brings lighting closer to the various users. According to Santen (2006), the lighting quality is assessed on the basis of the overall brightness at the chosen location, the distribution of light according to purpose, and the perceived well-being of the users at the chosen location. All this depends not only on the choice of a suitable lamp but also on the material and texture of the surfaces and the color of the light. Dark surfaces absorb much more than light surfaces; smooth surfaces reflect more than structured ones.

Illumination is not always expedient, especially when dangerous glow is present. Glow provides visual irritation and allows us to understand with the environment and react appropriately to events. Problems occur when conditions are so extreme that the human visual system cannot adapt. Too little or too much light is perceived as uncomfortable. In the worst cases, the response to a particular situation may be incorrect. Useless light, known as light pollution, is a particular form of environmental pollution. Environmental remediation begins once the source is eliminated. But night-light cannot be abolished simply to preserve nature.

15.2.2 Environmental Acceptance of the Light

Humans benefit from light to avoid discomfort, but the presence of artificial night-light in urban environments impacts ecosystems, animals, and vegetation. All major ecosystem functions are impaired, and biodiversity is damaged. Most living things are highly dependent on natural light. Even light reflected from illuminated surfaces causes sky glow and alters natural conditions. Some insects or birds are attracted to artificial light. Their behavior patterns, migration routes, and reproductive plans are affected.

Longcore and Rich (2004) found that artificial night-light alters the relationship between predators and prey. Predators might be more successful at hunting, but preys adapted to the dark might be confused by too much extra light.

It's not just direct light that seems to be problematic. The presence of the city increases the intensity of light in a larger radius. It manifests itself in the form of sky glow, which can be perceived from a great distance. Longcore and Rich (2004) conclude that this phenomenon may also alter the behavioral patterns of animals not directly associated with urban space.

Increased light has a completely different effect on plants. For green organisms, the following parameters are important: the value of the light spectrum, the light intensity, the light direction, and the duration of light exposure. Briggs (2006) states that night-light accelerates the plant growth cycle, resulting in faster seed germination, thicker leaf, faster stem growth, faster transition to flowering stage, and faster fruit ripening. Although these effects may even seem desirable, they often cause abnormalities in the vegetative parts of the plants.

15.2.3 Economic Efficiency Lighting

Choosing the right type of luminaire is as important as how much light is actually reflected into the environment. The technical characteristics make the luminaires interesting for the users. Nonshaded lighting elements are not suitable. The best choice is completely shaded luminaires. These shine evenly and have minimal glow. Suitable lighted streets are more attractive, and they are frequently used. The appearance of the luminaires is important as an aesthetic design element within urban structures. In practice, the qualitatively illuminated areas are more attractive and are therefore frequently used (Huber, 2006). Optimization of lighting towards quality improvement is imperative as it reduces operating costs.

15.2.4 User-Oriented Approach of Urban Artificial Night Lighting Management

The same lamp may produce a different effect. The absence or reduction of night lighting does not necessarily mean a low quality of living. At dusk, the negative effects of lighting are automatically reduced and energy costs are lowered. However, low costs are the key to successful sustainable development of the city. The starting point of the presented chapter is to consider how to achieve sustainable energy management of artificial night lighting in urban environments. In this context, the question is how to use light only where it is needed. How to find the optimal level between lighting "necessity," "intensity," and "cost."

In the technologically oriented aspect of energy management, illuminance is a measurable unit, expressed in lux. But sustainable management also includes the human dimension. We cannot talk about an absolute measurable light intensity here – the perception of the environment is perceived differently by different people. People perceive the illuminated environment differently, and our subjective perception depends on the spatial conditions. Focusing on a user-oriented approach, the following questions are addressed in the lighting of public urban areas:

(1) Is urban lighting necessary for users? And if so, why is it necessary?
(2) To what extent is it necessary to light public areas?
(3) Are we willing to invest material resources for lighting purposes?

There is a tendency to answer all these questions, and the main findings in this context are the following: Quality urban lighting means much more than providing a sufficient amount of light and minimizing energy consumption with the increase of renewable energy sources. It cannot be achieved as a permanent state, but only as a set of desirable characteristics of a complex system.

15.3 THE EXPERIMENT

The experiment and data analysis took place between 2016 and 2018. The experiment was based on a sample of 200 respondents, 100 men and 100 women, aged between 18 and 34 years. The respondents were anonymous. The interviews were conducted in small groups of up to 15 people. They lasted about half an hour. The interview location was a classroom. All participants were familiar with the research cases. The interviewer conducted the handling of the questionnaire passively. Each participant completed the questionnaire once. The questionnaire consisted of four questions. The second-largest Slovenian city was chosen as the research site. Maribor has an important central regional role in the traditional region of Lower Styria. The city preserves the awareness of its exceptional natural and geographical position. It is situated between the slopes of the Piramida and Kalvarija Hills and the Mestni vrh (city peak) to the north and the southern side of the Alps, Pekrska gorca, and Pohorje. The Drava leaves its alpine character here and becomes calmer as it enters the flat Pannonian plain. The Drava as a dividing line in the city has become an important focal point of the urban ambience. The city center on the left bank of the river consists of a series of squares in the old core of the historical center, churches, monuments, and historical facades. The right bank of the river has mainly residential and industrial functions.

As seen in Table 15.1, the selection of the studied cases focused primarily on basic urban typologies, such as street, square, and city park. The final selection of study cases was based on criteria such as urban characteristics, lighting purpose, lighting infrastructure, lighting type, lighting characteristics, lighting effect, lighting intensity, and color output. Based on the city matrix, 24 known cases were selected according to the principle of diversity. The majority of the cases are centrally located on the left bank of the river. It was assumed that central areas are better-known by the respondents. After the selection process, 13 locations (Gosposka Street, Poštna Street, Kneza Koclja Street, Lent – Old Town, Leon Štukelj Square, Castle Square, Main Square, Liberty Square, Slomšek Square, City Park Maribor, Market, Pohorje Ski Slope, and Clinical Centre) and 11 objects (pedestrian bridge, main bridge, university, theater, city hall, football stadium, Europark shopping center, Hills [Piramida, Kalvarija, Pekrska Gorca], Monument of the National Liberation War [NLW], Plague Monument, and A. M. Slomšek Monument). Respondents rated their lighting intensity and likability with a five-level rating scale.

15.4 STATISTICAL RESULTS AND DATA EVALUATION

Data obtained on the basis of a questionnaire were statistically processed and analyzed using SPSS Windows statistical program. Methods of descriptive statistics and inferential statistics were used.

TABLE 15.1
Criteria and Elements Used in Selecting Studied Cases

Selection Criteria	Elements
Urban characteristics	• Urban typology: street (main, side, transit), square (greenery on it, no greenery on it), green area (city park)
Illumination purpose	• Attracting attention, orientation ability, sense of safety, location aesthetic
Lighting infrastructure	• Freestanding lamp (on a pole), wall lamp, pendant lamp, floor lamp, interior lighting of the building (shop window), lighting of objects (information boards, billboards, road signs, etc.), light beam, etc.
	• One-sided installation of lights, two-sided installation of lights
Type of lights	• Mercury lamp, halogen lamp, sodium lamp, LED
Lighting characteristics	• Surface lighting for motor vehicles, surface lighting reserved for pedestrians and cyclists, decorative lighting of buildings
	• Direct and indirect distribution of luminous flux
	• Point and planar distribution of light
Lighting effect	• Increase safety, visual guidance (orientation), information, warning, decoration, etc.
Illuminance intensity	• Low, medium, high intensity (lux)
Color output	• Monochromatic, polychromatic

TABLE 15.2
Descriptive Statistical Evaluation of Individual Spatial Factors

Spatial Factors	$\bar{x}$	σ
a. Well-designed and maintained network of traffic routes	4.36	0.722
b. Functioning public transport system	4.10	0.967
c. Access to public services	4.38	0.705
d. The presence of greenery and larger green areas in the city	4.51	0.702
e. Well-maintained sports facilities	4.05	0.881
f. City squares as places of public life	3.87	0.986
g. Sense of safety	4.48	0.750
h. Illuminated surfaces	4.20	0.813
i. Presence of functional urban equipment	4.26	0.711
j. Beauty, attractiveness, and cleanliness of the city	4.48	0.641
k. Children's playgrounds	3.50	1.070
l. Parking lots	4.14	1.016
m. Garbage collection, regular cleaning of green areas	4.56	0.707

$\bar{x}$ is arithmetic mean, σ is standard deviation.

15.4.1 Users' Priorities Regarding Illumination Necessity

The guiding question of the first part of the study was: *Is urban lighting necessary for users? And if so, why is it necessary?* We determined which spatial factors are more or less important compared to others. Is spatial lighting necessary? How high does it rank on the scale of priorities? In Table 15.2, respondents rated the importance of different factors using a five-level assessment scale. On this scale, 1 represents extremely unimportant, and 5 represents extremely important. The first question was: *How important are the following factors for you?*

The results in Table 15.2 show that respondents rate the importance of individual spatial factors in the city relatively highly. The average values range from 3.50 to 4.56. The relatively lowest value score of 3.87 was given to the factor "*City squares as places of public life*"; the second-highest value was given to the factors "*Sense of safety*" and "*Beauty, attractiveness and cleanliness of the city.*" The factor "*Illuminated surfaces,*" which has a direct influence on them, ranked eighth out of 13 possibilities, which means that the respondents are not aware of the importance of this factor to this extent. They do not perceive a direct relationship between safety and the level of lighting. The standard deviations range from 0.641 to 1.070, and in most cases, the dispersion of responses is greater for those factors that have lower average scores. The only exception is the factor "*Parking lots.*"

By the second question we have established the importance of the illumination of each type of urban area. We have ascertained whether the one whose illumination is most important is also the most indispensable illumination. When can we speak of economic efficiency in this context? The respondents used the five-level evaluation scale to evaluate 15 typologically different areas and buildings that are most often the subject of lighting in urban areas. They figured out how important it is to light this area and where lighting could be easily missed if the cost of public lighting has to be reduced or fit with the supply from renewable energy sources. The second question was: *Assess how important lighting the listed areas is to you! If it were necessary to reduce lighting to reduce the cost of public lighting, where would it be most difficult to miss it?*

The results in Table 15.3 show that in most cases, lighting is an important spatial factor for each urban area type. The lowest average importance rating (2.15) was assigned to plant lighting. Also less important is surface lighting: outdoor lighting of shopping malls (2.77), lighting of facades of public buildings (2.85), and lighting of statues and monuments (2.95). The highest average score (3.90) is given to the lighting of squares. Lighting of the old town (3.81), the ski slope (3.77), and bridges (3.75) is also highly rated. The average ratings range from 2.15 to 3.90, depending on the subject of illumination.

Respondents would tend to miss the reduction of lighting compared to the importance of lighting, as the average ratings for the reduction of lighting are lower than its importance in all cases. They range from 2.00 to 3.76. Thus, the lowest value is missing in the case of plant lighting, where we also have the lowest value for the importance of lighting. In this case, respondents would be most likely to accept the reduction. However, they would be most likely to agree to the reduction of lighting in the downtown area, which also has the highest average value for the importance of lighting. All other cases are predominantly in between these two extremes. Of all the options, lighting of sacred buildings stands out the most. Respondents would be most affected by any reduction in this lighting (the average score for lack of light is one of the highest at 3.65), although they are aware that their lighting is not the most important (a score of 3.60 is not one of the highest).

TABLE 15.3
Descriptive Measurements of the Importance of Lighting Individual Types of Urban Areas and the Results of the t-Test and Correlation for Dependent Samples of the Pairs Day–Night

Subject of Illumination	Importance $\bar{x}$	σ	Missing Subject $\bar{x}$	σ	|Difference| $\bar{x}$	t	2p	r
a. Urban arterial roads lighting	3.66	0.917	3.23	1.114	0.430	6,237	0.000	0.559
b. Local road lighting	3.51	0.919	3.36	1.032	0.145	2,438	0.016	0.634
c. Lighting of public areas in the city center	3.97	0.763	3.76	0.905	0.215	3,709	0.000	0.528
d. Old town lighting	3.81	0.837	3.51	1.027	0.300	4,823	0.000	0.571
e. Square lighting	3.90	0.868	3.62	0.960	0.285	4,972	0.000	0.611
f. Park lighting	3.56	0.944	3.34	1.127	0.215	3,352	0.000	0.629
g. Outdoor lighting of shopping malls	2.77	1.121	2.51	1.169	0.255	3,730	0.000	0.644
h. Sports surfaces lighting	3.13	1.022	2.98	1.154	0.150	2,174	0.031	0.604
i. Facade of public buildings lighting	2.85	0.972	2.59	1.094	0.255	4,462	0.000	0.700
j. Plant lighting (trees, shrubs)	2.15	1.034	2.00	1.051	0.145	2,188	0.030	0.596
k. Sacral buildings lighting	3.60	1.121	3.65	1.247	0.065	−0,783	0.435	0.653
l. Parking area lighting	3.06	1.013	2.93	1.077	0.125	1,925	0.056	0.616
m. Statues and monuments lighting	2.97	0.972	2.58	1.077	0.395	5,877	0.000	0.574
n. Ski slope lighting	3.77	1.160	3.37	1.323	0.400	5,643	0.000	0.681
o. Bridge lighting	3.75	0.902	3.39	1.097	0.360	5,712	0.000	0.618

$\bar{x}$ is aritmetic mean; σ, standard deviation; t, value difference arithmetic test; $2p$, bidirectional level of statistical significance; r, Pearson product–moment correlation coefficient.

The results of the t-test show that in most cases, there is a statistically significant difference between the importance of lighting and the absence of reduction, which, in other words, means that in most cases the respondents would agree to a possible reduction. The only exception is the lighting of sacred buildings and parking lots, where respondents would mostly disagree with a reduction in lighting.

The results of the correlation coefficient r show that all cases are moderately correlated distribution functions. The coefficient ranges from 0.528 for the illumination of areas in the city center to 0.681 for the illumination of the ski slope. In this case, the difference in average values is also high (0.400), which is the second-largest difference.

15.4.2 Users' Priorities Regarding Illumination Intensity

The leading question of the research's central part was: *To what extent is it necessary to illuminate public areas?* Respondents assessed whether the lighting intensity affects the illumination likability. Are brighter surfaces more likable? The relationship between illumination intensity and likability was determined at ten sites using a five-level evaluation scale. The questionnaire question was: *Evaluate the illumination intensity and likability of the listed locations.*

Case study (CS) number: CS1, Gosposka Street; CS2, Gosposka Street; CS3, Poštna Street; CS4, Kneza Koclja Street; CS4, Lent – Old Town; CS5, Leon Štukelj Square; CS6, Castle Square; CS7, Main Square; CS8, Anton Martin Slomšek Square; CS9, Liberty Square; CS10, City Park Maribor; CS11, market; CS12, Pohorje ski slope; CS13, clinical center; CS14, pedestrian bridge; CS15, main bridge; CS16, university; CS17, theater; CS18, city hall; CS19, football stadium; CS20, Europark shopping center; CS21, hills (Piramida, Kalvarija, Pekrska Gorca); CS22, National Liberation War Monument; SC23, Plague Monument; CS24, Anton Martin Slomšek Monument.

The results in Table 15.4 show how illumination intensity affects likability. CS5, Leon Štukelj Square, received the highest rating of likability (4.09), and CS22, National Liberation War Monument at Castle Square, received the lowest (2.82). In terms of illumination intensity, CS20, Europark shopping center, received the highest average rating (4.25). Among the worst-illuminated locations was, again, CS21, NLW Monument (2.52), followed by CS10, city park (2.53). The correlation coefficient r ranges from 0.198 to 0.598, which means that in all cases, a slight to moderate correlation can be found between illuminance intensity and likability. In the sixth column of Table 15.4, a new ratio is introduced and named sustainability coefficient. It is expressed as the quotient of the average values between likability and illumination intensity (equation 1):

$$\text{Sustainability coefficient}(S) = \frac{\text{Likeability}(L)}{\text{Illumination Intensity}(I)} \quad (1)$$

The sustainability coefficient (S) is a subjective measure that has no physical basis, since both values, likability and

TABLE 15.4
Descriptive Measurements of the Illumination Intensity and Likability of the Locations and the Results of the t-Test and Correlation

Case Study (CS)	Likability (L) $\bar{x}$	σ	Illumination Intensity (I) $\bar{x}$	σ	$\frac{(L)}{(I)}$	|Difference| $\bar{x}$	t	2p	r
CS1	3.12	0.727	3.30	0.677	0.95	0.180	−2.656	0.009	0.198
CS2	3.16	0.073	2.87	0.780	1.10	0.290	4.441	0.000	0.512
CS3	3.49	0.902	3.71	0.850	0.94	0.215	−2.925	0.004	0.395
CS4	3.24	0.947	2.90	0.824	1.12	0.335	4.964	0.000	0.448
CS5	4.09	0.911	4.10	0.822	0.98	0.011	−0.141	0.888	0.283
CS6	3.34	0.803	3.28	0.727	1.02	0.067	1.214	0.226	0.534
CS7	3.25	0.831	3.40	0.830	0.96	0.145	−2.469	0.014	0.517
CS8	3.34	0.803	3.28	0.727	1.02	0.067	1.214	0.226	0.534
CS9	3.05	0.868	2.86	0.847	1.07	0.188	2.880	0.005	0.523
CS10	3.14	0.955	2.53	0.851	1.24	0.607	7.803	0.000	0.345
CS11	2.66	0.887	2.56	0.926	1.03	0.097	1.430	0.155	0.539
CS12	3.88	0,927	3.89	0.965	1.00	0.011	−0.147	0.883	0.416
CS13	3.00	0.765	3.33	0.778	0.90	0.329	−4.566	0.000	0.310
CS14	3.41	1.098	2.80	0.893	1,22	0.605	7.914	0.000	0.524
CS15	3.33	0.954	3.30	0.737	1.01	0.029	0.429	0.669	0.483
CS16	3.71	0.902	3.68	0.838	1.01	0.036	0.558	0.578	0.459
CS17	3.74	0.822	3.62	0.822	1.03	0.115	1.712	0.089	0.451
CS18	3.33	0.919	3.35	0.889	0,99	0.018	−0.268	0,789	0,543
CS19	3.52	0.994	3.60	1.058	0.98	0.082	−0.943	0.247	0.344
CS20	3.46	1.107	4.25	0.819	0.81	0.786	−9.036	0.000	0.228
CS21	3.42	0.986	2.78	1.017	1.23	0.644	7.510	0.000	0,339
CS22	2.84	0.902	2.52	0.038	1.13	0.325	4.928	0.000	0.598
CS23	3.20	0.903	3.21	0.949	1.00	0.006	−0.088	0.930	0.575
CS24	3.20	0.789	3.05	0.926	1.05	0.022	0.286	0.024	0.454

$\bar{x}$ is arithmetic mean; σ, standard deviation; t, value difference arithmetic test; $2p$, bidirectional level of statistical significance; r, Pearson product–moment correlation coefficient.

illumination intensity, are determined on the basis of average ratings of subject opinion. However, the coefficient expresses very well our expectation to achieve the highest possible sympathy with the lowest possible illumination intensity. Lighting intensity is also an economic category, because to achieve higher lighting intensity, a certain type of technology requires more energy, which has its price, which means higher costs.

The value of the sustainability coefficient is around 1. Anything above 1 is a favorable result, which means that, on average, Likability is rated higher than Illumination Intensity. Conversely, this means that we invest a disproportionate amount of energy for low likability.

The results of the sustainability coefficient range from 0.81 for the outdoor area of CS20, Europark shopping center, to 1.24 for CS10, city park. The CS10, city park, site is an open-space site that has a relatively high likability score (3.14) with a relatively low illuminance score (2.53). This result means that if we increased the light intensity, we would only lose on the sustainability coefficient, which can also be interpreted to mean that this site is adequately lit. In the case of CS22, NLW Monument, we have a relatively low result of average values for both likability and light intensity (2.84 and 2.52, respectively), although the sustainability coefficient is relatively favorable (1.13). This result means that we need to improve the likability of lighting. By increasing the lighting intensity, we only worsen the existing result. For an unpleasantly lit place, it is better to have a subdued light that is less noticeable and less disturbing to people and surroundings. In the case of the CS20, Europark shopping center, we are dealing with a relatively pleasant lighting of the site (3.46) with a very high estimated light intensity (4.25). A low sustainability coefficient means that this location is too strongly lit.

The results of the t-test show that there are statistically significant differences in both directions, depending on the case. In cases where the means are statistically significantly different in the direction of higher means for likability, the t-test results are positive (the difference in mean values is positive). These examples are: CS3, Poštna Street; CS6, Castle Square; CS8, Slomšek Square; CS9, Liberty Square; CS10, Maribor City Park; CS4, Lent; CS14, pedestrian bridge; CS15, main bridge; CS16, university; CS17, theater; CS11, market; CS21, hills; CS22, NLW Monument; and

CS24, A. M. Slomšek Monument. In all other cases, the average values for likability are lower than the average values for illuminance. From here, there are the negative values of t. In the cases where $2p > 0.05$, there are no statistically significant differences. These examples are the CS6, Castle Square, and CS8, Slomšek Square; CS15, main bridge; CS16, university; CS17, theater; CS11, market; CS12, Pohorje ski slope; and CS23, Plague Monument, which have $S \geq 1.00$ at the same time. CS12, Pohorje ski slope, and CS23, Plague Monument, are classified in this group, although the result of the t-test is negative, but due to rounding to two decimal places, $S = 1.00$ is calculated.

There are also examples (CS5, Leon Štukelj Square; CS18, city hall; and CS19, football stadium) where it is also $2p > 0.05$ and, at the same time, $S < 1.00$. So there are also statistically insignificant differences between the averages. In all those cases where we do not have statistically significant differences between the mean values for likability and lighting intensity, the coefficient S is very close to 1.00, so in our cases, it is between 0.98 and 1.03. However, this interval partially overlaps with the interval $S > 1.00$, so we can combine it into an interval within which we can still speak of **urban open space illumination compliance**. This is the interval $S \geq 0.98$. The limit $S = 0.98$ is a rounded value of the coefficient S, where we can still find no statistically significant difference between likability in illumination intensity. In the present study, all cases except five ($S < 0.98$) were evaluated as compliant: CS1, Gosposka Street; CS4, Kneza Koclja Street; CS7, main square; CS13, clinical center; and CS20, Europark shopping center.

In summary, the higher the value of the sustainability coefficient for a given location, the more sustainably that location is lit. In other words, this means that we have achieved a higher site likability with a lower lighting intensity. For all those case studies where $S \geq 0.98$ ($Sn \geq 1$), we say that the lighting compliance of the open space is achieved. The normalized value of Sn is introduced to set the boundary of the closed interval to the value of 1 to make it easier to express (equation 2).

$$\text{Normalised } S(Sn) = S + 0{,}02 \qquad (2)$$

This can also be written in the form as (equation 3):

$$Sn \begin{cases} \geq 1, \text{lighting complience is achieved} \\ < 1, \text{lighting complience is not achieved} \end{cases} \qquad (3)$$

In cases where Sn < 1, lighting compliance is not achieved. Illumination intensity is high in relation to likability. In such cases, we consume too much energy compared to the cases when lighting compliance is achieved.

15.4.3 Users' Priorities Regarding Illumination Costs

The guiding question of the last part of the research was: *Are we willing to invest material resources for lighting?*

TABLE 15.5
The Number of Answers and the Percentage of the Evaluation of the Importance of Considering the Costs When Planning Public Lighting

Consideration of Costs When Planning Public Lighting	f	f %
Irrelevant	3	1.5
A little important	11	5.5
Important	122	61.0
Very important	52	26.0
Extremely important	12	6.0
Together	**200**	**100.0**

f is number of responses; *f* %, percentage of responses.

Light intensity is strongly related to the cost of public lighting. Higher intensity or irrational direction of light usually means higher electricity consumption, and therefore higher costs. Consequently, we have to take into account the aspect that lower energy consumption requires smaller place for photovoltaic system or wind turbine and less impact on the environment and landscape. The question was: *Do you think that when planning public lighting, its cost should also be taken into account?*

Around 98.5% of respondents are aware that the cost of public lighting is important. Only 5.5% consider these costs to be not very important, 61.5% of the respondents consider these costs to be important, and 32.0% rated these costs as very or extremely important.

The most important contribution of this part of the survey is to show how the public evaluates the cost necessity aspect of public lighting. The public is aware of the costs and wants these costs to be minimized while seeking the greatest positive effects and purposes that we want lighting to achieve.

15.5 CONCLUSION AND DISCUSSION

Today the function of daylight can be replaced. Lighting is a specific environmental design that depends on which spatial components should be visible and emphasized, what effect the lighting should have, what kind of light is used, where lighting elements can be placed, and last but not the least, who is the user of the space (Santen, 2006). Public lighting is a technical discipline which, in addition to a rigorous technical approach based on measurable physical quantities, also requires consideration of a subjective approach in which the psychological aspect must not be neglected. This is understandable, because we introduce night-light precisely for the users, to satisfy their needs. It would be right to take the user, his needs, and the psychological aspect as a starting point (Gregory, R. L., 1998). However, all this requires the introduction of a subjective measure. In this context, the sustainability coefficient (S) is derived as a basic subjective measure that combines two subjective factors, such as likability (L) and illumination intensity (I).

This coefficient is related to energy consumption, which is a purely technically measurable quantity. The sustainability coefficient (S) thus combines the subjective aspect with the technical aspect, which together form a kind of holistic whole.

In the cities of the future, the share of green energy is increasing and the share of nuclear energy is decreasing. Green energy is still not an unlimited resource. A lot of energy is consumed for lighting public urban areas, so the question is whether we can consume this energy sustainably and whether we can produce it entirely locally with renewable energy sources such as solar and wind energy source. It is necessary to find the value where the effect of light and the cost of energy are in harmony. Sustainable energy management of the future means monitoring between illumination need, intensity, and cost.

Is urban lighting necessary for users? And if so, why is it necessary? Users clearly express their expectation of illuminated spaces, but public lighting is not at the forefront of their consciousness. With its critical view and user-centered approach, the study aims to encourage a more conscious use of lighting elements in environmental design in order to reduce light pollution and energy losses.

If we agree that artificial night-light should not be absent, the question arises: To what extent is it necessary to illuminate public areas? This chapter presents the newly defined sustainability coefficient (S) as a subjective evaluation quotient of likability and intensity of lighting. The higher the value of this coefficient for a place, the more sustainably the place is lit. This means that with a lower lighting intensity, a higher site likability is achieved. Based on statistical parameters, an interval of $S \geq 0.98$ ($S_n \geq 1$) was set. Within this interval, urban open-space illumination compliance can be achieved. The introduction of this coefficient and the determination of the lighting compliance value provide a quantitative measure of sustainable lighting management in the city.

Anything above 0.98 is a favorable result, which means that, on average, likability is rated higher than illuminance. The reverse is also possible, which means that we invest a disproportionate amount of energy for low level of likability.

An important aspect of sustainability is the inclusion of a time component. Multiple applications of the sustainability coefficient allow us to obtain comparable results before and after spatial interventions, such as a lighting retrofit. A single application at a specific location also provides useful results. The coefficient can be used as an evaluation tool during the implementation phase of a preliminary analysis, followed by the elaboration of a lighting strategy. The latter was demonstrated in several cases with different urban backgrounds. The study was conducted in the city of Maribor (Slovenia).

Are we willing to invest material resources for quality artificial night lighting? With thoughtful planning, it is possible to achieve a high level of lighting at a low cost. However, lower costs also mean a contribution to more efficient economic growth and, indirectly, to the sustainability of the energy economy. Artificial night lighting of the urban environment is much more than just providing sufficient light with reduced energy consumption.

To get a holistic picture of a city lighting, it is necessary to spread the research over a wider radio. More cases are needed. Coefficient testing at multiple locations with similar urban typology, such as streets, parks, and squares, would also provide new insights by comparing the results of the same typology. In order to get a proper knowledge of the larger geographical area, it is also recommended to extend the research to different geographical and cultural areas.

15.6 REFERENCES

Briggs, W. R. *Physiology of Plant Responses to Artificial Lighting, Ecological Consequences of Artificial Night Lighting.* (ed.) Longcore, T., Rich, C. Washington, DC: Island Press, 2006, pp. 389–411.

Dabbagh, E. The Effects of Color and Light on the Beautification of Urban Space and the Subjective Perception of Citizens. *International Journal of Engineering Science Invention (IJESI)*, 2019, 8(3), 20–25.

Gregory, R. L. *Eye and Brain, The Psychology of Seeing.* Oxford, UK: Oxford University Press, 1998.

Huber, C. Sicherheit durch Licht. *Str. und Verk*, 2006, 9, 6–9.

Longcore, T., Rich, C. Ecological Light Pollution. *Ecological Environment*, 2004, 2(4), 191–198.

Mizon, B. *Light Pollution: Responses and Remedies.* London: Springer, 2002.

Pellegrino, A. Lighting Control System to Improve Energy Performance and Environmental Quality of Buildings: Limits and Potentials. *V: Razsvetljava delovnih mest. Petnajsto mednarodno posvetovanje Razsvetljava*, 2006, Bled, 12–13. okt. 2006. Orgulan A. (ur.). Maribor, Slovensko društvo za razsvetljavo, 2006, 37–46.

Rebernik, N., Goličnik Marušić, B., Bahillo, A., Osaba, E. 4-dimensional Model and Combined Methodological Approach to Inclusive Urban Planning and Design for All. *Sustainable Cities and Society*, 2019, 44, 195–214. DOI: 10.1016/j.scs.2018.10.001

Rebernik, N., Szajczyk, M., Bahillo, A., Goličnik Marušić, B. Measuring Disability Inclusion Performance in Cities Using Disability Inclusion Evaluation Tool (DIETool). *Sustainability*, 2020, 12, 1378.

Rozman Cafuta, M. Open Space Evaluation Methodology and Three Dimensions Evaluation Model as a Base for Sustainable Development Tracking. *Sustainability*, 2015, 7(10), 13690–13712

Santen, C. *Light Zone City: Light Planning in the Urban Context.* Basel, Switzerland: Birkhaeuser, 2006.

16 A Review on Multiobjective Control Schemes of Conventional Hybrid DC/AC Microgrid

S. Mamatha and G. Mallesham

CONTENTS

- 16.1 Introduction 197
- 16.2 Why the Potential Grid Solution Is a Hybrid AC/DC Microgrid 198
  - 16.2.1 DC Power Transmission in the Existing AC Power Supply System 198
  - 16.2.2 AC Power System Load Evaluation 198
  - 16.2.3 Incorporating Renewable Energy Systems into Existing Power System Network 198
- 16.3 Conventional Hybrid DC–AC Microgrid Architecture 199
- 16.4 Idea of Management of Power and Control Scheme 199
  - 16.4.1 Islanded Mode of Operation in Conventional Hybrid DC/AC Microgrid 199
  - 16.4.2 Grid-Connected Operating Mode of the Proposed System 200
- 16.5 Transition from Islanded Mode to Grid-Connected Mode 201
  - 16.5.1 Transition to Grid-Connected Mode from Islanded Mode 202
  - 16.5.2 Transition to Islanded Operating Mode from Grid-Connected 202
- 16.6 Interlinking Converter Control 202
- 16.7 Conclusion 202
- 16.8 References 202

16.1 INTRODUCTION

In developing countries like India, population growth is increasing day by day; as well, electricity demand for the country is also increasing with the growth of population. This has a major effect on energy demand requirements. Some rural areas of India do not have continuous electricity. Because of this, many farmers are facing problems due to the lack of electricity supply in their fields for water pumping. Till April 2020, India's total installed capacity is 3,70,106 MW. Major contribution for electricity is from thermal power generating stations. As the conventional methods of power generation may lead to pollution, in the future there will be a problem of deterioration of fossil fuels. In addition, an increasing demand for electricity due to increased population will lead to alternative energy sources, such as renewable energy systems [1, 2].

In power system, the incorporation of distributed generators (DGs) is increasing day by day; hence, managing the power of the main grid and DGs has become a big task. In this concern, in power system network, microgrid has grown into a broadly valid concept to connect with distributed generators. The microgrid is defined as the set of distributed energy sources and storage devices, whereas in conventional power systems, AC microgrids are already established, and in particular, several surveys were recorded on the sharing of power between the parallelly connected energy sources.

As the majority of RES generate DC power, it is feasible to establish a DC grid in the power system network. Due to the increased demand for modern DC loads and to maintain the reliability of the DC load, to make the system cost-effective, many electrical engineers are proposing the idea of a DC microgrid. The idea is, by connecting DC loads to the DC grid, power conversion stages, such as AC to DC and DC to DC, may be reduced; because of this conversion, cost will be become less. Nevertheless, as of today, most of the existing power grids are working for AC supply only, and it is not anticipated that solely DC microgrids can occur exclusively in power grids. In recent studies, it has become much important to avail the benefits of both the grids by connecting them with an interlinking device which is called the bidirectional main converter. The main idea is to merge the two grids through a main converter, called bidirectional converter, having the capability to convert AC to DC and vice versa. By this new structure, power exchange is possible depending on the load demand.

As the nature of the load is changing day by day, there is a need for the distribution of power in the DC paradigm. AC microgrids are the foremost established microgrids as corresponds to the conventional power systems. But many of the renewable energy sources will generate DC power and need to be interconnected with the DC microgrid. Nowadays, heavy usage of DC loads, such as with laptops, computers, phones, printers, etc., has made researchers think regarding the aforesaid hybrid system. As advances have taken place in the power electronic sector, DC voltage regulation has become an easy task. By this new thought, DC loads and AC loads can be fed uninterrupted power as the power exchange is done between the AC and the DC subsystems, using power electronic converter interlinking both grids [1, 3–5].

16.2 WHY THE POTENTIAL GRID SOLUTION IS A HYBRID AC/DC MICROGRID

In the 1880s, Thomas Edison introduced the DC supply system, at the same time Tesla invented the AC supply system. In the 1890s, AC emerged as the potential power supply system. Because of the invention of the transformer, transfer of power became much easier. Does it mean DC power supply has disappeared? The answer is an unequivocal no. As advances took place in power electronics and control system technologies in traditional power system loads, due to the increase of RES in low-voltage distribution systems and increased demand of power electronics–based HVDC systems, researchers had to think again regarding the establishment of DC supply into the existing AC power supply systems [6–8].

16.2.1 DC Power Transmission in the Existing AC Power Supply System

As we know, in high-voltage AC transmission, many technological achievements have taken place. It does not represent that for the same voltage level of transmission, AC transmission is much superior to the DC transmission system. Also, in AC power systems, under the high-voltage rating and high-power-flow condition, there is a chance of blackouts of the complete system because of the AC transients present in the system. As AC transient stage is much longer than the DC transient stage, in high-voltage DC transmission, the aforementioned problem can be avoided [9–12].

16.2.2 AC Power System Load Evaluation

The power supply systems are established at the earliest to supply the loads, such as heating loads, lighting loads, and motor loads. DC loads and AC loads are manufactured at the early stages majorly depending on the existing power supply systems at that time. Because of this, all types of loads in the power system networks are adapted forcefully to AC power systems [11, 12]. Because of tremendous improvements in the power electronics sector, scientists have invented a device that will convert AC power supply to DC power supply, called a rectifier, used to connect DC loads to the AC network. In the existing power system, there is a drastic shift in the type of load applied to the system because of the tremendous increase in utilizing DC loads by the users. When we return to researching loads in modern power systems, we find that DC and AC loads with converters (ACwC) play a dominant role in most AC power systems. At present, the load used by the electricity consumer is not pure AC. Some of the domestic loads used presently are for servers, printers, video recorders, and TVs, etc., and all these are of DC-type loads; also, some of the domestic loads which are AC types, such as in refrigerators, air conditioners, industrial machinery, washing machines, etc., are presently using inverters instead of AC motors to save energy and for speed control.

Table 16.1 displays the types of loads used in the power system networks in the future. From the table, we conclude that in the future, most of the domestic and some of the industrial loads will be majorly DC type.

Table 16.2 displays the number of stages required to convert the type of power to satisfy the type of load.

16.2.3 Incorporating Renewable Energy Systems into Existing Power System Network

RES consists of power generating sources, such as solar panels, hydraulic power, wind turbines, fuel cells, geothermal energy sources, wave energy, tidal energy, etc. These sources are natural sources that cost nothing and are pollution-free. Incorporating these RES into the existing grid may lead to issues such as output power from solar panels is DC power, and power output of fuel cell is also DC, whereas from wind turbines, power output will be AC power or DC, depending on the way of construction. To connect RES with the existing utility grid, there is a need for a power electronics converter to convert the DC output generated by renewable energy systems to AC. Even though the output of the

TABLE 16.1
Power System Typical Loads

	Type of Load	AC	DC	ACWC
1	Electronics loads	Not used	Used	Not used
2	UPS and energy storage	Not used	Used	Not used
3	Electrochemical process	Not used	Used	Not used
4	Future motor driver	Used	Used	Used
5	Future air conditioner	Not used	Not used	Used
6	Electric arc furnace	Used	Used	Not used
7	Heating	Used	Used	Used

TABLE 16.2
Power Conversion Stages between Type of Source and Load

S. No.	System	Conversion Needed		
		AC Load	DC Load	ACwC Load
1	Alternating current source	No conversion needed	AC-to-DC conversion needed	AC-to-DC-to-AC conversion needed
	Direct current source	DC-to-AC conversion needed	DC-to-AC-to-DC conversion needed	DC-to-AC-to-DC-to-AC conversion needed
2	Alternating current source	AC-to-DC conversion needed	AC-to-DC -to-AC conversion needed	AC-to-DC-to-AC conversion needed
	Direct current source	DC-to-AC conversion needed	No conversion needed	DC-to-AC conversion Nneded

solar system or any RES which generates output power as DC, this power output cannot be directly connected with the existing grid without power conversion and synchronization. To synchronize the new power with the existing system, controlling of power electronics device used for power conversion is essential. When RES is linked to the existing grid, the battery energy management system must be maintained using a charging–discharging controller. With the help of the proposed hybrid system number of conversion, stages will be reduced and smooth power exchange will be possible from AC side to DC and DC side to AC [13, 14].

16.3 CONVENTIONAL HYBRID DC–AC MICROGRID ARCHITECTURE

Due to the limited fossil fuels available, and by considering the environmental issues raised due to the utilization of fossil fuels to generate electricity, many electrical engineers are paying concentration on the generation of electrical energy using RES and also on the efficiency of power generated by renewable sources. By connecting these RES into the existing AC grid, many numbers of power conversion stages are required to satisfy the load demand as well as the type of load. Table 16.2 shows various conversion techniques used for an individual AC microgrid or DC microgrid. By observing the number of conversion stages as shown in Table 16.2, there will be a question raised, such as, Can we reduce the conversion process, or can we reduce the additional equipment? This idea leads to the concept of a hybrid power system which includes both the AC link and DC link based on the already-existing infrastructure. Many electrical engineers are also working on the issue of adding DC microgrids in the distribution systems for better load management. At the distribution level, rearranging the existing utility grid as a hybrid grid may reduce the synchronization problem, and also, power conversion stages can be reduced [15].

The structure of the hybrid DC–AC microgrid is shown in Figure 16.1. In Figure 16.1, DC bus and AC bus are connected to distributed generators and storage devices, and both buses are connected with an interlinking converter which is a bidirectional converter, whereas a bidirectional converter or interlinking converter will increase the capacity and reliability of the system [16–20]. This structure requires better coordination between the grids to control voltage level and power. The major benefit of this hybrid system which is linked using a bidirectional converter is, it reduces the number of power conversion stages; due to this, cost of the system also reduces, and the overall efficiency of the system improves.

Even though the idea of a hybrid DC–AC microgrid is more feasible, it is much necessary to study and investigate the energy management and power balance between AC and DC buses. For better management of power between DC grid and AC grid, control of voltage and control of frequency are essential.

16.4 IDEA OF MANAGEMENT OF POWER AND CONTROL SCHEME

As the proposed system consists of many numbers of DGs containing multiple renewable energy sources, storage devices and AC loads and DC loads, which are connected to their respective AC and DC grids, must maintain the best control strategy for better power flow. As well, the coordination between DC and AC grids is necessary [20–24]. Figure 16.2 shows the architecture of a hybrid DC–AC microgrid connected with a bidirectional converter. The interlinking converter, which is connected between the DC grid and AC grid, has to work under the bidirectional power control mode. The proposed structure involves the following two modes of operations: (i) islanded mode and (ii) grid-connected mode. Figure 16.3 shows a brief view of the power management scheme used in the islanded mode of operation and grid-connected mode of operation.

16.4.1 ISLANDED MODE OF OPERATION IN CONVENTIONAL HYBRID DC/AC MICROGRID

In the proposed hybrid DC–AC microgrid, both AC microgrid as well as DC microgrid consist of multiple sources and loads. In this architecture, coordination between distributed generators, bidirectional or interlinking converters, and storage devices plays a crucial role. In this islanded mode of operation, by applying the voltage control methods, such as droop control, master–slave control, etc., we can control the operating voltage and frequency of the AC grid as well as the DC grid. In a DC subsystem, operating voltage can also be controlled by utilizing a power balance control

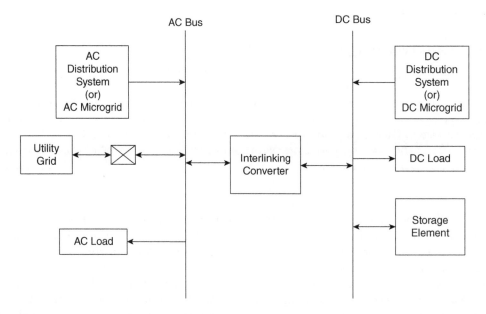

FIGURE 16.1 Conventional hybrid DC–AC microgrid system.

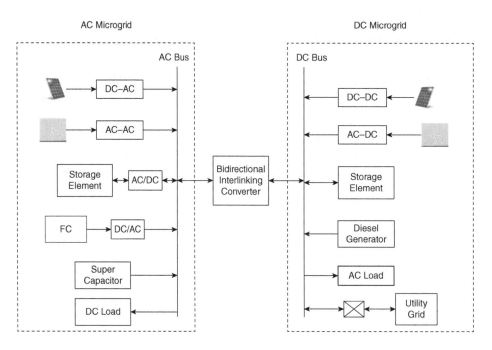

FIGURE 16.2 Architecture of hybrid system using bidirectional converter.

strategy. It is observed that in the stand-alone mode of operation, the bidirectional converter, which is also called interlinking converter, will play a dominant role in controlling the voltage and frequency and in managing power [25–28]. This bidirectional power flow converter can operate in the following three modes: (i) bidirectional converter operating to control AC microgrid power, (ii) bidirectional converter operating to control DC microgrid power, and (iii) bidirectional converter operating to control the power of both the microgrids. The major important thing in an islanded mode of operation is to incorporate good communication between the previous three cases.

Figure 16.4 shows the structure of parallelly connected interlinking converters connected between the two microgrids. In parallel connection, one converter can operate in voltage control of AC or DC system [29], whereas another converter can manage the power flow between AC and DC systems.

16.4.2 GRID-CONNECTED OPERATING MODE OF THE PROPOSED SYSTEM

In grid-connected mode, the bidirectional converter needs to operate in two methods, namely, transmitted and

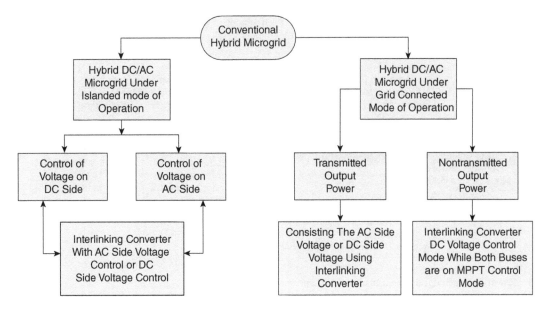

FIGURE 16.3 Power management scheme.

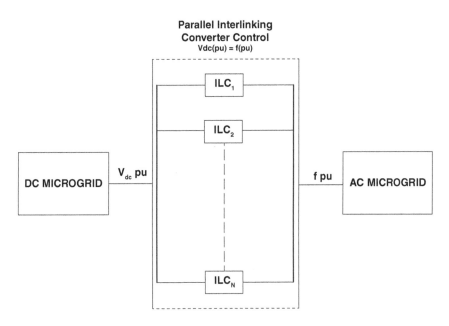

FIGURE 16.4 Hybrid AC/DC microgrid with parallel interlinking converters.

nontransmitted power method. In the first method, voltage of the DC microgrid [30–32] is set to some desired value, and the bidirectional converter needs to work for the voltage regulation on the DC side. This first method is the voltage regulation method. In the second method, transmitted power output can be produced by the combined operation of the AC bus and the DC bus, including the interlinking converter. The interlinking converter will work under power control in this mode of operation. DGs in the DC bus and the AC bus will operate on maximum power point in a grid-connected nontransmitted output power operation mode.

16.5 TRANSITION FROM ISLANDED MODE TO GRID-CONNECTED MODE

To reduce disturbances in frequency variation and voltage variation and to make a smooth power flow between AC and DC microgrids, the change from islanded mode to grid-connected mode is required. The control strategies for the change from islanded to grid-connected operations and the change from grid-connected to islanded mode are discussed separately here. Figure 16.5 provides the summary of this transition.

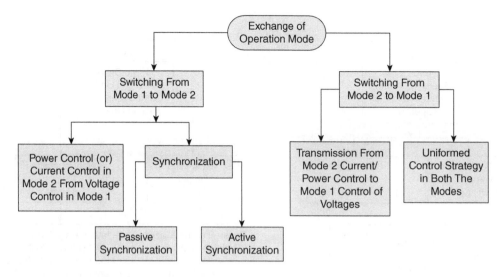

FIGURE 16.5 Power management during transition from one mode to another mode.

16.5.1 Transition to Grid-Connected Mode from Islanded Mode

In the transition from islanded operation mode to grid-connected operation mode, a major task is to make synchronization between voltages of the microgrid and the main grid before reconnection. There are mainly two types of synchronization methods to maintain synchronization between the microgrid and the main grid. Synchronization methods are (i) active synchronization and (ii) passive synchronization [33–37]. In some situations, the synchronization device is incorporated into the control strategies, whereas in others, the synchronization signals are generated by a separate synchronization unit to provide a microgrid to connect with the grid. The active synchronization methods can be divided into two types. In the first type, one or more distributed generators initiate the synchronization process, and the other distributed generators will follow them. And in the second type, in the synchronization process, simultaneously all distributed generators will participate. The first synchronization method is primarily used in microgrids, where some distributed generators operate on the current control mode, while others work on the voltage control mode. The second synchronization method is primarily used in the microgrids that operate on the voltage control mode for all distributed generators. After the synchronization, the microgrid will be connected to the main grid at zero crossing of voltage.

16.5.2 Transition to Islanded Operating Mode from Grid-Connected

In this case of transition, there are two classes of control strategies; first one is switching of control strategies from power/current control mode in grid-connected operation to voltage control mode in islanded mode operation, and second one is uniform control in both islanded mode and grid-connected mode. By using better control strategy in this transition mode, we can extract the best power output without keeping much stress on power electronic controllers used [38].

16.6 INTERLINKING CONVERTER CONTROL

For better power exchange between AC and DC subgrids, interlinking converter control is necessary. As shown in the Figure 16.4, by connecting the interlinking converters in parallel, a large amount of power exchange is possible. By connecting single interlinking converter, we can manage the exchange of power between the AC microgrid and the DC microgrid, but it might create overstressing in the interlinking converter and circulating power flow [39–43]. By connecting multiple interlinking converters between the AC subgrid and the DC subgrid with suitable novel droop control method, power sharing will be improved and it will reduce circulating powers.

16.7 CONCLUSION

For more than 100 years, power systems with AC have served the world according to the convenience of the user. But nowadays, there is rapid change and advancement in modern society; due to this, type of load used by the power consumer is also changing day by day. Because of the load demand, many electrical engineers are also trying to develop a hybrid power system which is user-friendly. In this chapter, one of the hybrid AC/DC microgrid architecture is discussed with its power management and control scheme.

16.8 REFERENCES

[1] Farzam Nejabatkhah and Yun Wei, "Overview of power management strategies of hybrid AC/DC microgrid," *IEEE Transactions on Power Electronics*, vol. 30, no. 12, Dec 2015
[2] Rouzbeh Reza Ahrabi, Yunwei Li, and Farzam Nejabatkhah, "Hybrid AC/DC network with parallel LCC-VSC

interlinking converters," *IEEE Transactions on Power Systems*, vol. 36, no. 1, Jan 2021

[3] Farzam Nejabatkhah, Yun Wei Li, and Hao Tian, "Power quality control of smart hybrid AC/DC microgrids: An overview," *IEEE Access*, vol. 7, 2019

[4] T. Ma, M. H. Cintuglu, and O. A. Mohammed, "Control of a hybrid AC/DC microgrid involving energy storage and pulsed loads," *IEEE Trans. Ind. Appl.*, vol. 53, no. 1, pp. 567–575, Jan 2017

[5] Y. Liu, X. Hou, X. Wang, C. Lin, and J. M. Guerrero, "A coordinated control for photovoltaic generators and energy storages in low-voltage AC/DC hybrid microgrids under islanded mode," *Energies*, vol. 9, no. 8, p. 651, 2016.

[6] A. A. Hamad, M. A. Azzouz, and E. F. El-Saadany, "A sequential power flow algorithm for islanded hybrid ac/dc microgrids," *IEEE Trans. Power Syst.*, vol. 31, no. 5, pp. 3961–3970, Sept 2016.

[7] M. Davari and Y. A. R. I. Mohamed, "Robust multi-objective control of VSC-based dc-voltage power port in hybrid ac/dc multi-terminal microgrids," *IEEE Trans. Smart Grid*, vol. 4, no. 3, pp. 1597–1612, Sept 2013.

[8] N. Eghtedarpour and E. Farjah, "Power control and management in hybrid ac/dc microgrid," *IEEE Trans. Smart Grid*, vol. 5, no. 3, pp. 1494–1505, May 2014

[9] A. A. Eajal, E. F. E. Saadany, and K. Ponnambalam, "Equal power sharing in islanded ac/dc hybrid microgrids," in *Proc. 2016 IEEE Elect. Power and Energy Conf. (EPEC)*, Ottawa, ON, Oct 2016, pp. 1–6.

[10] J. Zhang, D. Guo, F. Wang, Y. Zuo, and H. Zhang, "Control strategy of interlinking converter in hybrid AC/DC microgrid," in *Proc. 2013 Int. Conf. Renewable Energy Res. And Applicat. (ICRERA)*, Madrid, Oct 2013, pp. 97–102.

[11] M. Davari and Y. A. R. I. Mohamed, "Robust multi-objective control of VSC-based DC-voltage power port in hybrid ac/dc multi-terminal microgrids," *IEEE Trans. Smart Grid*, vol. 4, no. 3, pp. 1597–1612, Sept 2013.

[12] S. Bahrami, V. W. S. Wong, and J. Jatskevich, "Optimal power flow for AC-DC networks," in *Proc. 2014 IEEE Int. Conf. Smart Grid Commun. (smartgridcomm)*, Venice, Nov 2014, pp. 49–54.

[13] P. Wang, C. Jin, D. Zhu, Y. Tang, P. C. Loh, and F. H. Choo, "Distributed control for autonomous operation of a three-port AC/DC/DS hybrid microgrid," *IEEE Trans. Ind. Electron.*, vol. 62, no. 2, pp. 1279–1290, Feb 2015.

[14] X. Liu, P. Wang, and P. C. Loh, "A hybrid AC/DC microgrid and its coordination control," *IEEE Trans. Smart Grid*, vol. 2, no. 2, pp. 278–286, June 2011.

[15] E. Unamuno and J. Barrena, "Hybrid ac/dc microgrids – part I: Review and classification of topologies," *Renew. Sustain. Energy Rev.*, vol. 52, pp. 1251–1259, 2015.

[16] E. Unamuno and J. A. Barrena, "Hybrid ac/dc microgrids – part II: Review and classification of control strategies," *Renew. Sustain. Energy Rev.*, vol. 52, pp. 1123–1134, 2015.

[17] M. Rahman, M. Hossain, and J. Lu, "Coordinated control of threephase AC and DC type EV-ESSS for efficient hybrid microgrid operations," *Energy Conversion and Management*, vol. 122, pp. 488–503, 2016.

[18] S. D. Manshadi and M. Khodayar, "Decentralized operation framework for hybrid ac/dc microgrid," in *Proc. 2016 North American Power Symp. (NAPS)*, Denver, CO, Sept 2016, pp. 1–6.

[19] E. Unamuno and J. A. Barrena, "Primary control operation modes in islanded hybrid ac/dc microgrids," in *Proc. IEEE EUROCON 2015 – Int. Conf. Comput. As a Tool (EUROCON)*, Salamanca, Sept 2015, pp. 1–6.

[20] P. Caramia, G. Carpinelli, F. Mottola, and G. Russo, "An optimal control of distributed energy resources to improve the power quality and to reduce energy costs of a hybrid ac-dc microgrid," in *Proc. 2016 IEEE 16th Int. Conf. Environment and Elect. Eng. (EEEIC)*, Florence, June 2016, pp. 1–7.

[21] W. Guo, X. Han, C. Ren, and P. Wang, "The control method of bidirectional ac/dc converter with unbalanced voltage in hybrid microgrid," in *Proc. 2015 IEEE 10th Conf. Indus. Electron. and Applicat. (ICIEA)*, Auckland, June 2015, pp. 381–386.

[22] D. K. Dheer, S. Doolla, and A. K. Rathore, "Small signal modeling and stability analysis of a droop based hybrid ac/dc microgrid," in *Proc. IECON 2016–42nd Annu. Conf. IEEE Ind. Electron. Soc., Florence*, Oct 2016, pp. 3775–3780.

[23] C. Jin, J. Wang, K. L. Hai, C. F. Hoong, and P. Wang, "Coordination secondary control for autonomous hybrid ac/dc microgrids with global power sharing operation," in *Proc. IECON 2016–42nd Annu. Conf. IEEE Ind. Electron. Soc.*, Florence, Oct 2016, pp. 4066–4071.

[24] K. Sun, X. Wang, Y. W. Li, F. Nejabatkhah, Y. Mei, and X. Lu, "Parallel operation of bidirectional interfacing converters in a hybrid ac/dc microgrid under unbalanced grid voltage conditions," *IEEE Trans. Power Electron.*, vol. 32, no. 3, pp. 1872–1884, March 2017.

[25] P. C. Loh, D. Li, Y. K. Chai, and F. Blaabjerg, "Autonomous control of interlinking converter with energy storage in hybrid ac-dc microgrid," *IEEE Trans. Ind. Appl.*, vol. 49, no. 3, pp. 1374–1382, May 2013.

[26] L. Zhang, F. Gao, N. Li, Q. Zhang, and C. Wang, "Interlinking modular multilevel converter of hybrid ac-dc distribution system with integrated battery energy storage," in *Proc. 2015 IEEE Energy Conversion Congr. and Expo. (ECCE)*, Montreal, QC, Sept 2015, pp. 70–77.

[27] H. Xiao, A. Luo, Z. Shuai, G. Jin, and Y. Huang, "An improved control method for multiple bidirectional power converters in hybrid ac/dc microgrid," *IEEE Trans. Smart Grid*, vol. 7, no. 1, pp. 340–347, Jan 2016.

[28] P. Wang, X. Liu, C. Jin, P. Loh, and F. Choo, "A hybrid ac/dc microgrid architecture, operation and control," in *Proc. 2011 IEEE Power and Energy Soc. General Meeting*, San Diego, CA, July 2011, pp. 1–8.

[29] P. C. Loh, D. Li, Y. K. Chai, and F. Blaabjerg, "Autonomous operation of ac-dc microgrids with minimised interlinking energy flow," *IET Power Electron.*, vol. 6, no. 8, pp. 1650–1657, September 2013.

[30] N. Eghtedarpour and E. Farjah, "Power control and management in a hybrid ac/dc microgrid," *IEEE Trans. Smart Grid*, vol. 5, no. 3, pp. 1494–1505, May 2014. M. Davari and Y. A. R. I. Mohamed, "Robust multi-objective control of VSC-based DC-voltage power port in hybrid ac/dc multi-terminal microgrids," *IEEE Trans. Smart Grid*, vol. 4, no. 3, pp. 1597–1612, Sept 2013.

[31] M. Davari and Y. A. R. I. Mohamed, "Robust multi-objective control of VSC-based DC-voltage power port in hybrid ac/dc multi-terminal microgrids," *IEEE Trans. Smart Grid*, vol. 4, no. 3, pp. 1597–1612, Sept 2013.

[32] A. A. A. Radwan and Y. A. R. I. Mohamed, "Assessment and mitigation of interaction dynamics in hybrid ac/dc distribution generation systems," *IEEE Trans. Smart Grid*, vol. 3, no. 3, pp. 1382–1393, Sept 2012.

[33] A. Abdelsalam, H. Gabbar, and A. Sharaf, "Performance enhancement of hybrid ac/dc microgrid based d-facts," *Int. J. Elect. Power and Energy Syst.*, vol. 63, pp. 382–393, 2014.

[34] Z. Jiang and X. Yu, "Power electronics interfaces for hybrid DC and AC-linked microgrids," in *Proc. 2009 IEEE 6th Int. Power Electron. and Motion Control Conf.,* Wuhan, May 2009, pp. 730–736.

[35] Y. Liu, A. Escobar-Mejía, C. Farnell, Y. Zhang, J. C. Balda, and H. A. Mantooth, "Modular multilevel converter with high-frequency transformers for interfacing hybrid dc and ac microgrid systems," in *Proc. 2014 IEEE 5th Int. Symp. Power Electron. For Distributed Generation Syst. (PEDG),* Galway, June 2014, pp. 1–6.

[36] S. Bahrami, V. W. S. Wong, and J. Jatskevich, "Optimal power flow for AC-DC networks," in *Proc. 2014 IEEE Int. Conf. Smart Grid Commun. (smartgridcomm),* Venice, Nov 2014, pp. 49–54.

[37] P. C. Loh, D. Li, Y. K. Chai, and F. Blaabjerg, "Autonomous operation of hybrid microgrid with ac and dc subgrids," *IEEE Trans. Power Electron.,* vol. 28, no. 5, pp. 2214–2223, May 2013.

[38] L. Che, M. Shahidehpour, A. Alabdulwahab, and Y. Al-Turki, "Hierarchical coordination of a community microgrid with ac and dc microgrids," *IEEE Trans. Smart Grid,* vol. 6, no. 6, pp. 3042–3051, Nov 2015.

[39] Y. Xia, Y. Peng, P. Yang, M. Yu, and W. Wei, "Distributed coordination control for multiple bidirectional power converters in a hybrid ac/dc microgrid," *IEEE Trans. Power Electron.,* vol. 32, no. 6, pp. 4949–4959, June 2017.

[40] A. A. Eajal, E. F. E. Saadany, and K. Ponnambalam, "Equal power sharing in islanded ac/dc hybrid microgrids," in *Proc. 2016 IEEE Elect. Power and Energy Conf. (EPEC),* Ottawa, ON, Oct 2016, pp. 1–6.

[41] A. A. Eajal, E. F. El-Saadany, and K. Ponnambalam, "Inexact power sharing in AC/DC hybrid microgrids," in *Proc. 2016 IEEE Can. Conf. On Elect. And Comput. Eng. (CCECE),* Vancouver, BC, May 2016, pp. 1–5.

[42] P. J. Hart, R. H. Lasseter, and T. M. Jahns, "Symmetric droop control for improved hybrid ac/dc microgrid transient performance," in *Proc. 2016 IEEE Energy Conversion Congr. and Expo. (ECCE),* Milwaukee, WI, Sept 2016, pp. 1–8.

[43] P. Wang, L. Geol, X. Liu, and F. H. Choo, "Harmonizing AC and DC: A hybrid AC/DC future grid solution," *IEEE Power Energy Mag.,* vol. 11, no. 3, pp. 76–83, May–Jun 2013.

17 Recent Advancements in Solar Thermal Technology for Heating and Cooling Applications

Sunita Mahavar

CONTENTS

17.1 Introduction 205
17.2 Solar Thermal Technology 208
    17.2.1 Basic Working Principle 208
    17.2.2 Basic Energy Balance Equation 210
17.3 Heating Technology and Applications 211
    17.3.1 Low Temperature 211
    17.3.2 Medium Temperature 218
    17.3.3 High Temperature 219
17.4 Cooling Technology and Applications 221
    17.4.1 Solar Desiccant Systems (Open System) 221
    17.4.2 Solar Absorption Cooling (Closed System) 222
    17.4.3 Solar Adsorption Cooling 223
17.5 Conclusion 223
17.6 References 224

17.1 INTRODUCTION

In the current century, the estimated population growth is 7 billion to 11 billion people – that will result into huge increase in the global energy consumption of the world. Figure 17.1 depicts the total production and consumption of energy from year 1990 to 2017 for the various regions of the world [1]. Vast industrialization, prompt technological changes, and advancements are the major factors for the rapid increase in energy generation and consumption after year 2000. The primary and final energy demands are given in Table 17.1 for year 2020 and 2040. The primary energy need of the world will increase by 23%, while final energy will upturn about 18% from the present. Although research and development would be continued to improve the efficiency of the present technology to meet the energy demand, lack of sufficient energy resources may befall for the sustainable development of the present world for the future [1–3].

About 54% share of the total energy of the world is utilized by the industrial sector of the world, and the estimated growth of this consumption is 1.2% annually. The major share of this energy is met by fossil fuel consumption that can be altered from environment-friendly sources, viz, solar energy [4]. This shift would also lead to solve other challenges before the world, such as energy security, unemployment, climate change, etc. [5]. The sun is a supreme power on Earth; the energy radiated by the sun in just 14 s is equal to the energy received by the Earth in about 1,000 years. Figure 17.2 depicts solar potential of different territories of the world, and Figure 17.3 infers areas in black dots which may combinedly fulfill the world's total energy demand if efficiency conversion is only 8% for the solar systems [5–8]. Clearly, solar thermal energy, with its versatile technological development for low- to high-temperature applications, has distinguished relevance among all renewable energy sources. From water heating to space heating and from space cooling to power generation, all these energy demands can be successfully delivered using solar thermal technologies.

To design a solar thermal system, understanding of the fundamentals of the solar geometry is of prime necessity. The solar radiation received at any surface without scattering in the Earth's atmosphere is called beam or direct radiation (I_b), while the radiation received after being subjected to scattering is called diffuse radiation (I_d). Global radiation is mainly the sum of the beam and diffuse radiation (I_g). At any time, the beam flux on the surface should be converted into an equivalent value corresponding to the normal direction to the surface of a solar thermal system. In Figure 17.4, θ_i is the angle between an incident beam of flux I_{bn} and the normal to a plane surface, so the equivalent flux falling normal to the surface is $I_{bn} \cos \theta_i$. The angle θ_i depends on various angles, viz, latitude of the location (ϕ), the declination (δ), the surface azimuth angle (γ), the hour

DOI: 10.1201/9781003321897-17

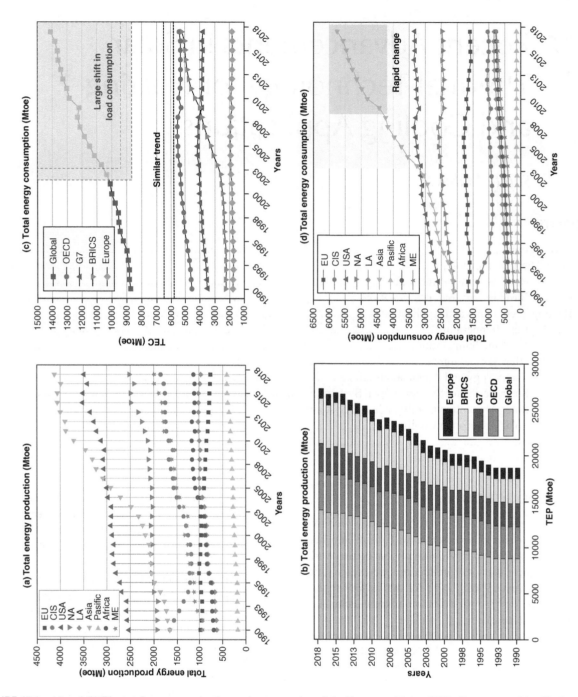

FIGURE 17.1 (a) and (b) The total energy production and consumption of the European Union (EU), Commonwealth of Independent States (CIS), USA, North America (NA), Latin America (LA), Pacific, Africa, and Middle East (ME). (b) and (c) The total energy production and consumption of Europe, Brazil, Russia, India, China, and South Africa (BRICS), the group of seven (G7), Organization for Economic Co-operation and Development (OECD), and global demand [1].

angle (ω), and the slope or tilt (β) of the surface, which are shown in Figures 17.5 and 17.6. The limits/variation of these angles are: the latitude varies from -90 to +90° (positive for the Northern Hemisphere and negative for the Southern Hemisphere); hour angle (ω) is 15° per hour (zero for local noontime, 12:00, +ve in forenoon, and -ve in afternoon); surface azimuth angle (γ) varies from -180 to +180°; and tilt angle (β) can vary from 0 to 180°. If the surface tilt angle is zero, then the incidence angle θ_i is equivalent to zenith angle. The declination angle can be obtained through following equation [9–11]:

$$\delta \text{ (in degrees)} = 23.45 \sin\left[\frac{360}{365}(284+n)\right] \quad (17.1)$$

Where n is the number of day of the year from January 1.

In terms of various angles described previously, the angle of incidence (θ_i) at any surface is given as:

TABLE 17.1
Primary and Final Energy Consumption of the Distinct Domains of the World [1]

Category	Primary Energy Consumption		Final Energy Consumption	
	2020	2040	2020	2040
Europe	1,858	1,650	1,354	1,193
Asia	6,318	8,770	4,372	5,776
Africa	749	1,301	590	970
ME	777	742	504	491
NA	2,562	2,347	1,764	1,568
LA	874	1,258	638	853
CIS	955	1,246	642	805
Pacific	149	172	105	120
World	14,243	17,487	9,970	11,775

FIGURE 17.2 Solar energy potential in the world [7].

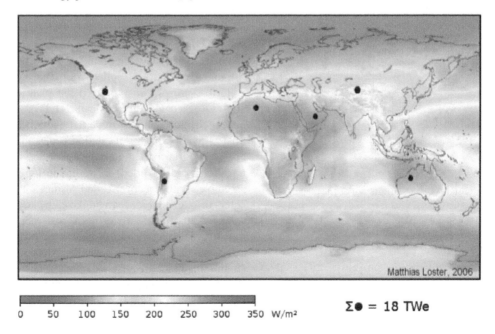

FIGURE 17.3 Premier solar potential areas in the world (black dot) [5, 8].

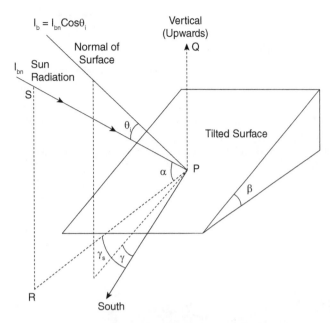

FIGURE 17.4 Beam radiation (I_b) falling on a surface (θ_i, β α, γ, γ_s are angle of incidence, tilt angle of surface, altitude of sun, surface azimuth angle, and solar azimuth angle, respectively) [11].

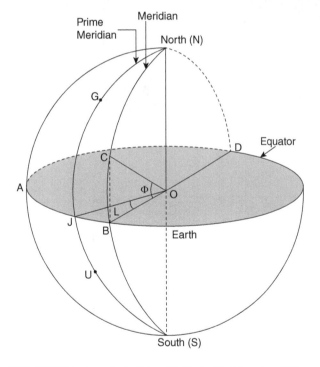

FIGURE 17.5 Longitude (L) and latitude (ϕ) of the place [11].

$$\cos\theta_i = \sin\phi\,(\sin\delta\cos\beta + \cos\delta\cos\gamma\cos\omega\sin\beta) \\ + \cos\phi\,(\cos\delta\cos\omega\cos\beta - \sin\delta\cos\gamma\sin\beta) \quad (17.2)\\ + \cos\delta\sin\gamma\sin\omega\sin\beta$$

The solar/local time on which the hour angle depends can be computed using the following equations:

$$Solar\ time = Standard\ time \pm 4\left(L_{st} - L_{loc}\right) \\ (\min) + E(\min) \quad (17.3)$$

E = 229.2(0.000075 + 0.001868 cos B − 0.032077 sin B − 0.014615 cos 2 B − 0.04089 sin 2B)

$$B = \frac{(n-1)360}{365}$$

Where L_{st} is the standard meridian and L_{loc} is the longitude of the location. In the first correction, the −ve and +ve signs correspond to the Eastern and Western Hemisphere, respectively.

17.2 SOLAR THERMAL TECHNOLOGY

The solar thermal energy has much protentional to meet heating (low to high temperature) as well as cooling (space and refrigeration) energy demands at domestic scale and small to large industrial scales. The thermal energy demand in households and industries for drying, water heating, cooking, dyeing, pasteurization, bleaching, chilling/cooling, etc. can be significantly met using solar thermal collectors. About 50% thermal energy demand of industries are in temperature range about 30°C–400°C [12]. For high-temperature applications, the solar concentrator technology can produce energy in the range of MW, and it provides an alternative of furnace oil-based heating systems [13–15]. The thermal efficiencies of various solar thermal collectors at different temperatures are given in Table 17.2. It infers that efficiencies of STSs vary about 30–80%, and clearly, the parabolic dish collector is the most efficient technology.

17.2.1 Basic Working Principle

On the basic working principle, all solar thermal systems are broadly divided into two categories, (i) nonconcentrating and (ii) concentrating collector, and further, a classification can be made as shown in Figure 17.7.

Nonconcentrating collectors. The nonconcentrating collectors are of two types: (i) flat plate collector (FPC) and (ii) evacuated tube collector (ETC). Basically, a flat plate collector consists of a flat metallic absorber plate (that is painted black), insulation (sides and bottoms), and one or two transparent glaze (glass or acrylic) cover on the top (horizontal or titled) for radiation input. All these are assembled within a cover called casing. When the radiation enters through the transparent glaze, the large portion of the solar radiation is absorbed by the absorber and the heat is transferred to liquid, air, or substance in the collector. In a flat plate solar water heater, the conductive metal tubes (containing fluid) are attached upon the absorbing plate. The metal cooking utensil (with water/food) is kept on absorber plate in a solar cooker. In a solar still, the absorber plate can directly contain a layer of water. The drying substance can be kept on some stand in case of a solar dryer, and in an air

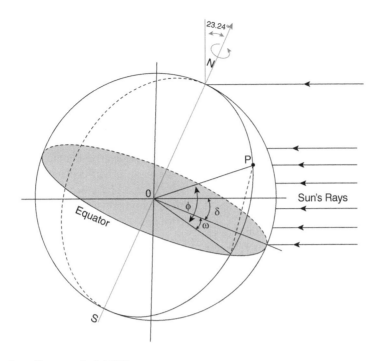

FIGURE 17.6 Representation of hour angle (ω) [11].

TABLE: 17.2
Thermal efficiecies of different solar collectors [4].

			Efficiency in Different Temperature Range (%)		
	Low	Medium	High		
Collector Type	(100°C–150°C)	(150°C–250°C)	300°C	400°C	500°C
Flat plate	34	-	-	-	-
Evacuated tubular	34	-	-	-	-
Compound parabolic	64	62	-	-	-
Linear Fresnel reflector	65	63	60	-	-
Parabolic trough	79	77	73	-	-
Heliostat	-	69	65	62	58
Parabolic dish	-	79	76	71	66

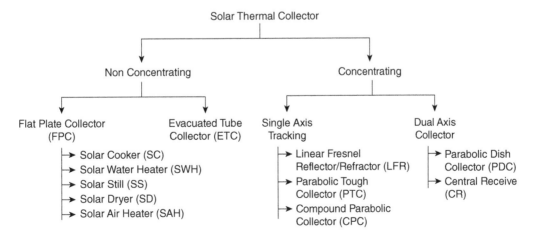

FIGURE 17.7 Classification of solar thermal collectors.

heater, a proper circulation of air can be made in the flat plate collector [9–11, 16].

The evacuated tube collector (ETC) is an array of cylindrical-shape transparent double tubes arranged parallel to each other and are attached to a storage tank on the top. The individual concentric glass tube is evacuated to minimize radiation losses. The inner tube is an absorber for this system that is painted black with the selective coating materials. The incident radiation is always perpendicular in an ETC for the absorbing surface and hence produces better efficiency than an FTC. Further, to increase heat transfer, a metallic absorber in the inner tube can also be incorporated in the design. These tubes can be of desired lengths and numbers as per their application for water heating, cooking, water purification, etc. For water heating, in both the systems (FPCs and ETCs), thermosyphon is the process through which water is collected in a storage tank kept at certain height of the system [17–18].

Concentrating collectors. A reflector or refractive surface, a metallic receiver, a support structure, and a tracking unit are the main components of a concentrating collector. The reflector (curved/ plane segmented) reflects the incident solar beam radiations on a focus point (line/point) to be received by a metallic receiver (linear/cavity). The receiver (coated with selective coating) absorbs the radiations and transfers the heat to the working fluid (dynamic/ stationary) inside the receiver. An accurate one- or two-axis tracking mechanism is required for all types of solar concentrators to receive reflected radiations at the focus point throughout the day. The concentrators can also be categorized on the basis of optical concentration and temperature generation.

17.2.2 Basic Energy Balance Equation

For a solar thermal collector, an energy balance equation is a relation between the absorbed energy rate, the utilized rate of heat, and the rate of thermal losses in a steady state [9–11]. The rate of utilized solar energy q_u is equal to the difference of the absorbed rate of solar energy q_a through absorber/receiver and the rate at which heat is lost from the collector to the surroundings through the basic heat transfer mechanism (conduction, convection, and reradiation). An energy balance is the following equation for a steady state:

$$q_u = q_a - q_l \quad (17.1)$$

This rate of utilized energy is the rate of heat utilized by the working substance that might be air, water, or any other fluid/ substance, depending on the nonconcentrator or concentrating collector application. This value can be experimentally determined for any collector using the thermal profile of the working load. The rates of absorbed heat and loss are:

$$q_a = A_a S \quad (17.2)$$

$$q_l = U_l A_c (T_{cm} - T_a) \quad (17.3)$$

For the nonconcentrator collector, the collector surface absorbs the radiation, and in nonconcentrating collector, the reflector surface reflects the radiation at the receiver. The area A_a is the aperture area of the collector for a nonconcentrating collector and the reflector area for a concentrating collector. The incident solar flux absorbed through the absorber/receiver is denoted by S. The treatment of S depends on absorber/receiver geometry. At any time, the solar flux absorbed by a nonconcentrator collector after passing through horizontal transparent collector surface is given by [10–11]:

$$S = I_b r_b (\tau\alpha)_b + \{I_d r_d + (I_b + I_d) r_r\} (\tau\alpha)_d \quad (17.4)$$

Here, the subscript b and d are used for *beam* and *diffuse*, respectively. I is the incident radiation on the horizontal transparent surface (W/m^2), r is the tilt factor, r_r is the radiation shape factor for the surface, τ is the transmissivity of the transparent glaze material, α is the absorptivity of the absorber, and $\tau\alpha$ is the transmissivity-absorptivity product. The overall transmissivity τ of a transparent surface can be computed by the equations given in references [9–10]. The equation for S for a concentrating collector is:

$$S = I_b \rho (\gamma\tau\alpha)_n K_{\gamma\tau\alpha} \quad (17.5)$$

Here, ρ is the specular reflectance (for reflector type) or transmittance (for refractor type) of the concentrator; α is the absorptivity of the receiver. If the receiver is covered with any transparent surface, then τ is the transmissivity of the receiver. The fraction of reflected radiation of the incident radiation on the concentrator is the intercept factor γ. The $K_{\gamma\tau\alpha}$ is incidence angle modifier that includes the effects of angle of incidence on the intercept factor.

The U_l is the overall loss coefficient in equation (2.3), T_{cm} is the average temperature of the collector (absorber/ receiver), and T_a is the ambient temperature. In case of a nonconcentrating collector, the rate of heat lost is the sum of the heat lost from the top (due to radiation and convection), from the bottom (conduction loss dominant), and from the sides (conduction loss dominant). In a concentrating collector, the rate of heat lost is the sum of heat lost due to convection and reradiation from the receiver surface and conduction through the receiver's support structure. In general, the rate of heat loss can be given as:

$$\frac{q_l}{A_c} = h_w (T_{cm} - T_a) + \epsilon\sigma (T_{cm}^4 - T_{sky}^4) + U_{cond} (T_{cm} - T_a) \quad (17.6)$$

The area of absorber/receiver is A_c. In the preceding equation, the first term is due to convection losses; h_w is the convective heat transfer coefficient between absorber/ receiver and the surrounding air. The second term is due to radiation losses; ϵ is the emissivity of the absorber/receiver, σ is the Stefan-Boltzmann constant, and T_{sky} is the sky temperature. The conduction loss coefficient (U_{cond}) is significant in a nonconcentrating collector; it depends on the thermal conductivity and thickness of the insulation. In a concentrating collector, if the receiver is covered through insulation, then this term may be considered; else, other losses (convection and reradiaton) are more dominating and this term can be neglected. For a concentrating collector, the ration of A_a/A_c is defined as area concentration ratio (C). For both collectors, the instantons collection efficiency (an important thermal performance parameter) can be given by:

$$\eta_i = \frac{q_u}{A_c I} \qquad (17.7)$$

For a nonconcentrating collector, I is the global radiation, while for a concentrator collector, this is only beam radiation [9].

17.3 HEATING TECHNOLOGY AND APPLICATIONS

After developing a basic understanding of the working principle and performance parameters of solar thermal systems, the research and development in solar thermal technology and the applications for low- to high-temperature use are briefly included in this section.

17.3.1 Low Temperature

Solar water heater (SWH). This is the most widely accepted solar thermal technology in many countries. In solar water heating, Hottel and Woertz have done the pioneer work in 1942; later, basic characteristic equations to calculate useful heat gain for a known inlet fluid temperature was introduced by Hottel-Whiller-Bliss, also known as the HWB equation [19, 20]. Various types of solar water heaters have been developed, tested, and implemented at domestic and industrial scales. The following classification can be made for the solar water heaters developed so far:

In the earlier work, Reddy and Kaushika [21] have investigated such systems with a transparent insulating material (TIM), which is incorporated within the glaze of the system. They have identified significant reduction in the heat losses of the system as an effect of TIM. The effect of a corrugated absorber surface on FPC is investigated by Kumar and Rosen [22]. The study infers an increase in thermal performance of the system due to more availability of the surface area. In some other study, Helal et al. [23] have done a design innovation by using three parabolic branches composed of a compound parabolic concentrator and placing an integrated storage tank at the focus. A passive FPC integrated with a collector storage system is a close-coupled system in that natural rise of heated water occurs by the thermosyphon flow. Kumar and Rosen [24] designed this system in Toronto, in that they used the concept of extended storage section. At an optimized ratio of waters in two storage tanks, a better thermal performance of the system is reported. In the direct- and indirect-circulation active FPC, a pump (Figure 17.9(a)) is used for circulation of fluid; that fluid is water for the direct system and a nonfreezing heat transfer fluid for the active system [6]. Chen et al. [25] tested an indirect active FPC with an ETFE foil (transparent polymer) between the glass cover and the absorber for different flow rate at the University of Denmark. A theoretical prediction of the system efficiency matches with the experimental results. The material coating at the absorber has also been tested widely. In one such study, a copper oxide layer grown in absorber using chemical treatment is tested by Alami and Aokal [26]. Almost 25% increase in the efficiency is found than in the conventional ones. In an ETC direct-flow pipe (Figure 17.9(b) left), there are two pipes inside the concentric transparent tubes, one for inlet and another for outlet [4], while a heat pipe ETC has a metallic plate attached to the absorber place that increases radiation absorption rate and gives better performance (temp. about 120°C) than the direct flow pipe (Figure 17.9(b) right). In the recent advancement, photovoltaic panels are also combined with solar water heater. Such combinations are shown in Figure 17.9(c) for different water flow conditions [27]. These efficient systems provide heat and electricity, simultaneously. Recently, a software tool called System Advisor Model (SAM) was introduced by Suresh and Rao [28] for coupling of solar

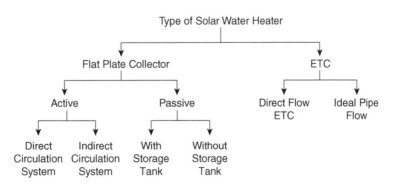

FIGURE 17.8 Classification of solar water heater systems.

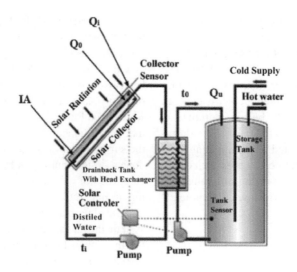

FIGURE 17.9(a) Indirect active FPC [6].

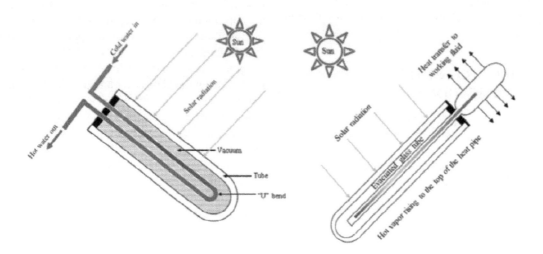

FIGURE 17.9(b) ETC direct flow (left) and heat pipe (right) [4].

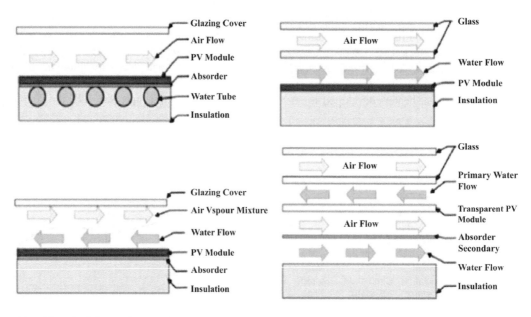

FIGURE 17.9(c) PV and FPC combination [27].

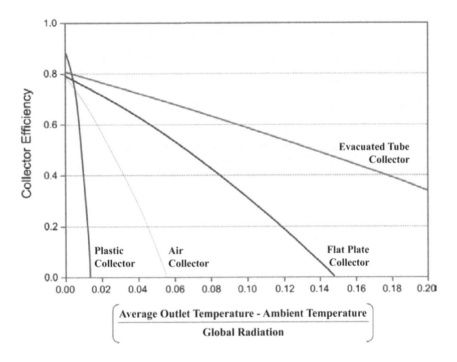

FIGURE 17.9(d)   Efficiencies of different collectors [18].

collectors for process heating. The thermal efficiencies of collectors with respect to ratio of difference of the outlet and ambient temperatures to the insolation value are illustrated by Brian [18] and are presented in Figure 17.9(d).

Solar cooker. The concept of solar cooking started in the seventeenth century. The design and development in this simple technology (that incorporates an FPC with cooking pots on absorber plate) have been much advanced till now, with better thermal performance, low cost, long durability, and less cooking time. The solar cookers can be divided into different categories, as inferred in Figure 17.9(e) [11, 29]. A detailed review on solar cooking can be found in [30–34]. A low-cost, single-family solar cooker that is designed and tested by Mahavar et al. [32] has a conventional design, but the new component materials make the system efficient at reasonable cost (Figure 17.9(f)). In low sunlight or in the dark, the cooking limitations of solar cookers cause less popularization of these systems at domestic and industrial scales. To mitigate this hindrance, an advance type of solar cookers has been designed and presented by many researchers. A comparative study of solar cooking in a parabolic dish cooker and in a booster mirror cooker is presented by Hosny and Ziyan [15]. The performance of parabolic dish cooker is observed better to decrease cooking time as well as for high-temperature cooking. The more-advanced types of cookers include heat backup using integrated electric system or thermal storage (latent or sensible) in the absence of sun-hours. Mahavar et al. have fabricated a solar-cum-electric cooker (SEC), shown in Figure 17.9(g), as per the analytical model presented in their paper [34]. A vast review is presented by Omara et al. [35] on phase-change materials (PCMs) as storage in solar cookers. The storage is applied not only on conventional systems but also on other advanced cookers – viz, Nayak et al. have integrated a storage system (acetanilide and stearic acid) with ETC, and Saini et al. used a PCM storage with a parabolic trough; the systems are shown in Figure 17.9(h) and (i), respectively [36–37].

Solar dryer. This technology is most popular in agriculture application for crop drying. The fossil fuel–powered dryers are high-temperature dryers which are costly and not environment friendly. So conventionally, agriculture products can be placed in the sun for open drying, but a controlled temperature and earlier drying are possible in a closed-chamber solar dryer. The solar drying systems can be classified as in Figure 17.9(j).

The basic elements of a direct-type solar dryer system are a flat plate collector and a load tray; those are shown in Figure 17.9(k) [39]. In an indirect solar dryer (Figure 17.9(l)), a drying chamber (having multiple absorber/load tray and a chimney) is attached to an FPC [38]. The air heated in collector may be ducted to the drying chamber either naturally by buoyant or using fans/pumps. The first one is passive indirect, and the other one is active indirect dryer. In mixed-typed solar dryers, the substance is heated through the direct absorption of solar radiation or by the preheated air coming from the solar air heater [16, 38–40].

El-Sebaii and Shalaby [40] have presented a detailed review on indirect and mixed-mode solar dryers. Figure 17.9(m) shows the industrial solar drying of banana at 1,000 kg capacity [41]. A mix-type 60% efficient solar dryer, shown in Figure 17.9(n), was designed by Almuhanna [42].

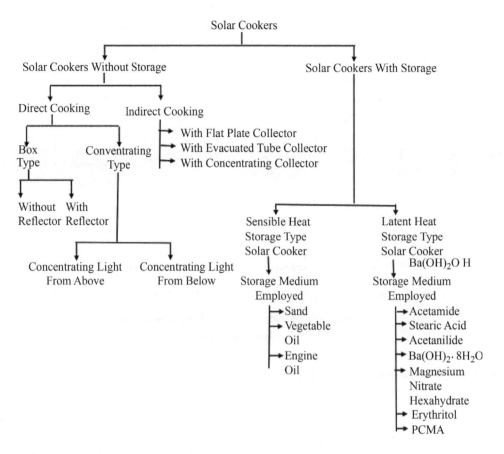

FIGURE 17.9(e) Classification of solar cookers [11, 29].

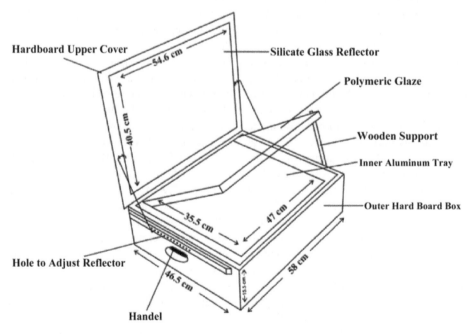

FIGURE 17.9(f) Schematic diagram of SFSC [33].

FIGURE 17.9(g) Solar-cum-electric cooker (SEC) [34].

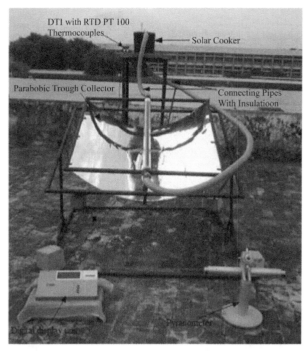

FIGURE 17.9(i) Solar cooking with parabolic trough [37].

FIGURE 17.9(h) Solar cooking with ETC [36].

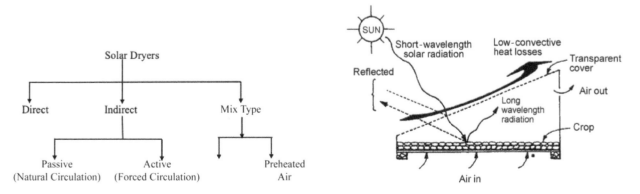

FIGURE 17.9(j) Classification of solar dryers [16, 38].

FIGURE 17.9(k) Direct solar dryer [39].

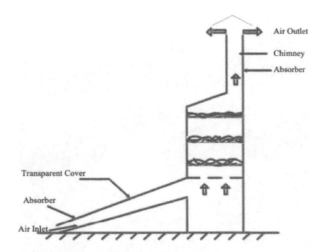

FIGURE 17.9(l) Indirect natural convection dryer [38].

FIGURE 17.9(m) Design and performance evaluation of a double-pass solar drier for drying of red chilli (*Capsicum annum L.*) [41].

FIGURE 17.9(n) Solar air heater (solar dryer, greenhouse type): (1) window (air intake), (2) extracting fan, (3) wire-netted floor, (4) air chamber, (5) fiberglass cover [42].

Solar air heaters. The usual applications of solar air heaters (SAHs) are for space heating and agriculture product drying [43–44]. A schematic view of an SAH experimental setup is shown in Figure 17.9(o). Esen et al. presented a modeling to compute efficiency of a new solar air heater (SAH) system by using least-squares support vector machine (LS-SVM) method [45]. A solar air heater coupled with photovoltaic panel is an effective method to lower down the temperature of PV panel to increase electric efficiency and air heating, simultaneously. In Figure 17.9(p), a PV/T air collector is depicted [46].

Solar distillation. Safe and clean drinking water is a necessity for every human. About 3.5 million people lose their life every year due to lacking water supply and improper sanitation [17]. The projected water scarcity in 2040, shown in Figure 17.9(q), is an alarming situation for many parts of the world [17, 47, 48]. The purification of saline water for various applications is done through many conventional and nonconventional techniques. Due to economical and eco-friendly techniques, solar distillation provides a viable solution for water desalination and purification [17]. Solar distillation systems are classified as passive- and active-type solar stills. A detailed classification is given in Figure 17.9(r). A passive solar (basin-type solar still) is an FPC in which a shallow, airtight basin is lined with a black material that contains the saline water [16–17, 47–49]. Depending on the glaze tilt, these may be single- or double-slope systems (Figure 17.9(s)) [49]. The evaporated water is collected through certain channels inside the system. In the active type of solar stills, an additional heat system to feed extra thermal energy to the basin for faster evaporation is integrated in the system. In the recent advancements, the role of nanofluids for water yield increase has been investigated by many researchers on passive solar stills. Some nanomaterials, viz, copper oxide (CuO), alumina (or aluminum oxide) (Al_2O_3), cuprous oxide (Cu_2O), graphite, etc., are used. Sharshir et al. [50] have tested the copper oxide and graphite powder in the still basin with varying depths. About 45% and 53% improvement in the efficiencies are reported. The effect of silica gel filled with graphite is investigated by Prasad et al. [51], and about 49% increase in efficiency is noted. The detailed review on active solar distillation and various means to increase heat transfer rate in solar stills can be found in [17, 52]. Prasad and Tiwari [53] have studied a double-effect solar distillation (forced circulation) unit that is coupled with a compound parabolic concentration (CPC) collector. A double-basin solar still (Figure 17.9(t)) under three conditions – (i) with black granite gravel, (ii) with ETC tubes, (iii) with both black granite gravel and ETC tubes – is tested by Panchal [54]. The experimental results show that the output increases by 56% when coupled with vacuum tubes and by 65% with vacuum tubes and black granite gravel. A combination of photovoltaic panel and active solar still (Figure 17.9(u)) has been studied by Kumar and Tiwari [55]; the study reports that the yield increased by more than 3.5 times than the passive

Advancements in Solar Thermal Technology: Heating and Cooling 217

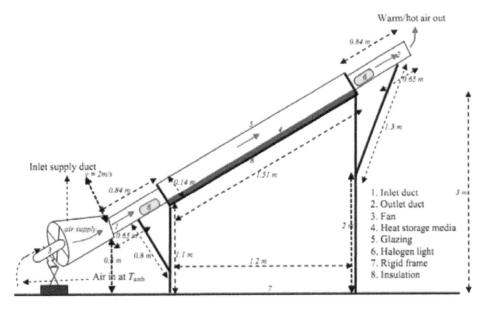

FIGURE 17.9(o) A schematic view of the experimental setup of solar air heater [43].

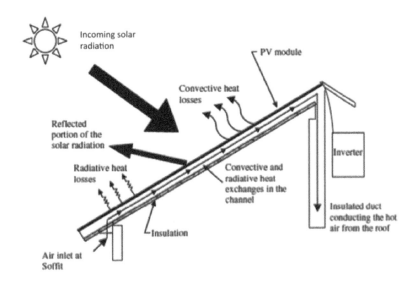

FIGURE 17.9(p) Schematic of a typical building-integrated PV/T air design [46].

FIGURE 17.9(q) Projected water stress in 2040 [47].

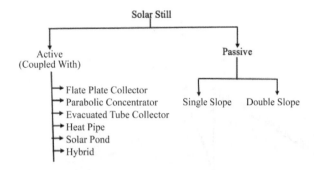

FIGURE 17.9(r) Classification of solar still.

FIGURE 17.9(t) Double-basin solar still with evacuated vacuum tubes [54].

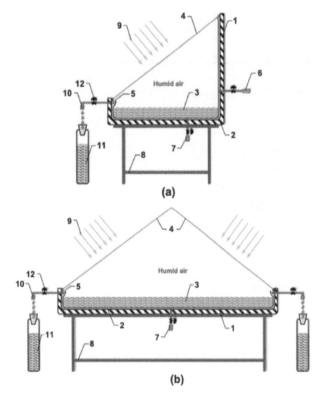

FIGURE 17.9(s) (a) Single- and (b) double-slope solar stills: (1) Insulation, (2) basin liner (3) saline water, (4) glass cover, (5) trough, (6) water supply, (7) drainage, (8) still supports, (9) solar radiation, (10) conduit for distillate collection, (11) desalinated water, (12) control valve [49].

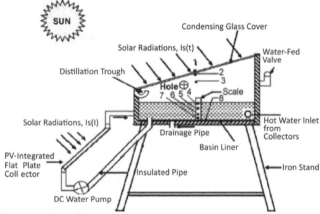

FIGURE 17.9(u) Schematic of hybrid (PV/T) active solar still [55].

solar still. A new hybrid desalination system consists of a wind turbine (WT), and an inclined solar water distillation integrated with the main solar still is also presented by Etawil and Zhengming [48].

17.3.2 Medium Temperature

The thermal energy demand below temperature 250°C that accounts for 60% share of total energy demand of industries can be supplied by solar concentrator technology. The basic working principle of a concentrating collector has been explained in Section 2.1. The classification of concentrating technology is also mentioned in that section.

Parabolic trough. This system requires single-axis tracking of the sun, and it collects energy through an absorber tube at the focal distance. This tube may be metallic or evacuated glass tube. A number of researchers have performed theoretical and experimental work on the development of

FIGURE 17.10(a) Solar parabolic trough collector system developed at IIT Madras [56].

FIGURE 17.10(c) Photo of constructed double-glazed ICS solar heater with CPC reflector [58].

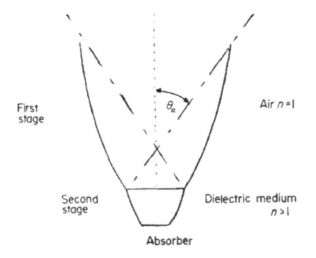

FIGURE 17.10(b) Two-stage CPC solar collector [57].

FIGURE 17.10(d) A Fresnel prototype system [62].

this technology for various applications. A parabolic trough collector that is investigated by Reddy et al. [56] is shown in Figure 17.10(a). The system efficiency is reported to be in the range of 64% to 67%.

Compound parabolic concentrating. An earlier CPC schematic is shown in Figure 17.10(b), which is a single-stage system described by Rabl [57]. Kessentini and Bouden [58] have investigated an integrated collector storage with CPC reflectors, as shown in Figure 17.10(c). Double glazing is used in the system that increased the thermal performance of the system due to reduced heat losses. This technology is getting the attention of researchers and industry developers due to no tracking mechanism, low heat losses, and good thermal efficiency. Pranesh et al. [59] have presented a detailed review on acceptance of CPC technology for domestic and industrial applications.

Linear Fresnel reflector (LFR). Around 1960, the pioneer work in linear Fresnel reflector (LFR) technology was done by Giovanni Francia. This thermal technology may be employed for power generation plants from a few kW to hundreds of MW [60]. An investigation of 250 kW LFR prototype, depicted in Figure 17.10(d), was performed by Beltagy et al. [61]. The plant has 14 lines of LFR, and it is tracking the sun on a north–south axis alignment. The efficiency of the system is about 40%. At large scale, Bishoyi and Sudhakar [60] have evaluated the performance of a 100 MW LFR solar thermal power plant (with thermal storage system) using system advisor model (SAM) software (NREL). Nevertheless, this technology can also be used at small scale – viz, Mokhtar et al. [61] have tested a linear Fresnel receiver (LFR) for water heating system.

17.3.3 High Temperature

Among all solar concentrator technologies, only parabolic dish concentrator and central receiver (heliostat) can produce high (400°C) to very high (2,000°C) temperature at focus point. Both the systems require dual-axis continuous tracking of the sun. The research and development in these technologies have reached an advanced stage due to their high potential in power generation. However, parabolic dish concentrator can be employed from small (solar still, solar cooking, etc.) to large (power generation) thermal applications. Omara and Eltawil [63] used an airtight solar still at the focus of a dish concentrator (SDC) and performed water distillation. The system performance was compared to conventional solar still and found to be 34% more efficient. The simulation of the thermal performance of a parabolic dish receiver system with argon gas as working fluid was done by Wang and Siddiqui [64]. A low-cost parabolic dish concentrator was designed by Hijazi et al. [65] for direct electricity generation. A 400 m² solar concentrator dish, shown in Figure 17.11(a), was installed at the Australian National University campus in 1994 [66]. A 20 m² prototype solar parabolic dish (Figure 17.11(b)) with cavity receiver was investigated by Reddy and Veershetty [67] at IIT Madras, India. Currently, the use of photovoltaic panels at focus point is also being tested in concentrator technology as like LFR and parabolic trough technologies. A heliostat field or central receiver has a large number of flat reflectors arranged

FIGURE 17.11(a) The big dish prototype at ANU, Australia [66].

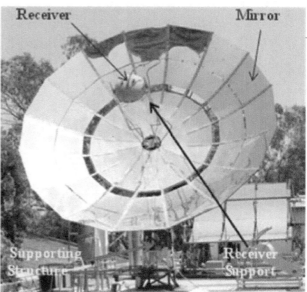

FIGURE 17.11(b) Solar parabolic dish collector at IIT Madrar, India [67].

FIGURE 17.11(c) Experimental arrangement of heliostat [68].

Advancements in Solar Thermal Technology: Heating and Cooling

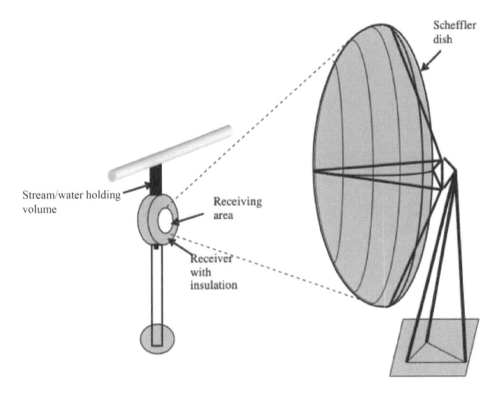

FIGURE 17.11(d) A Scheffler concentrator [13].

FIGURE 17.11(E) Arun solar concentrator [69].

at ground with dual-axis tracking mechanism. The receiver is centrally located above from the ground at the focus of all flat reflectors. Although this high-investment technology is empowered for power generation, researchers are also investigating the possibility for domestic-scale applications. Hayat et al. [68] have performed such small-scale experimental study; the setup is shown in Figure 17.11(c). A Scheffler concentrator (Figure 17.11(d)), which is a moving concentrator with fixed focus, and ARUN, which is a paraboloid dish with Fresnel reflectors (Figure 17.11(e)), are popular technologies in India for steam generation and medium-temperature industrial applications [13, 69].

17.4 COOLING TECHNOLOGY AND APPLICATIONS

Worldwide, the energy consumption for the air-conditioning and cold is rising year by year. The solar thermal systems can also provide cheap and clean energy solution to space cooling and refrigeration. The solar cooling systems can be classified into two main categories: (i) solar electric cooling and (ii) solar thermal cooling. In the solar electric cooling, the conventional electric vapor compressor of air conditioner is driven through electricity produced by the photovoltaic panels. The solar thermal cooling technology is called sorption technology, in which systems may be opened or closed and can be classified as inferred in the following text. To cool a space or building, an absorption or adsorption process is used to create chilled water that can be used directly in a conventional air handler unit [70].

17.4.1 SOLAR DESICCANT SYSTEMS (OPEN SYSTEM)

The solid and liquid desiccant cycles represent the open system. The liquid desiccant system has a higher thermal coefficient of performance (COP) than the solid desiccant system. The main components of this system are a desiccant-coated heat exchanger (DCHE), an evaporative cooler, and conventional cooling coils. In these systems, a desiccant

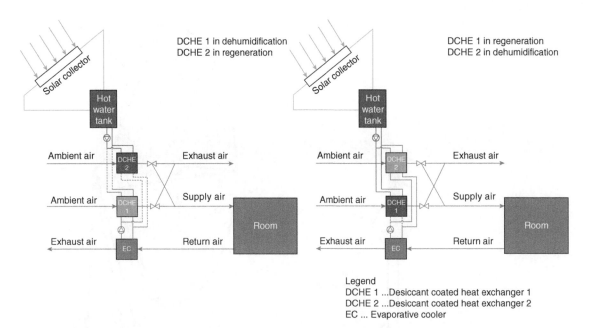

FIGURE 17.12 A solar-driven, desiccant-coated heat exchanger cooling system [72].

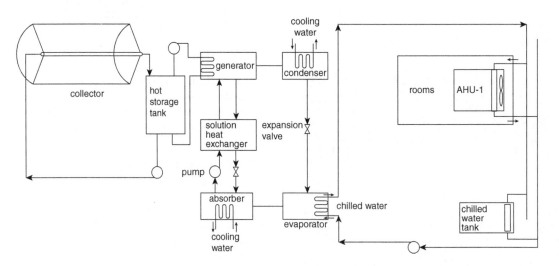

FIGURE 17.13(A) The basic principle of the solar absorption cooling system coupled with solar collector and storage [73].

material absorbs the humidity of the space [71]. Erkek et al. have done a simulation study of a solid desiccant–coated small-scale heat exchanger, shown in Figure 17.12 [72]. Firstly, ambient air passes through DCHE 1, absolute humidity decreases, and the adsorbent material adsorbs the water. The adsorption heat is ejected by means of cooling water passing through the DCHE 1. This process is dehumidification phase 1. Now, a regeneration phase 1 occurs, in which heated water through a solar collector at temperature 50°C to 75°C is circulated to DCHE 2. The humidity is released as exhaust air to the environment, while the desiccant coating layer is regenerated. This is a half cycle of the process; in the second half, dehumidification phase 2 occurs in the DCHE 2, and regeneration phase 2 is accomplished in the DCHE 1. Two solid desiccants, silica gel and aluminum fumarate, are considered by Erkek et al. [72] in their study.

Both are tested under two conditions: (i) water-cooled dehumidification and (ii) adiabatic dehumidification. The silica gel composite coating, over aluminum fumarate coating, is found more efficient.

17.4.2 Solar Absorption Cooling (Closed System)

Absorption cooling is the first and oldest form of air-conditioning and refrigeration. In absorption refrigeration, a mechanical compressor (used in vapor compression refrigeration cycle) is replaced by the thermal compressor (consists of an absorber and a generator) (Figure 17.13(a)). As shown in Figure 17.13(a), the refrigerant gets vaporized in the evaporator by taking heat from the surroundings. The vapor gets absorbed in the poor solution of the absorbent in the absorber [73]. Now the concentrated solution is pumped to the

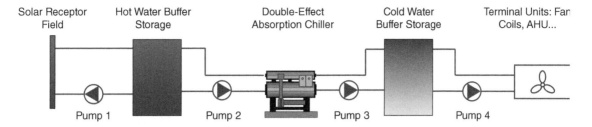

FIGURE 17.13(B) Double-effect absorption system employing a solar concentrating system [77].

generator. The generator is powered through a solar collector. Here, the refrigerant gets separated from the absorber and goes to the condenser. The extra heat of refrigerant is rejected in the condenser, and it flows back to the evaporator at low pressure through the expansion device. So far, two refrigerant/absorbent pairs are used: (i) $H_2O/LiBr$ (water refrigerant and LiBr absorbent) and (ii) NH_3/H_2O (ammonia refrigerant and water absorbent) [73–75]. Both the nonconcentrator solar collectors and the solar concentrator collectors can be used to power the generator in an absorption cooling system. Based on the generator temperature, the absorption system can be classified as single-effect (generator temp. 90°C), half-effect, and double-effect (generator temp. 140°C) solar absorption cycles. To improve the performance of a solar thermal cooling system, some novel concepts are proposed [76]. A comparison of a single-effect absorption chiller (powered through evacuated tube collectors) and a double-effect absorption chiller (powered by Fresnel reflective solar concentrating system) for use in a specific three-floor building is presented by Chemisana et al. [77]. The higher performance is observed in the second case, which also has a less investment cost (Figure 17.13(b)).

17.4.3 Solar Adsorption Cooling

This technology requires lower temperature than absorption, from 60°C to 120°C, and the corresponding COP is about 0.3–0.6. The adsorption process is a volumetric phenomenon, while the absorption process is a surface phenomenon. An adsorption system is a solid, porous surface with a large surface area and a large adsorptive capacity. In unsaturated surface, the vapor molecules interact with the surface and get adsorbed onto the surface [78]. As like absorption cycle, in these systems the refrigeration cycle has two sorption chambers, a condenser, and an evaporator. In the first step, the adsorbed water in silica gel is desorbed in hot adsorber (which is heated through solar thermal system). In the second step, the desorbed refrigerant (water) is condensed into liquid by rejecting heat through the cooling water supplied from a cooling tower. Then, the condensed water goes through the expansion valve to the evaporator, where it vaporizes under low partial pressure. This process produces low temperature in the evaporator surroundings. In the last, the vaporized water is again adsorbed in the silica gel at cold adsorber. Some solid-sorbent-and-refrigerant pairs are activated carbon–ammonia, activated carbon–ethanol, silica gel–water, and zeolite–water. Dzigbor and Chimphango [79] tested the activated carbon (AC) + NaCl (10–35.7% w/v) composite adsorbents paired with either high-purity (99.7%) or low-grade ethanol (60% ethanol/40% water) refrigerants to assess the potential of ethanol/water mixture as a refrigerant (Figure 17.14). The study showed that low-grade ethanol can be used as a potential alternative refrigerant in the adsorption system where pure ethanol is limited.

17.5 CONCLUSION

The versatile range of solar thermal systems provides a wide scope, from high-temperature (power generation) to cooling (refrigeration/air conditioning) applications. Besides all technical development and advancement, financial feasibility, social acceptance, and new market and government policies for solar thermal systems, still the contribution of this technology in the world energy production is limited; that reflects the challenges in the path of complete acceptability of STSs in comparison to the conventional energy systems. In the chapter conclusion, a few such parameters and future possibilities of improvements are addressed.

Technical limitation. The efficiencies of various thermal solar collectors are 30% to 80%, but under unfavorable weather conditions (cloudy, rainy, and windy day and at night), even the most efficient solar dish technology gets also technically vulnerable. Moreover, different geometrical (DNI, ambient temperature, etc.) conditions also affect the operating parameters and the system performance of a solar thermal system. The solution of such technical challenges lies in the suitable heat backup using solar thermal energy storage or conventional means. The large installation area and the proper maintenance requirement (skillful cleaning, repairing, and handling) are other technical issues that need to be addressed by the researchers in this field.

Financial feasibility. So far, for domestic and small industrial scale concern, the solar thermal systems are in limited practice. There is still a need for novel designs with inexpensive materials to bring these systems for more common uses. In the large industrial sectors, the long payback period at a high initial installation cost creates a hurdle for taking this technology over conventional ones. On the one hand, modern solar cooling technologies are still at the development stage, while, on the other hand, one of the oldest solar concentrator

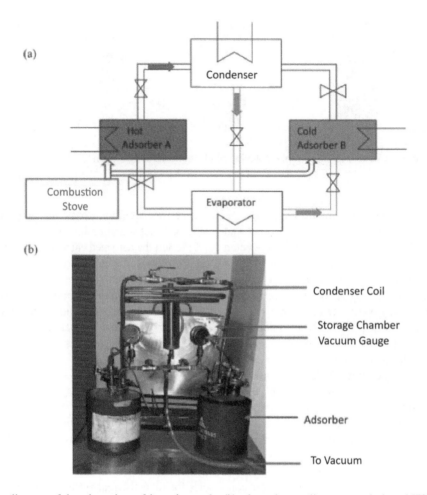

FIGURE 17.14 Flow diagram of the adsorption refrigeration cycle; (b) adsorption cooling system designed [79].

cooking technology still could not turn up as a good alternative to the conventional one due to high cost.

Policies and market issues. According to a report by the International Energy Agency, by 2050, the potential of solar heating industrial applications will be around 3,200 GW for low-temperature heating applications and 1,000 GW for solar cooling systems. Yet the unavailability of adequate policy, regulatory support, and promotional incentives prevents bringing this technology comparable to the conventional ones. Policy makers can make a bridge between government support and the market trends to promote this technology and bring it to the mainstream.

Environmental and safety issues. With great environmental benefits of reduced fossil fuel consumption and low carbon footprint, solar thermal energy is undisputably one of the cleanest technologies for the future of the world. Nevertheless, a few safety issues set challenges in the path of popularization; these are: (i) the large structures at MW thermal plants can cause noise and air pollution during the manufacturing and system installation; (ii) any leakage of thermal fluids can cause harm to the surroundings and the ecosystem; (iii) the high-optical-concentration need advances safety for the pant labor.

Social barriers. The large space requirement, inadequate efficiency for low-heat application, poor reliability, technical and financial issues, and lack of social awareness (about the benefits of solar thermal technology, marketing, etc.) are some barriers in the social and wide acceptability of the solar thermal technology for low to high thermal and cooling applications.

17.6 REFERENCES

[1] T. Ahmad, D. Zhang, A critical review of comparative global historical energy consumption and future demand: The story told so far, *Energy Reports* 6 (2020) 1973–1991.

[2] G.P. Beretta, World energy consumption and resources: An outlook for the rest of the century, *International Journal of Environmental Technology and Management* 7 (2007) 1–2.

[3] Y.W. Rufael, Coal consumption and economic growth revisited Yemane Wolde-Rufael, *Applied Energy* 87 (2010) 160–167.

[4] R.K. Kumar, N.V.V Chaitanya, S.K. Natarajan, Solar thermal energy technologies and its applications for process heating and power generation- A review, *Journal of Cleaner Production* 282 (2021) 125–296.

[5] E. Kabir, P. Kumar, S. Kumar, A.A. Adelodun, K.H. Kim, Solar energy: Potential and future prospects, *Renewable and Sustainable Energy Reviews* 82 (2018) 894–900.

[6] M.S. Hossain, R. Saidur, H. Fayaz, N.A. Rahim, M.R. Islam, J.U. Ahamed, M.M. Rahman, Review on solar water heater collector and thermal energy performance of circulating pipe, *Renewable and Sustainable Energy Reviews* 15 (2011) 3801–3812.

[7] T. Singh, M.A.A. Hussien, T.A. Ansari, K. Saoud, G. McKay, Critical review of solar thermal resources in GCC and application of nanofluids for development of efficient and cost effective CSP technologies, *Renewable and Sustainable Energy Reviews* 91 (2018) 708–719.

[8] National Aeronautics and Space Administration. Plotted from satellite data supplied by NASA Clouds and the Earth's Radiant Energy System (CERES); 2014.

[9] J.A. Duffie, W.A. Beckman, *Solar energy thermal processes*. New York: John Willy and Sons, 1974.

[10] S.P. Sukhatme, *Solar energy: Principles of thermal collection and storage*, 2nd ed. New Delhi: Tata McGraw-Hill Publication, 2001.

[11] S. Mahavar, *Modelling, development and testing of improved components for specific purpose solar thermal appliances*. Ph.D. Thesis, University of Rajasthan, Jaipur, India; 2013.

[12] R. Ramaiah and K.S.S. Shekar, Solar thermal energy utilization for medium temperature industrial process heat applications – A review, *IOP Conference Series: Material Science Engineering* 376 (2018) 012035.

[13] V. Sardeshpande, I.R. Pillai, Effect of micro-level and macro-level factors on adoption potential of solar concentrators for medium temperature thermal applications, *Energy for Sustainable Development* 16 (2012) 216–223.

[14] T. Kapoor, *Sun focus*, Patparganj Industrial Estate, New Delhi – 110 092, India issue 3, Jan 2014, SVS Press 116.

[15] Z. Hosny, A. Ziyan, Experimental investigation of tracking paraboloid and box solar cookers under Egyptian environment, *Applied Thermal Engineering* 18 (1998) 1375–1394.

[16] S. Mahavar, *Review of materials used in various solar thermal appliances. Solar engineering-I (applications)* Vol. 5, Ch. 6. Editors: Dr. Sri Sivakumar, Dr. Umesh Chandra Sharma & Dr. Ram Prasad. Texas: Studium Press LLC, 2015.

[17] K. Sampathkumar, T.V. Arjunan, P. Pitchandi, P. Senthilkumar, Active solar distillation – A detailed review, *Renewable and Sustainable Energy Reviews* 14 (2010) 1503–1526.

[18] N. Brian, Anatomy of a solar collector: Developments in materials, components and efficiency improvements in solar thermal collector systems, *Refocus* 7 (2006) 32–35.

[19] H. Hottel, A. Whiller, Evaluation of flat-plate collector performance, *Transactions of the Interanational Conference on the Use of Solar Energy* (1955).

[20] H. Hottel, B. Woertz, Performance of flat-plate solar-heat collectors, *Transaction of Americal Soceity of Mechanical Engineers (ASME)* 64 (1942).

[21] K.S. Reddy, N.D. Kaushika, Comparative study of transparent insulation materials cover systems for integrated-collector-storage solar water heaters, *Solar Energy Materials & Solar Cells* 58 (1999) 431–446.

[22] R. Kumar, M.A. Rosen, Thermal performance of integrated collector storage solar water heater with corrugated absorber surface, *Applied Thermal Engineering* 30 (2010) 1764–1768.

[23] O. Helal, B. Chaouachi, S. Gabsi, Design and thermal performance of an ICS solar water heater based on three parabolic sections, *Solar Energy* 85 (2011) 2421–2432.

[24] R. Kumar, M.A. Rosen, Integrated collector-storage solar water heater with extended storage unit, *Applied Thermal Engineering* 31 (2011) 348–354.

[25] Z. Chen, S. Furbo, B. Perers, J. Fan, E. Andersen, Efficiencies of flat plate solar collectors at different flow rates, *Energy Procedia* 30 (2012) 65–72.

[26] A.H. Alami, K. Aokal, Enhancement of spectral absorption of solar thermal collectors by bulk graphene addition via high-pressure graphite blasting, *Energy Conversion and Management* 156 (2018) 757–764.

[27] Z. Xingxing, Review of R&D progress and practical application of the solar photovoltaic/thermal (PV/T) technologies, *Renewable and Sustainable Energy Reviews* 16 (2012) 599–617.

[28] N.S. Suresh, B.S. Rao, Solar energy for process heating: A case study of select Indian industries, *Journal of Cleaner Production*, 151 (2017) 439–451.

[29] R.C. Punia, *Thesis: Modelling, design development and study of some non-concentrating community solar thermal appliances*. Thesis submitted to University of Rajasthan, Jaipur, India, 2013.

[30] T.U.C. Arunachala, A. Kundapur, Cost-effective solar cookers: A global review, *Solar Energy* 207 (2020) 903–916.

[31] S. Mahavar, P. Rajawat, V.K. Marwal, R.C. Punia, P. Dashora, Modeling and on-field testing of a solar rice cooker, *Energy* 49 (2012) 404–12.

[32] S. Mahavar, N. Sengar, P. Rajawat, M. Verma, P. Dashora, Design development and performance studies of a novel Single Family Solar Cooker, *Renewable Energy* 47 (2012) 67–76.

[33] S. Mahavar, P. Rajawat, R.C. Punia, N. Sengar, P. Dashora, Evaluating the optimum load range for box-type solar cookers, *Renewable Energy* 74 (2015) 187–194.

[34] S. Mahavar, N. Sengar, P. Dashora, Analytical model for electric back up power estimation of solar box type cookers, *Energy* 134 (2017) 871–881.

[35] A.A.M. Omara, A.A. Abuelnuor, H.A. Mohammed, D. Habibi, O. Younis, Improving solar cooker performance using phase change materials: A comprehensive review, *Solar Energy* 207 (2020) 539–563.

[36] N. Nayak, A. Jarir, H.A. Ghassani, Solar cooker study under Oman conditions for late evening cooking using stearic acid and acetanilide as PCM materials, *Journal of Solar Energy* 7 (2016) 314–328.

[37] G. Saini, H. Singh, K. Saini, A. Yadav, Experimental investigation of the solar cooker during sunshine and off-sunshine hours using the thermal energy storage unit based on a parabolic trough collector, *International Journal of Ambient Energy* 37 (2016), 597–608.

[38] A.G.M.B. Mustayen, S. Mekhilef, R. Saidur, Performance study of different solar dryers: A review, *Renewable and Sustainable Energy Reviews* 34 (2014) 463–470.

[39] A. Sharma, C.R. Chen, N.V. Lan, Solar-energy drying systems: A review, *Renewable and Sustainable Energy Reviews* 13 (2009) 1185–1210.

[40] A.A.E. Sebaii, S.M. Shalaby, Solar drying of agricultural products: A review, *Renewable and Sustainable Energy Reviews* 16 (2012) 37–43.

[41] S. Janjai, P. Intawee, J. Kaewkiewa, C. Sritus, V. Khamvongsa, A large-scale solar greenhouse dryer using polycarbonate cover: Modeling and testing in a tropical environment of Lao People's Democratic Republic, *Renew Energy* 36 (2011) 1053–1062.

[42] E.A. Almuhanna, Utilization of a solar greenhouse as a solar dryer for drying dates under the climatic conditions of the Eastern Province of Saudi Arabia part I: Thermal performance analysis of a solar dryer, *Journal of Agricultural Science* 4(3) (2012) 237–246.

[43] A. Saxena, N. Agarwal, G. Srivastava, Design and performance of a solar air heater with long term heat storage, *International Journal of Heat and Mass Transfer* 60 (2013) 8–16.

[44] M.A. Wazed, Y. Nukman, M.T. Islam, Design and fabrication of a cost effective solar air heater for Bangladesh, *Applied Energy* 87 (2010) 3030–3036.

[45] H. Esen, F. Ozgen, M. Esen, A. Sengur, Modelling of a new solar air heater through least-squares support

vector machines, *Expert Systems with Applications* 36 (2009) 10673–10682.

[46] K. Panagiota, J.C. Mohammad, S. Eric, Numerical, modelling of forced convective heat transfer from the inclined windward roof of an isolated low-rise building with application to photovoltaic/thermal systems, *Applied Thermal Engineering* 31 (2011) 1950–1963.

[47] F.E. Ahmed, R. Hashaikeh, N. Hilal, Solar powered desalination – Technology, energy and future outlook, *Desalination* 453 (2019) 54–76.

[48] M.A. Eltawil, Z. Zhengming, Wind turbine-inclined still collector integration with solar still for brackish water desalination, *Desalination* 249 (2009) 490–497.

[49] O. Bait, M.S. Ameur, Enhanced heat and mass transfer in solar stills using nanofluids: A review, *Solar Energy* 170 (2018) 694–722.

[50] S.W. Sharshir, G. Peng, L. Wu, N. Yang, F.A. Essa, A.H. Elsheikhd, I.T.M. Showgi, A.E. Kabeel, Enhancing the solar still performance using nanofluids and glass cover cooling: Experimental study, *Applied Thermal Engineering* 113 (2017) 684–693.

[51] P.R. Prasad, P. Pujitha, B.G.V. Rajeev, K. Vikky, Energy efficient solar water still, *International Journal of ChemTech Research* 3 (4) (2011) 1781–1787.

[52] H.K. Jani, K.V. Modi, A review on numerous means of enhancing heat transfer rate in solar- thermal based desalination devices, *Renewable and Sustainable Energy Reviews* 93-111.

[53] B. Prasad, G.N. Tiwari, Analysis of double effect active solar distillation, *Energy Conversion and Management* 37 (11) (1996) 1647–56. Elsevier BIOBASE, ScienceDirect.

[54] H.N. Panchal, Enhancement of distillate output of double basin solar still with vacuum tubes, *Journal of King Saud University – Engineering Sciences* 27 (2015) 170–175.

[55] S. Kumar, A. Tiwari, An experimental study of hybrid photovoltaic thermal (PV/T) active solar still, *International Journal of Energy Research* 32 (2008), 847–858.

[56] K.S. Reddy, K.R. Kumar, C.S. Ajay, Experimental investigation of porous disc enhanced receiver for solar parabolic trough collector, *Renewable Energy* 77 (2015) 308–319.

[57] A. Rabl, Comparison of solar concentrator, *Solar Energy* 18 (1976) 93–111.

[58] H. Kessentini, C. Bouden, Numerical simulation, design, and construction of a double glazed compound parabolic concentrators-type integrated collector storage water heater, *Journal of Solar Energy Engineering* 138 (2016).

[59] V. Pranesh, R. Velraj, S. Christopher, V. Kumaresan, A 50 year review of basic and applied research in compound parabolic concentrating solar thermal collector for domestic and industrial applications, *Solar Energy* 187 (2019) 293–340.

[60] D. Bishoyi, K. Sudhakar, Modeling and performance simulation of 100 MW LFR based solar thermal power plant in Udaipur India, *Resource-Efficient Technologies* 3 (2017) 365–377.

[61] G. Mokhtar, B. Boussad, S. Noureddine, A linear Fresnel reflector as a solar system for heating water: Theoretical and experimental study, *Case Studies in Thermal Engineering* 8 (2016) 176–186.

[62] H. Beltagy, D. Semmar, C. Lehaut, N. Said, Theoretical and experimental performance analysis of a Fresnel type solar concentrator, *Renewable Energy* 101 (2017) 782–793.

[63] Z.M. Omara, M. A. Eltawil, Hybrid of solar dish concentrator, new boiler and simple solar collector for brackish water desalination, *Desalination* 326 (2013) 62–68.

[64] M. Wang, K. Siddiqui, The impact of geometrical parameters on the thermal performance of a solar receiver of dish-type concentrated solar energy system, *Renewable Energy* 35 (2010) 2501–2513.

[65] H. Hijazi, O. Mokhiamar, O. Elsamni, Mechanical design of a low cost parabolic solar dish concentrator, *Alexandria Engineering Journal* 55 (2016) 1–11.

[66] K. Lovegrove, A. Zawadski, J. Coventy, Paraboloidal dish solar concentrators for multi-megawatt power generation, *Solar World Congress, Beijing*, September (2007) 18–22.

[67] K.S. Reddy, G. Veershetty, Viability analysis of solar parabolic dish stand-alone power plant for Indian conditions, *Applied Energy* 102 (2013) 908–922.

[68] H.M.A. Hayat, S. Hussain, H.M. Ali, N. Anwar, M.N. Iqbal, Case studies on the effect of two-dimensional heliostat tracking on the performance of domestic scale solar thermal tower, *Case Studies in Thermal Engineering* 21 (2020) 100681.

[69] S. Indora, T.C. Kandpal, Institutional cooking with solar energy: A review, *Renewable and Sustainable Energy Reviews* 84 (2018), 131–154.

[70] T.S. Ge, R.Z. Wang, Z.Y. Xu, Q.W. Pan, S. Du, X.M. Chen, T. Ma, X. Wu, X.L. Sun, J.F. Chen, Solar heating and cooling: Present and future development, *Renewable Energy* 126 (2018), 1126–1140.

[71] T.S. Ge, J.C. Xu, Ch. 3. Review of solar-powered desiccant cooling systems, in *Advances in Solar Heating and Cooling*. Woodhead Publishing, 2016, 329–379.

[72] T.U. Erkek, A. Gungor, H. Fugmann, A. Morgenstern, C. Bongs, Performance evaluation of a desiccant coated heat exchanger with two different desiccant materials, *Applied Thermal Engineering* 143 (2018) 701–710.

[73] L.A. Chidambaram, A.S. Ramana, G. Kamaraj, R. Velraj, Review of solar cooling methods and thermal storage options, *Renewable and Sustainable Energy Reviews* 15 (2011) 3220–3228.

[74] J.P. Praene, A. Bastide, L. Franck, G. François, H. Boyer, Simulation and optimization of a solar absorption cooling system using evacuated tube collectors, *9th REHVA world congress Wellbeing Indoors*, Helsinki, Finland (2007).

[75] D. Rusovas, S. Jaundalders, P. Stanka, Evaluation of solar sorption refrigeration system performance in Latvia, *16th International Scientific Conference Engineering for Rural Development* 24–26 (2017).

[76] T. Otanicar, R.A. Taylor, P.E. Phelan, Prospects for solar cooling – an economic and environmental assessment, *Solar Energy* 86 (5) (2012) 1287–1299.

[77] D. Chemisana, J.L. Villada, A. Coronas, J.I. Rosell, C. Lodi, Building integration of concentrating systems for solar cooling applications, *Applied Thermal Engineering* 50 (2013) 1472–1479.

[78] H.Z. Hassan, A.A. Mohamad, A review on solar-powered closed physisorption cooling systems, *Renewable and Sustainable Energy Reviews* (2012) 2516–2538.

[79] A. Dzigbor, A. Chimphango, Evaluating the potential of using ethanol/water mixture as a refrigerant in adsorption cooling system by using activated carbon – sodium chloride composite adsorbent, *International Journal of Refrigeration* 97 (2019) 132–142.

18 Developments in Wide-Area Monitoring for Major Renewables
Wind and Solar Energy

S. Behera and B. B. Pati

CONTENTS

18.1 Introduction: Wide-Area Measurement Systems (WAMS) 227
18.2 Need for Wide-Area Monitoring 228
    18.2.1 Transient Angle Instability 229
    18.2.2 Voltage Instability 229
    18.2.3 Frequency Instability 229
18.3 PMU Standard 229
18.4 Phasor Measurement Unit (PMU) 230
18.5 Phasor Data Concentrator (PDC) 230
18.6 Fundamentals of Phasors and Phasor Extraction Technique 231
18.7 Types of PMU 232
    18.7.1 Distribution PMU 232
    18.7.2 Mobile Device–Based Phasor Measurement Unit (MDPMU) 233
18.8 Challenges in PMU Implementation 233
18.9 PMU Placement 233
18.10 State Estimation 235
    18.10.1 Transmission State Estimation 235
    18.10.2 Distribution State Estimation 235
    18.10.3 Harmonic State Estimation 237
18.11 Event Detection/Islanding Detection with PMU 237
    18.11.1 What Is Islanding? 237
18.12 Systems Control 238
    18.12.1 Classification of Oscillation 238
18.13 Rotational Inertia 240
18.14 Topology Identification 241
18.15 Security of Power System 241
    18.15.1 What Is a Denial-of-Service (DoS) Cyberattack? 242
    18.15.2 Missing Data Recovery Methods 242
18.16 Synchrophasor in Microgrid 243
18.17 Conclusion 243
18.18 References 245

18.1 INTRODUCTION: WIDE-AREA MEASUREMENT SYSTEMS (WAMS)

Wide-area measurement systems (WAMS) in a smart grid measures and records variables in the power grid over a wide area and across traditional control boundaries, and then those measurements are used to improve grid stability and events through wide-area situational awareness and analysis of the data.

[Matic-Cuka and Kezunovic, 2014]

As the underlying purpose is monitoring, it can be termed wide-area monitoring systems. The fossil fuel crisis and environmental issues have turned power systems towards integrating wind and solar renewable power sources. But these generations are dependent on natural resources, specifically wind. So they constitute in different capacities, depending on the feasibility, and deliver as distributed generation (DG). The DGs may penetrate at transmission or distribution level, varying in capacity from kW to MW rating. The very nature of spatiotemporal variation makes these sources non-dispatchable. So the real-time picture of system dynamics is essential. Such a behavior adds unintentional disturbances in the power system, which increases the vulnerability of the power system to blackout. So wide-area monitoring to prevent and post facto analysis for corrective action are a mandate. As such, wide-area synchronized monitoring of both magnitude and phase angle is the first step that is required for fair control and protection. As such, demand for a smart, fast, reliable system with cutting-edge communication and fast computation framework is indispensable for WAMS. Thanks to the development of a phasor measurement unit (PMU), that has eased the synchronized measurement. As a result, many possible blackouts could be prevented by event detection, and the reliability of the power supply could be improved with an increase in power transmission. Specifically, the growing share of photovoltaic and wind power arouses many positive and negative possibilities. Based on the report of greening India by NREL, Berkeley Lab, and POSOCO, the use of HVDC links and monitoring by PMU has been discussed under different operational scenarios. But the effect of generators such as PV and wind are to be monitored, not only the lines.

18.2 NEED FOR WIDE-AREA MONITORING

Wide-area monitoring system (WAMS) is the essential base for protection and control as it gives an observation in large where the myopic view is incapable to envisage or avoid emergencies that may lead to system breakdown. So a trustworthy security estimation and optimized coordinated corrective action can moderate or prevent instabilities in large. The first task is, fast detection of disturbance, enhance power system reliability, operate near the stability limit, increase power transfer capacity noncompromising with security, and load curtailment. The conventional system, such as supervisory control and data acquisition (SCADA) or energy management system (EMS), results in inapt local control actions that approximate steady-state conditions. Synchronized measurement, followed by stability evaluation and algorithm to bring the system back to stable operation, is termed as wide-area monitoring and control (WAMC) system. It facilitates:

- Wider parametric view of the power system network
- Customized tasks on power systems, such as oscillation detection, voltage stability assessment of the transmission lines, and thermal stability

The wide-area power system instability is mainly due to three causes [Ashok et al., 2021]:

- Transient angle instability (loss of synchronism)
- Voltage instability (voltage collapse)
- Frequency instability (frequency collapse)

The severity and actions are represented in Figure 18.1.

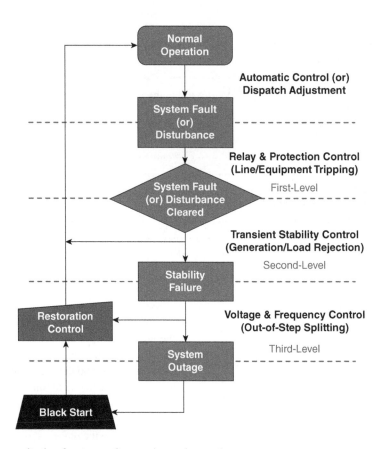

FIGURE 18.1 Wide-area monitoring for stages of protection and control.

Source: Ashok et al., 2021

18.2.1 Transient Angle Instability

When a cluster of generators in different areas accelerates at different speeds, this may lead to loss of synchronism. The transmission line fault causes oscillations in generator rotor angles. This then leads to inadequate coupling due to removal of transmission lines, lack of synchronizing power, and the generators in different areas missign synchronism. Subsequently, the angular change widens between groups, the voltage in two different areas will differ from standard reference, protective relays trip further lines, and cascade tripping occurs within a few seconds. The transient angle stability can be maintained by braking resistors or flexible AC transmission system (FACTS) devices and fast valve control to lessen generation.

Small-signal angle instability is frequent due to mismatch of loads and generation. This can increase rotor angle due to insufficient synchronizing torque, and the rotor oscillates at increased magnitude due to inadequate damping. Power system stabilizers (PSS) and FACTS devices can damp out small-signal angle oscillations.

18.2.2 Voltage Instability

Voltage instability occurs when in the system a disturbance or increase in load demand occurs and voltage drops beyond the acceptable limit. Thus, the system fails to keep the voltage profile within the stable limit. The instability is caused mainly by increased reactive power requirements in the system. Nevertheless, it takes several minutes for the voltage collapse to spread.

The emergency counteractions are:

- Resetting of generator voltage set point
- Fast redispatch of generation
- Automatic control of shunt switching
- Series compensation
- Blocking of tap changer of transformers

18.2.3 Frequency Instability

The power system operates within a thin range of frequency for stability and safe operation of the system as a whole. When the range is exceeded, frequency instability occurs. This is as a result of large disturbance, for example, major unbalance of generation and load or missing link of AC or DC. Emergency control and corrective actions to restore frequency stability are:
Trip of generators

- Load shedding, fast-valving to control generation
- HVDC power transfer control
- Controlled islanding of network (or) load

18.3 PMU STANDARD

According to IEEE Standard 2030 [IEEE guide, Sep. 2011], "PMU applications in distribution systems" are planned. The communications technology and interoperability between field devices for backup protection between SCADA and PMUs are required. Protocols IEC 61850 [IEC 61850] for timely communication and IEEE Standard 1815 [IEEE Standard 1815, 2012] are considered as relevant practices. The acceptable angle difference for restoration is prescribed in IEEE Standard 1547 for connecting distributed generation (DG) with electric power system [IEEE Standard, 2018]. PMUs constitute an integral part of WAMS. The PMU algorithms should satisfy within the accuracy and time during standard test conditions to take the benefit in real time for monitoring, control, and protection applications. For these, following the stipulated test procedure prescribed by the IEEE standard for synchrophasor measurement is very crucial. And these standards with the reconfiguration of the grid are gradually strengthened. Regarding this, a contrast of standards in 2011 and 2014 is presented in [Aalam and Shubhanga, 2020]. Also, the test procedures as per IEEE standard are enumerated and comprehensively described to apply the test to a PMU algorithm to authenticate. The steady-state and dynamic tests as per the standard are conducted using the Fourier filter (FF) algorithm, and outcomes are contrasted with the allowed limits. Then, the performance of the FF is confirmed using realistic signals in a system during fault conditions.

Out-of-band interference test/interharmonic test. This test is performed to ascertain whether the PMU algorithm's performance has deviated in the existence of out-of-band frequency signals. A frequency f is termed as out-of-band or interharmonic if:

$$|f - f_0| \geq F_{rr}/2$$

Where f_0 is the standard frequency and F_{rr} is the phasor reporting rate. This test is prescribed for measurement-class PMUs. Other limits are frequency error (FE) limit (Hz), rate of frequency error (RFE) limit (Hz/s), and total vector error (TVE) limit (%).

IEC/IEEE 60255–118–1. It states a synchronized phasor (synchrophasor), frequency, and rate of change of frequency measurements. It designates time stamping and synchronization requirements for the measurement of the aforementioned parameters.

IEEE C.37.118 standards. Communication of PMU data in a smart grid can be via transmission control protocol (TCP)/internet protocol (IP) that comprises protocols based on IEEE C37.118, substation automation protocols (IEC 61850), etc. in WAMS.

PMUs and phasor data concentrators (PDCs) constitute the two important physical apparatuses of a WAMS, and all of such units are time-synchronized using a universal global positioning satellite (GPS) clock. IEEE Standard C37.118.22 has a compact message format.

The packets are filtered and analyzed in the PDCs. The PDC time-synchronizes the phasor data collected from several PMUs and constructs a time-aligned output data stream

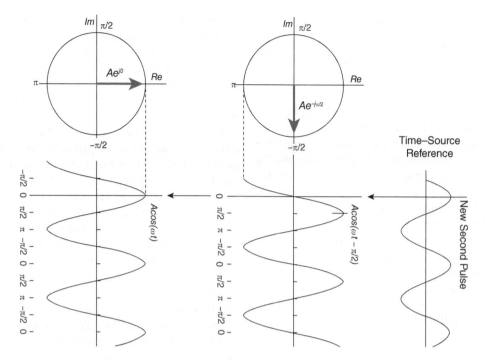

FIGURE 18.2 Concept of time-stamped phasors.

at each time stamp, ready to use by wide-area protection and control functions. Moreover, measurement requirements are quantity, resolution, accuracy, and latency [Usman and Faruque, 2019; Castello et al., 2014].

18.4 PHASOR MEASUREMENT UNIT (PMU)

The data that is vital to be recorded in synchronous measurement are:

1. Phasor values
2. Loads and generation
3. Other influential parameters

Phasors mean voltages and currents with both magnitude and angles, and these can be collected through PMU measurements. This envisages important system operation. These are used for state estimation and predictions. IEEE standard defines PMU as "an electronic telemetering apparatus that takes voltage signals and current signals obtained from current transformers and potential transformers as input and generates current phasor and voltage phasor, respectively, which are time-synchronized or time-stamped." The concept of time-stamped phasors is illustrated in Figure 18.2, where the two phasors are from two different locations.

The PMU converts the signals to digital form and data packets into [IEEE Standard C37.118] frames and sends the real-time measurements gathered from the grid to the PDC over a fast communication arrangement. PMU also generates frequency and rate of change of frequency (ROCOF) [Phadke and Thorp, 2008]. These data improve observability in the following ways. The transmission line trip can be observed by the change in voltage phasor separation in both magnitude and angle and similar variation in line current (MW and MVAR). The generator trip can be observed by the frequency drop, increased ROCOF, variation in angular separation, and drop in voltage magnitude. Load variations can be observed by continual high-frequency disturbances, phase angle separation, and voltage.

The reporting rate changes with system frequency – viz, for 50 Hz, it is 10–50 frames per second, whereas for 60 Hz, it is 10–60 frames per second. Frequency response can be quickly restored with a high-frequency data rate of 120/s than 10/s as the dynamics can be clear and observed fast.

18.5 PHASOR DATA CONCENTRATOR (PDC)

It can be regarded as the collector for storage synchrophasor technology (Figure 18.3). It receives the time-synchronized voltage and current data from multiple PMUs optimally located in a particular area. These PDCs then transfer data to the super-PDC, which sends information to the control room for monitoring, control, and protection for a WAMS. Kanabar et al. [2013] ideated PMU incorporation with PDC in a WAMS.

The synchrophasor technology (SPT) differs from SCADA technology in universal time synchronization. Whereas phasor angles are computed postprocessing data centrally at EMS in SCADA and reporting rate is slow once in 4–6 s, it is estimated locally (at the substation level or PMU) and reports as per the requirement in 10/12/15/20/30/60 frames per second.

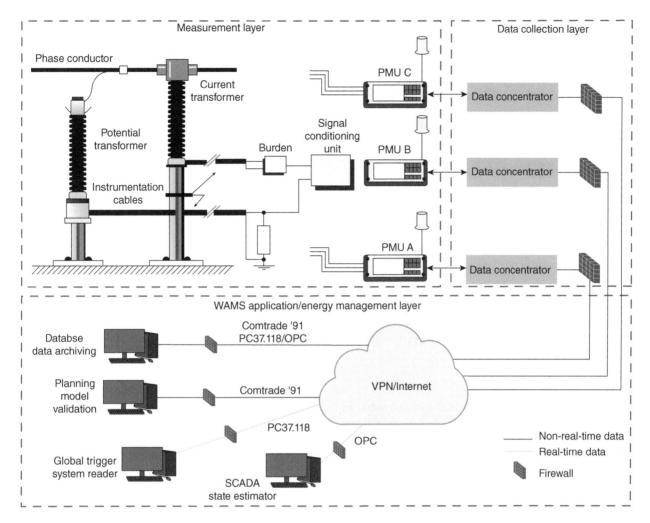

FIGURE 18.3 Synchrophasor measurement topology.

Source: Usman and Faruque, 2019

18.6 FUNDAMENTALS OF PHASORS AND PHASOR EXTRACTION TECHNIQUE

It is clear from the previous section that for WAMS, data loss may occur if queued longer, which is collected from the geographically dispersed location. Here the SCADA fails. This leads to inappropriate control or protection. This necessitates that data has time information or that is synchronous. The synchronous measurement helps in monitoring data per cycle, taking the voltage magnitude and phase angle and also tracking the changes in multiple locations. The difference in phase angle at two different buses indicates the stress in the grid. So it significantly supports real-time monitoring and postmortem of contingencies and blackouts. PMUs can provide such features, and the new technology is known as SPT [Phadke and Tianshu, 2018].

Phasor estimation algorithms in use are listed in [Mohanty et al., 2020], out of which one is the discrete Fourier transform (DFT), and the complete processing is shown in Figure 18.4.

The applications of SPT in WAMS are for offline/online applications that are meant for validation of power system model, state estimation, postdisturbance analysis, monitoring oscillations, and stability. The purpose is power system restoration by measurements and estimation via real-time control and adaptive protection.

In these aspects, the broad categories of basic functionalities of PMUs are as follows:

- System monitoring/state estimation
- Event recording
- Analysis system/load characteristics
- System controls

The synchrophasor-based applications target improved visibility and situational awareness, which can cover both inside and beyond the substation. For the transmission side, PMUs supplement wide-spread telemetry in vogue for power system quantities, voltage, current, active power, and reactive power nearly real-time. The increased resolution and precision of

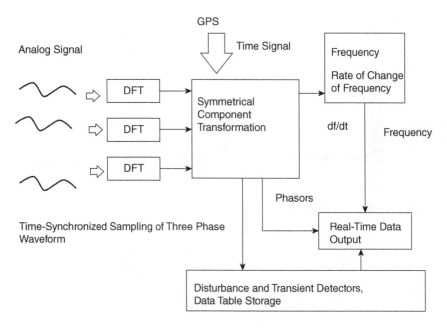

FIGURE 18.4 Block diagram of the internal structure of phasor estimator.

Source: Mohanty et al., 2020

SMT substantially advance conventional measurements, microscopically portraying the dynamics across the time dimension by time-stamped frequency measurements, even without any phasor differences. These dynamics are visualized in the relation of quantities over large distances across the network – viz, oscillations and their damping. Sometimes, the input noise that distorts the signal appears at the output of the PMU as data noise. This is one type of uncertainty added to the reported synchrophasor data.

18.7 TYPES OF PMU

The PMU as per the purpose of use can be classified broadly as per the synchrohasor measurement algorithm (SMA) per IEC/IEEE 60255–118–1 [Synchrophasor for power systems – Measurements, 2018; Bonian et al., 2019] as:

- Protection class, or P-PMU
- Measurement class, or M-PMU

Compared with the M-PMU, P-PMU needs a much faster response. This constraints data window width and inclusion of filter. Thereby, it can achieve a fast, dynamic response with disturbance rejection. Confirmation of standard tests and accuracy is required for both types.

Again, according to the use in the network:

- PMU in a transmission system
- Distribution PMU/micro-PMU

18.7.1 Distribution PMU

PMU that can be installed in distribution system (DS) is usually known as D-PMU, which helps in the microscopic observation of renewable variation that makes the system more reliable and resilient and improves power quality. Micro-PMU (μ-PMU), developed by the University of California (UC) in association with Power Standards Lab (PSL) and Lawrence Berkeley National Lab (LBNL), is suitable for DS [Von Meier et al., 2017].

The characteristics of μ-PMU with reference to PMUs in transmission are:

Increased measurement resolution
More accurate phase angle measurement
Reduced capital cost

At present, a countable number of pilot projects have installed D-PMU [Eto et al., 2016]. As stated by [Y. Liu et al. 2020], "[t]he applications of D-PMU for state estimation, voltage stability monitoring, fault detection and fault localization, are critical," and advances are multidisciplinary, including data mining.

Distribution network topology changes frequently due to isolation, the addition of generator or load, and restoration. So topological information is a prerequisite for appropriate action. It is done by lowering the state estimation residual [Monticelli, 2012; Singh et al., 2010]. DG contribution can be observed in real time by D-PMUs at the point of interconnection (POI) of DGs and the prosumer subsystem. This can also help net metering and billing. Reverse power flow is not encouraged as it raises the voltage at POI [Mortazavi et al., 2015; Hasheminamin et al., 2015]; hence, recording flow directions is a serious concern.

As per Glavic and Cutsem [2011], "[v]oltage stability assessment are two types: (i) single location measurement-based methods, and (ii) wide-area measurement-based methods." In such a case, PMU data–based method has

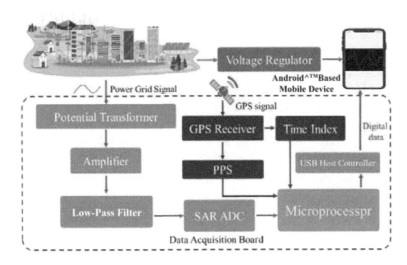

FIGURE 18.5 Structure of the MDPMU.
Source: Wang et al., 2020a

higher accuracy. Indeed, synchronized measurements from D-PMUs enabled a spectrum of methods for VSI prediction.

Dynamic feeder rating assessment can improve power flow as per real-time line statuses and ambient situations. With D-PMU, faithful recognition of events by tracking the fault in real time is possible by D-PMU instead of postmortem. Features extraction from D-PMU data – viz, total harmonic distortions (THD), ROCOF, and rate of changes of angle differences – helps detect islanding operation [Cardenas et al., 2014]. Low-frequency oscillations, though not prominent in DS, could be spread via substations to transmission level and can be detected from D-PMU data. Also, major fault decisions based on D-PMU data can island microgrids to operate independently [Li et al., 2018]. Restoration is also easy, as system frequency and voltage stability can be observed. Cross-correlation of synchronized voltage amplitudes only from different D-PMU can identify phases for the correct sequence at the time of restoration [Pezeshki and Wolfs, 2012]. Moreover, synchronized angle measurements, in addition, simplify the process [Wen et al., 2015].

18.7.2 Mobile Device–Based Phasor Measurement Unit (MDPMU)

The functional architecture of the MDPMU is depicted in Figure 18.5 [Wang et al., 2020a]. The analog measurement from the grid is processed to digital form and sent to and from an Android TM–based mobile via USB host controller in the wireless medium. Time synchronization with GPS is achieved by pulse-per-second (PPS) signal. It can be used for short-time phasor estimation computation in a smartphone. It is cheaper and can be deployed widely.

18.8 CHALLENGES IN PMU IMPLEMENTATION

The following are the main challenges [Maheswari et al., 2020]:

- Integration of SPT with SCADA
- Optimally placing PMUs
- Communication delays
- Speed of computing and storage
- Developing methods for exhaustive postmortem investigation

The local control requires the substation data and acts fast. A coordinated control in the neighborhood through a communication interface can achieve better control of the system and prevent cascaded events that can be decentralized and distributed. The SCADA system receives data from RTUs. EMS requires data from SCADA and smart meters and balances generation and load. The centralized control can implement EMS as well as protection in a wide area. The time frame and the nature of control are shown in Figure 18.6 to assess the need for synchrophasor measurements.

A "grid-interactive battery energy storage system (GI-BESS)," termed by [Torkzadeh et al., 2019], can participate in power oscillation damping control. This can be established by installing PMUs in locations at both ends of the line that is to be observed. The super-PDC stores all the data required for the control in a widespread network. The clear structure of the constituents is shown in Figure 18.7. It is also possible that the BESS can participate in the demand response market.

18.9 PMU PLACEMENT

WAMS for protection and control systems demand observability of the system. Placing the reduced number of PMUs, maintaining the observability but reducing the cost of PMU placement, is an optimization task. The selection of the bus depends on the purpose and structure of the system. For the transmission system, it can be at the highest interconnected bus or weak bus, whereas in DS it can be at the branching of radial lines.

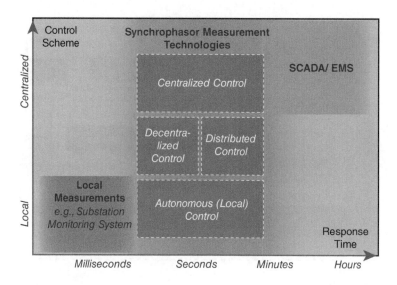

FIGURE 18.6 Control scheme–time frame–technology mapping.

Source: Torkzadeh et al., 2019

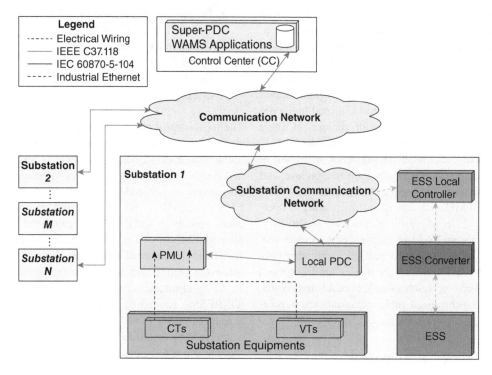

FIGURE 18.7 WAMS-based solution for control.

Source: Torkzadeh et al., 2019

As per [Guideline NERC, 2016], transmission system visualization is of three types:

- **System-wide:** PMUs deliberately placed at specific EHV bus
- **Interface-based:** PMUs placed on both sides of main transmission interfaces
- **Line-based:** PMUs placed at both ends of a circuit

Situational awareness can be deeper and corrected by widely monitoring frequency, voltage trend study, and power flow monitoring. However, alarms are yet to develop based on these measurements that prevent the worst conditions.

In a case study with PV and thermal generation in a DS monitoring for a fault occurrence, the observability demands placement of PMU in the intermediate bus rather than putting in all three buses [Ray and Mishra, 2020]. Variable energy resource (VER) integrated is to be observed. One of the purposes of installing PMU at the POI of a wind farm could be a check of its control mode of operations: power factor control (constant or range) or voltage control mode. This is possible by taking the measurements from the POI, such as voltage and active and reactive power. Section 6.1.3 of the ERCOT Nodal Operating Guide states that "[n]ew generation Facilities over 20 MVA aggregated at a single site placed into service after January 1, 2017," will be required to have a PMU installed. By this, damp out of oscillations could be done easily by curtailing the generation from VER [Guideline NERC, 2016]. This can support grid reliability. A multicriteria decision support (MCDS) approach and the analytical hierarchy process (AHP) have been used to rank the locations for PMU installation [Chatterjee et al., 2020]. The multiobjective is optimized by intelligent search technique (IST). The stated decision criteria are: "bus-to-bus connectivity, buses connected to a radial bus, less significant bus, load bus and generator bus connectivity, transformer connection, and bus with a compensating device for a maximum observability." Sometimes, a missing measurement may cost high to the wide-area measurement, and this can be overcome by this method of placement.

Some criteria defined by [Phanendrababu, 2020] are system observability index (SOI), restorable islands observability index (RIOI), critical bus observability index (CBOI), and critical line observability index (CLOI), which can be used for multicriteria PMU placement.

18.10 STATE ESTIMATION

State estimation is of three types:

- Transmission system state estimation (TSSE)
- Distribution system state estimation (DSSE)
- Harmonic state estimation (HSE)

State estimation methods are classified into nonparametric (i.e., DFT) and parametric (i.e., Kalman filter (KF)) methods.

18.10.1 Transmission State Estimation

Transmission system constitutes of long interconnected lines with EHV and HV lines and generator and load in the MW power capacity. DFT-based method estimation is simple and rejects harmonics to derive the phasors. But it has scalloping loss due to the main lobe of the FF for deviation of the fundamental frequency. The phase-locked loop (PLL)–based methods are also simple, but it performs well without disturbance, and it degrades its performance speed and accuracy on such occurrence. Dynamics and distorted waves challenge the precise and speedy estimation of the phasor. To deal with simultaneous occurrence of speed and accuracy an "enhanced flat window-based P class synchrophasor measurement algorithm (EFW-PSMA)" is suggested [Xue et al., 2019]. The EFW is developed by least-square (LS), which acts as a low-pass filter (LPF) to segregate the fundamental signal. Then, LS estimates the frequency and ROCOF. The advantage is its computational inexpensiveness and simple method. The TVE limits are within 0.3% for the most severe benchmark tests, as per the IEC/IEEE 60255–118–1.

Interharmonics are contributed by PV systems because of their maximum power point tracking control. These are sub-multiple of fundamental. They can degrade power quality and distance protection schemes. They affect estimation in PMUs. The effects of sub- and supersynchronous frequency signals on PMU measurements are studied in [Hao et al., 2016] and [Jin et al., 2017]. Generally, the DFT algorithm is used for state estimation, but it fails for interharmonics. So a "phasor signal processing model (PSPM) with matrix pencil (MP) algorithm" is suggested that segregates the harmonics. The model has been evaluated to conform to IEEE C.37.118 Standards [Prakash et al., 2020].

The effect of quadrature demodulation (QD) synchrophasor estimation algorithms (SEA) on real-time voltage stability evaluation has been studied [Thilakarathne et al., 2020] for a P-class PMU. In the QD method, the input signal is multiplied by a complex fundamental signal, as shown in Figure 18.8. A modified DFT algorithm (e.g., s-DFT) is also proposed that works accurately for real-time stability assessment.

18.10.2 Distribution State Estimation

A DS normally takes MV-to-LV lines with the generator and loads. Along with electrical quantities, DSSE has to estimate the inertia constant (H), which is a mechanical parameter, as it varies for DGs. To observe DS, more PMUs are to be deployed [Von Meier et al., 2018]. Further, the measurement should be robust to any missing or false information. So a multiarea state estimator (MASE) can be a better solution. A reduced number of PMU channels can reduce the cost of a DS. This can be achieved by the development of the estimation algorithm as a contributory. The impact of the measured data is gauged in terms of the performance of the SE, reliability of measurements, and effects of the pseudomeasurements.

A few of the performance indices to assess the performance are [Gol et al., 2012]:

1. Measurement system vulnerability (MSV) ratio

$$MSV = \frac{Number\ of\ critical\ measurements}{Total\ number\ of\ measurements}$$

An MSV ratio of less than 3% is recommended.

2. The pseudomeasurement ratio (PMR)

$$PMR = \frac{Number\ of\ pseudo-measurements}{N-1}$$

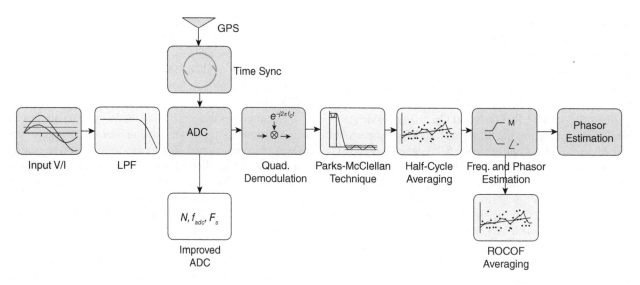

FIGURE 18.8 Enhanced P-class q-demodulation-based state estimation.

Source: Thilakarathne et al., 2020

Where *N* is the islands that can be observed. Multiple pseudomeasurements corrupt good measurements.

3. State estimation accuracy (SEA)

$$SEA = \{maximum\ vaiance\ of\ state\ estimates\}$$

$SEA < 10^{-6}$ is recommended.

[Gol et al., 2012] presented a method to place the critical measurements in a system to make the whole system observable. The algorithm can be used to place multiple measurements in each iteration, and it uses binary integer programming. It was shown that the performance of the two-step multiarea state estimator was comparable to the performance of the standard one-step multiarea state estimator in terms of accuracy even if the number of PMU channels is considered finite. Also, it was found that the increased number of communication channels resulted in improved estimation results (but mostly up to two to three channels). Thus, there is no need for installing higher-channel PMUs in the distribution network. However, more channels offer more availability of measurements and increased reliability as they provide measurement redundancy, but with increased cost [Fatima et al., 2020].

With high VERs in the system, the dynamic line rating to increase the load ability of the line within thermal stability is another estimation task. This can be done in real time, taking weather data, the sag of wires, and electrical loading [Singh et al., 2017]. Instead of measuring the temperature of the line, real-time estimates of line resistance can map to this value [Singh et al., 2018]. This is possible for both aerial lines and underground cables. As the line length is small, the magnitude and phase errors of the instrument transformers may affect the estimation modifying PMU measurements at both the end nodes. The assumption that the line is exposed to the same weather condition holds for its short length.

The state estimation techniques are singular value decomposition (SVD), principal component analysis (PCA) for dimension reduction of the data. Other data handling is also involved in different stages for feature extraction, feature selection, data compression, and event detection [Thasnimol and Rajathy, 2020]. The growing installation of PMUs and the increased reporting rates need compression of data to store but retain accurate important information. The compression methods can be either lossy or lossless [Sayood, 2012]. In lossless compression, important information is preserved by exploring the statistics of the data and using efficient bit-wise encoding techniques to compress it. Some of the methods are:

Deflate, Bzip2, Lempel-Ziv 77 (LZ77), Lempel-Ziv-Markov algorithm (LZMA), and the Szip [Klump et al., 2010], [Top and Breneman, 2013]. But the lossless method has a low compression ratio (CR). About a thousand PMUs can send data to PDC in the TB range. Whereas the lossy methods have high CR. Some of the methods are:

Discrete wavelet transformation (DWT) [Santoso et al., 1997], [Ning et al., 2011], improved DWT [Hamid and Kawasaki, 2002; Khan et al., 2015; Khan et al., 2016], exception compression (EC)-swing door trending (SDT) [Zhang et al., 2015], etc. The PCA can reduce dimensions to achieve early event detection [Xie et al., 2014] but is effective for frequency variation in data. A multiscale PCA model executes a spatial cluster analysis, then multiple PCA models compress the clusters. This two-stage method works well without the loss of significant information under normal and outage conditions. The cross-entropy and the SVD, after identifying data patterns, compress the data to an awesome small size and preserves the accuracy of disturbance records inside bounds [Wang et al., 2020b].

A real-time deep neural network–based nonlinear state estimator (DNNSE) with "analysis on the wire" framework is effective for large-size systems by µPMU measurements with highly dispersed VERs [Basumallik et al., 2020]

The states and transient reactance of a distributed synchronous generator–based combined heat and power unit in a DS have been estimated online, taking data from μ-PMU located at the POI. A constrained LS algorithm is tested under faults, islanding, and load/capacitor switching [Dutta et al., 2020]. However, the extended KF approach does not predict the states with accuracy in a DS. This is because the dynamics of distribution states are fast. So a fast algorithm should compute states in a short time within the time-varying stream of measured data. This demands reduction in computation time, and this has been done by Hessian calculation instead of matrix inverse to minimize the error of estimation [Song et al., 2019].

18.10.3 Harmonic State Estimation

Power electronic interface for renewable DERs contributes significantly to harmonics proportion in DS. The bad effects of harmonics on power quality include overheating of distribution lines, torque pulsation, and damage connected devices. D-PMUs can serve both DSSE and HSE. A design to estimate harmonics is presented in [Liao et al., 2007], and D-PMUs applied for harmonic synchrophasors are presented in [Carta et al., 2009].

18.11 EVENT DETECTION/ISLANDING DETECTION WITH PMU

18.11.1 What Is Islanding?

Islanding is "a situation where a portion of the grid and the local loads connected to the DG is isolated from the main grid by an opening of the circuit breaker (CB) at the POI," as stated by [Dutta et al., 2018a].

When islanded, the DG feeds the local load. The islanding can be intended (planned) or unintended (unintentional or unplanned). Unintended islanding breaches voltage and frequency stability and safety of utility personnel during restoration. So detection of islanding to isolate the DG from the main grid is essential. "Non-detection zone (NDZ) is the range of power mismatches (both active and reactive) of DG supply and the local load," as stated by [Pouryekta, 2018]. As per the IEEE 1547–2003 standard, "after islanding detection, the DG should isolate within two seconds after the utility disconnection" [IEEE Application Guide, 2009]. The local active islanding detection scheme (IDS) has low NDZ, but it is slow, is DG-specific, and degrades the power quality, whereas local passive IDSs hold large NDZ and IDS might fail, with false alarms and trips, as a result of inappropriate threshold set. Stringent limits cause false islanding, and a loose boundary imparts slow detection of the islanding.

Further, islanding also affects the normal protection coordination of the DS. So adaptive protection coordination is required. PMU data can be used efficiently in digital relays. For that, islanding instant is essential. The main design problems associated with IDS using PMU data [Subramanian and Loganathan, 2020; Chandak et al., 2018] are:

Choice of thresholds
Existence of NDZ

Inverse relation of islanding detection time on power mismatch deliberately for low power mismatches.

Modify the control of the inverter.

Some of the islanding scenarios are:

(a) $P_{DG} > P_{LL}$
(b) $P_{DG} < P_{LL}$
(c) $P_{DG} = P_{LL}$

And nonislanding scenarios (NIS) are:

(a) Faults
(b) Load shedding
(c) Switching of industrial motor loads

Where P_{DG} and P_{LL} are power from DG and local load.

Dutta et al. [2018b] proposed "[p]hase angle difference (PAD) between positive and zero sequence components to decide the islanding conditions." The method claims to be robust and has no NDZ. An IDS that uses frequency and phase angle from the μ-PMU at the reference substation and DG terminals calculates a cumulative sum of frequency difference (CMFD) and PAD.

The Pearson correlation coefficient of aforementioned quantities computed in real time generates a time-bound threshold. This accurately classifies islanding and nonislanding events due to load changes and faults. It subsides the requirement of any frequency and phase-angle-based thresholds. Again, the detection time does not depend on power difference and NDZ limits within 0.01 Hz. The system has two PV generators in a radial DS [Kumar and Jena, 2020].

An unintended IDS that is highly efficient under zero-power-mismatch conditions and satisfies the limit of trip time has been developed in an DS with wind and PV plants that take parameters from μ-PMU. The calculation of the difference between voltage phasors, ROCOF, and rate of change of PAD of grid and island is used. The μ-PMU-based IDS proposed [Subramanian and Loganathan, 2020] voltage-phase angle difference (VPAD) as a criterion that reduces NDZ and tripping time.

The moving-window PCA [Radhakrishnan et al., 2020] was generalized for distinguishing a wide-area disturbance from an islanding event by Guo et al. [2015] even for low-inertia DGs. The phase difference between positive-sequence voltage and current has been taken to decide the threshold in a phase-angle-based PCA IDS [Muda and Jena, 2018]. The IDS is effective for power mismatch and NISs.

When multiple microgrids with wind power generators but constant load are considered with islanding detection with two -PMU installed, the load reliability is improved with fast islanding detection after training. "Islanding detection monitoring factor (IDMF) and rate of change of inverse hyperbolic cosecant function of voltage (ROCIHCF)," the voltage at the POI for VER is computed, as stated by [Micky et al., 2021]. For faults, the disconnection of the DG or continue generation in the island based on the probability of power balance (PoB) and the probability of islanding duration in the submicrogrids work under the uncertainties of RES. Event prediction after training is done by the decision tree (DT) method. Event detection that leads to frequency disturbances, such as generator trips, load disconnections, line trips, etc., by deep learning to build a convolutional neural network (CNN) model is presented [Wang et al., 2020c]. For this, the ROCOF and the relative angle shift (RAS) are the inputs of the CNN. The decision time is reduced and suits practical scenarios.

A drawback of the PMU-based islanding detection scheme (IDS) is that as a remote method, its reliability depends on the communication latency and needs passive IDS as a backup [Appasani and Mohanta, 2018]. Packet loss, noise, and bias errors need calibration of such devices to ensure the reliability of the PMU, whereas NDZ can be reduced by a higher sampling rate of PMU from 24 to 120/cycle, affecting data communication. The number of channels limits phasors that are measured – viz, one channel is for one pair of voltage and current phasor. In a study for the channel requirement for IDS of the grid-tied PV plant, a three-channel PMU was installed at POI to measure node voltage phasor and three connected line currents. Only the voltage-phase angle of bus 1 was sufficient to detect the islanding of the PV side. Hence, only one-channel PMU is needed in such a case for efficient IDS that later includes PV and segregates the grid [Khair et al., 2020].

18.12 SYSTEMS CONTROL

The contribution of PMUs in DS is reviewed [Sanchez-ayala et al., 2013]. Control of DS with radial feeder depends on the capacity and stiffness. So integrating more DG demands a change of operation modes of distribution feeders. One such can be a closed-loop of two or more radial feeders to reduce impedance at POI and allow more DG to penetrate [Chen et al., 2008]. As the next generation of DS has to incorporate PEVs, real-time active control can benefit from the wide-area measurement from PMU. This will also deal with the intermittence of RESs. The rapid growth of rooftop PV impacts weak feeders, causes power quality issues, disturbs voltage control and protection, and creates stability issues. Large penetration can be dealt with dynamic volt-VAr compensation via DG inverters or D STATCOMs. With proper monitoring, control, protection, and automation systems in place, a closed-loop operation can increase reliability and efficiency, reduce voltage drops and losses, reduce voltage sensitivity to DG output variations, and promote more effective use of existing feeder capacity. This is futuristic to apply to PEV (plugin electric vehicle) integration and can increase penetration levels by modification of feeder capacity by looping.

The synchrophasor data–driven dynamic models have been used to design control algorithms such that seamless disconnection and reconnection of a microgrid is possible, maintaining active power flow to the load [Konakalla et al., 2019]. The advantage of such control is effective when microgrid power flow modifies according to changing operational situations.

The increasing penetration of nonsynchronous VER in the grid complicates the frequency stability. High -ve deviation and fast change can lead to islanding and load shedding. In the case of +ve deviation, it can cause overloading, voltage collapse, and blackouts. In a case study of data from PMUs in terms of ROCOF, the NS-VER does not cause any problem for 3/4 penetration level [Xypolytou et al., 2018].

WAMS by optimal PMU placement is an integral part to monitor low-frequency oscillation (LFO), and postmortem analysis counteract sit by the installation of power system stabilizer (PSS) and system protection schemes (SPS) to improve power oscillation damping (POD) [Kumar et al., 2018]. A voltage control problem may be handled in real time without knowledge of the model parameters. The model-free approach is particularly desired since the increasing integration of renewable energy sources means that the electric grid is becoming increasingly complex [Okekunle, 2020].

The electrical utility is liable to low-frequency oscillations (LFO) [Agarwal and Kumar, 2020] that are inherent. When damping reduces and if some power system control excites the oscillatory modes, it is observed.

18.12.1 Classification of Oscillation

1. Interarea mode: 0.05–0.3 Hz
2. Intra-area mode: 0.4–1.0 Hz
3. Interplant mode: 1.0–2.0 Hz
4. Intraplant mode: 1.5–3.0 Hz
5. Control mode: not defined
6. Torsional mode: 10–46 Hz

These are natural modes and frequency changes with the system topology, operating conditions, generation, and load. These push the system to a small-signal stability limit. Then oscillations grow further for negative damping, which may lead to blackouts.

Oscillatory stability tracking, analysis, and control have been reviewed [Meegahapola et al., 2020] in association with PMU data. Emerging big-data analytics on PMU data could be a future direction. The success of control depends on timely action, and for this role of hardware of PMU, SE algorithm and communication are crucial. Since the model-based methods fail to estimate the modal properties of oscillations, such as oscillatory frequency, damping ratio, and mode shape, data-based estimation is used. Estimations suitable for the ringdown (postfault) data are Prony analysis [Hauer et al., 1990], minimal realization [Kamwa et al., 1993], eigensystem realization [Sanchez-Gasca and Chow, 1999], Fourier transformation, Hilbert–Hung transformation [Messina and Vittal, 2006], matrix pencil [G. Liu et al., 2007], wavelet transformation (WT), variable projection [Borden and Lesieutre, 2014], fast Fourier transform (FFT), Wigner–Ville distribution (WVD), S-transform (ST), estimation of signal parameters by rational invariance technique (ESPRIT) [Ashok et al., 2021], and phase-locked loop. Estimation methods suitable for normal data are spectral analysis [Hauer and Cresap, 1981], the Yule–Walker method [Wies et al., 2003], frequency domain decomposition [Liu and Venkatasubramanian, 2008], and the autoregressive moving average exogenous (ARMAX) model [Dosiek and Pierre, 2013]. The recursive method (least mean squares adaptive filtering [Wies et al., 2004] and robust recursive least squares [Zhou et al., 2007]) and subspace system identification [Kamwa et al., 1996] are appropriate for both normal and postfault cases.

Again, the estimation of mode shape delivers valuable information for damping control. The recent methods are the continuous modal parameter estimator [Wilson et al., 2003], PCA [Anaparthi et al., 2005], Prony analysis, moment-matching method [Freitas et al., 2008], matrix pencil [Jin et al., 2017], Kalman filtering [Chaudhuri and B. Chaudhuri, 2011], and phase-locked loop method for ringdown data, as well as cross-spectrum analysis [Trudnowski, 2008], frequency domain decomposition [Kamwa et al., 1993], channel-matching method [Dosiek et al., 2009], transfer function [Zhou et al., 2009], and ARMAX model [Dosiek and Pierre, 2013] for ambient data. Stochastic subspace system identification [Jiang et al., 2016] is applicable for data of both cases. Prony analysis, frequency domain decomposition, PLL method, ARMAX model, matrix pencil, and subspace system identification can be employed for both mode and mode shape estimation, but subspace system identification accommodates both fault and normal case; however, it needs high computation effort [Meegahapola et al., 2020].

Besides LFO, new oscillation modes induced by the VERs are power resonance due to the interaction between renewables and voltage source converters (VSCs), HVDC controllers, and forced oscillations due to intermittent sources, as shown in Figure 18.9. The existing monitoring and estimation tools based on modal analysis require a large volume of data and do not disclose how the damping control propagates.

Data is termed as "big data" if it has characteristics [De Mauro et al., 2016] – volume, variety, and velocity – and the data from PMUs have all these characteristics. About trillions of samples are accumulated annually. The advantage of the data-based method in WAMS is that it provides a realistic solution to handle the serious threats and challenges due to the significant incorporation of RESs to system operational planning. Data quality and cybersecurity are very important for data-based analysis tools to be reliable. Mixed causes of oscillation are to be properly treated. However, only countable applications of data analytics techniques for oscillatory stability monitoring and analysis are reported and expect future growth with the open end.

Dynamics of VERs influence systems under coherent oscillation. This is identified by the spectrum similarity approach [Yadav et al., 2019a]. The Stockwell transform is effective in extracting many time–frequency (TF) features that depict a pair-wise comparison between synchrophasor frequency signals. Then mean shift spectral clustering is

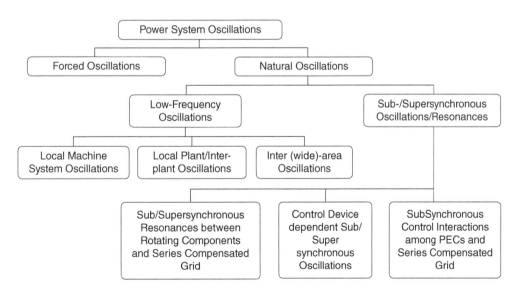

FIGURE 18.9 Classification of oscillations.

Source: Meegahapola et al., 2020

used to cluster the buses with similar TF spectrum tested in a model of an actual transmission system of India grid. Each generator in a coherent area (CA) exhibits almost-analogous amplitude and phase participation in a common set of interarea modes (0.1–1 Hz), showing a natural difference for coherency detection. But such generators also join in intra-area oscillations (1–3 Hz) of different degrees, and the coherency detection becomes a complex task.

Interarea oscillation can be damped by optimal placement of a power system stabilizer that works on PMU data for WAMS. Further, the effect of delay in data communication is compensated [Prakash et al., 2019]. Power angle is an indication of power flow and stress on the transmission system. Real-time monitoring of area angle is feasible from synchrophasor measurements to improve awareness and alert if exceeds the limit. An estimator that adapts major topology change and with a reduced number of PMU deployment for area angle measurement is proposed [Ju et al., 2020] that works in real time.

18.13 ROTATIONAL INERTIA

Large-scale PV plants that can improve damping oscillations without the need for curtailment have been studied [Silva-Saravia et al., 2020]. The projected step-down modulation (SDM) control method modulates active power such that on oscillation event, the PV panel voltage is controlled to shift from maximum power point (MPP) temporarily. This power differential is utilized for active power modulation to mitigate oscillation. So it has three control modes: prefault mode, activation mode, and recovery mode. The operation is at MPP for the first and last modes, and in the activation mode, the SDM works.

Synchrophasors measurement for PSS in coordination with an optimized load frequency control loop in order to tackle the undamped local and wide-area oscillatory problems is presented [Mekki and Krichen, 2020]. This gives a mixed PSS with both local and remote measurements. In the presence of wind power penetration at a different level, the mixed PSS performance is better. Series compensation improves system transmission capacity, but the doubly fed induction generator (DFIG) converters in wind farms introduce subsynchronous control interaction (SSCI) that challenges stability. These wind farms (WFs) operate asynchronously. At the natural resonance frequency, as the slip is negative, the effective resistance of the WFs-integrated power system decreases. To tackle such problems, a PMU data–based SSR damping control (SSRDC) is designed by controlling equivalent impedance that improves the SSR oscillation damping [Mahish and Pradhan, 2019]. The impedance offers sufficient damping for SSR along with the thyristor-controlled series capacitor (TCSC) control. It has been tested for variation in wind speed and the number of operational DFIGs. For a growing communication delay up to 0.8 s, the magnitude of the frequency response is limited to 0.1 Hz. But the magnitude suffers beyond this frequency. A large duration of delay beyond this degrades the performance of SDC. A "synchrophasor data-based distributed droop control (SDDC)" for the grid-integrated WFs, to improve primary frequency response (PFR), is stated in [Mahish and Pradhan, 2019a]. The PMU data is used to obtain load-generation mismatch. Then, the ratio of power reserve (RPR) is computed for the WFs. Each generator shares power as per the RPR and their wind speed in a WF. The droop is calculated based on the power share to generate the allocated share. The SDDC works effectively for a load disturbance, variation in wind speed at a WF, and a synchronous generator outage.

A high percentage of wind power in the generation mix degrades the inertia and raises frequency stability concerns. But the power electronic converter control associated with it can add virtual inertia. The impact can be observed by inertial estimation based on the swing equation and PMU data; it is applied to the system model of the southeast zone of Mexico with large WFs. This takes the bus angle variation at generation buses streamed from PMUs and active power measurements. It is accurate for the first oscillation and robust for variations in the system to estimate the inertia in the generation units [Beltran et al., 2018]. However, the load dynamics sacrifice its precision as it uses the initiation time of perturbation to calculate the peak value of the first angular oscillation. For frequency stability, the ARMAX model using a WAMS [Lugnani et al., 2020] for inertia estimation is proposed.

Data mining of the stochastic sources can be useful to check the influence on the output power.

Data mining algorithms applied on the PMU measurements made on WFs interfaced to the distribution and the transmission grids are analyzed. The model takes active and reactive power generation (P_{Wind}, Q_{Wind}) at POI of WF and their set values $P_{Settings}$, $Q_{Settings}$ respectively, which are operational decisions based on measurements and calculation from selective PMUs. The algorithm puts forth a novel early warning signal (EWS) and diagnoses the structure of critical changes for transmission systems and wind power operators as a situational awareness (SA) indicator [Cotilla-Sanchez et al., 2013]. So the emergency can be prevented by reducing wind power or increasing reactive power supply, and thus retaining a normal state [Klarić et al., 2018].

The system inertia and its map in real time can be assessed by the PMUs-based method proposed in [Tuo and Li, 2020]. It is derived using electrical distance and clustering algorithm, and high- and low-inertia areas are estimated. Due to the degradation of system frequency response, conventional methods are not fast enough to halt a frequency deviation. Frequency regulation becomes much more important for future low-inertia power systems; in addition, it becomes more difficult to determine the regulation reserve requirement. To address this challenge, many frequency control schematics have been developed. The synthetic governor control method reserves the wind power generation by working in the overspeed zone instead of maximum power point tracking (MPPT) [Jin et al., 2017; Thongam and Ouhrouche, 2011]. Wind power plant inertia control exploits the kinetic energy stored in wind blades and turbines and provides a synthetic inertial frequency response in seconds [Muller et al., 2002]. The virtual inertia method presented in [Arani and El-Saadany, 2013] is similar to the kinetic inertia of a synchronous generator to recover the system dynamics. Wind energy conversion system has low inertia; however, it can participate in frequency response if it has a margin from maximum power operation and works in an underloaded condition. A distributed synchronized droop control of the wind farm taking information on optimum power share ratio (PSR) and the time-synchronized data of frequency of POI inertia of power system is proposed to improve primary frequency response. However, the delay or false data in communication may degrade the performance [Mahish and Pradhan 2019c].

System inertia estimation methods based on the ROCOF measurements may suffer high noise and bias. The frequency response under different renewable generation penetration levels is first tested. Then, an index based on electrical distance is used to estimate the inertia distribution over the entire grid. Butterworth filter is introduced to mitigate the impact of noise-induced measurement errors [Tuo and Li, 2020]. To reduce the bias from the location of measurements relative to the location of in-feed loss, disturbances on different buses over an observation window are combined, then a clustering algorithm based on electrical distance is utilized to accurately estimate the location of the centroid of inertia (COI) suitable for measurements. Areas with different inertia distribution levels are proposed to provide useful information to generation dispatch and frequency control. To evaluate the impact of renewable penetration on inertia distribution, another emulation was conducted under a scenario of a 20% wind penetration level. A significant excursion of COI location can be observed due to the installation of wind plants. It shows that the COI location shifts towards the area where many synchronous generators are located and synchronized online.

A case study of the Great Britain (GB) power system is presented [Ashton et al., 2014]. First, an event is detected from transients, and the estimate of inertia is carried out in presence of wind energy, summing up regional inertia estimates.

18.14 TOPOLOGY IDENTIFICATION

Topology identification (TI) in distribution networks is important for frequent modifications. Insufficient measurement infrastructure for the vast network increases the difficulty. TI is performed jointly based on both fundamental and harmonic synchrophasor measurements. Moreover, an analysis is carried out to pick the number and the location of harmonic sources and sensors warranted for total observability. The study of TI in radial networks is done as DSs are typically of such a topology [Chen et al., 2020]. The TI is formulated and then solved by mixed-integer linear programming (MILP). Pseudopower injection [Farajollahi et al., 2019; Gandluru et al., 2019; Tian et al., 2015] and pseudocurrent injection [Abur et al, 1995] methods may cause incorrect topology identification. Harmonic measurements are used to resolve the aforementioned issue [Tian et al., 2015]. The harmonic current phasors differ from fundamental current phasors as the former appears on the load side rather than the substation side and not all the loads contribute significant harmonic pollution.

The resiliency of electric DS is vital as erratic weather and cyber threats are growing. Natural disasters landfall and movement are predicted, and linked paths are forecasted nowadays sufficiently in advance with accuracy and allow to take pre-event reconfiguration to minimize disruption of supply in DS [Pandey et al., 2020]. Proactive control in real time is possible with D-PMU, with a high reporting rate, as high as 120 samples/second. The accuracy D-PMUs in output data should be higher compared to the PMUs to catch the frequent changes in system states. This check can be done either prior to deployment or by anomaly detection by data mining approaches after installation.

An ensemble technique is used which uses unsupervised machine learning [Zhou et al., 2018] for such purposes. A synchrophasor anomaly detection (SyncAD) tool has been developed [Xu et al., 2016]. The algorithm for such anomaly detection is shown in Figure 18.10.

Thus, before the extreme condition, the "shift-and-shed" algorithm recognizes critical loads and recommends a switching operation to supply these loads. The rate of change of angle across both feeders and substations is observed. When the frequency across both of them is lower than the threshold, to prevent trip of the ROCOF relays, the algorithm hands over the information to the "resilient reconfiguration algorithm." By minimum spanning tree (MST), the switching sequence to maintain power is generated. The resiliency is improved as the lost service is reduced.

18.15 SECURITY OF POWER SYSTEM

The role of the synchrophasor communication system (SPCS) is crucial in aggregating the synchrophasor data from the different geographical locations of PMUs to the PDC [Appasani and Mohanta 2020]. WAMS is pivoted on the availability of correct and timely communication of synchrophasor measurements for the algorithms for protection, control, or islanding to operate within the stipulated time in a cyberphysical system of the power system. The PMUs ease real-time grid situational awareness and observability, estimate states, and assess stability online.

Reliability in power system protection considers the four features:

1. Dependability – defined as the degree of certainty that the relays will definitely function correctly for the faults for which they are designed to function.
2. Security – the measure of the certainty that the relays will not function for any fault not designed to operate/normal condition.

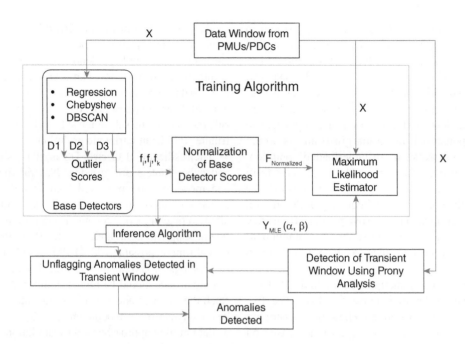

FIGURE 18.10 Anomaly detection system.
Source: Pandey et al., 2020.

3. Selectivity – the capability to affect the minimum action when performing its proposed function.
4. Robustness – the capability to function acceptably over the full range of both steady state and dynamics.

There exists a trade-off between dependability and security. The protection should be "dependable" in a normal state; however, when the system is highly stressed (due to a fault, line trip, or generator outage), bias toward "security" is desired. To "adapt" the security dependability balance situation awareness can be updated in real time, for which PMUs are of great help.

18.15.1 What Is a Denial-of-Service (DoS) Cyberattack?

It is unethical hacking such that the data is unavailable to the PDC from PMUs so that system dynamics could not be visualized in real time. This may result in the failure of time-critical protection. The hacker may also modify the data as it goes undetected and open critical circuit breakers with malefic intent and jeopardize the whole power system within a very short time before analyzing.

So the detection of DoS can make the system function resiliently to mitigate the negative effect of such effect, maintaining system stability. To mitigate the impact of such attack to protect resiliency, a modified WAMS structure with the added functions is proposed [Chawla et al., 2020]:

- Data unavailability detector (DUD)
- DoS mitigation (DM)
- Root-cause analyzer (RCA)

The DUD detects not a number (NaN) (due to DoS or lack of communication) from the data at PDC; it passes it through DM and raises an alarm instead of passing to synchrophasor-based control/protection algorithm. The critical circuit breakers are blocked from an operation and passed to RCA simultaneously to categorize if it is DoS or lack of communication. The DM module recovers the missing data, and the complete data is passed to WAMS application algorithms. This assists resiliency.

18.15.2 Missing Data Recovery Methods

Missing data is recovered by pattern classification by imputation based on statistical and machine learning methods as data misses during real-time testing. Machine learning–based k-nearest neighbors (KNNs) are applied. Deep-learning techniques like recurrent neural networks (RNN) and other regenerative classes of algorithms can also be deployed, as they take the temporal and spatial information in the data to estimate missing values. The mean of ten nearest neighbors in the historical data replaces the missing data. To apply in real time, KNN is applied after training 15 clusters of data with no missing values, and the cluster centroids are stored. When a data stream with missing values is spotted, the closest centroid from the available data and the corresponding values from this centroid replace the missing PMU data. Then this-rectified stream is passed to the SE algorithm (normal/stressed) to classify for awareness of the condition. Also, it goes to protection/application algorithms. However, the complexity, time, and memory requirements of these detection and recovery algorithms should match with the timeline of preventive and corrective actions.

The synchrophasors data transfer are susceptible to a single point of failure, such as near the PDC. A blockchain-based decentralization in a peer-to-peer system for synchrophasor networks enhances defense against cyberattacks [Bhattacharjee et al., 2020]. Data manipulation attacks can be identified by bloom filter, and Merkle tree verifies that it is from a PMU. Then, the blockchain consensus can recon the security of the protection framework.

A different approach to assess reliability online and improve synchrophasor data communications is proposed [Seyedi et al., 2020]. Loss of successive data packets due to communication failure can be detected by observing the reporting rate of PMUs and the delay thresholds by analysis of the time of arrival (TOA) and time of departure (TOD) of data packets. An LTE mobile communication for PMUS in the distribution feeder is proposed. The data latency can be predicted, and priority data can be delivered and thereby minimize the missing data effect.

A system protection system (SPS) functions to provide a safety backup for the electrical network during unplanned/extreme/N-1-1 contingency conditions or when operational constraints are violated. SPS could be more cost-effective than network modification, but some risks hurdle the SPS. These are risk of mismatch of remand response, interference with other SPS, increase in complex analysis and coordination, and inappropriate execution of planned automation.

The failure of SPS is mainly due to:

1. Hardware failure
2. Incorrect logic in the algorithm
3. Software failure
4. Manual mistakes

Recognizing the importance of SPS in any large grid, its misoperation or failure would cost a lot. The installation of SPSs in the Indian grid is discussed in [Agarwal and Kumar, 2020]. These constitute wide-impact SPSs and local-impact types [CIGRE 38.02.19, 2001]. Wide-impact SPSs are those where the action is visualized at locations at distance and, at the same time, when its linked SPS operates. While in the local-impact SPS, the action is confined and it provides instant action. An automated system should activate within milliseconds to a few seconds in response to contingencies. This monitoring of SPS utilizes PMUs to distinguish the change of normalcy, and the preplanned automatic remedial action is taken to establish security and reliability. The fast action of the SPS keeps the parameters of the system bounded and prevents unstable conditions. The Indian grid has 11 HVDC (four HVDC back-to-back, three interregional HVDC bipoles, three intraregional HVDC bipoles, and one multiterminal HVDC), with an AC network with PMUs installed. PDCs are shown in Figure 18.11 [Shukla et al., 2019].

18.16 SYNCHROPHASOR IN MICROGRID

Microgrid protection should be dependable for internal faults but should be reliable for external faults to minimize unintended interruptions and serve the local load, which is the basic requirement in a microgrid. A protection strategy can be dependable for high-resistance fault if both magnitude and angle of the impedance are considered instead of the magnitude alone as the threshold. Such a strategy has been derived from the positive-sequence voltage and current measurements from PMU to detect the fault with a planar threshold of both magnitude and angle [Sharma and Samantaray, 2020]. The discrimination of fault and no-fault for severe no-fault cases, including capacitor switching, disconnection of DG, grid-islanding resynchronization, removal of the section, and starting of the motor, is tested for a microgrid in both grid-connected and islanding condition and external and internal fault [Sharma and Samantaray, 2020]. The observation of system dynamics and deriving voltage stability index (VSI) from μ-PMUs in a radial distribution feeder is, the service to a critical load of the hospital could be maintained in a practical case study in India considering a cyclone [Kumar and Jeena, 2020].

Real-time accurate classification of an event using PMU data requires that it should work with a small batch of samples for fast completion of the task. So time series methods suffer in performance in real-time protection decisions as it occurs in millisecond. The data processing time can be reduced by taking only transient conditions for feature extraction. To classify the disturbance on real-time DT, KNN [Brahma et al., 2017], support vector machines (SVMs) [Seethalekshmi et al., 2012], etc. suffer from accuracy despite prior training and testing [Xin et al., 2018]. Compared to the traditional machine learning classifiers, DNN provides better accuracy performance for a large volume of data with lower testing time. A completely data-driven approach with a 1:2 ratio of renewable to conventional generation is done. A diffusion-type kernel density estimator (DKDE) depicts the feature in the shape of 3-D voltage and frequency distribution along time in terms of probability density functions (PDFs) [Yadav et al., 2019b]. Then these features are used to train a multilayered deep neural network for event classification.

The data-driven secondary voltage control (SVC) using PMU data is designed for decentralized control in the closed loop [Nascimento et al., 2020]. This utilizes the reactive power capability of generators in the system and cuts the requirement for additional reactive power sources.

18.17 CONCLUSION

There has been significant recent growth in the renewable energy sector worldwide, and indications suggest that the pace of development is only going to increase. Domination of the large- and small-scale renewable generation makes a highly diverse and decentralized energy system. As nonsynchronous wind and solar generators are weather-dependent,

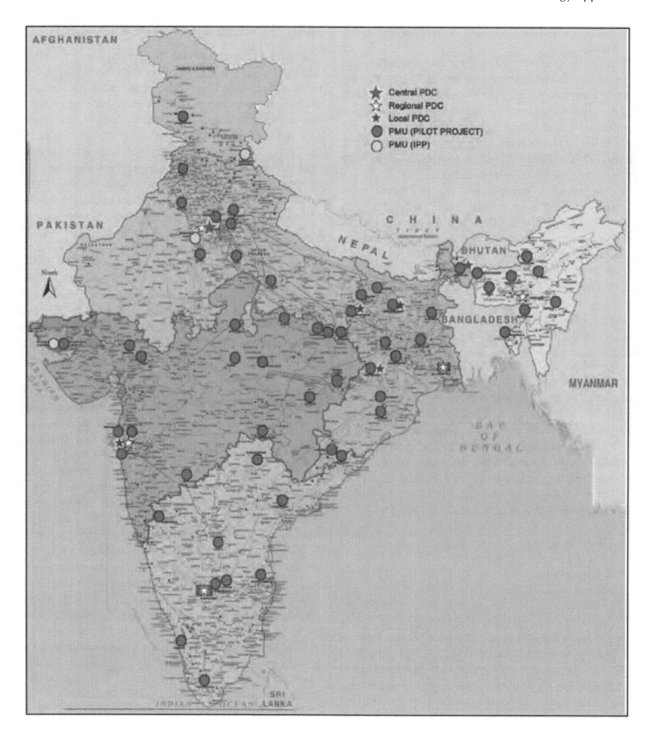

FIGURE 18.11 WAMS network in India in 2016.

Source: Shukla et al., 2019

they cannot offer the same baseload stability that thermal and hydrogenators can. Increased variable renewable generation, combined with the exponential growth of distributed energy resources and an aging fleet of thermal generators, has also led to increased variations in both load and supply in the grid. Delay in system connection and curtailment is a regular feature. Wide-area monitoring is required to control all these aforementioned drawbacks and can make a reliable as well as efficient system. Availability of synchrophasor measurements has drastically supported this monitoring and take-control-and-protection decisions. The smartness of the grid has improved, but the reliability and security of the data communication are a new role-player for the data-assisted decision.

18.18 REFERENCES

M. K. Aalam and K.N. Shubhanga, "Synchrophasor measurement standard comparison and testing of an FF-based PMU," *IEEE International Conference on Industry 4.0 Technology (I4Tech)*, pp. 115–120, February 2020.

Abur, D. Shirmohammadi, and C.S. Cheng, "Estimation of switch statuses for radial power distribution systems," in: *Proceedings of IEEE International Symposium on Circuits and Systems*, Vol. 2, Seattle, WA, USA, 1995.

K. Agarwal and C.X. Kumar, "Oscillation detection and mitigation using synchrophasor technology in the Indian power grid," in: *Power System Grid Operation Using Synchrophasor Technology*, pp. 195–216. Springer, Cham, 2020.

K.K. Anaparthi, B. Chaudhuri, N.F. Thornhill, and B.C. Pal, "Coherency identification in power systems through principal component analysis," *IEEE Transanctions on Power System*, Vol. 20, No. 3, pp. 1658–1660, Aug. 2005.

B. Appasani and D.K. Mohanta, "A review on synchrophasor communication system: Communication technologies, standards and applications," *Protection and Control of Modern Power System*, Vol. 3, No. 1, pp. 1–17, 2018.

B. Appasani and D.K. Mohanta, "Uncertainty analysis and risk assessment for effective decision-making using wide-area synchrophasor measurement system," in: *Decision Making Applications in Modern Power Systems*. Academic Press, pp. 63–88, 2020.

M.F.M. Arani and E.F. El-Saadany, "Implementing virtual inertia in dfig-based wind power generation," *IEEE Transactions on Power Systems*, Vol. 28, pp. 1373–1384, May 2013.

V. Ashok, A. Yadav and A.Y. Abdelaziz, "A Comprehensive Review on Wide-Area Protection, Control and Monitoring Systems," in: Haes H. Alhelou, A.Y. Abdelaziz, P. Siano (eds) *Wide Area Power Systems Stability, Protection, and Security. Power Systems*. Springer, Cham, pp. 1–43, 2021.

P.M. Ashton, C.S. Saunders, G.A. Taylor, A.M. Carter, and M.E. Bradley, "Inertia estimation of the GB power system using synchrophasor measurements," *IEEE Transactions on Power Systems*, Vol. 30, No. 2, pp. 701–709, 2014.

S. Basumallik, Y. Liu, D. Katramatos, and S. Yoo, "Real-time distribution state estimation with massive μPMU streaming data," in: *IEEE Power & Energy Society Innovative Smart Grid Technologies Conference (ISGT)*, pp. 1–5, February 2020.

O. Beltran, R. Peña, J. Segundo, A. Esparza, E. Muljadi and D. Wenzhong, "Inertia estimation of wind power plants based on the swing equation and phasor measurement units," *Applied Sciences*, Vol. 8, No. 12, p. 2413, 2018.

Bhattacharjee, S. Badsha, A.R. Shahid, H. Livani, S. Sengupta, "Block-phasor: A decentralized blockchain framework to enhance security of synchrophasor," in: *IEEE Kansas Power and Energy Conference (KPEC)*, pp. 1–6, Jul 2020.

S. Bonian, K. Martin, A. Goldstein, and B. Dickerson, "Synchrophasor measurements for power system monitoring and control under the standard IEC/IEEE 60255–118–1," in: *IEEE Innovative Smart Grid Technologies-Asia (ISGT Asia)*, pp. 655–660, May 2019.

A.R. Borden and B.C. Lesieutre, "Variable projection method for power system modal identification," *IEEE Trans. Power Syst.*, Vol. 29, No. 6, pp. 2613–2620, Nov. 2014.

S. Brahma, R. Kavasseri, H. Cao, N.R. Chaudhuri, T. Alexopoulos, and Y. Cui, "Real-time identification of dynamic events in power systems using PMU data, and potential applications – Models, promises, and challenges," *IEEE Trans. Power Del.*, Vol. 32, No. 1, pp. 294–301, Feb. 2017.

J. Cardenas, G. Mikhael, J. Kaminski, and I. Voloh, "Islanding detection with phasor measurement units," *67th Annual Conference for Protective Relay Engineers*, College Station, TX, USA, April 2014.

L. Carta, N. Locci and C. Muscas, "A PMU for the measurement of synchronized harmonic phasors in three-phase distribution networks," *IEEE Trans. Instrum. Meas.*, Vol. 58 pp. 3723–3730, 2009.

P. Castello, J. Liu, C. Muscas, P.A. Pegoraro, F. Ponci and A. Monti, "A fast and accurate PMU algorithm for P+ M class measurement of synchrophasor and frequency," *IEEE Transactions on Instrumentation and Measurement*, Vol. 63, No. 12, pp. 2837–2845, 2014.

S. Chandak, M. Mishra, S. Nayak, and P.K. Rout, "Optimal feature selection for islanding detection in distributed generation," *IET Smart Grid*, Vol. 1, No. 3, pp. 85–95, 2018.

S. Chatterjee, P.K. Ghosh, B.K. Saha Roy, "PMU-based power system component monitoring scheme satisfying complete observability with multicriteria decision support," *International Transactions on Electrical Energy Systems*, Vol. 30, No. 2, e122, 23 Feb 2020.

N.R. Chaudhuri, and B. Chaudhuri, "Damping and relative mode-shape estimation in near real-time through phasor approach," *IEEE Transactions on Power Systems*, Vol. 26, No. 1, pp. 364–373, Feb. 2011.

Chawla, P. Agrawal, A. Singh, B.K. Panigrahi, K. Paul, and B. Bhalja, "Denial-of-service resilient frameworks for synchrophasor-based wide area monitoring systems," *Computer*, Vol. 53, No. 5, pp. 14–24, May 2020. IEEE Systems Journal.

L. Chen, M. Farajollahi, M. Ghamkhari, W. Zhao, S. Huang, and H. Mohsenian-Rad, "Switch status identification in distribution networks using harmonic synchrophasor measurements," *IEEE Transactions on Smart Grid*, Vol. 12, No. 3, pp. 2413–2424, 2020.

T.H. Chen, W.C. Yan, Y.D. Cai, and N.C. Yang, "Voltage variation analysis of normally closed-loop distribution feeders interconnected with distributed generation", in: *Proceedings of 4th IASTED Asian Conference on Power and Energy Systems*, Apr 2008.

CIGRE 38.02.19 System protection schemes in power networks, June 2001.

Communication Networks and Systems in Substations, IEC 61850.

E. Cotilla-Sanchez, P. Hines, and C. Danforth, "Predicting critical transitions from time series synchrophasor data," in: *Proceedings of the 2013 IEEE Power and Energy Society General Meeting*, Vancouver, BC, Canada, July 2013.

De Mauro, M. Greco, and M. Grimaldi, "A formal definition of big data based on its essential features," *Libr. Rev.*, Vol. 65, No. 3, pp. 122–135, Apr. 2016.

L. Dosiek and J.W. Pierre, "Estimating electromechanical modes and mode shapes using the multichannel ARMAX model," *IEEE Transactions on Power System*, Vol. 28, No. 2, pp. 1950–1959, May 2013.

L. Dosiek, J.W. Pierre, D.J. Trudnowski, and N. Zhou, "A channel matching approach for estimating electromechanical mode shape and coherence," *IEEE Power and Energy Society General Meeting*, pp. 1–8, 2009.

R. Dutta, P. Kundu, and A.K. Srivastava, "Dynamic state and parameter estimation of distributed generators using constraint recursive least square algorithm and μ-PMU data," in: *IEEE International Conference on Power Electronics, Smart Grid and Renewable Energy (PESGRE2020)*, pp. 1–5, January 2020.

S. Dutta, P.K. Sadhu, M.J.B. Reddy, and D.K. Mohanta, "Shifting of research trends in islanding detection method – A comprehensive survey," *Protection and Control of Modern Power System*, 2018a, Vol. 3, No. 1, pp. 1–20.

S. Dutta, P.K. Sadhu, M.J.B. Reddy, and D.K. Mohanta, "Smart inadvertent islanding detection employing p-type PMU for an active distribution network," *IET Generation Transmission & Distribution*, Vol. 12, pp. 4615–4625, 2018b.

J.H. Eto, E. Stewart, T. Smith, M Buckner, H. Kirkham, F. Tuffner, and D. Schoenwald, "Scoping study on research and development priorities for distribution-system phasor measurement units," 2016. Available: https://prod.sandia.gov/techlib-noauth/access-control.cgi/2016/163546r.pdf.

M. Farajollahi, A. Shahsavari, and H. Mohsenian-Rad, "Topology identification in distribution systems using line current sensors: An MILP approach," *IEEE Transaction on Smart Grid*, Vol. 11, No. 2, pp. 1159–1170, Aug 2019.

K. Fatima, M. Sefid, M.A. Anees, and M. Rihan, "An investigation of the impact of synchrophasors measurement on multi-area state estimation in active distribution grids," *Australian Journal of Electrical and Electronics Engineering*, Vol. 17, No. 2, pp. 122–131, 2020.

F.D. Freitas, N. Martins, and L.F.J. Fernandes, "Reliable modeshapes for major system modes extracted from concentrated WAMS measurements processed by a SIMO identification algorithm," *2008 IEEE Power and Energy Society General Meeting – Conversion and Delivery of Electrical Energy in the 21st Century*, IEEE, 2008.

Gandluru, S. Poudel, and A. Dubey, "Joint estimation of operational topology and outages for unbalanced power distribution systems," *IEEE Transaction on Power System*, Vol. 35, No. 1, pp. 605–617, Aug 2019.

M. Glavic and T.V. Cutsem, "A short survey of methods for voltage instability detection," *IEEE Power and Energy Society General Meeting*, San Diego, CA, USA, July 2011.

M. Gol, A. Abur, and F. Galvan, "Metrics for success: Performance metrics for power system state estimators and measurement designs," *IEEE Power and Energy Magazine*, Vol. 10, No. 5, pp. 50–57, 2012.

Guideline, N.R. *PMU Placement and Installation*. NERC, Atlanta, GA, USA, 2016.

Y. Guo, K. Li, D.M. Laverty, and Y. Xue, "Synchrophasor-based islanding detection for distributed generation systems using systematic principal component analysis approaches," *IEEE Transaction on Power Delivery*, Vol. 30, pp. 2544–2552, 2015.

E. Hamid and Z. Kawasaki, "Wavelet-based data compression of power system disturbances using the minimum description length criterion," *IEEE Transaction on Power Delivery*, IEEE, Vol. 17, No. 2, pp. 460–466, 2002.

L.I.U. Hao, B.I. Tianshu, X. Chang, G.U.O., Xiaolong, W.A.N.G. Lu, C.A.O. Chuang, Y.A.N. Qing, and L.I. Jinsong, "Impacts of subsynchronous and supersynchronous frequency components on synchrophasor measurements," *Journal of Modern Power Systems and Clean Energy*, Vol. 4, No. 3, pp. 362–369, 2016.

M. Hasheminamin, V.G. Agelidis, V. Salehi, R. Teodorescu, and B Hredzak, "Index-based assessment of voltage rise and reverse power flow phenomena in a distribution feeder under high PV penetration," *IEEE Journal of Photovoltaics*, Vol. 5, No. 4, pp. 1158–1168, 2015.

J. Hauer and R. Cresap, "Measurement and modeling of Pacific AC intertie response to random load switching," *IEEE Transaction on Power Apparatus and Systems*, Vol. PAS-100, No. 1, pp. 353–359, Jan. 1981.

J.F. Hauer, C.J. Demeure, and L.L. Scharf, "Initial results in prony analysis of power system response signals," *IEEE Transaction on Power System*, Vol. 5, No. 1, pp. 80–89, 1990.

IEEE Application Guide for IEEE Std 1547(TM), IEEE Standard for Interconnecting Distributed Resources with Electric Power Systems; IEEE Std 1547.2–2008; IEEE: Piscataway, NJ, USA, 2009; pp. 1–217.

IEEE Guide for Smart Grid Interoperability of Energy Technology and Information Technology Operation with the Electric Power System (EPS), End-use applications, and loads, *IEEE Standard 2030–2011*, Sep. 2011.

IEEE Standard for Electric Power Systems Communications – Distributed Network Protocol (DNP3), IEEE Standard 1815, 2012.

IEEE standard for interconnection and interoperability of distributed energy resources with associated electric power systems interfaces, IEEE Standards Coordinating Committee, 2018.

IEEE/IEC 60255-118-1-2018 – IEEE/IEC International Standard – Measuring relays and protection equipment – Part 118-1: Synchrophasor for power systems – Measurements, 2018.

T. Jiang, Y. Mu, H. Jia, N. Lu, H. Yuan, J. Yan, and W. Li, "A novel dominant mode estimation method for analyzing inter-area oscillation in China southern power grid," *IEEE Transaction on Smart Grid*, Vol. 7, No. 5, pp. 2549–2560, Sep. 2016.

T. Jin, S. Liu, R.C.C. Flesch, and W. Su, "A method for the identification of low frequency oscillation modes in power systems subjected to noise," *Applied Energy*, Vol. 206, pp. 1379–1392, Nov. 2017.

W. Ju, I. Dobson, K. Martin, K. Sun, N. Nayak, I. Singh, H. Silva-Saravia, A. Faris, L. Zhang and Y. Wang, "Real-time area angle monitoring using synchrophasors: A practical framework and utility deployment," *IEEE Transactions on Smart Grid.*, Vol. 12, No. 1, pp. 859–870, 2020.

Kamwa, G. Trudel, and L. Gerin-Lajoie, "Low-order blackbox models for control system design in large power systems," *IEEE Transanction on Power System*, Vol. 11, No. 1, pp. 303–311, 1996.

Kamwa, R. Grondin, E.J. Dickinson, and S. Fortin, "A minimal realization approach to reduced-order modelling and modal analysis for power system response signals," *IEEE Transaction on Power System*, Vol. 8, No. 3, pp. 1020–1029, 1993.

M. Kanabar, M.G. Adamiak and J. Rodrigues, "Optimizing wide area measurement system architectures with advancements in phasor data's concentrators (PDCs)," in: *IEEE Power and Energy Society General Meeting*, 2013.

Khair, M. Zuhaib, and M. Rihan, "Effective utilization of limited channel PMUs for islanding detection in a solar PV integrated distribution system," *Journal of the Institution of Engineers (India): Series B*, pp. 1–12, 2020.

J. Khan, S.M.A. Bhuiyan, G. Murphy, and M. Arline, "Embedded-zerotree-wavelet-based data denoising and compression for smart grid," *IEEE Transaction on Industrial Applications*, Vol. 51, No. 5, pp. 4190–4200, 2015.

J. Khan, S.M.A. Bhuiyan, G. Murphy, and J. Williams, "Data denoising and compression for smart grid communication," *IEEE. Transaction Signal Information Processing Network*, Vol. 2, No. 2, pp. 200–214, 2016.

M. Klarić, I. Kuzle, and N. Holjevac, "Wind power monitoring and control based on synchrophasor measurement data mining," *Energies*, Vol. 11, No. 12, 3525, 2018.

R. Klump, P. Agarwal, J. Tate, and H. Khurana, "Lossless compression of synchronized phasor measurements," *IEEE PESGM*, Providence, USA, Jul 2010.

S.A.R. Konakalla, A. Valibeygi, and R.A. de Callafon, "Microgrid dynamic modeling and islanding control with synchrophasor data," *IEEE Transactions on Smart Grid*, Vol. 11, No. 1, pp. 905–915, 2019.

C. Kumar, M.V. Rao, P. Seshadri, V. Pandey, C. Ghangrekar, S. Chitturi, V.K. Shrivastava, and A. Gartia, "Detection of LFO and evaluation of damping improvement using synchrophasor Measurement," in: *IEEE PES Asia-Pacific Power and Energy Engineering Conference (APPEEC)*, pp. 701–706, Oct 2018.

D.S. Kumar, J.S. Savier, and S.S. Biju, "Micro-synchrophasor based special protection scheme for distribution system automation in a smart city," *Protection and Control of Modern Power Systems*, Vol. 5, No. 1, Dec 2020, pp. 1–4.

G.P. Kumar and P. Jena, "Pearson's correlation coefficient for islanding detection using micro-PMU measurements," *IEEE Systems Journal*, Vol. 15, No. 4, 2020, pp. 5078–5089.

J. Li, Y. Liu, and L. Wu, "Optimal operation for community-based multi-party microgrid in grid-connected and islanded modes," *IEEE Transaction on Smart Grid*, Vol. 9, No. 2, pp. 756–765, 2018.

H. Liao, "Power system harmonic state estimation and observability analysis via sparsity maximization," *IEEE Transaction on Power System*, Vol. 22, No. 1, pp. 15–23, 2007.

G. Liu, J. Quintero, and V.M. Venkatasubramanian, "Oscillation monitoring system based on wide area synchrophasors in power systems," *2007 iREP Symposium – Bulk Power System Dynamics and Control – VII. Revitalizing Operational Reliability*, pp. 1–13, 2007.

G. Liu and V. Venkatasubramanian, "Oscillation monitoring from ambient PMU measurements by frequency domain decomposition," *2008 IEEE International Symposium on Circuits and Systems*, pp. 2821–2824, 2008.

Y. Liu, L. Wu, and J. Li, "D-PMU based applications for emerging active distribution systems: A review," *Electric Power Systems Research*, Vol. 179, 106063, 2020.

Y. Liu, S. You, and Y. Liu, "Study of wind and PV frequency control in U.S. power grids: EI and TI case studies," *IEEE Power and Energy Technology Systems Journal*, Vol. 4, pp. 65–73, Sep. 2017.

L. Lugnani, D. Dotta, C. Lackner and J. Chow, "ARMAX-based method for inertial constant estimation of generation units using synchrophasors," *Electric Power Systems Research*, Vol. 180, 106097, 2020.

M. Maheswari, N.S. Vanitha, and N. Loganathan, "Wide-area measurement systems and phasor measurement units," *Wide Area Power Systems Stability, Protection, and Security*, pp. 105–126, 2020.

P. Mahish and A.K. Pradhan, "Mitigating subsynchronous resonance using synchrophasor data based control of wind farms," *IEEE Transactions on Power Delivery*, Vol. 35, No. 1, pp. 364–376, 2019.

P. Mahish and A.K. Pradhan, "Synchrophasor data based distributed droop control in grid integrated wind farms to improve primary frequency response," in: *2019 8th International Conference on Power Systems (ICPS)*, pp. 1–5, December 2019a.

P. Mahish and A.K. Pradhan, "Distributed synchronized control in grid integrated wind farms to improve primary frequency regulation," *IEEE Transactions on Power Systems*, Vol. 35, No 1, pp. 362–373.08760440, 2019c.

B. Matic-Cuka and M. Kezunovic, "Islanding detection for inverter- based distributed generation using support vector machine method," *IEEE Transaction on Smart Grid*, Vol. 5, No. 6, pp. 2676–2686, 2014.

L.G. Meegahapola, S. Bu, D.P. Wadduwage, C.Y. Chung and X. Yu, "Review on oscillatory stability in power grids with renewable energy sources: Monitoring, analysis, and control using synchrophasor technology," *IEEE Transactions on Industrial Electronics*, Vol. 68, No. 1, pp. 519–531, Jan 2020.

N. Mekki and L. Krichen, "Coordinated designs of fuzzy PSSs and load frequency control for damping power system oscillations considering wind power penetration," in: *Wide Area Power Systems Stability, Protection, and Security*. Springer, Cham, pp. 167–188, 2020.

A.R. Messina and V. Vittal, "Nonlinear, non-stationary analysis of interarea oscillations via hilbert spectral analysis," *IEEE Transaction on Power System*, Vol. 21, No. 3, pp. 1234–1241, Aug. 2006.

R.R. Micky, R. Sunitha, and S. Ashok, "Techno-economic analysis of WAMS based islanding detection algorithm for microgrids with minimal PMU in smart grid environment," in: *Wide Area Power Systems Stability, Protection, and Security*. Springer, Cham, pp. 499–522, 2021.

M. Mohanty, R. Kant, A. Kumar, D. Sahu, and S. Choudhury, "A brief review on synchro phasor technology and phasor measurement unit," *Advances in Electrical Control and Signal Systems*, pp. 705–721, 2020.

Monticelli, *State Estimation in Electric Power Systems: A Generalized Approach*. Springer Science Business Media, LLC, New York, NY, USA, 2012.

H. Mortazavi, H. Mehrjerdi, M. Saad, S Lefebvre, D Asber and L Lenoir, "A monitoring technique for reversed power flow detection with high PV penetration level," *IEEE Trans. Smart Grid*, Vol. 6, No. 5, pp. 2221–2232, 2015.

H. Muda and P. Jena, "Phase angle-based PC technique for islanding detection of distributed generations," *IET Renewable Power Generation*, Vol. 12, pp. 735–746, 2018.

S. Muller, M. Deicke, and R.W.D. Doncker, "Doubly fed induction generator systems for wind turbines," *IEEE Industry Applications Magazine*, Vol. 8, pp. 26–33, May 2002.

M.M. Nascimento, R.T. Bernardo, and D. Dotta, "Data-driven secondary voltage control design using PMU measurements," in: *IEEE Power & Energy Society Innovative Smart Grid Technologies Conference (ISGT)*, IEEE, pp. 1–5, February 2020.

J. Ning, J. Wang, W. Gao, et al, "A wavelet-based data compression technique for smart grid," *IEEE Transactions on Smart Grid*, Vol. 2, No. 1, pp. 212–218, 2011.

D.J. Okekunle, "Synchrophasor-based predictive control considering optimal phasor measurement unit placements methods" (Doctoral dissertation, University of Edinburgh), 2020.

S. Pandey, S. Chanda, A.K. Srivastava, and R.O. Hovsapian, "Resiliency-driven proactive distribution system reconfiguration with synchrophasor data," *IEEE Transactions on Power Systems*, Vol. 35, No. 4, pp. 2748–2758, 2020.

H. Pezeshki and P. Wolfs, "Correlation based method for phase identification in a three phase lv distribution network," *22nd Australasian Universities Power Engineering Conference (AUPEC)*, Bali, Indonesia, September 2012.

A.G. Phadke and J.S. Thorp, *Synchronized Phasor Measurements and Their Applications*, Vol. 1. Springer, New York, 2008.

A.G. Phadke and B.I. Tianshu, "Phasor measurement units, WAMS, and their applications in protection and control of power systems," *Journal of Modern Power Syst. Clean Energy*, Vol. 6, No. 4, pp. 619–629, 2018.

N.V. Phanendrababu, "Optimal selection of phasor measurement units," in: *Wide Area Power Systems Stability, Protection, and Security*. Springer, Cham, pp. 127–166, 2020.

Pouryekta, V.K. Ramachandaramurthy, N. Mithulananthan, and A. Arulampalam, "Islanding detection and enhancement of microgrid performance," *IEEE Systems*, Vol. 12, No. 4, pp. 3131–3141, 2018.

T. Prakash, S.R. Mohanty, and V.P. Singh, "PMU-Assisted Zone-3 Protection Scheme for PV Integrated Power Systems Immune to Interharmonics," *IEEE Systems Journal*, Vol. 14, No. 3, pp. 3267–3276, 2020.

T. Prakash, V.P. Singh, S.R. Mohanty, "A synchrophasor measurement based wide-area power system stabilizer design for inter-area oscillation damping considering variable time-delays," *International Journal of Electrical Power & Energy Systems*, Vol. 105, pp. 131–41, 2019.

R.M. Radhakrishnan, A. Sankar, and S. Rajan, "Synchrophasor based islanding detection for microgrids using moving window principal component analysis and extended mathematical morphology," *IET Renewable Power Generation*, Vol. 14, No. 12, pp. 2089–2099, 2020.

P. Ray and D.P. Mishra, "Introduction to condition monitoring of wide area monitoring system," in: *Soft Computing in Condition Monitoring and Diagnostics of Electrical and Mechanical Systems*. Springer, Singapore, pp. 71–89, 2020.

Report by the National Renewable Energy Laboratory (NREL), Lawrence Berkeley National Laboratory (Berkeley Lab), Power System Operation Corporation (POSOCO), and the United States Agency for International Development (USAID), *Greening the Grid: Pathways to Integrate 175 Gigawatts of Renewable Energy into India's Electric Grid*, Vol. I, National Study, NREL.

G. Sanchez-Ayala, J.R. Agüerc, D. Elizondo, and M. Lelic, "Current trends on applications of PMUs in distribution systems," in: *2013 IEEE PES Innovative Smart Grid Technologies Conference (ISGT)*, pp. 1–6, February 2013.

J.J. Sanchez-Gasca and J.H. Chow, "Performance Comparison of three Identification Methods for the Analysis of Electromechanical Oscillations," *IEEE Transaction on Power System*, Vol. 14, No. 3, pp. 995–1002, 1999.

S. Santoso, E. Powers, and W. Grady, "Power quality disturbance data compression using wavelet transform methods," *IEEE. Transaction on Power Delivery*, Vol. 12, No. 3, pp. 1250–1257, 1997.

K. Sayood, *Introduction to Data Compression*. Elsevier, India, 4th edn. 2012.

K. Seethalekshmi, S.N. Singh, and S.C. Srivastava, "A classification approach using support vector machines to prevent distance relay maloperation under power swing and voltage instability," *IEEE Trans. Power Del.*, Vol. 27, No. 3, pp. 1124–1133, Jul. 2012.

Y. Seyedi, H. Karimi, C. Wetté, and B. Sansò, "A new approach to reliability assessment and improvement of synchrophasor communications in smart grids," *IEEE Transactions on Smart Grid*, Vol. 11, No. 5, pp. 4415–4426, 2020.

N.K. Sharma and S.R. Samantaray, "A composite magnitude-phase plane of impedance difference for microgrid protection using synchrophasor measurements," *IEEE Systems Journal*, Vol. 15, No. 3, pp. 4199–4209, Jun 2020.

R. Shukla, R. Chakrabarti, S.R. Narasimhan, and S.K. Soonee, "Indian power system operation utilizing multiple HVDCs and WAMS," in: *Power System Grid Operation Using Synchrophasor Technology*. Springer, Cham, pp. 403–432, 2019.

H. Silva-Saravia, H. Pulgar-Painemal, L.M. Tolbert, D.A. Schoenwald, and W. Ju, "Enabling utility-scale solar PV plants for electromechanical oscillation damping," *IEEE Transactions on Sustainable Energy*, Vol. 12, No. 1, pp. 138–147, 2020.

N. Singh, H. Hooshyar, and L. Vanfretti, "Feeder dynamic rating application for active distribution network using synchrophasors," *Sustainable Energy, Grids and Networks*, Vol. 10, pp. 35–45, 2017.

R. Singh, E. Manitsas, B.C. Pal, G. Strbac, "A recursive Bayesian approach for identification of network configuration changes in distribution system state estimation," *IEEE Trans. Power Syst.*, Vol. 25, No. 3, pp. 1329–1336, 2010.

R.S. Singh, S. Cobben, M., Gibescu, H. van den Brom, D. Colangelo, and G. Rietveld, "Medium voltage line parameter estimation using synchrophasor data: A step towards dynamic line rating," in: *2018 IEEE Power and Energy Society General Meeting (PESGM)*, pp. 1–5, August 2018.

J. Song, E. Dall' Anese, A. Simonetto, and H. Zhu, "Dynamic distribution state estimation using synchrophasor data," *IEEE Transactions on Smart Grid*, Vol. 11, No. 1, pp. 821–831, 2019.

K. Subramanian and A.K. Loganathan, "Islanding detection using a micro-synchrophasor for distribution systems with distributed generation," *Energies*, Vol. 13, No. 19, 5180, 2020.

C.M. Thasnimol, and R. Rajathy, "The paradigm revolution in the distribution grid: The cutting-edge and enabling technologies," *Open Computer Science*, Vol. 10, No. 1, pp. 369–395, 2020.

C. Thilakarathne, L. Meegahapola, and N. Fernando, "Real-time voltage stability assessment using phasor measurement units: Influence of synchrophasor estimation algorithms," *International Journal of Electrical Power & Energy Systems*, Vol. 119, 105933, 2020.

J.S. Thongam and M. Ouhrouche, "MPPT control methods in wind energy conversion systems," in: *InTech*, chapter 15, pp. 339–360, 2011.

Z. Tian, W. Wu, and B. Zhang, "A mixed integer quadratic programming model for topology identification in distribution network," *IEEE Transactions on Power Systems*, Vol. 31, No. 1, pp. 823–824, 2015.

P. Top and J. Breneman, "Compressing phasor measurement data," *North American Power Symposium (NAPS)*, Manhattan, USA, 2013.

R. Torkzadeh, M. Eliassi, P. Mazidi, P. Rodriguez, D. Brnobić, K.F. Krommydas, A.C. Stratigakos, C. Dikeakos, M. Michael, R. Tapakis, and V. Vita, "Synchrophasor based monitoring system for grid interactive energy storage system control," in *The International Symposium on High Voltage Engineering*, pp. 95–106, Aug 2019.

D.J. Trudnowski, "Estimating electromechanical mode shape from synchrophasor measurements," *IEEE Trans. Power Syst.*, Vol. 23, No. 3, pp. 1188–1195, Aug. 2008.

M. Tuo and X. Li, "Dynamic estimation of power system inertia distribution using synchrophasor measurements," *arXiv*, preprint arXiv:2006.11520, 2020.

M.U. Usman and M.O. Faruque, "Applications of synchrophasor technologies in power systems," *Journal of Modern Power Systems and Clean Energy*, Vol. 7, No. 2, pp. 211–226, 2019.

Von Meier, E. Stewart, A. McEachern, M. Andersen, and L. Mehrmanesh, "Precision micro-synchrophasors for distribution systems: A summary of applications," *IEEE Transactions on Smart Grid*, Vol. 8, No. 6, pp. 2926–2936, 2017.

Von Meier, U.C. Berkeley, K. Brady, et al., "Synchrophasor monitoring for distribution systems: Technical foundations and applications a white paper by the NASPI distribution task team," NASPI: Berkeley, CA, USA, 2018.

W. Wang, C. Chen, W. Yao, K. Sun, W. Qiu, and Y. Liu, "Synchrophasor data compression under disturbance conditions via cross-entropy-based singular value decomposition," *IEEE Transactions on Industrial Informatics*, Vol. 17, No. 4, pp. 2716–2726, 2020b.

W. Wang, K. Sun, C. Chen, Y. He, S. You, W. Yao, J. Dong, C. Zeng, X. Deng, and Y. Liu, "Advanced synchrophasor-based application for potential distributed energy resources management: Key technology, challenge and vision," in: *IEEE/IAS Industrial and Commercial Power System Asia (I&CPS Asia)*, 2020a, pp. 1120–1124, Jul 2020.

W. Wang, H. Yin, C. Chen, A. Till, W. Yao, X. Deng, and Y. Liu, "Frequency disturbance event detection based on synchrophasors and deep learning," *IEEE Transactions on Smart Grid*, Vol. 11, No. 4, pp. 3593–3605, 2020c.

M.H.F. Wen, R. Arghandehy, A.V. Meiery, et al., "Phase identification in distribution networks with micro-synchrophasors," *IEEE Power & Energy Society General Meeting*, Denver, CO, USA, July 2015.

R.W. Wies, J.W. Pierre, and D.J. Trudnowski, "Use of ARMA block processing for estimating stationary low-frequency electromechanical modes of power systems," *IEEE Transaction on Power System*, Vol. 18, No. 1, pp. 167–173, Feb. 2003.

R.W. Wies, J.W. Pierre, and D.J. Trudnowski, "Use of least mean squares (LMS) adaptive filtering technique for estimating low-frequency electromechanical modes in power systems," *2004 IEEE PES GM*, Vol. 2, pp. 1863–1870, 2004.

D.H. Wilson, K. Hay, and G.J. Rogers, "Dynamic model verification using a continuous modal parameter estimator," *2003 IEEE Bologna Power Tech Conference Proceedings*, Vol. 2, pp. 233–238, 2003.

L. Xie, Y. Chen, and P. Kumar, "Dimensionality reduction of synchrophasor data for early event detection: Linearized analysis," *IEEE Transaction on Power System*, Vol. 29, No. 6, pp. 2784–2794, 2014.

Y. Xin, L. Kong, Z. Liu, Y. Chen, Y. Li, H. Zhu, M. Gao, H. Hou, and C. Wang, "Machine learning and deep learning methods for cybersecurity," *IEEE Access*, Vol. 6, pp. 35365–35381, 2018.

Y. Xu, C.-C. Liu, K. Schneider, F. Tuffner, and D. Ton, "Microgrids for service restoration to critical load in a resilient distribution system," *IEEE Transactions on Smart Grid*, Vol. 9, No. 1, pp. 426–437, 2016.

H. Xue, Y. Cheng, and M. Ruan, "Enhanced flat window-based synchrophasor measurement algorithm for P class PMUs," *Energies*, Vol. 12, No. 21, 4039, 2019.

E. Xypolytou, W. Gawlik, T. Zseby, and J. Fabini, "Impact of asynchronous renewable generation infeed on grid frequency: Analysis based on synchrophasor measurements," *Sustainability*, Vol. 10, No. 5, 1605, 2018.

R. Yadav, A.K. Pradhan, and I. Kamwa, "A spectrum similarity approach for identifying coherency change patterns in power system due to variability in renewable generation," *IEEE Transactions on Power Systems*, Vol. 34, No. 5, pp. 3769–3779.08663402, (2019a).

R. Yadav, S. Raj, and A K. Pradhan, "Real-time event classification in power system with renewables using kernel density estimation and deep neural network," *IEEE Transaction on Smart Grid*, Vol. 10, No. 6, pp. 6849–6859, Nov. 2019b.

F. Zhang, L. Cheng, X. Li, Y. Sun, W. Zhao, and W. Zhao, "Application of a real-time data compression and adapted protocol technique for WAMS," *IEEE Transactions on Power Systems*, Vol. 30, No. 2, pp. 653–662, 2015.

N. Zhou, Z. Huang, L. Dosiek, D. Trudnowski, and J.W. Pierre, "Electromechanical mode shape estimation based on transfer function identification using PMU measurements," *IEEE Power & Energy Society General Meeting*, IEEE, 2009.

N. Zhou, J.W. Pierre, D.J. Trudnowski, and R.T. Guttromson, "Robust RLS methods for online estimation of power system electromechanical modes," *IEEE Transactions on Power Systems*, Vol. 22, No. 3, pp. 1240–1249, Aug. 2007.

M. Zhou, Y. Wang, A.K. Srivastava, Y. Wu, and P. Banerjee, "Ensemble based algorithm for synchrophasor data anomaly detection," *IEEE Transactions on Smart Grid*, Vol. 10, No. 3, pp. 2979–2988, 2018.

19 Solving Issues of Grid Integration of Solar and Wind Energy Models by Using a Novel Power Flow Algorithm

R. Satish, K. Vaisakh, and Almoataz Y. Abdelaziz

CONTENTS

19.1 Introduction	251
19.2 Modeling of Network Components	252
19.2.1 Lines	252
19.2.2 Loads	252
19.2.3 Capacitor Banks	253
19.2.4 Three-Phase Transformer	253
19.3 Modeling of Solar and Wind Energy Models	253
19.3.1 Synchronous Generator	254
19.3.2 Induction Generator	254
19.3.3 Power Electronic Converter	254
19.4 Voltage Unbalance	254
19.5 Algorithm for Developing BUS_NUM and BRANCH_NUM Matrices	254
19.6 Three-Phase PFA with Multiple Solar and Wind Energy Models	256
19.7 Results and Discussions	258
19.7.1 IEEE-13 Bus URDN	258
19.7.1.1 Case Studies on IEEE-13 Bus URDN	260
19.7.1.2 Discussions	260
19.7.2 Case Studies and Discussions on IEEE-34 Bus URDN	260
19.8 Conclusions	265
19.9 References	265

19.1 INTRODUCTION

The renewable energy sources like photovoltaic, wind fuel cells, etc. use synchronous/induction generators or induction generators combined with power electronic converters (PECs) or only PECs to transfer AC power to the grids. Distribution networks are unbalanced due to unsymmetrical spacing between the conductors and unequal distribution of single-phase and two-phase loads among the three phases of the network. The severity of VU problem with the high penetration of single-phase photovoltaic systems into secondary radial distribution networks is presented in [1]. Authors of [2] propose a mitigation strategy for restricting the VU depending on the distributed batteries included in grid-interfaced rooftop PV systems. The potential capability of residential PV inverters is investigated in [3] to develop a distributed reactive power compensation scheme for voltage regulation in three-phase four-wire unbalanced low-voltage distribution networks. In [4], a novel decentralized plug-in electric vehicle reactive power discharging approach is presented, where the reactive power is discharged at selected nodes for further reduction of voltage unbalance. Authors of [5] proposed a novel algorithm for power angle control (PAC) which aims to simplify the control algorithm and obtain a fast dynamic response. It also extends the capability of PAC to compensate voltage unbalance with or without phase angle jump in a simple way. In [6], a three-phase power flow method for real-time distribution system is proposed, which gives some initial discussions on a PV node concept in three-phase power flow based on compensation method. In [7], a novel current injection-based Newton–Raphson power flow algorithm with new PV bus representation for improving the convergence characteristics is presented. In this algorithm, rectangular coordinates are used to represent all nonlinear current injection equations. The connection of distributed generation with the utility grid is specified in IEEE-1547 standard [8]. Induction generator can be modeled as PQ bus in steady-state cases [9, 10]. Synchronous generator based on its excitation control [11, 12] can be modeled as either PV or PQ buses or static voltage characteristic model (SVCM). Based on the control scheme used, the PECs can be modeled as either PV or PQ node [9, 12]. In [13], demand-side management techniques (load shifting and peak clipping) are used to maximize the profit for retailer company by reducing total demand at peak hours and achieve an optimal daily load schedule using linear programming (LP) and genetic algorithm (GA). In [14], a novel functional and robust PSOGSA optimizer is demonstrated to discover the optimal allocations of DG units for enhancing voltage stability in addition to minifying the power loss and operating cost. Authors of [15–17] propose an algorithm for optimal sizing and placement of distributed generation in unbalanced distribution networks. In [18], a three-phase power flow algorithm for URDN with multiple installations of DGs and distribution static synchronous compensator (D-STATCOM) is proposed.

This chapter proposes a novel three-phase PFA for URDN with multiple integrations of solar and wind energy models using BUS_NUM and BRANCH_NUM matrices. These matrices store the information of bus numbers and branch numbers of newly divided sections of URDN. These sections are created by using the radial topology in URDN. The PFA uses the basic principles of circuit theory, and they can be easily understood. Both PV and PQ models of solar and wind energy sources are effectively incorporated into the proposed three-phase power flow algorithm using compensation-based method. The work is organized in the following order. Modeling of network components is discussed in Section 19.2. Modeling of solar and wind energy models is discussed in Section 19.3. The VU is discussed in Section 19.4. Algorithm for developing BUS_NUM and BRANCH_NUM matrices is presented in Section 19.5. Three-phase PFA with multiple solar and wind models is presented in Section 19.6. The results and discussions with several case studies on IEEE-13 and IEEE-34 bus URDN are presented in Section 19.7. The concluding remarks are discussed in Section 19.8.

19.2 MODELING OF NETWORK COMPONENTS

Distribution network consists of major components like lines (overhead and underground), transformers, capacitor banks, and loads. A brief modeling of these components is presented in this section.

19.2.1 Lines

For a four-wire overhead line, the primitive impedance matrix is of 4×4 size, and for a four-wire underground line, this matrix is of 7×7 size. For underground circuits that do not have the additional neutral conductor, the primitive impedance matrix will be 6×6 size. The elements in the primitive matrix are obtained with Carson's equations [19]. The Kron reduction will reduce the primitive matrices to phase matrices of size 3×3 in a grounded neutral system. Figure 18.1 represents a three-phase line model of overhead (or underground) line which is connected between the buses j and k. The shunt admittance of the line is neglected for the distribution voltage levels.

In Figure 18.1, $[Z_{abc}]_{jk}$ is the phase impedance matrix for the branch jk. For the two-phase and single-phase line sections, the phase matrices will have entries zeros in place of missing phases.

$$[Z_{abc}]_{jk} = \begin{bmatrix} Z_{aa} & Z_{ab} & Z_{ac} \\ Z_{ba} & Z_{bb} & Z_{bc} \\ Z_{ca} & Z_{cb} & Z_{cc} \end{bmatrix}_{jk} \quad (1)$$

From Figure 18.1, the voltages of the buses j and k are related as:

$$\begin{bmatrix} V_a \\ V_b \\ V_c \end{bmatrix}_k = \begin{bmatrix} V_a \\ V_b \\ V_c \end{bmatrix}_j - \begin{bmatrix} Z_{aa} & Z_{ab} & Z_{ac} \\ Z_{ba} & Z_{bb} & Z_{bc} \\ Z_{ca} & Z_{cb} & Z_{cc} \end{bmatrix}_{jk} \cdot \begin{bmatrix} I_a \\ I_b \\ I_c \end{bmatrix}_{jk} \quad (2)$$

19.2.2 Loads

The three-phase line currents serving the different types of spot loads and distributed loads are presented in Table 19.1. The detailed discussion of Table 19.1 is presented in [19, 20].

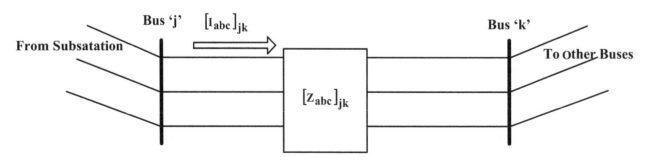

FIGURE 19.1 Three-phase line segment between buses j and k.

Solving Issues of Grid Integration of Energy Models

TABLE 19.1
Load Modeling

	Wye Connection	Delta Connection
Voltage and specified load at bus	$\begin{bmatrix} \|V_{an}\|\angle\delta_a \\ \|V_{bn}\|\angle\delta_b \\ \|V_{cn}\|\angle\delta_c \end{bmatrix} \begin{bmatrix} \|S_a\|\angle\theta_a \\ \|S_b\|\angle\theta_b \\ \|S_c\|\angle\theta_c \end{bmatrix}$	$\begin{bmatrix} \|V_{ab}\|\angle\delta_{ab} \\ \|V_{bc}\|\angle\delta_{bc} \\ \|V_{ca}\|\angle\delta_{ca} \end{bmatrix} \begin{bmatrix} \|S_{ab}\|\angle\theta_{ab} \\ \|S_{bc}\|\angle\theta_{bc} \\ \|S_{ca}\|\angle\theta_{ca} \end{bmatrix}$
Phase currents serving constant power load	$\begin{bmatrix} IL_a \\ IL_b \\ IL_c \end{bmatrix} = \begin{bmatrix} \left(\frac{S_a}{V_{an}}\right)^* \\ \left(\frac{S_b}{V_{bn}}\right)^* \\ \left(\frac{S_c}{V_{cn}}\right)^* \end{bmatrix}$	$\begin{bmatrix} IL_{ab} \\ IL_{bc} \\ IL_{ca} \end{bmatrix} = \begin{bmatrix} \left(\frac{S_{ab}}{V_{ab}}\right)^* \\ \left(\frac{S_{bc}}{V_{bc}}\right)^* \\ \left(\frac{S_{ca}}{V_{ca}}\right)^* \end{bmatrix}$
Phase currents serving the constant impedance load	$\begin{bmatrix} ZL_a \\ ZL_b \\ ZL_c \end{bmatrix} = \begin{bmatrix} \frac{\|V_{an}\|^2}{S_a^*} \\ \frac{\|V_{bn}\|^2}{S_b^*} \\ \frac{\|V_{cn}\|^2}{S_c^*} \end{bmatrix}$ $\begin{bmatrix} IL_a \\ IL_b \\ IL_c \end{bmatrix} = \begin{bmatrix} \frac{V_{an}}{ZL_a} \\ \frac{V_{bn}}{ZL_b} \\ \frac{V_{cn}}{ZL_c} \end{bmatrix}$	$\begin{bmatrix} ZL_{ab} \\ ZL_{bc} \\ ZL_{ca} \end{bmatrix} = \begin{bmatrix} \frac{\|V_{ab}\|^2}{S_{ab}^*} \\ \frac{\|V_{bc}\|^2}{S_{bc}^*} \\ \frac{\|V_{ca}\|^2}{S_{ca}^*} \end{bmatrix}$ $\begin{bmatrix} IL_{ab} \\ IL_{bc} \\ IL_{ca} \end{bmatrix} = \begin{bmatrix} \frac{V_{ab}}{ZL_{ab}} \\ \frac{V_{bc}}{ZL_{bc}} \\ \frac{V_{ca}}{ZL_{ca}} \end{bmatrix}$
Phase currents serving the constant current load	$\begin{bmatrix} IL_a \\ IL_b \\ IL_c \end{bmatrix} = \begin{bmatrix} \|IL_a\|\angle(\delta_a - \theta_a) \\ \|IL_b\|\angle(\delta_b - \theta_b) \\ \|IL_c\|\angle(\delta_c - \theta_c) \end{bmatrix}$	$\begin{bmatrix} IL_{ab} \\ IL_{bc} \\ IL_{ca} \end{bmatrix} = \begin{bmatrix} \|IL_{ab}\|\angle(\delta_{ab} - \theta_{ab}) \\ \|IL_{bc}\|\angle(\delta_{bc} - \theta_{bc}) \\ \|IL_{ca}\|\angle(\delta_{ca} - \theta_{ca}) \end{bmatrix}$
Line currents entering the load	$\begin{bmatrix} IL_a \\ IL_b \\ IL_c \end{bmatrix}$	$\begin{bmatrix} IL_a \\ IL_b \\ IL_c \end{bmatrix} = \begin{bmatrix} 1 & 0 & -1 \\ -1 & 1 & 0 \\ 0 & -1 & 1 \end{bmatrix} \cdot \begin{bmatrix} IL_{ab} \\ IL_{bc} \\ IL_{ca} \end{bmatrix}$
Distributed loads		Create a dummy node at 1/4 length from the sending end, and 2/3 of lode is connected there. At the end of the line section, 1/3 of load is connected.

19.2.3 Capacitor Banks

The modeling of capacitor banks is presented in Table 19.2.

19.2.4 Three-Phase Transformer

The voltage and current relationships between the primary and secondary sides for different transformer connections are presented in [21].

19.3 MODELING OF SOLAR AND WIND ENERGY MODELS

Solar systems convert solar energy into electricity, and like fuel cells, their output DC power is converted via an inverter-to-grid compatible AC power.

The grid-connected wind turbines are divided into fixed- and variable-speed groups [22, 23]. In the first group, a propeller through a gear box rotates the rotor of a squirrel cage induction generator, which is directly connected to the grid. In the second group, a doubly fed induction generator or a synchronous generator, either permanent magnet or conventional one, is used [9]. The AC output power of these units is converted via a power electronic–based rectifier and inverter-to-grid compatible AC power.

In accordance with the aforementioned, the primary energy of solar and wind systems may be injected to the grid via either a synchronous or asynchronous electric machine which is directly connected to the grid, a combination of an electric machine and a power electronic interface, or only via a power electronic interface. If electric machine is directly connected to the grid, its operation determines

TABLE 19.2
Capacitor Banks Modeling

	Wye Connected	Delta Connected
Specified bus voltage and reactive power	$\begin{bmatrix} \|V_{an}\|\angle\theta_a \\ \|V_{bn}\|\angle\theta_b \\ \|V_{cn}\|\angle\theta_c \end{bmatrix}, \begin{bmatrix} Q_a \\ Q_b \\ Q_c \end{bmatrix}$	$\begin{bmatrix} \|V_{ab}\|\angle\theta_{ab} \\ \|V_{bc}\|\angle\theta_{bc} \\ \|V_{ca}\|\angle\theta_{ca} \end{bmatrix}, \begin{bmatrix} Q_{ab} \\ Q_{bc} \\ Q_{ca} \end{bmatrix}$
Phase currents serving the capacitor bank	$[B_{abc}] = \begin{bmatrix} \dfrac{Q_a}{\|V_a\|^2} \\ \dfrac{Q_b}{\|V_b\|^2} \\ \dfrac{Q_c}{\|V_c\|^2} \end{bmatrix}$	$[B_{abc}] = \begin{bmatrix} \dfrac{Q_{ab}}{\|V_{ab}\|^2} \\ \dfrac{Q_{bc}}{\|V_{bc}\|^2} \\ \dfrac{Q_{ca}}{\|V_{ca}\|^2} \end{bmatrix}$
	$\begin{bmatrix} IC_a \\ IC_b \\ IC_c \end{bmatrix} = \begin{bmatrix} j \cdot B_a \cdot V_{an} \\ j \cdot B_b \cdot V_{bn} \\ j \cdot B_c \cdot V_{cn} \end{bmatrix}$	$\begin{bmatrix} IC_{ab} \\ IC_{bc} \\ IC_{ca} \end{bmatrix} = \begin{bmatrix} j \cdot B_{ab} \cdot V_{ab} \\ j \cdot B_{bc} \cdot V_{bc} \\ j \cdot B_{ca} \cdot V_{ca} \end{bmatrix}$
Line currents entering the capacitor bank	$\begin{bmatrix} IC_a \\ IC_b \\ IC_c \end{bmatrix}$	$\begin{bmatrix} IC_a \\ IC_b \\ IC_c \end{bmatrix} = \begin{bmatrix} 1 & 0 & -1 \\ -1 & 1 & 0 \\ 0 & -1 & 1 \end{bmatrix} \cdot \begin{bmatrix} IC_{ab} \\ IC_{bc} \\ IC_{ca} \end{bmatrix}$

the model of solar and wind energy for power flow studies. In other cases, the characteristics of the interface control circuit determine the solar and wind energy model. These models are extracted in the following.

19.3.1 Synchronous Generator

Based on the excitation system, synchronous generators are of two types [11]. The first one is the fixed-excitation voltage type, and the second one is regulating-excitation voltage type. In regulating-excitation voltage type, the excitation can be controlled in two ways. One is maintaining the constant terminal voltage (voltage control mode), and the other is maintaining the constant power factor (power factor control mode). In PFAs, the synchronous generator which maintains constant terminal voltage is considered as PV bus, and the synchronous generator which maintains constant power factor is treated as PQ bus. The synchronous generator with fixed excitation voltage is treated as SVCM [24].

19.3.2 Induction Generator

The active and reactive powers developed by the induction generator are functions of both slip and bus voltage [10]. Assuming the active power developed as constant and neglecting the very low dependency of reactive power with the slip, the induction generator can be modeled as SVCM, and further, since the bus voltage is nearer to 1 p.u in steady-state cases, the reactive power is also constant. Hence, the induction generator is modeled as PQ bus in PFAs.

19.3.3 Power Electronic Converter

The modeling of PECs in PFAs depends on the control method employed in converter control circuit. The PEC is modeled as PV bus if the control circuit in converter is designed to control both P and V independently [12]. If it is designed to control P and Q independently, then it is modeled as PQ bus [9].

19.4 VOLTAGE UNBALANCE

Voltage unbalance is a common problem in distribution networks. The increase in voltage unbalance can cause overheating and derating of all induction motor–type loads and can cause network problems, such as maloperation of protective relays generation of noncharacteristic harmonics from power electronic loads [25]. There have been several methods of definition and elucidation of voltage unbalance factor (VUF) [26]. The true definition of VU is the ratio of negative-sequence voltage component (V_2) to the positive-sequence voltage component (V_1). The VUF% is given by equation (3). International standards of allowable VU limits are presented in [1].

$$VUF\% = \left(\frac{V_2}{V_1}\right) \bullet 100 \qquad (3)$$

19.5 ALGORITHM FOR DEVELOPING BUS_NUM AND BRANCH_NUM MATRICES

After the numbering of buses and branches is done, as mentioned in [27], the following steps are to be followed to write a software code to divide the distribution network

Solving Issues of Grid Integration of Energy Models

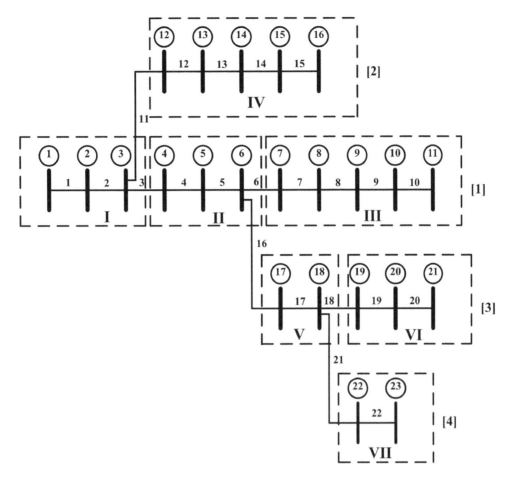

FIGURE 19.2 Divided sections for sample distribution network.

into different sections, as shown in Figure 18.2, for the given sample distribution network.

1. From the distribution network shown in Figure 18.2, form Table 19.3.
2. Start with BN = 1. Read the RE of BN, that is, 2. Then check how many times this 2 appears in SE row. From the previous table, it appears one time. That means bus 2 is the sending end for only one branch. Fill these RE 2 and BN 1 in two different variables (BUS_NUM and BRANCH_NUM) as the first row and first column elements. Then increase the column number by one.
3. Increase the BN (i.e., BN = 2). Read the RE of BN, that is, 3. Then, as in step 1, check for the appearance of 3 in SE row. The bus 3 appears two times. That means from the bus 3, two branches are leaving. Then fill these RE 3 and BN 2 into the same variables as the first row and present column elements. Name this row elements as "Section I." Now, increase the row number by one and set the column number to one.
4. Similarly, increase the BN. Read the RE of BN. Then check for the appearance of this RE in SE row. If it appears one time, then fill these RE and BN values as the present row and present

TABLE 19.3
Branch Numbering of Distribution Network in Figure 2

Branch Number (BN)	Sending Bus (SE)	Receiving Bus (RE)
1	1	2
2	2	3
3	3	4
4	4	5
5	5	6
6	6	7
7	7	8
8	8	9
9	9	10
10	10	11
11	3	12
12	12	13
13	13	14
14	14	15
15	15	16
16	6	17
17	17	18
18	18	19
19	19	20
20	20	21
21	18	22
22	22	23

column elements of variables BUS_NUM and BRANCH_NUM. Then increase the column number by one. And then repeat step 4. If it appears for no time or more than one time in SE row, then fill the corresponding RE and BN values as present row and present column elements. Then identify this row as a section. Then increase the row number by one and set the column number to one. And repeat step 4.

The previous steps are repeated until the BN value reaches the last branch number. At the end, the BUS_NUM and BRANCH_NUM matrices are obtained as follows:

$$\text{BUS\_NUM} = \begin{bmatrix} 1 & 2 & 3 & 0 & 0 \\ 4 & 5 & 6 & 0 & 0 \\ 7 & 8 & 9 & 10 & 11 \\ 12 & 13 & 14 & 15 & 16 \\ 17 & 18 & 0 & 0 & 0 \\ 19 & 20 & 21 & 0 & 0 \\ 22 & 23 & 0 & 0 & 0 \end{bmatrix} \begin{matrix} \rightarrow \text{Section I} \\ \rightarrow \text{Section II} \\ \rightarrow \text{Section III} \\ \rightarrow \text{Section IV} \\ \rightarrow \text{Section V} \\ \rightarrow \text{Section VI} \\ \rightarrow \text{Section VII} \end{matrix}$$

$$\text{BRANCH\_NUM} = \begin{bmatrix} 1 & 2 & 0 & 0 & 0 \\ 3 & 4 & 5 & 0 & 0 \\ 6 & 7 & 8 & 9 & 10 \\ 11 & 12 & 13 & 14 & 15 \\ 16 & 17 & 0 & 0 & 0 \\ 18 & 19 & 20 & 0 & 0 \\ 21 & 22 & 0 & 0 & 0 \end{bmatrix} \begin{matrix} \rightarrow \text{Section-I} \\ \rightarrow \text{Section-II} \\ \rightarrow \text{Section-III} \\ \rightarrow \text{Section-IV} \\ \rightarrow \text{Section-V} \\ \rightarrow \text{Section-VI} \\ \rightarrow \text{Section-VII} \end{matrix}$$

19.6 THREE-PHASE PFA WITH MULTIPLE SOLAR AND WIND ENERGY MODELS

After the BUS_NUM and BRANCH_NUM matrices are developed for the URDN, the following steps illustrate the iterative procedure for PFA:

1. The voltages at all busses are assigned as substation bus voltage.

$$\begin{bmatrix} V_a \\ V_b \\ V_c \end{bmatrix} = \begin{bmatrix} 1\angle 0° \\ 1\angle -120° \\ 1\angle 120° \end{bmatrix} \text{p.u} \tag{4}$$

2. Find the load current at all buses with the specified load power, type of load, and voltage at the bus.
3. The power flow program should start with collecting current at bus 23 (the tail bus in Section VII, as in BUS_NUM); thereby, find the current in branch 22 (the tail branch in Section VII, as in BRANCH_NUM). Then proceed towards the head bus (i.e., bus 22) and head branch (i.e., branch 21) in finding the bus currents and branch currents, respectively. This is illustrated with a sample section that consists of three buses, as shown in Figure 18.3.

$$\lfloor I_{abc} \rfloor_k = \lfloor IL_{abc} \rfloor_k + \lfloor Ish_{abc} \rfloor_k + \lfloor IC_{abc} \rfloor_k \tag{5}$$

$$\lfloor I_{abc} \rfloor_{jk} = \lfloor I_{abc} \rfloor_k \tag{6}$$

$$\lfloor I_{abc} \rfloor_j = \lfloor I_{abc} \rfloor_{jk} + \lfloor IL_{abc} \rfloor_j + \lfloor Ish_{abc} \rfloor_j + \lfloor IC_{abc} \rfloor_j \tag{7}$$

$$\lfloor I_{abc} \rfloor_{ij} = \lfloor I_{abc} \rfloor_j \tag{8}$$

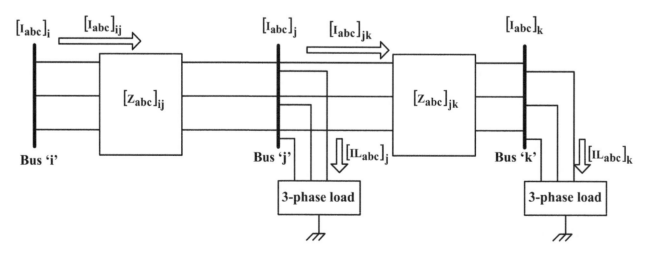

FIGURE 19.3 A sample section consists of three busses.

$$\left[I_{abc}\right]_i = \left[I_{abc}\right]_{ij} \quad (9)$$

Where:

$\left[I_{abc}\right]_k$ is three-phase current at bus k.

$\left[I_{abc}\right]_{jk}$ is three-phase current in branch jk.

$\left[IL_{abc}\right]_k$ is three-phase load current at bus k.

$\left[Ish_{abc}\right]_k$ is three-phase current drawn by shunt admittance at bus k.

$\left[IC_{abc}\right]_k$ is three-phase current drawn by capacitor bank at bus k, if any.

4. Then go to Section VI and repeat the above step in finding head bus current and head branch current. Likewise, move up to Section I and find current at head bus in Section I (bus 1) and head branch in Section I (branch 1).
5. Now start with bus 1 (head bus in Section I) and proceed towards the tail bus in finding the bus voltages with equation (2).
6. Then go to Section II and proceed towards the tail bus in finding the bus voltages. Likewise, move up to bus 23 (tail bus in Section VII).
7. Repeat steps 3 to 6 until the voltage magnitude mismatches at all buses in successive iterations are below the tolerance limit.

$$\left|\left[V_{abc}\right]_i^r - \left[V_{abc}\right]_i^{r-1}\right| \leq \left[\varepsilon_{abc}\right] \quad (10)$$

Where r is the iteration number.

8. Select the locations of solar and wind energy sources.
9. Then, for the outside γ^{th} iteration, check the type of solar and wind energy source available.
10. If the solar and wind energy source is modeled as PQ at bus, say, j, then the current injected by the solar and wind energy source is obtained with the specified power rating and bus voltage by equation (11).

$$\left[I_{G,abc}\right]_j^\gamma = \begin{bmatrix} \left(S_{G,a}/V_a\right)^* \\ \left(S_{G,b}/V_b\right)^* \\ \left(S_{G,c}/V_c\right)^* \end{bmatrix}_j^\gamma \quad (11)$$

11. If the solar and wind energy sources are modeled as PV buses, then calculate the voltage mismatches at all the PV buses with equation (12).

$$\begin{bmatrix} \Delta V_a \\ \Delta V_b \\ \Delta V_c \end{bmatrix}^\gamma = \begin{vmatrix} V_a^{sp} \\ V_b^{sp} \\ V_c^{sp} \end{vmatrix} - \begin{vmatrix} V_a^{cal} \\ V_b^{cal} \\ V_c^{cal} \end{vmatrix}^\gamma \forall \text{PV buses} \quad (12)$$

Where $\left[\Delta V\right]^\gamma$ is the voltage mismatch matrix. And its size is $3 \bullet n \times 1$.

12. If the voltage mismatches are not less than the specified tolerance value, then incremental current injections at PV buses in order to maintain the specified voltages is calculated with equation (13).

$$\left[\Delta I\right]^\gamma = \left[ZPV\right]^{-1} \bullet \left[\Delta V\right]^\gamma \quad (13)$$

$\left[ZPV\right]$ can be formed by observing the following numerical properties of its entries [6]. The diagonal entry ZPV_{pp} is equal to the modulus of the sum positive sequence impedance of all line sections between PV buses p and the substation bus. Since all the lines are three-phase line sections, the size of ZPV_{pp} is 3×3. If any two PV nodes, p and q, have different paths from root node, then the off diagonal entry ZPV_{pq} (size 3×3) is zero. If p and q share a piece of common path to the root node, then ZPV_{pq} is equal to the modulus of the sum-positive sequence impedance of all line sections on this common path. The dimension of $\left[ZPV\right]$ is equal to $3 \bullet n \times 3 \bullet n$.

13. If the reactive power generations were unlimited by solar and wind energy sources, then the incremental reactive current to be injected at the jth PV bus is obtained as equation (14).

$$\begin{bmatrix} \Delta I_{G,a} \\ \Delta I_{G,b} \\ \Delta I_{G,c} \end{bmatrix}_j^\gamma = \begin{bmatrix} |\Delta I_a| \bullet \left(\cos\left(90^0 + \delta_{v,a}\right) + j^* \sin\left(90^0 + \delta_{v,a}\right)\right) \\ |\Delta I_b| \bullet \left(\cos\left(90^0 + \delta_{v,b}\right) + j^* \sin\left(90^0 + \delta_{v,b}\right)\right) \\ |\Delta I_c| \bullet \left(\cos\left(90^0 + \delta_{v,c}\right) + j^* \sin\left(90^0 + \delta_{v,c}\right)\right) \end{bmatrix}_j^\gamma \quad (14)$$

Where $\delta_{v,a}$, $\delta_{v,b}$, and $\delta_{v,c}$ are the voltage angles at the jth PV bus.

14. Now, the current in branch ij from Figure 18.4 is obtained from equation (15).

$$\begin{bmatrix} I_a \\ I_b \\ I_c \end{bmatrix}_{ij}^\gamma = \begin{bmatrix} IL_a \\ IL_b \\ IL_c \end{bmatrix}_j^\gamma - \begin{bmatrix} \Delta I_{G,a} \\ \Delta I_{G,b} \\ \Delta I_{G,c} \end{bmatrix}_j^\gamma \quad (15)$$

With $\left[V_{abc}\right]_i^\gamma$ and $\left[I_{abc}\right]_{ij}^\gamma$, the reactive power flow in the line $\left[Q_{abc}\right]_{ij}^\gamma$ is evaluated. Then, incremental reactive current injection is obtained with equation (16).

$$\begin{bmatrix} \Delta Q_{G,a} \\ \Delta Q_{G,b} \\ \Delta Q_{G,c} \end{bmatrix}_j^\gamma = \begin{bmatrix} QL_a \\ QL_b \\ QL_c \end{bmatrix}_j^\gamma - \begin{bmatrix} Q_a \\ Q_b \\ Q_c \end{bmatrix}_{ij}^\gamma \quad (16)$$

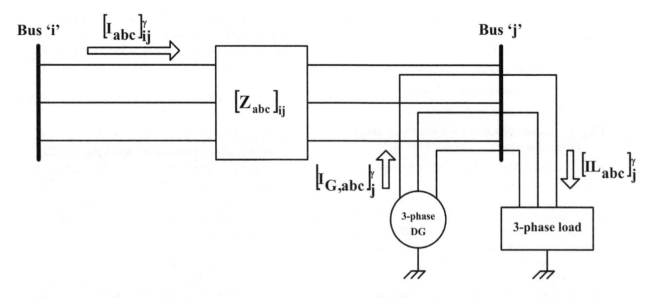

FIGURE 19.4 A sample two nodes in the system with wind energy source placed at bus j.

The reactive power generation needed at j^{th} PV bus is obtained with equation (17).

$$\begin{bmatrix} Q_{G,a} \\ Q_{G,b} \\ Q_{G,c} \end{bmatrix}_j^\gamma = \begin{bmatrix} Q_{G,a} \\ Q_{G,b} \\ Q_{G,c} \end{bmatrix}_j^{\gamma-1} + \begin{bmatrix} \Delta Q_{G,a} \\ \Delta Q_{G,b} \\ \Delta Q_{G,c} \end{bmatrix}_j^\gamma \quad (17)$$

The complex power generation at PV bus is obtained with equation (18).

$$\begin{bmatrix} S_{G,a} \\ S_{G,b} \\ S_{G,c} \end{bmatrix}_j^\gamma = \begin{bmatrix} P_{G,a} \\ P_{G,b} \\ P_{G,c} \end{bmatrix}_j + j \bullet \begin{bmatrix} Q_{G,a} \\ Q_{G,b} \\ Q_{G,c} \end{bmatrix}_j^\gamma \quad (18)$$

Where $\lfloor P_{G,abc} \rfloor_j$ is the specified real power generation of the solar and wind energy sources unit at j^{th} bus.

15. With the complex power obtained in equation (18), the current injected by the solar or wind energy source is calculated with equation (11).
16. If the reactive power generations by solar and wind energy sources were limited, then reactive power limits must be checked first. The total three-phase reactive power needed at j^{th} PV node is the sum of reactive power generations of three phases, as in equation (19), and is compared with the reactive power generation limits of the solar or wind energy sources.

$$\left(Q_G\right)_j^\gamma = \left(Q_{G,a}\right)_j^\gamma + \left(Q_{G,b}\right)_j^\gamma + \left(Q_{G,c}\right)_j^\gamma \quad (19)$$

If $Q_{j,min} \leq \left(Q_G\right)_j^\gamma \leq Q_{j,max}$, then set complex power generation is as in equation (18)

If $\left(Q_G\right)_j^\gamma \leq Q_{j,min}$, then set $\left(Q_G\right)_j^\gamma = Q_{j,min}$ and $\left(Q_{G,a}\right)_j^\gamma = \left(Q_{G,b}\right)_j^\gamma = \left(Q_{G,c}\right)_j^\gamma = Q_{j,min}/3$.

If $\left(Q_G\right)_j^\gamma \geq Q_{j,max}$, then set $\left(Q_G\right)_j^\gamma = Q_{j,max}$ and $\left(Q_{G,a}\right)_j^\gamma = \left(Q_{G,b}\right)_j^\gamma = \left(Q_{G,c}\right)_j^\gamma = Q_{j,max}/3$.

These reactive power generations are combined with active power generations at PV buses, and the current injections by solar and wind energy sources are obtained with equation (11).

17. Now, set $\gamma = \gamma + 1$ and repeat the steps from 9 with the current injections at solar and wind energy source buses.
18. Power flow will be stopped when it attains convergence criterion.
19. The power loss in all the branches is obtained with equation (20).

$$\begin{bmatrix} SLoss_a \\ SLoss_b \\ SLoss_c \end{bmatrix}_{ij} = \begin{bmatrix} (V_a)_i \cdot (I_a)_{ij}^* \\ (V_b)_i \cdot (I_b)_{ij}^* \\ (V_c)_i \cdot (I_c)_{ij}^* \end{bmatrix} - \begin{bmatrix} (V_a)_j \cdot (I_a)_{ij}^* \\ (V_b)_j \cdot (I_b)_{ij}^* \\ (V_c)_j \cdot (I_c)_{ij}^* \end{bmatrix} \quad (20)$$

20. Find the total power loss of the network, and VUF% at each bus is evaluated using equation (3).

19.7 RESULTS AND DISCUSSIONS

19.7.1 IEEE-13 Bus URDN

The accuracy of the proposed three-phase PFA without integrations of solar and wind energy sources is tested on IEEE-13 bus URDN. The data for IEEE-13 bus feeder

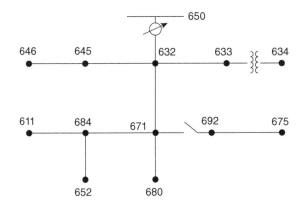

FIGURE 19.5 IEEE 13 bus URDN.

shown in Figure 18.5 is taken from [28]. The base values selected for the network are 5,000 kVA and 4.16 kV. The convergence tolerance is 10^{-4}. The PFA is converged after five iterations. The power flow solution is presented in Table 19.4 in comparison with the IEEE results. The maximum mismatches for voltage magnitudes and angle are found to be 0.0005 p.u and 0.010° respectively and are of insignificance. Therefore, the test results almost match the IEEE results in terms of accuracy. The power loss of the network in comparison with IEEE results is presented in Table 19.5.

TABLE 19.4

Voltage Solution for IEEE-13 Bus URDN

Bus	Phase	Calculated Results	IEEE Results [28]	Error in Voltage Mag.	Error in Voltage Ang.
650	a	1∠0°	1∠0°	0.0000	0.00
	b	1∠−120°	1∠−120°	0.0000	0.00
	c	1∠120°	1∠120°	0.0000	0.00
RG	a	1.0625∠0°	1.0625∠0°	0.0000	0.00
	b	1.0500∠−120°	1.0500∠−120°	0.0000	0.00
	c	1.0687∠120°	1.0687∠120°	0.0000	0.00
632	a	1.0210∠−2.49°	1.0210∠−2.49°	0.0000	0.00
	b	1.0420∠−121.72°	1.0420∠−121.72°	0.0000	0.00
	c	1.0175∠117.83°	1.0170∠117.83°	−0.0005	0.00
671	a	0.9900∠−5.30°	0.9900∠−5.30°	0.0000	0.00
	b	1.0529∠−122.34°	1.0529∠−122.34°	0.0000	0.00
	c	0.9778∠116.03°	0.977∠116.02°	0.0001	−0.01
680	a	0.9900∠−5.30°	0.9900∠−5.30°	0.0000	0.00
	b	1.0529∠−122.34°	1.0529∠−122.34°	0.0000	0.00
	c	0.9778∠116.03°	0.977∠116.02°	0.0001	−0.01
633	a	1.0180∠−2.55°	1.0180∠−2.56°	0.0000	0.01
	b	1.0401∠−121.77°	1.0401∠−121.77°	0.0000	0.00
	c	1.0148∠117.82°	1.0148∠117.82°	0.0000	0.00
634	a	0.9940∠−3.23°	0.9940∠−3.23°	0.0000	0.00
	b	1.0218∠−122.22°	1.0218∠−122.22°	0.0000	0.00
	c	0.9960∠117.35°	0.9960∠117.34°	0.0000	−0.01
645	b	1.0328∠−121.90°	1.0329∠−121.90°	0.0001	0.00
	c	1.0155∠117.86°	1.0155∠117.86°	0.0001	0.00
646	b	1.0311∠−121.98°	1.0311∠−121.98°	0.0000	0.00
	c	1.0134∠117.90°	1.0134∠117.90°	0.0000	0.01
692	a	0.9900∠−5.30°	0.9900∠−5.31°	0.0000	0.01
	b	1.0529∠−122.34°	1.0529∠−122.34°	0.0000	0.00
	c	0.9778∠116.03°	0.9777∠116.02°	−0.0001	−0.01
675	a	0.9835∠−5.55°	0.9835∠−5.56°	0.0000	0.01
	b	1.0553∠−122.52°	1.0553∠−122.52°	0.0000	0.00
	c	0.9759∠116.04°	0.9758∠116.03°	−0.0001	−0.01
684	a	0.9881∠−5.32°	0.9881∠−5.32°	0.0000	0.00
	c	0.9758∠115.92°	0.9758∠115.92°	0.0000	0.00
611	c	0.9738∠115.78°	0.9738∠115.78°	0.0000	0.00
652	a	0.9825∠−5.24°	0.9825∠−5.25°	0.0000	0.01

TABLE 19.5
Power Loss in IEEE-13 Bus URDN

Phase	Calculated Power Loss		IEEE Results [28]	
	Active (kW)	Reactive (kVAR)	Active (kW)	Reactive (kVAR)
a	39.13	152.62	39.107	152.585
b	-4.74	42.27	-4.697	42.217
c	76.59	129.69	76.653	129.850
Total	110.98	324.57	111.063	324.653

TABLE 19.6
Case Studies on IEEE-13 Bus URDN

Case Study	Description
Case 1	• Removed voltage regulator between buses 650 and 632. • Removed Capacitor banks at buses 675 and 611.
Case 2	• 3-phase solar cell, with control circuit of its converter, controlled independently P and Q placed at bus 634, is modeled as PQ bus with capacity P = 300 kW and Q = 197 kVAR for each phase. • 3-phase wind energy with voltage control mode placed at bus 675 is modeled as PV bus with P = 260 kW per phase and three-phase reactive power limits: $100 \leq Q \leq 650$ kVAR.

TABLE 19.7
Power Loss for Different Case Studies on IEEE-13 Bus URDN

Phase	Case 1		Case 2	
	Active (kW)	Reactive (kVAR)	Active (kW)	Reactive (kVAR)
A	39.48	194.82	7.9867	87.6800
B	−2.03	47.13	4.0219	10.1590
C	109.88	191.59	56.8390	88.4051
Total	147.33	433.54	68.84	186.24

19.7.1.1 Case Studies on IEEE-13 Bus URDN

The impacts of solar and wind energy source integrations on IEEE-13 bus URDN are analyzed with different case studies as presented in Table 19.6. The total active and reactive power loss on the network for different case studies are presented in Table 19.7. The voltage solution and VUF % for case studies 1 and 2 are presented in Table 19.8.

19.7.1.2 Discussions

From Table 19.7, the total active and reactive power losses are observed to be reduced in case 2. From Table 19.8, it is observed that in case 1, the maximum value of VUF% at bus 675 is observed to be 2.75, and minimum fundamental voltage on the network is 0.8651 p.u at bus 611 for c-phase. The results of this study are not desirable. In case 2, the minimum voltage of the network is improved to 0.8990 at bus 611 for c-phase, and the maximum value of VUF% at bus 675 is reduced to 2. The improvement in p.u voltage profile of case 2 with respect to case 1 is shown in Figure 18.6.

19.7.2 CASE STUDIES AND DISCUSSIONS ON IEEE-34 BUS URDN

The data for IEEE-34 bus URDN, shown in Figure 18.7, is given in [28]. The base values selected for the system are 2,500 kVAR and 4.9 kV. The convergence tolerance is taken as 10^{-4}. The load flow program is converged after two iterations. The rating and location of solar and wind energy sources for different case studies are presented in Table 19.9. The p.u voltage solution and VUF% for case studies 1 and 2 are presented in Table 19.10 and Table 19.11, respectively. Figure 18.8 compares the voltage profiles of different case studies. Figure 18.9 compares the power loss in different case studies. Figure 18.10 compares the maximum VUF% in different case studies. From Table 19.10, it is observed that the maximum value of VUF% at bus 860 is 1.30, and minimum fundamental voltage on the network is 0.7837 p.u at bus 890 for a-phase. From Table 19.11, with the integrations of solar and wind energy models, the maximum value of VUF% at bus 890 is reduced to 0.98, and the minimum voltage of the network is improved to 0.9009 at bus 860 for a-phase.

Solving Issues of Grid Integration of Energy Models

TABLE 19.8
Voltage Solution and VUF% for Different Case Studies on IEEE-13 Bus URDN

Bus	S. No.	Ph	Case I $\|V_{p.u}\|\angle\deg$	Case I VUF%	Case II $\|V_{p.u}\|\angle\deg$	Case II VUF%
650	1	a	1∠0°	0	1.0000∠0°	0
	2	b	1∠−120°		1.0000∠−120°	
	3	c	1∠120°		1.0000∠120°	
632	4	a	0.9498∠−2.74°	1.27	0.9730∠−1.68°	0.83
	5	b	0.9839∠−121.68°		0.9973∠−120.48°	
	6	c	0.9300∠117.80°		0.9540∠118.98°	
671	7	a	0.9109∠−5.89°	2.58	0.9429∠−4.10°	1.89
	8	b	0.9875∠−122.20°		1.0045∠−120.42°	
	9	c	0.8717∠115.95°		0.9056∠117.88°	
680	10	a	0.9109∠−5.89°	2.58	0.9429∠−4.10°	1.89
	11	b	0.9875∠−122.20°		1.0045∠−120.42°	
	12	c	0.8717∠115.95°		0.9056∠117.87°	
633	13	a	0.9466∠−2.82°	1.28	0.9749∠−1.64°	0.82
	14	b	0.9819∠−121.73°		1.0006∠−120.40°	
	15	c	0.9271∠117.79°		0.9574∠119.06°	
634	16	a	0.9207∠−3.60°	1.35	0.9950∠−1.00°	0.76
	17	b	0.9624∠−122.24°		1.0249∠−119.60°	
	18	c	0.9064∠117.21°		0.9828∠119.94°	
645	19	b	0.9745∠−121.86°	−	0.9880∠−120.66°	-
	20	c	0.9283∠117.82°		0.9523∠119.00°	
646	21	b	0.9729∠−121.93°	−	0.9863∠−120.73°	-
	22	c	0.9264∠117.86°		0.9503∠119.05°	
692	23	a	0.9109∠−5.89°	2.58	0.9429∠−4.10°	1.89
	24	b	0.9875∠−122.20°		1.0045∠−120.42°	
	25	c	0.8717∠115.95°		0.9056∠117.87°	
675	26	a	0.9025∠−6.07°	2.75	0.9379∠−4.19°	2.00
	27	b	0.9887∠−122.30°		1.0080∠−120.44°	
	28	c	0.8678∠116.06°		0.9047∠118.08°	
684	29	a	0.9093∠−5.95°	−	0.9411∠−4.15°	-
	30	c	0.8684∠115.91°		0.9023∠117.84°	
611	31	c	0.8651∠115.83°	−	0.8990∠117.77°	-
652	32	a	0.9041∠−5.87°	−	0.9358∠−4.07°	-

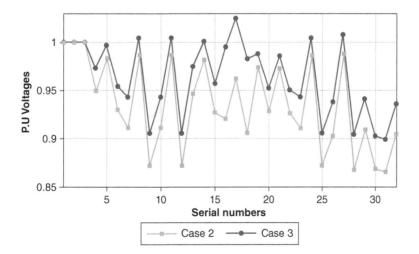

FIGURE 19.6 Comparison of p.u voltage profiles in different case studies on IEEE-13 bus URDN.

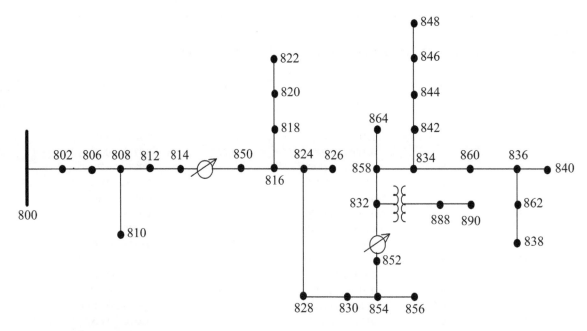

FIGURE 19.7 IEEE-34 bus URDN.

TABLE 19.9
Case Studies on IEEE-13 Bus URDN

Case Study	Description
Case 1	• Removed voltage regulators between buses 614 and 650 and 852 and 832.
	• Removed capacitor banks at buses 844 and 848.
Case 2	• 3-phase solar cell, with control circuit of its converter controlled independently P and Q, placed at bus 848, is modeled as PQ bus with capacity P = 150 kW and Q = 99 kVAR for each phase.
	• 3-phase wind energy with voltage control mode placed at bus 890 is modeled as PV bus with P = 130 kW per phase and three-phase reactive power limits $50 \leq Q \leq 325$ kVAR.

TABLE 19.10
Voltage Solution and VUF% for Case 1 on IEEE-34 Bus URDN

Bus No.	S. No.	Ph	$|V_{p.u}|\angle$deg	VUF %	Bus No	S. No.	Ph	$|V_{p.u}|\angle$deg	VUF %
800	1	a	1.0000∠ 0°			46	a	0.8156∠0.9764°	1.29
	2	b	1.0000∠−120°	0	834	47	b	0.8526∠−120.97°	
	3	c	1.0000∠ 120°			48	c	0.8560∠119.90°	
802	4	a	0.9972∠−0.00°			49	a	0.8155∠0.98°	1.29
	5	b	0.9980∠−120.03°	0.02	842	50	b	0.8525∠−120.96°	
	6	c	0.9981∠119.10°			51	c	0.8559∠119.91°	
806	7	a	0.9953∠−0.00°			52	a	0.8151∠0.99°	1.29
	8	b	0.9967∠−120.05°	0.03	844	53	b	0.8519∠−120.97°	
	9	c	0.9969∠119.99°			54	c	0.8554∠119.91°	
808	10	a	0.9604∠−0.07°			55	a	0.8149∠1.00°	1.28
	11	b	0.9746∠−120.38°	0.33	846	56	b	0.8513∠−120.96°	
	12	c	0.9742∠119.94°			57	c	0.8552∠119.91°	
812	13	a	0.9197∠−0.15°			58	a	0.8149∠1.00°	1.28
	14	b	0.9488∠−120.77°	0.70	848	59	b	0.8513∠−120.96°	
	15	c	0.9478∠119.89°			60	c	0.8552∠119.91°	
814	16	a	0.8875∠−0.22°		810	61	b	0.9975∠−120.03°	-
	17	b	0.9284∠−121.10°	1.01	818	62	a	0.8864∠−0.22°	-
	18	c	0.9269∠119.84°		820	63	a	0.8593∠−0.24°	-

Bus No.	S. No.	Ph	$\|V_{p.u}\|\angle\deg$	VUF %	Bus No	S. No.	Ph	$\|V_{p.u}\|\angle\deg$	VUF %
850	19	a	0.8875∠−0.22°		822	64	a	0.8558∠−0.24°	−
	20	b	0.9284∠−121.10°	1.01	826	65	b	0.9168∠−121.10°	−
	21	c	0.9268∠119.84°		856	66	b	0.8952∠−121.06°	−
816	22	a	0.8872∠−0.21°			67	a	0.7860∠−1.10°	1.27
	23	b	0.9280∠−121.10°	1.01	888	68	b	0.8242∠−122.92°	
	24	c	0.9266∠119.84°			69	c	0.8264∠118.05°	
824	25	a	0.8772∠−0.05°			70	a	0.7837∠−1.11°	1.29
	26	b	0.9170∠−121.10°	1.03	890	71	b	0.8221∠−122.96°	
	27	c	0.9170∠119.84°			72	c	0.8241∠118.02°	
828	28	a	0.8764∠−0.04°		864	73	a	0.8191∠0.90°	−
	29	b	0.9161∠−121.10°	1.03	860	74	a	0.8151∠0.90°	1.30
	30	c	0.9162∠119.84°			75	b	0.8521∠−120.96°	
830	31	a	0.8566∠0.26°			76	c	0.8556∠119.90°	
	32	b	0.8959∠−121.06°	1.11		77	a	0.8149∠0.99°	1.29
	33	c	0.8968∠119.86°		836	78	b	0.8517∠−120.97°	
854	34	a	0.8561∠0.27°			79	c	0.8554∠119.90°	
	35	b	0.8954∠−121.06°	1.11		80	a	0.8149∠0.99°	1.29
	36	c	0.8964∠119.86°		840	81	b	0.8517∠−120.97°	
852	37	a	0.8221∠0.82°			82	c	0.8554∠119.90°	
	38	b	0.8599∠−120.99°	1.26		83	a	0.8149∠0.99°	1.29
	39	c	0.8625∠119.90°		862	84	b	0.8517∠−120.97°	
832	40	a	0.8221∠0.82°			85	c	0.8554∠119.90°	
	41	b	0.8599∠−120.99°	1.26	838	86	b	0.8515∠−120.97°	−
	42	c	0.8625∠119.90°						
858	43	a	0.8191∠0.90°						
	44	b	0.8565∠−120.98°	1.28					
	45	c	0.8595∠119.90°						

TABLE 19.11
Voltage Solution and VUF% for Case 2 on IEEE-34 Bus URDN

Bus No.	S. No.	Ph	$\|V_{p.u}\|\angle\deg$	VUF %	Bus No	S. No.	Ph.	$\|V_{p.u}\|\angle\deg$	VUF %
800	1	a	1.0000∠0°	0		46	a	0.9072∠1.26°	0.97
	2	b	1.0000∠−120°		834	47	b	0.9353∠−120.34°	
	3	c	1.0000∠120°			48	c	0.9433∠120.27°	
802	4	a	0.9984∠0.01°	0.02		49	a	0.9072∠1.26°	0.97
	5	b	0.9990∠−120.02°		842	50	b	0.9352∠−120.34°	
	6	c	0.9992∠120.00°			51	c	0.9433∠120.27°	
806	7	a	0.9973∠0.01°	0.03		52	a	0.9073∠1.26°	0.97
	8	b	0.9984∠−120.02°		844	53	b	0.9351∠−120.34°	
	9	c	0.9987∠120.01°			54	c	0.9433∠120.27°	
808	10	a	0.9768∠0.08°	0.29		55	a	0.9084∠1.26°	0.96
	11	b	0.9890∠−120.19°		846	56	b	0.9357∠−120.35°	
	12	c	0.9897∠120.08°			57	c	0.9443∠120.26°	
812	13	a	0.9530∠0.15°	0.61		58	a	0.9085∠1.26°	0.96
	14	b	0.9779∠−120.38°		848	59	b	0.9358∠−120.35°	
	15	c	0.9792∠120.17°			60	c	0.9445∠120.26°	
814	16	a	0.9342∠0.22°	0.86	810	61	b	0.9987∠−120.02°	−
	17	b	0.9692∠−120.53°		818	62	a	0.9332∠0.22°	−
	18	c	0.9709∠120.25°		820	63	a	0.9072∠0.20°	−
850	19	a	0.9342∠0.22°	0.86	822	64	a	0.9039∠0.20°	−
	20	b	0.9692∠−120.53°		826	65	b	0.9634∠−120.52°	−
	21	c	0.9709∠120.25°		856	66	b	0.9540∠−120.46°	−
816	22	a	0.9340∠0.22°	0.86		67	a	0.9012∠1.41°	0.98
	23	b	0.9690∠−120.53°		888	68	b	0.9298∠−120.24°	
	24	c	0.9707∠120.25°			69	c	0.9366∠120.44°	

(*Continued*)

TABLE 19.11 (Continued)

Bus No.	S. No.	Ph	$\|V_{p.u}\|\angle$deg	VUF %	Bus No	S. No.	Ph.	$\|V_{p.u}\|\angle$deg	VUF %
824	25	a	0.9302∠0.37°	0.86		70	a	0.9009∠1.44°	0.97
	26	b	0.9636∠-120.52		890	71	b	0.9294∠-120.22°	
	27	c	0.9671∠120.25			72	c	0.9362∠120.46°	
828	28	a	0.9298∠0.39	0.86	864	73	a	0.9088∠1.20°	-
	29	b	0.9633∠-120.51°		860	74	a	0.9068∠1.27°	0.98
	30	c	0.9667∠120.25°			75	b	0.9348∠-120.34°	
830	31	a	0.9223∠0.66°	0.90		76	c	0.9429∠120.27°	
	32	b	0.9544∠-120.46°			77	a	0.9066∠1.27°	0.97
	33	c	0.9592∠120.26°		836	78	b	0.9344∠-120.34°	
854	34	a	0.9221∠0.67°	0.90		79	c	0.9428∠120.27°	
	35	b	0.9542∠-120.46°			80	a	0.9065∠1.27°	0.97
	36	c	0.9590∠120.26°		840	81	b	0.9344∠-120.34°	
852	37	a	0.9102∠1.16°	0.95		82	c	0.9428∠120.27°	
	38	b	0.9391∠-120.35°			83	a	0.9066∠1.27°	0.97
	39	c	0.9464∠120.29°		862	84	b	0.9344∠-120.34°	
832	40	a	0.9102∠1.16°	0.95		85	c	0.9428∠120.27°	
	41	b	0.9391∠-120.35°		838	86	b	0.9342∠-120.35°	-
	42	c	0.9464∠120.29°						
858	43	a	0.9088∠1.20°	0.96					
	44	b	0.9373∠-120.35°					—	
	45	c	0.9450∠120.28°						

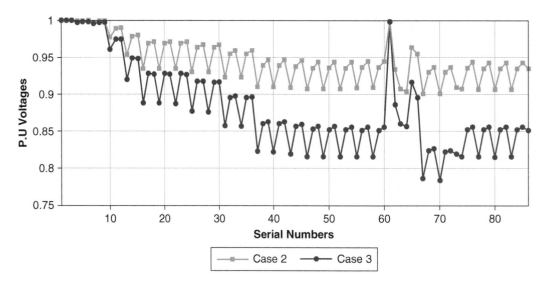

FIGURE 19.8 Comparison of p.u voltage profiles in different case studies on IEEE-34 bus URDN.

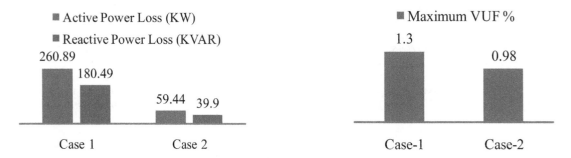

FIGURE 19.9 Comparison of power loss in different case studies on IEEE-34 bus URDN.

FIGURE 19.10 Comparison of maximum VUF% in different case studies on IEEE-34 bus URDN.

19.8 CONCLUSIONS

The BUS_NUM and BRANCH_NUM matrices developed in this chapter make the three-phase PFA simple. The three-phase PFA without solar and wind energy source integration is tested for accuracy on IEEE-13 bus URDN. It is found that the proposed three-phase PFA is accurate and has fast convergence. The impacts of multiple solar and wind energy sources on voltage profile, power loss, and VUF% are analyzed with the proposed three-phase PFA. The proposed algorithm is fast, accurate, and easy to understand and effectively handles both PV and PQ modeling solar and wind energy models. With the integration of these devices, there is an improvement in voltage profile, reduction in power loss, and reduction in severity of VU.

19.9 REFERENCES

[1] F.J. Ruiz-Rodriguez, J. C. Hernandez, F. Jurado., 'Voltage unbalance assessment in secondary radial distribution networks with single-phase photovoltaic systems', *Electrical Power and Energy Systems*, 2015, 64, pp. 646–654.

[2] Ahmed M. M. Nour, Ahmed A. Helal, Magdi, M. El-Saadawi, Ahmed Y. Hatata, 'A control scheme for voltage unbalance mitigation in distribution network with rooftop PV systems based on distributed batteries', *Electrical Power and Energy Systems*, 2021, 124, pp. 1–18.

[3] Mehdi Zeraati, Mohamad Esmail Hamedani Golshan, Josep M. Guerrero, 'Voltage quality improvement in low voltage distribution networks using reactive power capability of single-phase PV inverters', *IEEE Transactions on Smart Grid*, 2019, 10(5), pp. 5057–5065.

[4] Nasim Jabalameli, Xianging Su, Arindam Ghosh, 'Online centralized charging coordination of PEVS with decentralized var discharging for mitigation of voltage unbalance', *IEEE Power and Energy Technology Systems*, 2019, 6(3), pp. 152–161.

[5] Omid Abdoli, Eskandar Gholipour, Rahmat-Allah Hooshmand, 'A New approach to compensating voltage unbalance by UPQC-based PAC', *Electric Power Components and Systems*, 2019, pp. 1–13.

[6] D. Shirmohammadi, S. Carol, A. Cheng, 'A three phase power flow method for real time distribution system analysis', *IEEE Transactions on Power System*, 1995, 10, pp. 671–679.

[7] Abhishek Kumar, Bablesh Kumar Jha, Devender Singh & Rakesh Kumar Misra, 'A new current injection based power flow formulation' *Electric Power Components and Systems*, 2020, pp. 1–13.

[8] Digambar M. Tagare, 'IEEE standard for interconnecting distributed resources with electric power systems', *IEEE Standard 1547-2003*, 2003, pp. 1–16.

[9] M. H. Nehrir, C. Wang, S. R. Shaw, 'Feul cell: Promising devices for distributed generation', *IEEE Power Energy Magazine*, 2006, 4, pp. 47–53.

[10] S. Naka, T. Genji, Y. Fukuyama., 'Practical equipment models for fast distribution power flow considering interconnection of distributed generators', *Power Engineering Society Summer Meeting Conference Proceedings*, 15th -19th July 2001, pp. 1007–1012.

[11] Arturo Losi, Mario Russo, 'Dispersed generation modeling for object-oriented distribution load flow', *IEEE Trans. Power Delivery*, 2005, 20, pp. 1532–1540.

[12] H. Chen, J. Chen, D. Shi, X. Duan, 'Power flow study and voltage stability analysis for distribution systems with distributed generation', *IEEE Power Engineering Society General Meeting*, 18th -22nd 2006, pp. 1–8.

[13] Ahmed M. Ibrahim, Mahmoud A. Attia, Almoataz Y. Abdelaziz, 'A DSM approach for distribution systems with high wind power penetration', *Electric Power Components and Systems*, 2020, pp. 1–14.

[14] Mohamed A. Tolba, Ahmed A. Zaki Diab, Artem S. Vanin, Vladimir N. Tulsky, Almoataz Y. Abdelaziz, 'Integration of renewable distributed generation in distribution networks including a practical case study based on a hybrid PSOGSA optimization algorithm', *Electric Power Components and Systems*, 2019, pp. 1–14, 2019.

[15] A.Y. Abdelaziz, Y.G. Hegazy, Walid El-Khattam, M.M. Othman, 'Optimal allocation of stochastically dependent renewable energy based distributed generators in unbalanced distribution networks', *Electric Power Systems Research*, 2015, 119, pp. 34–44.

[16] M. M. Othman, Walid El-Khattam, Yasser G. Hegazy and Almoataz Y. Abdelaziz, 'Optimal placement and sizing of distributed generators in unbalanced distribution systems using supervised big bang-big crunch method', *IEEE Transactions on Power Systems*, 2015, pp. 1–9.

[17] M.M. Othman, Walid El-Khattam, Y.G. Hegazy, Almoataz Y. Abdelaziz, 'Optimal placement and sizing of voltage controlled distributed generators in unbalanced distribution networks using supervised firefly algorithm', *Electrical Power and Energy Systems*, *2016*, 82, pp. 105–113.

[18] R. Satish, P. Kantarao, K. Vaisakh, 'A new algorithm for impacts of multiple DGs and D-STATCOM in unbalanced radial distribution networks', *International Journal of Renewable Energy Technology*, 2021, 12(3), pp. 221–242.

[19] W. H. Kersting, *Distribution system modeling and analysis*, Fourth ed., CRC Press, 2017.

[20] Khushalani and Schulz, 'Unbalanced distribution power flow with distributed generation', *IEEE/PES Transactions on Distribution Conference*, 2006, pp. 301–306.

[21] T. -H. Chen, M. -S, Chen, T. Inoue, P. Kotas, and E. A. Chebli, 'Three-phase cogenerator and transformer models for distribution system', *IEEE Transactions on Power Delivery*, 1991, 6(4), pp. 1671–1681.

[22] K.C. Divya, P.S. Nagendra Rao, 'Models for wind turbine generating systems and their application in load flow studies', *Electric Power Systems Research*, 2006, 76, pp. 844–856.

[23] A.P. Agalgaonkar, S.V. Kulkarni, S.A. Khaparde, 'Impact of wind generation on loss and voltage profile in a distribution system', *Conference on Convergent Technologies for Asia-Pacific Region*, 2003, 2, pp. 775–779.

[24] Ryuto Shigenobu, Akito Nakadomari, Ying-Yi Hong, Paras Mandal, Hiroshi Takahashi and Tomonobu Senjyu, 'Optimization of voltage unbalance compensation by smart inverter', *Energies*, 2020, 13, pp. 1–22.

[25] Farhad Shahnia, Ritwik Majumder, Arindam Ghosh, et al., 'Voltage imbalance analysis in residential low voltage

distribution networks with rooftop PVs', *Electric Power System Research*, 2011, 81, pp. 1805–1814.

[26] Asheesh K. Singh, G. K. Singh, R. Mitra., 'Some observations on definitions of voltage unbalance', *39th North American Power Symposium (NAPS)*, September/October 2007, pp. 473–479.

[27] D. Das, H. S. Nagi and D. P. Kothari., 'Novel method for solving radial distribution networks', *IEE Proceedings-Generation, Transmission and Distribution*, 1994, 141, pp. 291–298.

[28] Radial Distribution Test Feeders, http://sites.ieee.org/pes-testfeeders/resources

20 Multifunctional PV-Integrated Bidirectional Off-Board EV Battery Charger Targeting Smart Cities

Rajesh Kumar Lenka, Anup Kumar Panda, Venkataramana Naik N, Laxmidhar Senapati, and Nishit Tiwary

CONTENTS

20.1 Introduction 267
20.2 EV Battery Charger Configuration 268
20.3 Control Strategy..268
    20.3.1 Islanding Operation..268
    20.3.2 Grid-Connected Operation 271
20.4 Simulation Results and Discussion...273
20.5 Conclusion 278
20.6 References...278

20.1 INTRODUCTION

Increasing environmental pollution and limiting petroleum fuels have resulted in two major changes in the world [1]. The first change is adopting renewable energy sources (RESs) by replacing conventional energy sources, and the second is adopting electric vehicles (EV) by replacing combustion engine–based vehicles [2]. The important factors that affect the EV sector are low-cost storage technology, efficient motor drive, charging stations, smart control algorithms, and government laws [3]. A survey reported in [4] states that renewable energy contributes only 20% of the total generated electricity, and conventional energy contributed 60%. Hence, even if conventional vehicles are completely replaced by electric vehicles (EVs), that does not solve the problem related to pollution and energy safety. As a result, various efforts are made recently to build RES-based EV charging stations [5–7]. The first alternative is to install the large RES generating unit at remote places and transmit the power to the EV charging station. However, this involves great planning, massive investment, and commissioning of the new transmission lines, which further incur more power losses. The second alternative is to install the small RES-based generating unit locally at public places, like parking areas, shopping malls, etc., and use it for EV charging. The second one is more beneficial as it involves less capital investment and reduces transmission loss by using underutilized space [8].

In [9], Alharbi et al. have proposed a PV-integrated EV fast charger. This work mainly focuses on optimizing the charging time of the fast charger and mitigating its effects on the utility grid. In [10], a wind energy–integrated EV charger is proposed for energy balancing. The proposed approach uses the stored energy in EVs to maintain energy balance in the wind generating system. In this scenario, optimum sizing of the EV batteries for energy balance in PV- and wind-integrated distributed generation system is discussed in [11]. Considering EV charging in PV- and wind-integrated microgrid, a model predictive control is proposed for both bidirectional converters in [12]. A PV battery system (PBS) for EV charging is proposed in [13]; however, EV charging is not possible without an auxiliary battery. An EV charger by utilizing the parking space for PV generation is proposed in [14]. This work focuses on developing the charging scheduling for optimal utilization of the PV generation unit. The proposed PV-based EV charger in [15] develops charging scheduling to minimize charging costs. Similarly, in [16] the charging cost is minimized by using a PBS-integrated EV charger. The available literature on PV-integrated EV chargers mostly focuses on optimal charging scheduling, charging cost minimization, optimal utilization of renewable energy, and optimization of PV and battery unit size. Moreover, the vehicle-to-grid reactive and active power exchange, power factor correction, and grid ancillary services provider operation of the charger are less explored. Therefore, this chapter presents a PV-integrated multifunctional off-board bidirectional EV battery charger that not only charges the EV batteries but also provides grid ancillary services like reactive power compensation and grid current harmonic compensation.

In [17], the proposed EV charger is designed to provide reactive power support (RPS) to the grid. However, to achieve uninterruptible RPS, the DC link voltage is regulated by EV batteries. Hence, the batteries are exposed to unwanted DC link ripples that affect its life. Moreover, it goes under more

DOI: 10.1201/9781003321897-20

charging and discharging cycles that reduce battery life. Similarly, in [18, 19], the charger is used to serve the grid with reactive power support. However, using EV batteries to regulate DC link voltage reduces its performance and life. In this chapter, the DC link voltage is regulated by using the grid instead of EV batteries. The application of the EV charger to mitigate grid current harmonics by operating as an active power filter is introduced in [20]. Further, in [20, 21, 22] and [23], the study is extended by developing new control strategies and topologies for achieving harmonic compensation by using an EV charger. To improve the harmonic compensation capability of the charger, a multilevel (three-level) converter is used in charger topology [21]. The multilevel converters have additional components as compared to conventional converters but have low component stress and losses. Furthermore, it has high efficiency, reduced size and switching frequency, and smaller and inexpensive filter and is suitable for high-power EV chargers [24]. In [22], during harmonic compensation, the EV batteries are used to sustain the DC link voltage, and the proposed control algorithm uses a phase-locked loop (PLL) for grid synchronization. However, the use of PLL in grid voltage phase estimation leads to slower transient response. In [23], second-order generalized integrator (SOGI) is used instead of PLL to improve transient performance. However, the dynamic performance of the SOGI during severe phase change results is larger transient in frequency estimation. The EV charger control function is developed only to serve the grid by providing harmonic compensation but not able to charge/discharge the EV along with providing this service.

This chapter presents a PV array and grid-integrated multifunctional off-board bidirectional EV battery charger that not only charges the EV batteries but also supplies the residential load simultaneously in a grid-connected and islanding operation. The proposed EV charger consists of a front-end multilevel bidirectional AC–DC converter and a back-end DC–DC bidirectional converter. The proposed EV charger has two operating modes; first, in the absence of a grid, it operates as an islanding PV system, and the second is grid-connected operation. In grid-connected operation, the charger controller enables discharging of EV-stored energy to the grid for revenue generation. Furthermore, the proposed charger control algorithm enables the EV charger to provide grid ancillary services, like reactive power and harmonic compensation. In islanding operation, the charger supplies the residential load in the vehicle-to-home mode. For achieving this, a resonant filter–based active component extraction (RF-ACE) control algorithm is used. Furthermore, for a seamless transition from islanding mode to grid-connected mode, adaptive notch filter (ANF) is used to synchronize the point of common coupling voltage with the grid voltage. The performance of the proposed control algorithm is thoroughly studied and evaluated in MATLAB/Simulink. The obtained results demonstrate the effectiveness of the proposed algorithm for achieving satisfactory operation of the charger during both islanding and grid-connected mode.

20.2 EV BATTERY CHARGER CONFIGURATION

The off-board EV battery charger configuration is presented in Figure 20.1. The charger consists of a front-end multilevel cascaded H-bridge bidirectional converter (CHBDC) which synchronizes the point of common coupling (PCC) voltage with the grid. The CHBDC operates the charger in both islanding and grid-connected operation. Furthermore, it is used to supply harmonic current and reactive current demand of the load.

The CHBDC topology includes a toroidal core transformer which provides galvanic isolation. Furthermore, in the toroidal core transformer facilitates, the converter works with a single DC excitation voltage. Therefore, the use of voltage matching sensors is not required to maintain equal power distribution among the modules [25]. In each phase, three H-bridge modules are attached in a cascaded manner with the cascaded transformer connected to add up the voltage to produce the desired voltage level. The transformer secondary winding in each phase is joined in cascaded and fed to the PCC through a filter inductor. The added advantages with the presented H-bridge configuration are very low charging current ripples, both current and voltage control capabilities, and provision of galvanic isolation [25]. The grid-facing CHBDC is coupled to the power grid through the L-filter to improve the converter output voltage quality. The multilevel structure requires additional component and control circuitry, which increases cost and complexity. However, lower component stress and losses improve converter efficiency. Furthermore, multilevel structure leads to improved performance of charger in harmonic compensation operation with small and inexpensive filter. The back-end bidirectional DC–DC converter (BBDC) is used to control the charging/discharging of the EV batteries. For fast charging, constant current (CC) and constant voltage (CV) charging of EV batteries is carried out by controlling BBDC. A boost converter is connected at the PV unit output to achieve optimal operation of PV unit. The charger is connected to the grid through an inductor filter and power electronic switch. The power electronic switch is controlled by synchronizing pulse command S.

20.3 CONTROL STRATEGY

The charger control algorithm is aimed at the efficient operation of the charger in grid-connected and islanding conditions. Furthermore, the proposed controller enables the charger to provide load reactive and harmonic current in grid-connected operation. The controller strategy adopted in this chapter is presented in Figure 20.2. The control algorithm is designed by considering different scenarios in both grid-connected and islanding conditions.

20.3.1 Islanding Operation

In islanding operation, the charger controller is designed to charge the EV in CC/CV mode and supply the residential load by taking the energy from the PV unit. As the residential

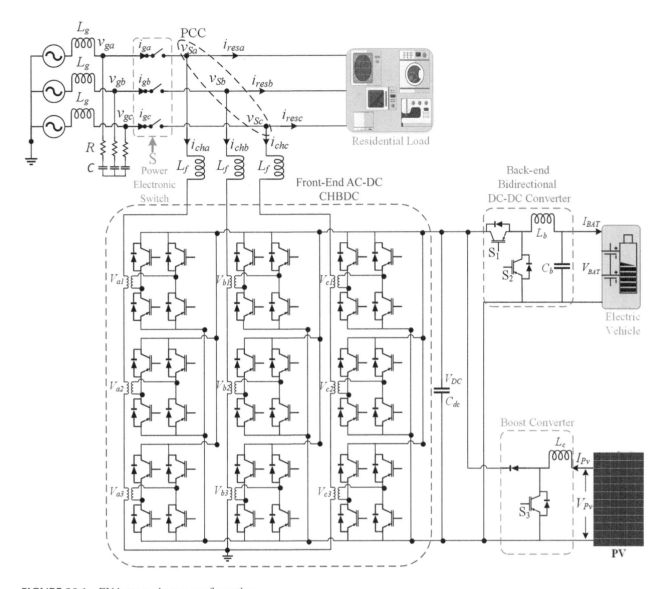

FIGURE 20.1 EV battery charger configuration.

load is an AC nonlinear load, the front-end CHBDC works as an inverter. The control algorithm is developed by considering the intermittent nature of the solar generating unit and discussed in the following.

Case I: When PV generation is more than or equal to the sum of residential load requirement and EV charging power, $(P_{PV} \geq P_{ev} + P_{res})$.

In this case, the PV generation power is enough to supply residential load along with charging the EV batteries. The BBDC is controlled to control the charging current of the EV. This chapter adopts the constant voltage (CV) and constant current (CC) charging of EV batteries. In the initial state of charging, the reference battery charging current is set to proper power level under constant current until battery voltage touches the rated permitted voltage level fixed by the manufacturer. After that, the battery charging is carried out at a maximum voltage level with the decreasing current until the current reached its rated cutoff level and battery voltage reaching its maximum level. During CC/CV charging, the BBDC works as a buck converter by controlling the switch S_1 to regulate battery charging current (I_{BAT}) and voltage (V_{BAT}). The control block diagram of BBDC during the EV battery charging is presented in Figure 20.2. The variables are the same as in Figure 20.1.

The charger topology does not have auxiliary battery storage to regulate DC link voltage. Therefore, the boost converter of the PV unit is used to regulate the DC link voltage by controlling the duty of S_3, as shown in Figure 20.2. During operation, the PV unit generates power following the combined demand of residential load and EV charging requirements. Hence, the maximum power point (MPP) of PV unit is governed by the loading conditions. Whenever PV power generation is not enough

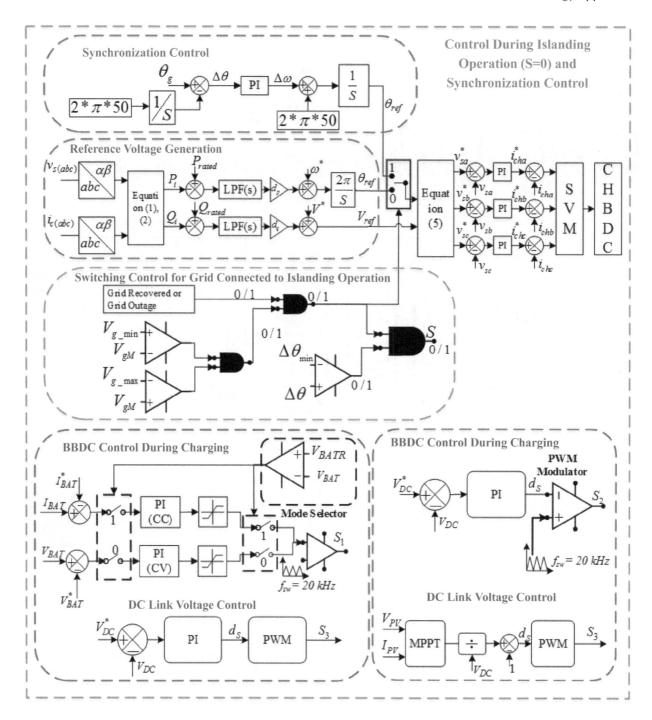

FIGURE 20.2 Charger control during islanding operation with the synchronization control.

to supply the residential load demand, the load is reduced to maintain the adequate operation of the charger.

Case II: When PV generation is less than the residential load demand, $(P_{PV} < P_{res})$.

In this case, the EV charger discharges the EV battery–stored energy to fulfill the residential load demand. The BBDC operates in boost mode to control discharging current (I_{BAT}) and voltage (V_{BAT}) by controlling the switching of S_2, as depicted in Figure 20.2. Furthermore, the DC link voltage is also controlled by BBDC. The boost converter draws maximum power available in the PV unit and operates in MPP by controlling the switching of the boost converter. The MPPT algorithm used here is adapted from [26].

In both cases, the CHBDC uses the PV input power to supply the AC nonlinear residential load. In this chapter, real power–frequency (P–f) control and reactive power–voltage (Q–V) control is adopted for CHBDC control. The instantaneous real and reactive power is calculated as given thus.

$$P_i = \frac{1}{2}\left[\left(v_{s\alpha} * i_{c\beta}\right) + \left(v_{s\beta} * i_{c\alpha}\right)\right] \quad (20.1)$$

$$Q_i = \frac{1}{2}\left[\left(v_{s\beta} * i_{c\alpha}\right) - \left(v_{s\alpha} * i_{c\beta}\right)\right] \quad (20.2)$$

Where $\left(v_{s\alpha}, v_{s\beta}\right)$ and $(i_{s\alpha}, i_{s\beta})$ are the $\alpha\beta$ transformation of $v_{s(abc)}$ and $i_{c(abc)}$, respectively.

Using the instantaneous power and P–f control and Q–V control, the reference sinusoidal voltages are estimated as shown in Figure 20.2. The mathematical representation to calculate the phase and voltage magnitude of the reference voltage is given as:

$$\omega_{ref} = \omega^* + d_p * LPF(s) * \left(P_{rated} - P_i\right) \quad (20.3)$$

$$V_r = V^* + d_q * LPF(s) * \left(Q_{rated} - Q_i\right) \quad (20.4)$$

With the reference phase θ_{ref} calculated by integrating reference frequency and reference magnitude, the reference sinusoidal voltages are obtained as:

$$v_{sa}^* = V_{ref}\sin(\theta_{ref}), v_{sb}^* = V_{ref}\sin(\theta_{ref} - 2\pi/3), v_{sc}^* = V_{ref}\sin(\theta_{ref} + 2\pi/3) \quad (20.5)$$

The measured voltages (v_{sa}, v_{sb}, v_{sc}) are compared with reference voltages $(v_{sa}^*, v_{sb}^*, v_{sc}^*)$, and a PI controller generates the reference current as shown in Figure 20.2. With the actual current and generated reference current, space vector modulation (SVM) generates gating pulses for CHBDC in islanding operation.

During islanding operation, the solar PV generation may or may not always fulfill the load demand. Furthermore, the EV owner may generate some revenue by selling the EV-stored energy to the grid. In both cases, the PCC voltages need to be synchronized with the grid voltages to avoid transients during power exchange. Therefore, the controller estimates grid voltage phases by using adaptive notch filter (ANF) and PCC voltage phases by using P–f control. The error between the phases is minimized by a PI controller which generates a frequency correction factor $\Delta\omega$. After getting the $\Delta\omega$ the changed reference frequency for sinusoidal reference voltage generation is changed to:

$$\omega_n = \omega_{ref} + \Delta\omega \quad (20.6)$$

Now, with changed reference frequency, the generated reference voltage minimizes the phase error. When the phase error reaches the minimum phase error requirement for grid synchronization stated in IEEE Standard 1547, the synchronizing switch gated (S = 1).

20.3.2 Grid-Connected Operation

In grid-connected operation, the charger provides load-reactive and harmonic current to make grid current sinusoidal and maintain UPF operation. In general, the residential load is highly nonlinear and draws harmonic current from the grid. The charger controller provides load harmonics and reactive current by using CHBDC to improve the grid power quality. To achieve this, the active load current component is extracted by using the resonant filter–based active component estimation (RF-ACE) technique. The resonant filter extracts the fundamental current component from the residential nonlinear load current. The transfer function of the resonant filter is given in equation 20.7. The performance of the resonant filter completely depends on the selection of w_{bp}. Bode plot of the resonant filter for different values of w_{bp} is plotted in Figure 20.3. The magnitude and phase plot shows the filter imposes zero phase shift and unit magnitude at the fundamental grid frequency $f^* = 50$ Hz.

$$\text{RF}(s) = \frac{2 * w_{bp} * s}{s^2 + (2 * w_{bp} * s) + (2 * \pi * f^*)^2} \quad (20.7)$$

Furthermore, it can be observed that the smaller values of w_{bp} ensure better selectivity of the filter. On the other hand, the larger values of w_{bp} ensure improved dynamic response of the filter. So the selection of the value of w_{bp} is a trade-off between dynamic response and selectivity of the filter. Here, the value of w_{bp} is selected to be 40.

After getting the fundamental current component, the active component can be estimated by sample and hold the fundamental current component at zero crossings of the 90° shifted synchronizing voltage template. This chapter adopts ANF-based synchronizing voltage template generation technique as depicted in Figure 20.4. The ANF for grid synchronization is first proposed in [27]. The ANF works effectively irrespective of the system disturbances as compared to the conventional PLL.

The dynamic equations of the ANF [28] are expressed as:

$$\ddot{x} + \theta^2 x = 2\zeta\theta e(t)$$
$$\dot{\theta} = -\gamma x \theta e(t)$$
$$e(t) = u(t) - \dot{x} \quad (20.8)$$

Where $u(t)$ is the input signal, θ is the estimated frequency of the input signal, and ζ and γ are two real positive constants. The selections ζ and γ decide the speed and estimation accuracy of the ANF. For fundamental component frequency ω_1 with amplitude U_1, the ANF has a unique periodic orbit located at [28]:

$$\begin{pmatrix} x \\ \dot{x} \\ \theta \end{pmatrix} = \begin{pmatrix} -\dfrac{U_1}{\omega_1}\cos(\omega_1 t + \varphi_1) \\ U_1 \sin(\omega_1 t + \varphi_1) \\ \omega_1 \end{pmatrix} \quad (20.9)$$

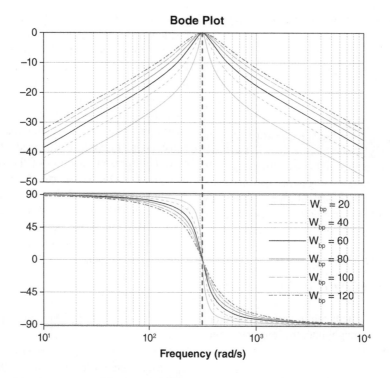

FIGURE 20.3 Frequency domain response of the resonant filter.

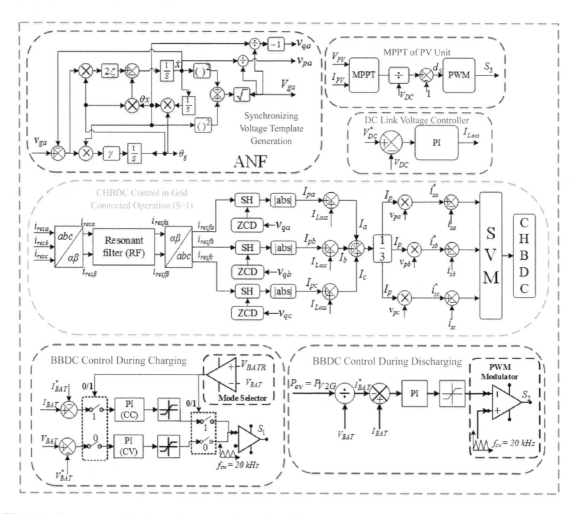

FIGURE 20.4 Charger control during grid-connected operation with ANF-based synchronization.

With this, the in-phase synchronizing voltage template (v_p) and 90° shifted synchronizing voltage template (v_q) can be estimated as:

$$U_1 = \sqrt{(\theta x)^2 + (\dot{x})^2}$$
$$v_p = \frac{\dot{x}}{U_1}, v_q = -\frac{\theta x}{U_1} \quad (20.10)$$

The detailed ANF structure is presented in Figure 20.4 to estimate v_{pa} and v_{qa} from the a-phase grid voltage v_{ga}. Similarly, the synchronizing voltage templates for b-phase and c-phase can be estimated by using ANF. After estimating each phase maximum voltage magnitude V_{ga}, V_{gb}, V_{gc}, the amplitude of the grid voltage can be calculated by using equation (20.11).

$$V_{gM} = \sqrt{\frac{(V_{ga} * v_{pa})^2 + (V_{gb} * v_{pb})^2 + (V_{gc} * v_{pc})^2}{3/2}} \quad (20.11)$$

With the fundamental frequency current component and 90° shifted synchronizing voltage template (v_q), the active component of each phase I_{pa}, I_{pb} and I_{pc} can be estimated as given in Figure 20.4. In grid-connected operation, the CHBDC regulates the DC link voltage. Furthermore, the losses incurred in the charging/discharging of the DC link capacitor are supplied by the grid. The current corresponding to the losses I_{Loss} is estimated by using a PI controller, as shown in Figure 20.4. With this, the active component of each phase is given as:

$$I_a = I_{pa} + I_{Loss}, I_b = I_{pb} + I_{Loss}, I_c = I_{pc} + I_{Loss} \quad (20.12)$$

Using equation (20.12), the active current component per phase can be calculated as given thus:

$$I_p = \frac{I_a + I_b + I_c}{3} \quad (20.13)$$

Now, the sinusoidal current that corresponds to the active component can be estimated as:

$$i_{pa} = I_p * v_{pa}, i_{pb} = I_p * v_{pb}, i_{pc} = I_p * v_{pc} \quad (20.14)$$

Using sinusoidal current that corresponds to the active component i_{pa}, i_{pb}, i_{pc}, the reference CHBDC currents are estimated as:

$$i^*_{ca} = i_{ga} - i_{pa}, i^*_{cb} = i_{gb} - i_{pb}, i^*_{cc} = i_{gc} - i_{pc} \quad (20.15)$$

The CHBDC reference currents $i^*_{ca}, i^*_{cb}, i^*_{cc}$ are compared with the measured output currents i_{ca}, i_{cb}, i_{cc}, and the error signal fed to the space vector modulation to generate gating pulses for CHBDC.

20.4 SIMULATION RESULTS AND DISCUSSION

In this chapter, a MATLAB simulation model of the proposed EV battery charger is designed to verify its steady-state and transient performance with the proposed control technique. The parameter specification used to design the charger simulation model is listed in Table 20.1. The steady-state performance of the charger during the islanding operation is depicted in Figure 20.5. The PCC voltages and nonlinear residential load currents are harmonically distorted with %THD of 3.68% and 26.32%, respectively. The PV unit supplies the residential load and simultaneously provides EV charging current.

In this case, the total load demand, including residential load, and EV charging are less than the PV generation; hence, it operates in MPP. The steady-state performance of the charger in grid-connected mode is depicted in Figure 20.6. In this operation, the PV unit works at MPP at all times, and the sudden change in load demand is supplied by the grid. The grid and the PV unit share proportionally the residential load demand. At 0.1S, the charger compensates the grid current harmonics by providing harmonic current (i_h), as shown in Figure 20.6. The distorted grid voltage quality

TABLE 20.1
Simulation Parameter Specifications

Parameters	Specifications
Charger apparent power	12.6 kVA
CHBDC filter	$L_f = 2.5$ mH (25 A)
BBDC elements	$L_b = 3.7$ mH, $C_b = 660\,\mu F$
Grid impedance (Z_s)	$R_s = 0.1\,\Omega$, $L_s = 1.6\,mH$,
DC link capacitor (C_{DC})	2,200 μF/500 V
Transformer (CHBDC)	1 kVA, 1-ϕ, toroidal core
Supply system	230 V rms, 50 Hz
EV battery	Nominal voltage = 240 V, 35 Ah, $V_{BATR} = 0.9 \times 240$ V
Transformer (front-end converter)	1 KVA, 1-phase toroidal core transformer
Nonlinear load	Diode bridge rectifier with R–L load

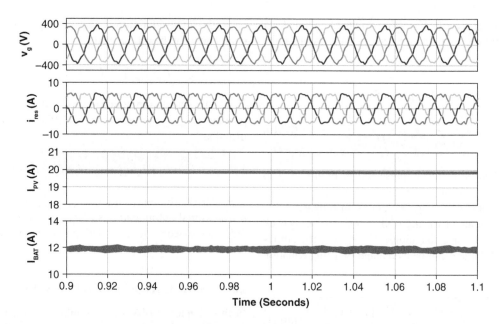

FIGURE 20.5 Steady-state performance of the charger in islanding operation.

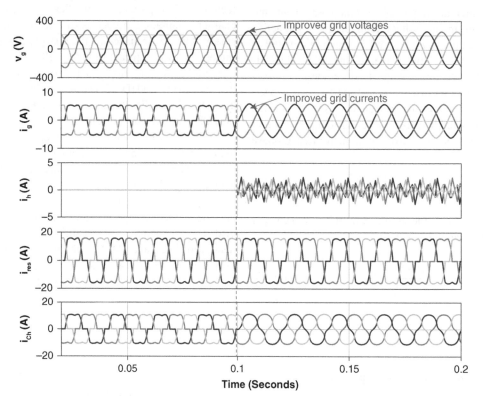

FIGURE 20.6 Steady-state performance of the charger in grid-connected operation.

also improved due to reduced harmonic voltage drop at grid impedances. The grid voltage %THD improved from 4.67% to 2.13%, and grid current %THD improved from 24.16% to 3.24%.

Figure 20.7 shows the transient performance of the charger in islanding operation when suddenly a residential load is connected to the PV unit. Previously, the PV unit generation was only used to charge the EV batteries. At 1S, the residential load is connected to the system; hence, the PV generation increases to supply the residential load without affecting the EV charging.

The performance charger during varying solar irradiance is depicted in Figure 20.8. In all cases, the residential load is kept constant. Figure 20.8(a) shows the charger performance with the increase in solar irradiance. An increase in irradiance increases the EV charging current as the

PV-Integrated Bidirectional Off-Board EV Battery Charger

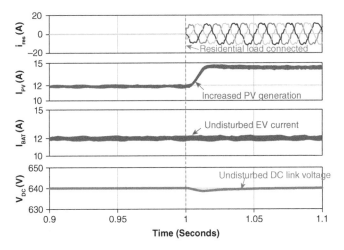

FIGURE 20.7 Transient performance of the charger in islanding operation.

residential load is kept constant. Similarly, a decrease in solar irradiance decreases EV charging current, as shown in Figure 20.8(b). Figure 20.8(c) shows the performance of the charger when solar irradiance becomes zero $(P_{PV} = 0)$. In this case, the EV discharges to supply the residential load, as shown in Figure 20.8(c). The EV current after PV generation becomes zero goes negative to discharge the EV batteries and supply the residential load.

In grid-connected operation, the PV unit operates in MPP at all times. Hence, the total residential load is shared by the grid and the PV unit during a load change. Initially, the grid and PV generations are combined to supply the residential load, as shown in Figure 20.9. At 0.05S, the charger provides harmonic current (i_h) to improve the grid voltage and current quality. The grid voltage %THD improved to 2.31% from 4.98%, and grid current %THD improved to 3.23% from 25.63%. During this operation, the residential

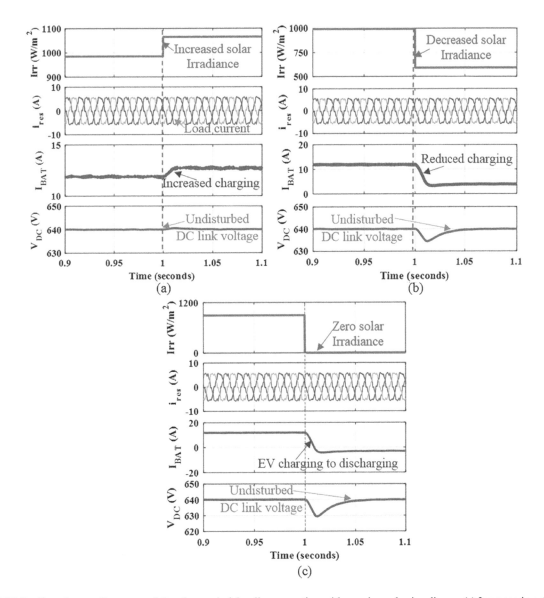

FIGURE 20.8 Transient performance of the charger in islanding operation with varying solar irradiance: (a) Increase in solar irradiance, (b) decrease in solar irradiance, (c) decrease in solar irradiance to zero.

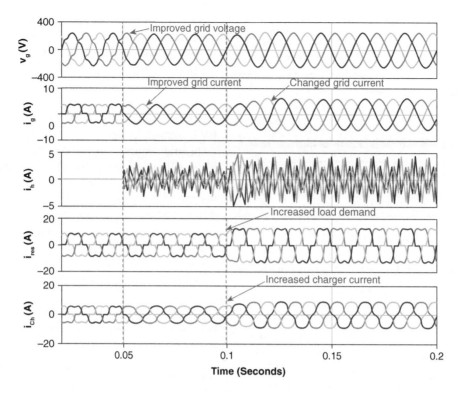

FIGURE 20.9 Performance of the charger in grid-connected operation with a sudden increase in residential load.

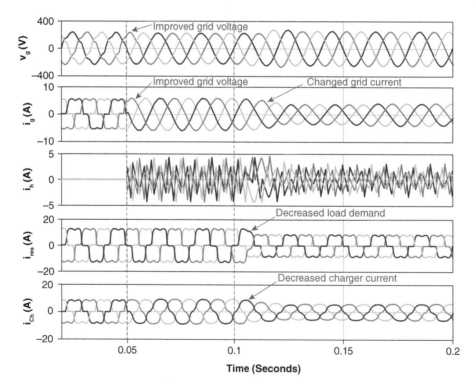

FIGURE 20.10 Performance of the charger in grid-connected operation with a sudden decrease in residential load.

load is increased at 0.1S. With an increase in load demand, the grid current increases to supply additional load requirement, and the charger provides the harmonic compensation to the grid current, as shown in Figure 20.9. Similarly, the dynamic performance of the charger with a sudden decrease in residential load is depicted in Figure 20.10. With a decrease in load demand, the grid current is reduced to maintain power balance, as shown in Figure 20.10.

PV-Integrated Bidirectional Off-Board EV Battery Charger

In the absence of the PV generation, the charger draws charging current from the grid and the residential load fully supplied by the grid. Furthermore, the charger can discharge the EV-stored energy to the grid and operates in the vehicle-to-grid mode. In both cases, the CHBDC and BBDC control the charging and discharging current and maintain UPF at PCC. The performance of the charger during EV charging in the absence of PV generation is depicted in Figure 20.11. At first, the residential load demand and EV charging current are supplied by the grid. The charger current (i_{ch}) is sinusoidal and in phase with PCC voltage. The grid current (i_g) is harmonically distorted due to nonlinear residential load. At 0.3S the charger starts grid current harmonic compensation operation by providing load harmonic current. With this, the grid current %THD improved from 25.12% to 3.21%. The charger current is harmonically distorted as it supplies harmonic current to improve grid current power quality. Figure 20.12 shows

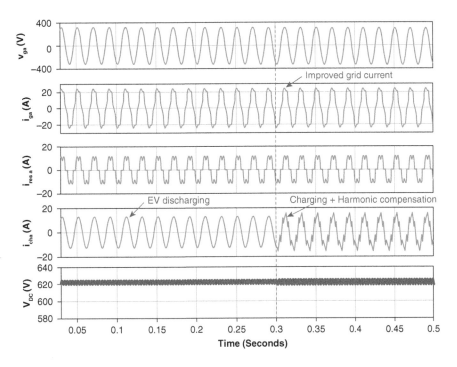

FIGURE 20.11 Performance of the charger in grid-connected operation with $P_{PV} = 0$ and EV charging.

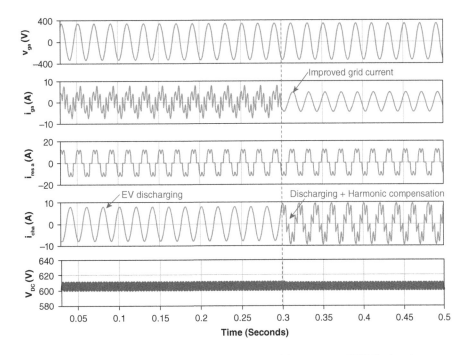

FIGURE 20.12 Performance of the charger in grid-connected operation with $P_{PV} = 0$ and EV discharging.

the performance of the charger during EV discharging along with grid current harmonic compensation. At first, the charger discharges the EV batteries to supply the grid; hence, the charger currents are out of phase with PCC voltage. At 0.3S, the charger delivers harmonic current to improve the grid current power quality. With this, the grid current %THD improves from 26.14% to 4.11%.

20.5 CONCLUSION

In this chapter, a three-phase PV-integrated off-board multifunctional EV battery charger is presented. The performance of the charger in both grid-connected and islanding operations for EV charging and supplying the residential load is validated by test results. The obtained results have verified the ability of the charger to provide grid current harmonic compensation and improve PCC voltage quality. In islanding operation, the charger controller controls the PV unit boost converter and BBDC for uninterrupted supply to the residential load. In grid-connected operation, the proposed RF-ACE effectively estimates active load current component from residential load current. The obtained results in grid-connected operation have verified that the charger always maintains grid voltage and current power quality as per IEEE Standard 519. Moreover, the performance of the charger in both transient and steady-state conditions verify the effectiveness of the proposed controller and its stability. The use of ANF minimizes the system transients during grid synchronization. Furthermore, the obtained results have validated the stable operation of the charger during different dynamic conditions. Therefore, the presented charger is a favorable candidate for EV charging, uninterruptible supply to the residential load, grid voltage correction, improving grid current power quality, and stability.

20.6 REFERENCES

[1] International Energy Agency- Renewables 2O18 – Analysis and Forecasts to 2023. [Online]. Available: https://webstore.iea.org/download/summary/2312?fileName=English-Renewables-2018-ES.pdf.

[2] M. Popescu, J. Goss, D. A. Staton, D. Hawkins, Y. C. Chong and A. Boglietti, "Electrical Vehicles – Practical Solutions for Power Traction Motor Systems," *IEEE Trans. Ind. Applicat.*, vol. 54, no. 3, pp. 2751–2762, May–June 2018.

[3] N. Chang, M. A. A. Faruque, Z. Shao, C. J. Xue, Y. Chen and D. Baek, "Survey of Low-Power Electric Vehicles: A Design Automation Perspective," *IEEE Design & Test*, Early Access.

[4] International Energy Agency-World Energy Outlook 2017. [Online]. Available: www.iea.org/weo2017/

[5] P. M. de Quevedo, G. Muñoz-Delgado and J. Contreras, "Impact of Electric Vehicles on the Expansion Planning of Distribution Systems considering Renewable Energy, Storage and Charging Stations," *IEEE Trans. Smart Grid*, Early Access

[6] A. Verma and B. Singh, "A Solar PV, BES, Grid and DG Set Based Hybrid Charging Station for Uninterruptible Charging at Minimized Charging Cost," in *IEEE Ind. Applicat. Soc. Annual Meeting (IAS)*, Portland, OR, 2018, pp. 1–8.

[7] V. Monteiro, J. G. Pinto and J. L. Afonso, "Experimental Validation of a Three-Port Integrated Topology to Interface Electric Vehicles and Renewables with the Electrical Grid," *IEEE Trans. Ind. Informat.*, vol. 14, no. 6, pp. 2364–2374, June 2018.

[8] M. S. Islam, N. Mithulananthan and K. Y. Lee, "Suitability of PV and Battery Storage in EV Charging at Business Premises," *IEEE Trans. Power Syst.*, vol. 33, no. 4, pp. 4382–4396, July 2018.

[9] W. Alharbi and K. Bhattacharya, "Flexibility Provisions from a Fast Charging Facility Equipped with DERs for Wind Integrated Grids," *IEEE Trans. Sustainable Energy*, Early Access.

[10] A. Tavakoli, M. Negnevitsky, D. T. Nguyen and K. M. Muttaqi, "Energy Exchange Between Electric Vehicle Load and Wind Generating Utilities," *IEEE Trans. Power Sys*, vol. 31, no. 2, pp. 1248–1258, 2016.

[11] K. Khalid Mehmood, S. U. Khan, S. Lee, Z. M. Haider, M. K. Rafique and C. Kim, "Optimal Sizing and Allocation of Battery Energy Storage Systems with Wind and Solar Power DGs in a Distribution Network for Voltage Regulation Considering the Lifespan of Batteries," *IET Renewable Power Generation*, vol. 11, no. 10, pp. 1305–1315, 2017.

[12] N. Saxena, I. Hussain, B. Singh and A. L. Vyas, "Implementation of a Grid-Integrated PV-Battery System for Residential and Electrical Vehicle Applications," *IEEE Trans. Ind. Electron.*, vol. 65, no. 8, pp. 6592–6601, Aug. 2018.

[13] Y. Zhang and L. Cai, "Dynamic Charging Scheduling for EV Parking Lots With Photovoltaic Power System," *IEEE Access*, vol. 6, pp. 56995–57005, 2018.

[14] Q. Yan, B. Zhang and M. Kezunovic, "Optimized Operational Cost Reduction for an EV Charging Station Integrated with Battery Energy Storage and PV Generation," *IEEE Trans. Smart Grid,* Early Access.

[15] G. R. Chandra Mouli, J. Schijffelen, M. van den Heuvel, M. Kardolus and P. Bauer, "A 10kW Solar-Powered Bidirectional EV Charger Compatible with Chademo and COMBO," *IEEE Trans. Power Electron.*, Early Access.

[16] K. Chaudhari, A. Ukil, K. N. Kumar, U. Manandhar and S. K. Kollimalla, "Hybrid Optimization for Economic Deployment of ESS in PV-Integrated EV Charging Stations," *IEEE Trans. Ind. Informat.*, vol. 14, no. 1, pp. 106–116, Jan. 2018.

[17] U. R. Prasanna, A. K. Singh and K. Rajashekara, "Novel Bidirectional Single-phase Single-Stage Isolated AC – DC Converter With PFC for Charging of Electric Vehicles," *IEEE Transactions on Transportation Electrification*, vol. 3, no. 3, pp. 536–544, Sept. 2017.

[18] D. B. Wickramasinghe Abeywardana, P. Acuna, B. Hredzak, R. P. Aguilera and V. G. Agelidis, "Single-Phase Boost Inverter-Based Electric Vehicle Charger With Integrated Vehicle to Grid Reactive Power Compensation," *IEEE Transactions on Power Electronics*, vol. 33, no. 4, pp. 3462–3471, April 2018.

[19] M. Restrepo, J. Morris, M. Kazerani and C. A. Cañizares, "Modeling and Testing of a Bidirectional Smart Charger for Distribution System EV Integration," *IEEE Transactions on Smart Grid*, vol. 9, no. 1, pp. 152–162, Jan. 2018.

[20] M. C. B. P. Rodrigues, H. J. Schettino, A. A. Ferreira, P. G. Barbosa and H. A. C. Braga, "Active Power Filter Operation of an Electric Vehicle Applied to Single-phase Networks," *2012 10th IEEE/IAS International Conference on Industry Applications*, Fortaleza, Brazil, 2012, pp. 1–8.

[21] A. Mortezaei, M. Abdul-Hak and M. G. Simoes, "A Bidirectional NPC-based Level 3 EV Charging System with Added Active

Filter Functionality in Smart Grid Applications," *2018 IEEE Transportation Electrification Conference and Expo (ITEC)*, Long Beach, CA, USA, 2018, pp. 201–206.

[22] V. Monteiro, J. G. Pinto and J. L. Afonso, "Operation Modes for the Electric Vehicle in Smart Grids and Smart Homes: Present and Proposed Modes," *IEEE Transactions on Vehicular Technology*, vol. 65, no. 3, pp. 1007–1020, March 2016.

[23] A. Verma and B. Singh, "Multi-Objective Reconfigurable Three-Phase Off-Board Charger for EV," *IEEE Transactions on Industry Applications*, vol. 55, no. 4, pp. 4192–4203, July-Aug. 2019.

[24] M. Yilmaz and P. T. Krein, "Review of Battery Charger Topologies, Charging Power Levels, and Infrastructure for Plug-In Electric and Hybrid Vehicles," *IEEE Transactions on Power Electronics*, vol. 28, no. 5, pp. 2151–2169, May 2013.

[25] A. R. Dash, A. K. Panda, R. K. Lenka, and R. Patel, "Performance Analysis of a Multilevel Inverter Based Shunt Active Filter with RT-EMD Control Technique under Ideal and Non-ideal Supply Voltage Conditions," *IET Generation, Transmission & Distribution*, vol. 13, no. 18, pp. 4037–4048, 2019.

[26] F. Liu, S. Duan, F. Liu, B. Liu and Y. Kang, "A Variable Step Size INC MPPT Method for PV Systems," *IEEE Trans. Ind. Electron.*, vol. 55, no. 7, pp. 2622–2628, July 2008.

[27] M. Mojiri and A. R. Bakhshai, "An Adaptive Notch Filter for Frequency Estimation of a Periodic Signal," *IEEE Transactions on Automatic Control*, vol. 49, no. 2, pp. 314–318, Feb. 2004.

[28] M. Mojiri, M. Karimi-Ghartemani and A. Bakhshai, "Time-Domain Signal Analysis Using Adaptive Notch Filter," *IEEE Transactions on Signal Processing*, vol. 55, no. 1, pp. 85–93, Jan. 2007.

21 Integration of Wind, Solar, and Pumped Hydro Renewable Energy Sources in Rayalaseema Region
A Case Study

Y. V. Siva Reddy, T. Bramhananda Reddy, and E. Sanjeeva Rayudu

CONTENTS

21.1 Introduction 281
21.2 Present Generation Scenario in Andhra Pradesh 281
21.3 Solar Power 281
21.4 Wind Power 283
21.5 Pumped Hydroelectric Storage (PHES) Plants 284
    21.5.1 Pumped Hydroelectric Storage in Rayalaseema 284
    21.5.2 Need for Pumped Storage and Integration 284
21.6 Need for Integration of Renewable Energy Sources in Rayalaseema Region 285
21.7 Case Study 285
21.8 Conclusions 285
21.9 References 285

21.1 INTRODUCTION

The per capita growth of any country is better explained by the per capita energy consumption of that country. The Indian economy is growing at a rapid pace due to increasing industrialization, development of irrigation facilities, urbanization, and well-resourced infrastructure in villages and the like. Hence, demand for power, which is one of the basic inputs for the economic growth of a country, is therefore on the rise in India. The chief sources of energy which are utilized for generation of electricity are fuel in all forms, that is, solid, liquid, and gaseous; water energy; and nuclear energy. The other sources of energy are sun (solar, photovoltaic, etc.), wind, and tides. With the decline in fossil fuels, renewable sources like wind and solar are being extensively tapped to meet energy requirements [1]. Owing to the intermittent nature of these sources, it is mandatory to have some other form of energy integrated with these energy generation so as to meet the daily load cycle. This book chapter will explore the possibilities of setting up such storage in the form of pumped hydro storage in Rayalaseema region, where solar and wind are abundantly available.

21.2 PRESENT GENERATION SCENARIO IN ANDHRA PRADESH

Andhra Pradesh, located centrally in the peninsular part of India, plays a major role in the operation of the southern grid. Before the bifurcation of the state, majority of the energy generation was done through thermal and hydro. But after bifurcation, the share of hydro energy had decreased, and moreover, the major hydro energy production is done through the Srisailam plant; due to scarcity of water for irrigation, water is not used for power production and is used only for agriculture. Though there is a large establishment of gas-powered plants, it is unable to generate required power due to the nonavailability of fuel. The state does not have coal fields of its own and imports coal from other states. This has led to the establishment of wind and hydro to a large extent in this region. The Rayalaseema region comprises Ananthapur, Chittor, Kadapa, and Kurnool Districts and is located in the southern part of the state.

By November 2020, Andhra Pradesh is the sixth largest power-generating state in the country. At present, the total installed generation capacity is 20,334 MW. Out of it, 6,160 MW is from thermal, 1,798 MW is from hydro, 4,055 MW is through gas and CGS, and 8,321 MW is from renewable energy sources (wind and solar forming a major portion of it). The following shows the generation mix of the state:

The peak demand for electricity of Andhra Pradesh has been on the rise at a fast pace, and for the year 2021–2022, it has been estimated as 11,843 MW. Though hydro power provides an environment-friendly peaking power, only 1,673.6 MW of hydro power capacity has been installed in AP so far.

21.3 SOLAR POWER

The state is potentially rich in photovoltaic power on its marginally productive lands. The state has a total

Map of currently operating and proposed solar power plants in the Indian state of Andhra Pradesh. Green are currently operating and blue are proposed.

Map of currently operating and proposed wind power plants[81] in the Indian state of Andhra Pradesh

FIGURE 21.1 Maps showing the location of the existing/proposed solar and wind plants.

Source: Courtesy of NREDCAP.

TABLE 21.1
Generation Mix in the State of Andhra Pradesh

Source	2014–2015	2015–2016	2016–2017	2017–2018	2018–2019	2019–2020*
Hydel-Genco	1,671	1,671	1,671	1,783	1,798	1,798
Thermal-Genco	3,148	3,378	3,948	4,321	5,010	5,010
CGS	2,017	2,100	1,801	2,331	2,403	2,530
Gas	1,205	1,343	1,621	1,599	1,525	1,525
Thermal-IPP	184	1,455	1,626	1,671	1,671	1,150
Subtotal	**8,225**	**9,948**	**10,667**	**11,705**	**12,407**	**12,013**
Wind	1,055	1,453	3,802	4,162	4,179	4,080
Solar	144	579	1,880	2,148	3,029	3,649
Other	451	553	558	585	585	5,92
Subtotal	**1,650**	**2,585**	**6,240**	**6,895**	**7,793**	**8,321**

TABLE 21.2
Installed Capacity Versus Peak Demand in the State of Andhra Pradesh

Year	Installed Capacity	Peak Demand	Q3	Q2	Q1	Base Demand
2016–2017	16,653	7,965	6,648	6,202	5,681	3,800
2017–2018	18,046	8,983	7,061	6,521	5,956	4,535
2018–2019	19,160	9,453	7,819	7,266	6,652	3,888
2019–2020*	20,334	10,207	10,207	9,118	10,170	4,000

Date: 3-18-2020

TABLE 21.3
Existing Solar Plants in Andhra Pradesh

S. No.	Solar Plant	Capacity (MW)
1	Anantapuramu-I	1,500
2	Kadapa	1,000
3	Kurnool	1,000
4	Anantapuramu-II	500
5	Hybrid Solar and Wind	160

TABLE 21.4
Proposed Solar plants in Andhra Pradesh

S. No.	Location/Solar Park Name	Bid Capacity (MW)	Capacity Won (MW)	Final Tariff (INR/kWh)
1	Chakrayapet Ultra Mega Solar Park, Chakrayapet Mandal, YSR Kadapa District	600	600	2.48
2	Thondur Ultra Mega Solar Park, Thondur Mandal, YSR Kadapa District	400	400	2.49
3	M. Kambaladinne Ultra Mega Solar Park, Mylavaram Mandal, YSR Kadapa District	600	600	2.49
4	Pendlimarri Ultra Mega Solar Park, Pendlimarri Mandal, YSR Kadapa District	600	600	2.49
5	Rudrasamudram Ultra Mega Solar Park, Donakonda Mandal, Prakasam District	600	600	2.49
6	CS Puram Ultra Mega Solar Park, CS Puram Mandal, Prakasam District	600	600	2.58
7	Uruchintala Ultra Mega Solar Park, Tadipatri Mandal, Anantapuram District	600	600	2.48
8	Kambadur Ultra Mega Solar Park, Kambadur Mandal, Anantapuram District	300	300	2.47
		600	600	2.48
		600	300	2.49
9	Mudigubba Ultra Mega Solar Park, Mudigubba Mandal, Anantapuram District	600	600	2.49
10	Kolimigundla Ultra Mega Solar Park, Kolimigundla Mandal, Kurnool District	600	600	2.48

installed photovoltaic power capacity of 3,750 MW as of December 2020. The Andhra Pradesh government has proposed to set up its own generating units to produce 10,000 MW of solar power to meet the energy requirements of the agriculture sector.

The state is on the anvil to add 10,000 MW of solar power capacity so as to provide free power supply to the agriculture sector during daytime. Out of 10,000 MW, 6,400 MW capacity located at ten sites were gone for bidding. The average winning tariffs are Rs 2.49 per unit. The state government would provide the land on lease (its own land and farmers' land), giving state guarantee for the timely payment for the power purchased. Thus, it ensures reliable daytime power to the farmers and also reduces the burden on the government, thereby making the functioning of the sector sustainable. All the identified sites are located in the Rayalaseema region, a backward where the rainfall is very less. This is a win-win situation for the farmers, the state government, and the private players, because the farmers will get an amount of Rs 25,000/per acre for leasing heir land, the state government will get power at cheaper rates, and power companies need not purchase land (which may drastically increase the cost of installation). This low tariff of Rs 2.48 would result in savings of nearly Rs 3,800 crore in the very first year and is likely to increase over the years, considering the fact the present tariff of Rs 2.48 for the energy from the 6,400 MW solar power project would remain constant over the next 30 years, whereas the cost associated with power from some of the other sources is likely to go up [2]. The sanctioned solar plant was conceived on a BOT basis and would therefore be transferred to the state after the expiry of the PPA.

21.4 WIND POWER

Wind, a natural source, is an indirect form of solar energy. Nearly 1% of the solar radiation reaching the Earth is converted into wind energy. Generating power through wind has several advantages, like low repayment period, less operational and maintenance cost, and moreover, it produces clean energy. Due to policies of the central and state governments across the country, the energy produced in this form has tremendously increased in the last decade.

Andhra Pradesh has taken the lead in this and has created a separate nodal agency, called NREDCAP (New and Renewable Energy Development Corporation of AP), which has taken up the tasks of wind mapping and monitoring wind projects. The state of Andhra Pradesh has a large capacity of windy sites suitable for setting up of wind plants, and majority of them are located in the Rayalaseema region.

It has already commissioned around 4,080 MW of wind energy, and nearly 7,500 MW of establishment is in pipeline and is shown in Table 21.5.

21.5 PUMPED HYDROELECTRIC STORAGE (PHES) PLANTS

The pumped storage plant is a combination of two reservoirs, one at lower level, and another at higher/upper level. The energy is stored in the upper-level reservoir during off-peak hours, which may be during the day or night, when excess energy is available than the actual demand. The energy stored is then utilized during peak hours of the day, during which the cost of energy is high.

To enable this process, we need to install reversible turbines in the plant such that the generating as well as pumped mode operations can be performed satisfactorily. During this entire process of generation and pumped modes, there is a net loss in energy due to hydraulic and electrical losses in the pumped and generation modes, respectively. Hence, pumped storage plants are not 100% efficient, and the overall efficiency of the plant ranges between 70% and 80%, depending on various factors.

21.5.1 Pumped Hydroelectric Storage in Rayalaseema

The two major rivers of the Indian subcontinent of Krishna and Godavari flow in the region of Andhra Pradesh and interlinking of rivers (Godavari–Krishna in the lower region and Krishna upper region–Penna has been completed). The Krishna–Penna link is through Pothireddypadu Head Regulator, located in Kurnool District of the Rayalaseema region. This link paved the way for the establishment of nearly 25 reservoirs in the Rayalaseema region. The reservoir system and the geography of the Rayalaseema region can be better utilized for the establishment of pumped storage plants, and Table 21.6 shows the possible locations of pumped storage plants as provided by NREDCAP.

21.5.2 Need for Pumped Storage and Integration

The pumped storage system is essential because of the following facts:

1. To fully utilize the differential costs of power during peak and off peak, that is, during generating and pumped modes of operation.
2. Pumped storage plants help in maintaining frequency regulation by balancing the load.
3. It also helps in load following, wherein the generator is ready to take the load during any load disturbance.
4. It reduces the spinning reserve required and also helps in controlling the reactive when operating in generating mode and taking no load.

During peak time and also during nights, energy is produced from these plants, and during daytime, when the solar energy is abundantly available, it may be used for pumping water from lower to upper reservoir. In another way, this helps in storage of surplus available in the form wind and solar, which are available at cheaper rates.

From the previous year's data, the load duration curve of a particular day may be easily predictable, and the resources available for meeting the same may also be known in advance. Due to the large penetration of the renewable

TABLE 21.5
Proposed Wind Producers in Andhra Pradesh

1	Suzlon and Axis	3274.6
2	Suzlon Energy	209.10
3	Axis Wind Energy	105.00
4	Wind World (India)	229.60
5	Gamesa Wind Turbines	840.55
6	Greenko Wind	385.60
7	Hetero Wind Power	208.50
8	Mytra Enery	221.25
9	Regen Power Tech	229.00
10	Ecoren Energy	494.40
11	Helios Infra Tech	180.00
12	Inox Wind	168.00
13	Hero Wind Energy	240.00
14	Acciona Energia	200.00
15	Other Developers	545.34
	Total:	7,530.94

TABLE 21.6
Proposed Pumped Storage Plants in Andhra Pradesh

S. No.	Name of PHES	Power Potential in MW
1	Gandikota PHES1	28,000
2	Gandikota PHES1	600
3	Paidipalem PHES1	1,850
4	Paidipalem PHES2	2,750
5	Buggavanka	600
6	Annamayya	1,150
7	Mylavaram	14,000
8	Brahmam sagar	13,000
10	Teluguganaga Subsidiary Reservoirs	2,600
11	OWK	4,700
12	OWK	800
13	Gorakallu PHES	12,500
14	Velugodu PHES	7,800
15	Mid Pennar	2,600
16	Chitravathi	500
17	Kalyani	370

TABLE 21.7
Economic Analysis of Pumped Hydro Energy Storage Plant

S. No.	Item	Expenditure in Crores of Rupees	Income in Crores of Rupees
1	Operation and maintenance expenses (including insurance) at 3.5% of the project cost in the first year, with 4.77% escalation every year	100	
2	Pumping charges at Rs 3/KWh (government is getting solar power at Rs 2.5/KWh)	331.89	
3	Depreciation charges	100	
4	Interest on working capital	10	
5	Total income on energy sold at Rs 7.5/Kwh		663.79
6	Total	541.89	663.79
7	Income	**121.11**	
8	Interest on income at 6.5%		7.87
9	Net income	**128.98**	

energy in power sector, there is a need for regulation and reserve capacity.

The energy requirements are increasing year after year, and with the introduction of the electric vehicles, the energy requirements in the next ten will be three times of the present requirements. To meet energy shortages and to integrate solar and wind, we need to establish pumped storage plants [3].

21.6 NEED FOR INTEGRATION OF RENEWABLE ENERGY SOURCES IN RAYALASEEMA REGION

The state has commissioned nearly 8,300 MW of solar and wind energy, and nearly 98% of this is from the Rayalaseema region, as is also evident from the solar and wind maps provided in the previous section. Moreover, the proposed plants will also be established in this region only, and tenders have been finalized for 6,500 MW of solar power. With these increased amounts of wind and solar generation systems, there is a greater degree of uncertainty. The daily load cycle may be reasonably predictable as a daily trend, but with uncertain variable output from wind generation, there is a need for added regulation and reserve capacity. Increased wind and solar penetration creates a need for greater regulation capacity and faster regulation ramping capability. Hence, establishment of a battery bank or setting up any form of storage is essential [4]. But as this is very costly, we need to explore other forms of storage, and the natural storage is through pumped storage plants.

21.7 CASE STUDY

For evaluating the effect of integration of solar wind and pumped hydro, a case study is done for a 500 MW pumped storage plant in association with solar and wind plants. The operating time of the plant is 5.5 hours per day.

Installed capacity = 500 MW.

Annual energy generation @ 95% availability = 885.06 GWh (after deducting T&D losses and auxiliary usage). Energy required for pumping with 80% efficiency = 1106.325GWh.

Average cost of establishing a 500 MW pumped storage plant @Rs 3Cr per MW = Rs 1,500 Cr.

Interest on capital during construction = Rs 200 Cr.

Payback period of the plant = (1700/128.98) = **13.18 years**. The average life of the project is 40 years. Hence, it is highly economical to set up a pumped storage plant [5]. If we take into account the other aspects, like load following, spinning reserve, frequency regulation, and reactive power control, this can be more profitable and hence reduced payback period.

21.8 CONCLUSIONS

The Rayalaseema region of Andhra Pradesh is bestowed with high potential for power production through natural and renewable resources, wind and solar. The state government has already approved tenders for the establishment of 6,500 MW of solar power in this region. The solar and wind energy in particular are highly intermittent in nature, thereby disturbing the grid network. This region is also known for its large network of reservoirs, which are underutilized, and the harvested energy from this is very poor. There is a large scope for the establishment of pumped storage plants in this region, which can solve integration as well as power problems.

21.9 REFERENCES

[1] Development and manufacturing of solar and wind energy technologies in Ethiopia: Challenges and policy implications, Mulualem G. Gebreslassie, *Energy*, May 2021, Pages 107–118

[2] Cost-based decision-making in middleware virtualization environments, Kaushik Dutta Debra and Vander Meer, *European Journal of Operational Research*, Volume 210, Issue 2, 16 April 2011, Pages 344–357

[3] A robust aggregate model and the two-stage solution method to incorporate energy intensive enterprises in power system unit commitment, Hongyang Jin, Zhengshuo Li, Hongbin Sun, Qinglai Guo, Runze Chen and Bin Wang, *Journal of Energy*, Volume 15 November 2017, Pages 1364–1378

[4] Comprehensive modeling and joint optimization of ice thermal and battery energy storage with provision of grid services, Hu Wuhua, Chin Choy Chai, Wenxian Yang and Yu Rongshan, *IEEE Region 10 International Conference TENCON*, IEEE XPLORE, ISSN. 2159–3450, December 2017.

[5] Viability of power distribution in India, challenges and way forward, Soumya Deep Das and R. Srikanth, *Energy Policy*, December 2020, Pages 1–11

22 Photovoltaic-Based Hybrid Integration of DC Microgrid into Public Ported Electric Vehicle

S. Pragaspathy, V. Karthikeyan, R. Kannan, N. S. D Prakash Korlepara, and Mr. Bekkam Krishna

CONTENTS

22.1 Introduction 287
    22.1.1 AC Microgrid 288
22.2 DC Microgrid 288
22.3 Preparatory Architecture of a Hybrid DC Microgrid System 289
22.4 Photovoltaic-Based Hybrid Integration of DC Microgrid into Public Ported Electric Vehicle 289
22.5 Solar PV System 289
22.6 Modeling of Solar PV Cell 289
22.7 DC/DC Converters for DC Microgrid 290
22.8 MPPT Operation in High-Gain DC–DC Converter 291
    22.8.1 Adaptive P&O Algorithm 291
22.9 Charging Methods of Battery Electric Vehicle 292
22.10 Slow Charging 293
22.11 Semifast or Medium Charging 294
22.12 Fast Charging 294
22.13 Bidirectional Converters in BEV and Microgrid Integration 294
22.14 Bidirectional DC–DC Converters 294
22.15 Bidirectional AC–DC Converters 296
22.16 Power Management Strategy for BEV Charger 297
    22.16.1 Case 1 297
    22.16.2 Case 2 297
    22.16.3 Case 3 298
22.17 Measured Results and Discussion 299
    22.17.1 Case 1 300
    22.17.2 Case 2 300
    22.17.3 Case 3 301
22.18 Conclusions 301
22.19 References 302

22.1 INTRODUCTION

The renewable technology is deployed gradually as a consequence of fast depletion of fossil fuels and its aftereffect with the environment [1]. The substantial crisis in electricity market is influenced to exploit renewable energy with large attraction and interest. Similarly, like the dispute now in support of conventional and alternate energy, "war of currents" is the debate in the beginning of twentieth century [2]. The argument was on the nature of electricity in generation, transmission, utilization, and all its fundamental aspects to be appropriate for delivery. Nikola Tesla and George Westinghouse stood behind alternating current (AC), but Thomas Alva Edison, their opponent, promoted direct current (DC) as the ample solution to the needs of the day. Even though both AC and DC have their specific task in the respective applications, the latter one was restricted to a voltage boundary in generation, and the impact of potential drop was a critical issue which makes it utilize the load locally [3].

The phenomena of stepping up the voltage in transferring power over long distances and stepping down the same for user requirement could be a possible venture in AC. In addition, the invention of AC induction machine by Galileo Ferraris and Nikola Tesla in 1887 was a significant shift in the phase which pushed the AC to win the "war of currents" as the renowned electrical system to generate and power the world. This remarkable execution is the milestone in the history of electricity which enables centralized AC power generation, transmission, distribution, and utilization worldwide. Thereafter, the conventional method of using coal, oil, and natural gas has gained a prominent place in the market of electricity generation [4]. The slow and continuous exposure of our environment to the inexorable greenhouse gas emission is undoubtedly harming the society, owing to the usage of fossil fuels, and on the other hand, escalating energy demand and aged infrastructure of the existing transmission and distribution system hassle the power delivery systems. The government, stakeholders, and researchers realized the need for nonconventional energy to provide a clean, better, and reliable source of power for a sustainable life on Earth [5]. Even though integration of renewable energy into the public grid becomes easier in the distributed generation systems, increasing penetration results in the protection and security issues which will raise the dispute on grid quality, safety, and reliability. "Microgrid," the subject matter of this chapter, is the identified technology to resolve the issues in energy integration and security [6].

Microgrid is a clustered network of energy generation, storage, and load in a confined locale, whereas *smart grid* is an intelligent microgrid with advanced communication technology for the control and monitoring of electrical network within the supply chain. Microgrid is a low- and medium-voltage distribution network serving the localized demand by interfacing the distributed energy sources to local substations passing through the power converters [7]. It facilitates the surplus power generation to feed the utility grid and receive back from them at times entailed. Also, the microgrid can operate autonomously, detaching from the utility network when the condition dictates due to system failure/fault. The classifications of microgrid based on their bus supply are AC microgrid, DC microgrid, and hybrid (AC/DC, DC/AC) microgrid systems. An AC microgrid is preferred mostly since it is the traditional power system, but the development in microgrid technology infrastructure and increase in practice of DC equipments slowly shift the attention towards DC microgrid.

Most of the renewable power generations are in DC form, which requires electronic converters to deliver for the AC microgrids. For instance, fluctuating AC output of a wind turbine requires back-to-back converters to synchronize the wind power with AC grid system [8]. Domestic and residential needs, such as in television, desktops, battery chargers, and energy-efficient lights that function in DC power, have an inbuilt rectifier unit in the package. Likewise, in industrial applications and deployment of BEV technology, AC–DC–AC conversion stages are prominent for the adjustable-speed drive mechanisms. These multiple engineering stages in all aspects certainly reduce the reliability of the power network. The overall efficiency can be improved by sorting out the number of conversion stages by way of implementing DC grid system. The sophisticated concept of microgrid, in tandem with modern power electronics, revives again Edison's exact vision of power delivery [9].

22.1.1 AC Microgrid

In AC microgrid systems, distributed generators and loads are connected via power electronic converters in accordance with the nature of supply they generate or consume. Energy storage devices are the DC loads connected to AC bus via bidirectional DC/AC converters. Utility grid interacts with the AC bus at the point of common coupling (PCC), which is regulated by an active switch **[10]**. AC microgrids are classified as high-frequency AC and line-frequency AC based on the operating voltage, whereas single- and three-phase systems pertain to the number of phases. The insertion of sources like solar PV and fuel cells in AC microgrid requires inverter to connect the bus, which reduces the overall efficiency and increases the net cost.

22.2 DC MICROGRID

The intense use of low-voltage electronic DC loads and the increase in the exploitation of renewable sources, such as solar PV and fuel cells, in the recent years have provoked the research more towards DC microgrid. The distributed DC power is structured by interconnecting the renewable sources and energy storage at the DC link of the converter interface and congregating the loads, respectively. The design is simpler than AC microgrid since it eliminates the rectifier and inverter sections; thereby, commutation loss is reduced and tracking of voltage magnitude and frequency is not required. The energy storage devices can be connected to the DC bus directly or via bidirectional converters that depend upon their requirements and applications **[11]**. For instance, unstable voltage across the battery, owing to the state of charge (SoC), temperature, and battery chemistry, will not only affect the lifetime of a battery but also generate inrush current and harms the stability of the DC bus voltage when connected directly. Therefore, it is suggested to interface a battery via bidirectional converter to the DC bus. The converter topology can alleviate the fluctuation issues and provide scope to incorporate more batteries irrespective of their SoC characteristics. The significance of DC microgrid over AC microgrid includes:

a) Transition towards renewable energy reduces CO_2 emission, and further drop can be achieved by increasing the power system efficiency in transmission and distribution sectors. DC microgrid with fewer number of power converters can be the substitution or alternative.
b) DC microgrid offers no skin effect, and thereby, a small cable can be used to transfer the same amount of current.
c) The unidirectional property of DC allows easy controlling of power flow by means of power electronic converters embedded in the system.
d) DC microgrid is secured with uninterrupted power supply during islanding mode of operations. The assignment is intricate in AC bus, owing to

incongruity between grid code requirement and seamless switch.

e) There is no task of reactive power stabilization and synchronization issues for grid-connected renewable systems in DC microgrid, which further reduces the complexity and enhances the reliability and stability of the overall system. Therefore, the renowned attributes of the DC microgrid system makes it an important subject of the current science.

22.3 PREPARATORY ARCHITECTURE OF A HYBRID DC MICROGRID SYSTEM

The distributed network needs the integration of BEVs along with renewable energy sources to satisfy the local demand and, moreover, distribute the excess generation to the utility grid without any reactive power stabilization [12]. The standards of BEV charging station needs to be improved by building multiport charging and smart real-time monitoring infrastructure. The charging station powered by solar PV and renewable energy connected to a DC microgrid must follow the clause for network protection and grounding. The configuration of architecture facilitates the microgrid to operate either in grid-connected or islanding mode [13]. The islanding mode is preferred during external disruptions or faulty situations. The controlling competency of the microgrid system stands unique among the conventional distributed networks. Bidirectional DC–DC converter has a key role in controlling the power flow between DC microgrid and BEV/energy storage unit, whereas bidirectional DC–AC converter controls the power flow between DC microgrid and utility grid [14]. Regulations of voltage within the microgrid and power flow across the charging terminals are controlled based on the requirement.

EV charging stations are ensured to be efficient alongside the energy management strategies. The modular design and moderate energy density of the energy storage system in rechargeable-type batteries are more suited for grid-connected applications. The control system incorporates a nonlinear voltage balance (NLVB) algorithm in the conventional PID controller to exercise power generation and delivery at the BEV charging stations in both steady and dynamic conditions of DC bus voltage. Conversely, a separate control strategy is required for the integration of renewable energy source and also for balancing the local demands [15]. Renewable energy in this distributed network reduces the power stress, and the investment spent over its integration is retaliated within the stipulated time. In this proposed chapter, PV-integrated hybrid DC microgrid experiences an unbalance in power level due to the variation in solar irradiation. Maximum power is obtained from the solar PV at every instant of time by the closed-loop monitoring of incoming solar insolation using adaptive perturb-and-observe (P&O) algorithm [16].

22.4 PHOTOVOLTAIC-BASED HYBRID INTEGRATION OF DC MICROGRID INTO PUBLIC PORTED ELECTRIC VEHICLE

The solar PV system is connected with DC bus to feed the power into BEV respectively. But due to interrupted energy availability in nature, power generation from solar PV gets varied, and therefore, the electric vehicle fails to charge at uniform rate. In order to overcome the aforementioned drawback, this particular book chapter presents a hybridization of solar PV with AC grid system interconnected with electric vehicle. Therefore, it helps to supply uninterrupted power at a uniform charging rate. The complete block diagram of photovoltaic-based hybrid integration of DC microgrid into public ported electric vehicle is portrayed in Figure 22.1.

22.5 SOLAR PV SYSTEM

A photovoltaic unit straightly converts the solar irradiance into photonic energy, despite the different material used for its building. Semiconductor materials are the most common platform to yield monocrystalline, polycrystalline, and thin-film PV technologies in the global market [17]. Solar cells preferring gallium arsenide are higher in efficiency and cost as well as go with some unique and space applications. PV cells are arranged in a matrix version within an assorted area. Several solar panels are connected in serial and parallel fashion to supply the ample voltage and current [18].

22.6 MODELING OF SOLAR PV CELL

A simplified electrical equivalent circuit of solar PV cell is represented in Figure 22.2. It has a light-generated DC current source (I_{ph}), diode (D), internal series (R_{se}) and parallel (R_{sh}) resistance [19]. The total current driven out from a solar cell depends upon the current flowing towards the diode (I_D) and series resistor (I_{se}). The characteristic equation of solar cell is given by (1) and (2):

$$I_{pv} = I_{ph} - I_D - I_{sh} \quad (1)$$

$$I_{pv} = I_{ph} - I_0[\exp(\frac{q(V_{pv} + I_{pv}R_s)}{nkt} - 1)] - \frac{V_{pv} + I_{pv}R_s}{R_{sh}} \quad (2)$$

Where I_0 is the reverse saturation current, V_{pv} is the terminal voltage, n is the ideality factor, q is elementary charge, k is the Boltzmann's constant, and t is the junction temperature, respectively.

The value of series and parallel resistors and current flowing through it depends on the dimensions of the PV cell. There exists a nonlinear relationship between the terminal voltage and the output current, depending on the existing conditions of solar irradiance and temperature. The practical I–V curve depicted in Figure 22.3 determines the output power of a PV cell. Maximum voltage (V_{oc}) can be obtained

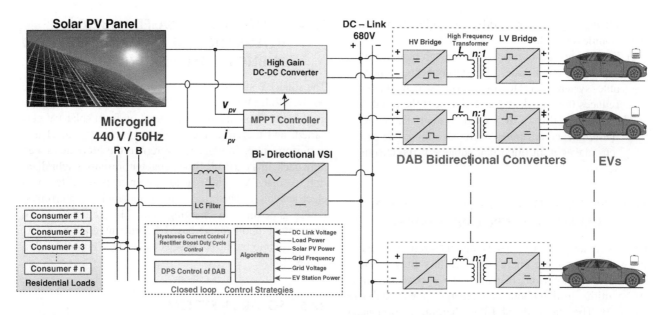

FIGURE 22.1 Structure of photovoltaic-based hybrid integration of DC microgrid into public ported electric vehicle.

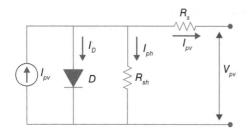

FIGURE 22.2 Equivalent circuit model of PV cell.

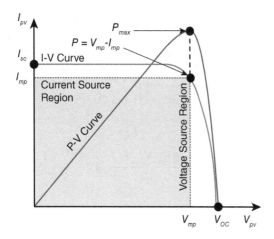

FIGURE 22.3 Practical I–V and P–V curves of a PV cell.

during open circuit, whereas the maximum current (I_{sc}) at short circuit, but neither this condition generates power. Maximum power can be achieved only when the terminal voltage and output current read V_{mp} and I_{mp}, respectively. Therefore, the PV panel can be controlled electrically or mechanically at every instant of time to track the maximum power operating point (MPOP).

22.7 DC/DC CONVERTERS FOR DC MICROGRID

DC microgrid cannot feed the load directly at the desired voltage levels. Regulation of voltage by DC/DC converter plays a prominent role in microgrid applications [20]. Reliable wavelet coefficient, low cost, and size are the additional essential requirements of the converter proffered in microgrids. The input voltage can be boosted in boost converter (Figure 22.4b) and are stepped down in buck converter (Figure 22.4a).

Buck-boost converter can do both the operations together and so are preferred widely (Figure 22.4c). Topologies of isolated DC–DC converters, such as fly-back and full-bridge types, are used in high-power applications. Recent-day converters, like two-phase interleaved boost converter (Figure 22.4d), quadratic boost converter (Figure 22.4f), and DC/DC converter based on voltage-lifting techniques, have high voltage conversion ratios with reduced switching stresses (Figure 22.4g). Modified voltage-lift technique based nonisolated boost converter is the high-gain converter with only one active switch, three diodes, and five storage components (Figure 22.4e).

From the standpoint of the mentioned elements, efficiency of the modified voltage-lift technique–based converter surpasses other converters in continuous and discontinuous current conduction mode. The load voltage carries small ripple content due to the low output capacitance at extensive load cycles. Time constant of the output capacitor is very small at light load, whereas higher for larger load, resulting in ripples, and consecutively reduces the output voltage. Closed-loop controller with a modified voltage lifting techniques implemented in this converter fixes the issues for various load circumstances.

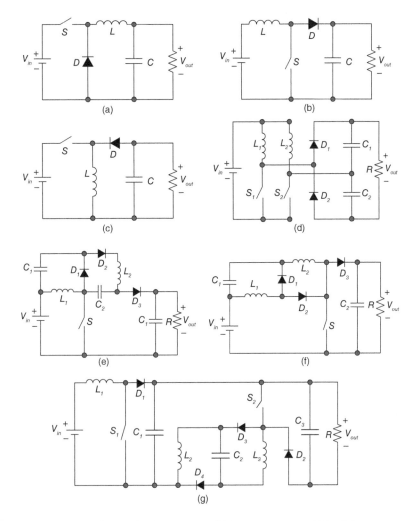

FIGURE 22.4 DC–DC converter topologies: (a) buck converter, (b) boost converter, (c) buck-boost converter, (d) two-phase interleaved boost converter with parallel input series output, (e) modified voltage-lift technique–based nonisolated boost DC–DC converter, (f) quadratic boost converter, and (g) DC–DC boost converter based on voltage-lift technique.

22.8 MPPT OPERATION IN HIGH-GAIN DC–DC CONVERTER

In the distributed energy network, there is no microgrid without the integration of renewable energy sources. Aforementioned in previous section, the intermittent nature of this inevitable resource can give maximum power for the existing conditions. Maximum power can be extracted through modified voltage-lift technique–based nonisolated boost DC–DC converter by tracking MPOP using adaptive P&O algorithm [21].

22.8.1 Adaptive P&O Algorithm

MPOP of the PV system is changing instantaneously with respect to the variation in atmospheric conditions. The main aspiration of the MPOP tracking system is to follow the region where the peak power swings. Respective maximum power can be delivered by the solar PV system for the instantaneous changing climatic conditions and pattern (which means for the vast range of electrical input), and so it is demanded. Solar radiation (photons) directly influences the PV module current, and the temperature rise drops the panel voltage. Tracking of the maximum power is governed by the algorithm which is essentially controlling the driving pulse of employed DC–DC converter between the load and PV modules. The algorithm receives the current and voltage of the module as feedback and integrates with duty cycles of the proposed DC–DC converter that will make the system capture the possible peak power at that time [22]. The maximum power in some application is regulated by PI controllers. Among various power-tracking algorithms, perturb-and-observe is extensively the preferred method and can be easier to make use of. The step size of the algorithm oscillates at every interval of time in the process of tracking peak power, respectively.

Plummeting the step size eventually retards the MPPT operation of solar photovoltaic system. If the output power is increased while enhancing the duty ratios, the operating region is said to be in uphill. Therefore, the duty ratios are supposed to increase in the same direction so as to increase the power. However, perturbation should be reversed to

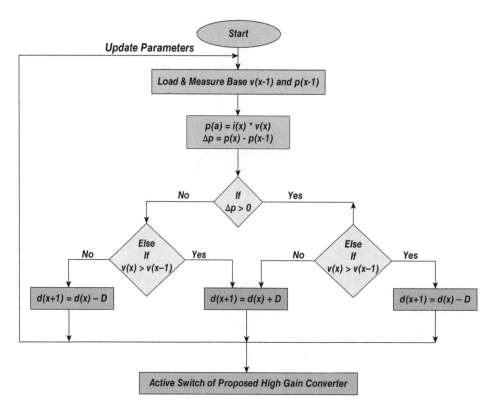

FIGURE 22.5 Flowchart of adaptive P&O algorithm.

TABLE 22.1
Condition for Positive and Negative Perturbation

Previous Perturbation	$\dfrac{\Delta P}{\Delta D}$	Next Perturbation
> 0	> 0	> 0
> 0	< 0	< 0
< 0	> 0	< 0
< 0	< 0	> 0

reach the MPOP when there is decline in power. Now the operating region is said to be in downhill. The panel voltage and current are the feedback parameters used to compute the duty ratios, and a digital signal processor is employed as the controller which is most suited for the P&O method. The algorithm is unsuccessful sometimes, owing to the abruptly changing atmospheric conditions. When the operating point is shifted from one point to another due to forward perturbation and also if the power decreases because of that action, then it should be reverse-perturbed.

The approach flowchart of adaptive P&O algorithm is illustrated in Figure 22.5, and perturbation conditions are revealed in Table 22.1. Meanwhile, increase in solar irradiance may lift the power curves and, in parallel, raise the power for same perturbation. Therefore, the operating point fails to achieve MPOP in the altered power curve, and it is continued for changing irradiance level. It is essential for a power-tracking algorithm to ensure the operation of solar photovoltaic system in MPOP under various conditions. Adaptive P&O method prevents the deviation from MPOP to the core.

22.9 CHARGING METHODS OF BATTERY ELECTRIC VEHICLE

Energy storage device is the most significant constituent of a microgrid system. Lifetime, cost, state of charge, and discharge of the energy storage system play a prominent role behind the technology innovation of BEV and to increase their market penetration. BEV range is likely to be estimated from the energy density and charging time of battery, which is also the main constraint in the market share of BEVs over conventional-fueled vehicle [23]. Moreover, the aforesaid parameters are firmly related to the attributes of a battery charger. Thereby, the charging architecture turns into an essential unit of BEV technology and plays a vital role in its widespread adoption. Battery chargers underwent a series of research as a consequence of continuous innovations in the evolution of BEV. They are mostly classified based on their power rating and charging times. Onboard and off-board chargers are the further classification based on their location [24].

Onboard chargers are fixed inside the vehicle and often give flexible operation, as the system requires least external component close to the power terminal. Onboard chargers with additional dedicated power converter and onboard chargers with existing integrated converter are commonly used to draw the battery power and deliver it to the BEV

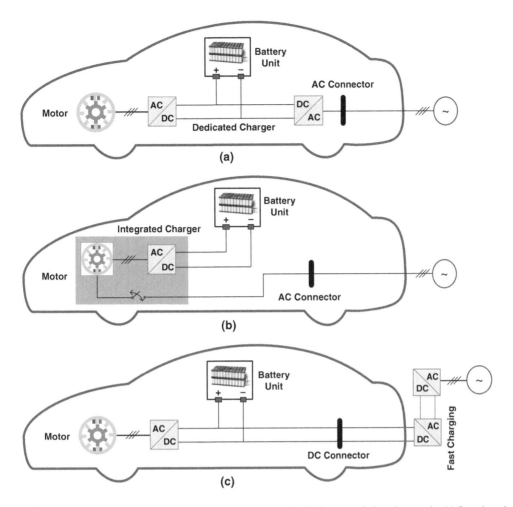

FIGURE 22.6 BEV charging infrastructure: (a) dedicated charging mode, (b) integrated charging mode, (c) fast charging mode.

motor. Dedicated-type converters are the slow chargers, as they are small in power rating due to the space limitation and therefore preferred for long time or overnight-charging applications. On the other hand, integrated chargers are high in power rating, and as well, the charging time is reduced.

They can be employed for both slow and medium fast charging applications. Off-board charger includes a built-in charging feature that usually dwells outside the vehicle. Therefore, there is no limitation identified with size and weight as seen in their onboard counterpart. Vehicle is connected directly from the DC source, and the high-power capability allows for lesser charging times. As there is no restriction in size and power levels, off-board chargers cannot be employed in domestic applications and are recommended only for public utilities like charging stations. The charging infrastructure of BEV is unfolded in Figure 22.6.

22.10 SLOW CHARGING

Dedicated onboard chargers are simple and easily accessible devices that perform an exclusive task in the charging applications. Stabilized DC voltage is generated in the first stage, and regulation of battery charging current is executed in the second stage. The galvanic isolation is imposed in the utility network pertaining to the safety regulations. The charger proposed in [25] comprises of dual channel booster, which controls the power flow and converts the alternating supply to DC as well. The control is realized by changing the pulse duration of converter, and also, the circuit inflicts lower current ripple that certainly permits the huge drop in the size of passive components like inductor and capacitor. Rating of the electronic switch can also be reduced as the current is partitioned between the two boosters. Incorporated high-frequency converter in the second stage reduces the conduction loss to a large extent, and thereby, the overall efficiency of the charger is increased respectively. The isolated configuration is replaced by resonant topology in [26]. In that order, the power flow control comprises of rectification and boost conversion stages. Switching loss due to high-frequency operation is reduced by inserting an exclusive frequency controller. The effective power factor of the circuit is close to unity, and the efficiency is reported at 93%. An alternate move where the power flow control is shifted to low-voltage end to ease the size of DC link capacitor is proposed in [27]. The charger is reverse-configured, and the back-end converter provides fixed and smooth switching frequency. Improving the voltage regulation and wiping out the harmonics are performed through the power flow control, respectively. The discussion portrays that the slow chargers do not associate with low-frequency transformers as they are complex in size and volume. The uniqueness of slow charger is galvanic isolation, and it is realized through high-frequency transformers. But the isolation requires cascaded converter stages that ensue in various losses and thereby affect the efficiency.

22.11 SEMIFAST OR MEDIUM CHARGING

Integrated onboard chargers do not have converter stages; instead, they employ a drive inverter and battery charger. The integrated onboard chargers do no support charging while BEV is in driving mode, and vice versa. This configuration eliminates the excessive cost, weight, and volume imposed on BEVs, as a single converter is used for both purposes. The rating of converter is ideal to the motor power, and charging is supposed to be executed with the same power. Indeed, power level (driving and charging) is greater than the onboard dedicated chargers. Meanwhile, the performance of the chargers is analyzed based on the integration, voltage, and torque ripples during the driving and charging operations. The exclusive use of semiconductor devices in the charging circuit board decides the level of integration. There are several topologies with maximum and partial integration proposed so far. As stated earlier, minimizing the number of components certainly reduces the size, weight, and overall cost of the charging unit. On the other hand, driving and charging the vehicle at the same time is mere impossible in this topology. Rectifier and interleaved buck-boost units are integrated in a nonisolated onboard charger [28]. The ripple values are reduced to an extent due to the two channel interleaved configurations. Furthermore, proposed topology ensures the regulation of voltage level to a wide range. The effective power factor of the circuit is close to unity, and efficiency is reported at 97%. Bidirectional and buck-boost converters are brought together for an exclusive design of onboard battery charger. Regeneration is performed during buck mode, whereas the battery discharges to the BEV through boost operation. The charging of battery is made possible in buck-boost action of the converter. The module is restricted for the aforesaid modes to operate at the same time.

22.12 FAST CHARGING

Dedicated and integrated version of onboard charger is discussed briefly in the preceding section. Onboard chargers are uncomplicated and mounted inside the battery-operated vehicle. The charging architecture offers flexible exchange of power with least onboard components [29]. Despite the fact, size and volume highly influence the power ratings, and thereby, the charging time is increased. Furthermore, implementation of fast and ultrafast charging infrastructure helps the aspiring community towards the adoption of BEVs in the larger scale. Subsequently, it replaces the number of batteries and batteries with larger capacity employed in the conventional slow charging configurations. The analysis put forth here clearly indicates that the onboard chargers could not be the effective solution for fast charging, owing to the size, weight, and cost, respectively. The off-board chargers are realized to be suitable for public charging stations to feed more number of BEVs. As similar to the fuel station, batteries of EVs are charged in a short duration, and the charging levels are monitored and recorded. The research and development alongside propose so many emerging battery technologies to facilitate fast and ultrafast charging as a realistic possibility for the current and future society.

22.13 BIDIRECTIONAL CONVERTERS IN BEV AND MICROGRID INTEGRATION

The role of bidirectional converter in the integration of BEVs and renewable energy has become popular now. It is the only solution for long-standing issues in connection to the distributed energy storage systems. Power electronic converters are the inevitable features of present-day microgrid systems, without which there is no penetration, integration, and interconnection of any form of energy sources. The exchange of power in the proposed system connecting the DC bus is carried out through bidirectional DC–DC and bidirectional AC–DC converters, respectively [30].

22.14 BIDIRECTIONAL DC–DC CONVERTERS

DC microgrid is supplied by the solar PV system, as shown in Figure 22.2. The power supplied to the microgrid can be stored in the distributed energy storage device as well charges the battery-operated electric vehicle. Isolated bidirectional dual active bridge DC–DC converter (IBDABC) is used as an interfacing module between the source and storage unit of the proposed system. The high-voltage and low-voltage sides of IBDABCs are isolated from each other to enable smooth power flow control and flexible and reliable operations. Soft switching methods are employed in the bidirectional forum to enhance the converter performance during the exchange of power. The converters are suffered with circulating currents due to phase shifting issues [31]. Dual phase shifting control techniques are deputed for IBDABCs to reduce the circulating circuit, and it is found to be more appropriate than other phase-shifting phenomena. Phase shifting is executed with two-level AC voltage in triple-phase shift modulation technique, whereas in extended phase shifting, two-level and three-level AC voltage is developed in their respective low-voltage and high-voltage bridges. Three-level AC voltage is the output across the bridges of dual phase-shift modulation, and this technique also limits the negative spikes developed in the IBDABCs. Therefore, the proposed chapter experiments the closed-loop control of dual phase shift modulation in IBDABC, pertaining to the integration of BEVs in DC microgrid platform. The design configuration of IBDABC is depicted in Figure 22.7.

Consider the proposed IBDABC is operated under two different modes during the exchange and retrieval of power between the microgrid, distributed energy storage system, and parallel-configured BEVs. If the phase shift is leading in forward-conduction mode, then it is lagging for

Photovoltaic-Based Hybrid Integration of DC Microgrid

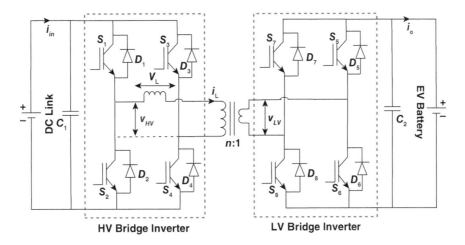

FIGURE 22.7 Circuit configuration of IBDABC.

reverse-conduction mode, and vice versa. The operating modes are stated as follows: (a) $0 \leq d_2 \leq d_1$ and (b) $0 \leq d_1 \leq d_2$, where d_1 and d_2 are inner and outer phase-shift duty ratios, respectively. Therefore, the inductor current is expressed as (3):

$$\frac{di_L(t)}{d_t} = \frac{(V_{in}(t) - nV_a(t))}{L} \quad (3)$$

The duty ratios are adjusted in such a way that their count $(d_1 + d_2)$ should not surpass or be equal to the value 1. But in general, scale of duty ratios (d_1 and d_2) is set between 0 and 1. Assume that the power is transferred to the parallel-connected BEVs and energy storage units from the solar PV source (or DC microgrid) through IBDABCs. While doing so, duty ratios (d_1) are tuned to shift the phase of V_{LV} by $d_2\pi$ with reference to the rising point of V_{HV}. As a result of this control, power transfer to charge the energy storage system is achieved respectively. The power flow operating direction is interchanged during the reverse conduction mode. Duty ratios are adjusted to shift the phase of V_{LV} in advance to V_{HV} by $d_2\pi$. Here, V_{LV} leads V_{HV}. The waveform portrayed in Figure 22.8 exemplifies the operation of IBDABC under dual phase shift conditions.

Bidirectional power flow between the source (solar PV/microgrid) and energy storage unit (BEV/distributed energy storage) is controlled by regulating the duty ratios d_1 and d_2. Furthermore, the magnitude of inductor voltage is subjected to the values of duty ratios respectively. The waveform similarity between the operating modes ($0 \leq d_2 \leq d_1$ and $0 \leq d_1 \leq d_2$) is affected during the time period T_2 and T_3. The expression of current for the instant $0 \leq d_2 \leq d_1$ is obtained as follows (4) to (6):

$$i_{L1} = \frac{n(V_a)}{2Lf_S}[k(1-d_1)-(d_1+4d_2-1)] \quad (4)$$

$$i_{LP} = \frac{n(V_a)}{2Lf_S}[k(1-d_1-4d_2)+(d_1-1)] \quad (5)$$

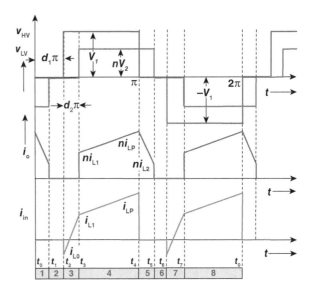

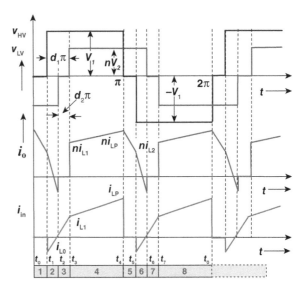

FIGURE 22.8 Operating waveforms of DAB converter under dual phase shift modulation (a) $0 \leq d_2 \leq d_1$ and (b) $0 \leq d_1 \leq d_2$.

$$i_{L2} = i_{L0} + \frac{n(V_a)}{2Lf_S}[k(1-d_1)-(d_1-1)] \quad (6)$$

Similarly, expression of current for the instant $0 \le d_1 \le d_2$ is derived as follows (7) and (8):

$$i_{L0} = \frac{n(V_a)}{2Lf_S}[k(d_1-1)-(4d_2-d_1-1)] \quad (7)$$

$$i_{L2} = \frac{n(V_a)}{2Lf_S}[k(4d_2+d_1-1)-(d_1-1)] \quad (8)$$

The losses of the system are a common factor to bring down the efficiency and also produce heat. The conduction and switching losses are two important losses in the power converters. As mentioned previously, the DAB converter operates under higher frequency of operation; thereby, switching losses are more dominant than conduction losses of the converter. Therefore, it is essential to analyze the switching losses of the DB converter, and a short description is addressed here.

Switching losses. The turn ON and OFF timing are responsible in producing the switching losses. It is inherently dependent on rise, delay, and fall time of the MOSFET during turning on and off time. Apart from this, the number of times ON and OFF action takes place depends on the frequency; therefore, operating frequency has a huge role in the production of switching losses. And also, it depends on the device current and voltage across the device during on and off states of the device. The switching losses for both bridges can be written as:

$$P_{switch\_HV} = 4fV_{sw\_HV}I_{sw\_HV}(T_{on}+T_{off}) \quad (9)$$

$$P_{switch\_LV} = 4fV_{sw\_LV}I_{sw\_LV}(T_{on}+T_{off}) \quad (10)$$

By using the two previous equations, the switching losses can be estimated under any operating conditions.

22.15 BIDIRECTIONAL AC–DC CONVERTERS

The idea in the exchange of power between the microgrid to BEVs and vice versa becomes a more attractive application due to the existence of bidirectional AC–DC converters [32]. In addition to power transfer, converters are involved in achieving the receipt of reactive power compensation and peak power shielding.

The three-phase bidirectional AC–DC converters employed in the course of power transfer between the active microgrid and BEVs are supposed to be more efficient, and henceforth, creative topologies with advanced controllers are explored to enhance the efficiency. Furthermore, the issues in power quality, output ripples, and distortions are forestalled through the means of novel configuration independently. The circuit topology of bidirectional AC–DC converter experimented in BEV-based microgrid environment is portrayed in Figure 22.9. The three-phase bidirectional AC–DC converter comprises of six active switches (S_1–S_6), with a diode connected in parallel across each (D_1–D_6). The supply is given through the inductors, which act as filters. Stable voltage is achieved through the presence of a DC link capacitor. There are two modes of operation for the bidirectional converter. As a rectifier, it transfers the power from AC bus to the DC bus during the first mode, and it performs the reverse operation as an inverter in the second mode, which makes the power flow from DC bus to AC bus respectively.

An adaptive control algorithm is imposed on the power converter for the bidirectional power flow, which conceives the static demeanor of the circuit based on the viable switching conditions. In addition, switching state enables the power flow control by exploiting the distinct nature of filter inductance. The pulse to trigger the active switch is generated appropriately from the error measured between the actual and reference values. Switches from the same leg

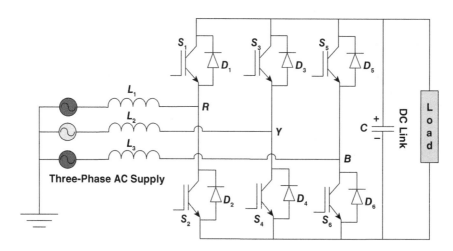

FIGURE 22.9 Circuit topology of three-phase bidirectional AC–DC converter.

Photovoltaic-Based Hybrid Integration of DC Microgrid

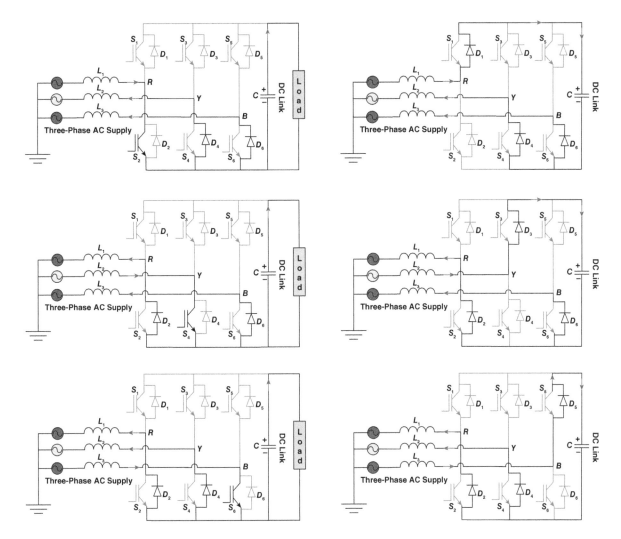

FIGURE 22.10 Modes of power flow in three-phase bidirectional AC–DC converter.

are operated in an alternative sequence so as to avoid the short-circuit problems. The power flow paths of three-phase bidirectional AC–DC converter under various switching states are illustrated in Figure 22.10.

22.16 POWER MANAGEMENT STRATEGY FOR BEV CHARGER

Initially, the chargers are connected to the BEVs, and then the power is set by the consumer. Depending on the customer requirement, power is being consumed from the DC link side through dual active bridge converter. However, the solar PV not at all generates the power based on the requirement of BEV, and therefore, a power management strategy has to be implemented in the proposed system to exercise the generated power accordingly. The operation is explained through the flowchart, which is clearly depicted in Figure 22.11. Firstly, the number of electric vehicles connected to the public port is estimated, and set power is immediately adjusted in the algorithm. Then, the solar PV power generation is monitored and estimated at every instant. The operation can be classified into the following three cases.

22.16.1 CASE 1

At every cycle of operation, both the powers are certainly compared to identify the difference in power. If there is no deviation from the load requirement and generation, algorithm can even be not in operation. But practically, this condition happens once in a while. Other than this condition, the difference in power is being calculated, and when the solar power generation is not equal to the BEV requirement, the DC link voltage will vary. In order to balance the power and also to maintain the DC link voltage, the excess power has to be delivered or supplied to the DC bus respectively.

22.16.2 CASE 2

In case 2, consider the power derived from solar PV is not sufficient to supply the BEV, and therefore, an additional power is required/drawn to satisfy the demand. As per the algorithm, the channel is set as 1 in DAB converter, and it means forward direction, and the reference power to the grid is set by the difference in power. Now, the bidirectional DC–AC converter performs in rectification mode.

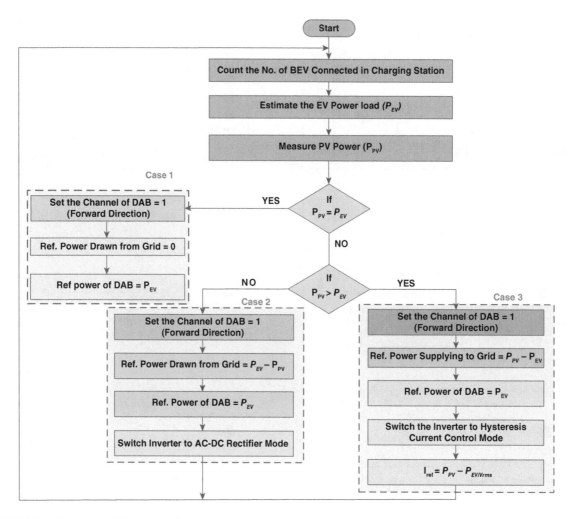

FIGURE 22.11 Flowchart of the proposed system.

All the gating signals get turned off, and consequently, the grid power is directly available at the DC link side. In the aforementioned theory, maximum power is delivered at the DC link voltage, and excess power required by the BEV is received from the grid automatically. In this particular way, the total power is mapped throughout the day while raising this condition respectively.

22.16.3 Case 3

An excess power will be available whenever the solar PV–generated power is higher than the BEV requirement (case 3). Now, the DC link voltage starts to rise and leads to system failure due to this excess power. In order to overcome this drawback, solar PV power, BEV power, and excess power must be balanced. According to the algorithm, reference power is calculated, and the DAB power is set as channel 1 to deliver the power in forward direction. Now, three-phase bidirectional converter must be performed in three-phase inverting mode. To satisfy the power or maintain the DC link voltage, an accurate reference current must be calculated. The reference current is determined using the following expressions:

$$I_{hys\_ref} = \left(\frac{V_{dclink} - V_{dclink\_act}}{V_{dclink}}\right) \frac{P_{grid\_ref}}{V_{rms}} \quad (11)$$

Where

$$P_{grid\_ref} = P_{PV} - P_{EV} \quad (12)$$

The grid power is being estimated from the following function:

$$P_{grid} = V_{rms} I_{hys\_act} \quad (13)$$

The accuracy and power balancing approach of the proposed solar PV–fed hybrid structure depend on the proportional and integral gains of K_p and K_i, and their mathematical relations are given as:

$$r = K_p e(t) + K_I \int e(t)dt \quad (14)$$

Using (11), the reference current is generated and applied to the hysteresis current controller. In hysteresis current controller, the band limit is set as 10% of the reference current,

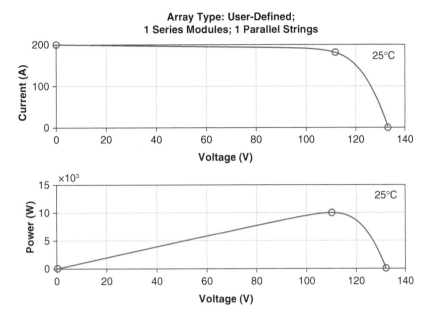

FIGURE 22.12 IV and PV characteristics of solar PV system – 15 kW$_{Peak}$ at 25°C.

respectively. Therefore, the actual sensed grid-injected current is compared with reference current plus band values. In accordance with the logic behind hysteresis controller, the diagonal pairs get ON and OFF in the three-phase voltage source inverter. The actual grid current is increased while the switches are turned ON, and vice versa. So the actual grid current gets oscillated within the boundary values of the band limit. Due to this, the DC link voltage is also maintained within the range, and hence power is balanced.

22.17 MEASURED RESULTS AND DISCUSSION

To validate the proposed research, the simulation study was carried out using PSCAD/EMDTC software. The results are clearly observed and well presented in this chapter. A 15 kW$_{Peak}$ capacity of solar panel is assembled in simulation platform. It is operated at the temperature of 25°C and also collected various sample data to plot the IV and PV characteristics of the modeled solar PV. The performance of an individual solar array module is studied and plotted in the graph, respectively. One series module and one parallel string data are collected and plotted as shown in Figure 22.12. It is justified that the nonlinear characteristics are similar to the theoretical approach. The rated values of short-circuit current and open-circuit voltage are 200 A and 130 V. However, maximum power is obtained while operating the solar PV module voltage of around 110 V at 25°C. Moreover, this 110 V is not constant or fixed voltage, but it varies with respect to atmospheric temperature. Each module power is around 3 kW$_{Peak}$, and almost 5 modules are connected in parallel to verify the feasibility of the proposed operation.

The adaptive P&O algorithm is implemented in the proposed system to track the maximum power under any environmental conditions. Several MPPT algorithms have been reported in recent literature. But P&O algorithm is most popularly used for solar PV applications. However, it has a drawback of slower response under sudden dynamic conditions. In order to improve the performance of tracking maximum power from the solar PV system, the step time has to be reduced. Even with reduced step time operation, the response becomes poorer, while change in irradiation occurs at sudden rate. In order to overcome this drawback, step time must be adjusted according to the solar irradiation and temperature changes. During large step variation in the environmental conditions, the step time becomes so high, and therefore, the maximum power point is located within few sampling time intervals. When the maximum power point is located around the exact point, then step time gets reduced to smaller value to maintain the accuracy; thereby, it results in lesser oscillations around the MPOP. After adopting this MPPT algorithm to DC–DC converter, the DC link voltage is adjusted to the desired value. It is obtained from the closed-loop control operation executed by MPPT algorithm. The DC link voltage tracks the reference voltage, and it is clearly shown in Figure 22.13. It is clear that it has less ripple content and ensures negligible oscillations on the whole operation eventually.

In the aforementioned statement, power is fed to the BEV using DAB converter by adjusting the phase angle plus PWM approach. According to the reference power set by the customer or operator, the phase angle and PWM signal is automatically set by the closed-loop section of the DAB converter. The DAB converter allows the power to BEV by keeping certain phase angle delay and PWM signal for the given power. The measured results are observed and presented in Figure 22.14. Based on the instantaneous voltage that appears across the inductor, the inductor current gets charged and discharged to transfer the power. In all the three cases discussed in the previous section, the DAB

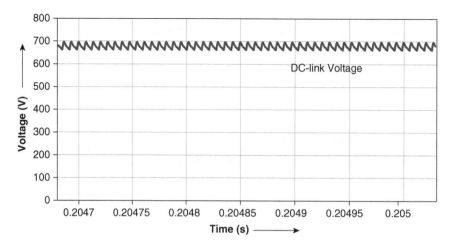

FIGURE 22.13 DC link voltage.

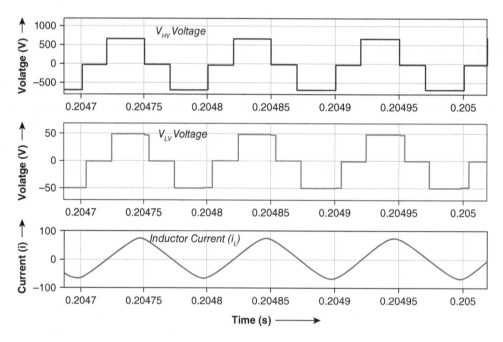

FIGURE 22.14 Steady-state voltage and current of DAB converter.

converter operates under forward condition. Therefore, the voltage at LV side is always behind the voltage at HV side to charge the BEV respectively. To validate the performance of the proposed hybrid system, simulation study was carried out under all three conditions and the measured results are exposed pertaining to the observation.

22.17.1 Case 1

When solar PV–generated power is approximately equal to the BEV charging power, no error power exists. Therefore, there is no command signal generated to supply or feed the power at grid level. At 800 W/m², the solar PV–generated power is 10 kW and BEV charging power is also 10 kW. The grid power is observed to be almost zero, which is clearly shown in **Figure 22.15.**

22.17.2 Case 2

Similarly, when the solar PV–generated power becomes lower than the required BEV charging power, the extra demanded power is computed and the bidirectional converter operates in rectification mode as per the algorithm. Due to the inverter turn OFF gating signals, additional power required is automatically consumed from grid and supplies to the BEV charging port eventually. To validate this condition, 12 kW solar PV power is generated and almost 15 kW of BEV charging power is consumed in the proposed system. Now, as per algorithmic approach, additional 3 kW of power is automatically consumed from the grid to satisfy the demand of BEV charging port. This result has been captured and presented in Figure 22.16.

Photovoltaic-Based Hybrid Integration of DC Microgrid

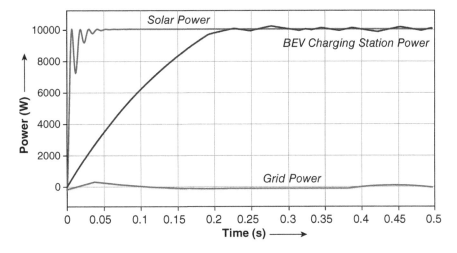

FIGURE 22.15 Simulation results of solar PV power, EV charging power, and grid power in case 1.

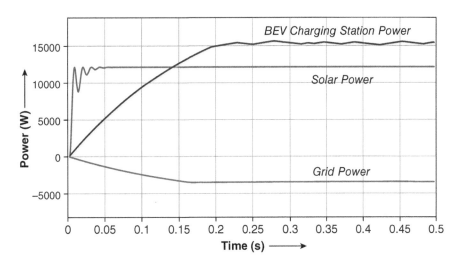

FIGURE 22.16 Simulation results of solar PV power, EV charging power, and grid power in case 2.

22.17.3 Case 3

Now, consider the solar PV–generated power is reduced due to the change in solar irradiation (400 W/m² approximately) and the corresponding power generation is about 7.5 kW. And also, the BEV charging power gets reduced to 5 kW, then the deviated power is almost 2.5 kW. Therefore, the additional power generated must be injected into the grid to maintain the DC link voltage at desired values. With the help of hysteresis current controller, the excess power is injected into the grid. Using expression (1), the reference current is computed and the respective gating signals are generated. At this juncture, solar module power, BEV charging power, and the grid-injected power are clearly captured and shown in Figure 22.17.

The waveform of voltage and current portrayed in Figure 22.18 ensures the quality of power injected into the grid, respectively. From all these attained results, it is concluded that the proposed solar PV–fed hybrid system is one of the most suitable solutions for BEV charging environment.

22.18 CONCLUSIONS

Solar PV–based hybrid integration of DC microgrid into public ported BEV is proposed in this chapter. The concept of microgrid in tandem with renewable integration has become the essential research of the present decade since the microgrid technology fueled by the renewable source is the modern-day development substituted to seize the challenges faced by the domestic grid in supply infrastructure and energy capacity, respectively. The basic architecture and components of a distributed network along with the associated issues of a microgrid system are revealed in this chapter clearly. The scope for solar energy is higher among other renewable resources due to its huge global availability potential, and therefore, it is preferred in the proposed investigation. The modeling of a solar cell is performed, and maximum power is achieved via adaptive P&O algorithm. The charging patterns of energy storage unit are illustrated in this chapter as they are considered to be the significant constituent of BEV and microgrid system. Converters used for integrating and establishing the

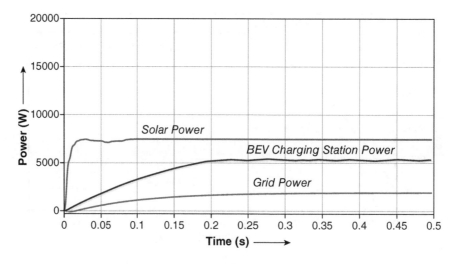

FIGURE 22.17 Simulation results of solar PV power, EV charging power, and grid power in case 3.

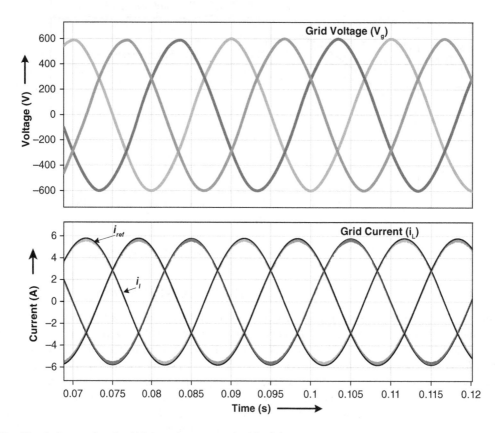

FIGURE 22.18 Simulation results of grid-injected current and grid voltage.

respective voltage across the loads are examined for high gain and versatile operation. Stability of the DC microgrid is analyzed under different loading conditions, and the power generated is exercised under various operating conditions. Validation of the discussion is achieved by experimenting 15 kW$_{Peak}$ solar panel at 25°C in the PSCAD/EMTDC software tool.

22.19 REFERENCES

[1] Pragaspathy, S., Baskaran, A. Mitigation of uncertainties in wind-powered renewable systems for environmental assets. *Polish Journal of Environmental Studies*. 2017 Jan 1; 26 (1).

[2] McPherson, S.S. *War of the Currents: Thomas Edison vs Nikola Tesla*. Twenty-First Century Books; 2012 Nov 1.

[3] Porter, S.F., Denkenberger, D., Mercier, C., May-Ostendorp, P., Turnbull, P. Reviving the war of currents: Opportunities to save energy with DC distribution in commercial buildings. In. *Proc. ACEEE Summer Study Energy Efficiency Buildings*, 2014 Aug (Vol. 85, p. 3).

[4] Höök, M., Tang, X. Depletion of fossil fuels and anthropogenic climate change – A review. *Energy Policy*. 2013 Jan 1; 52: 797–809.

[5] Dincer I. Renewable energy and sustainable development: A crucial review. *Renewable and Sustainable Energy Reviews*. 2000 Jun 1; 4 (2): 157–75.

[6] Rahbar, K., Chai, C.C., Zhang, R. Energy cooperation optimization in microgrids with renewable energy integration. *IEEE Transactions on Smart Grid*. 2016; 9 (2): 1482–1493.

[7] Khodayar, M.E., Barati, M., Shahidehpour, M. Integration of high reliability distribution system in microgrid operation. *IEEE Transactions on Smart Grid*. 2012 Sep 28; 3 (4): 1997–2006.

[8] Hosseinzadeh, M., Salmasi, F.R. Robust optimal power management system for a hybrid AC/DC micro-grid. *IEEE Transactions on Sustainable Energy*. 2015 Apr 1; 6 (3): 675–687.

[9] Kumar, D., Zare, F., Ghosh, A. DC microgrid technology: System architectures, AC grid interfaces, grounding schemes, power quality, communication networks, applications, and standardizations aspects. *IEEE Access*. 2017 Jun 1; 5: 12230–12256.

[10] Jayachandran, M., Ravi, G. Predictive power management strategy for PV/battery hybrid unit based islanded AC microgrid. *International Journal of Electrical Power & Energy Systems*. 2019 Sep 1; 110: 487–496.

[11] Shen, Lei, et al. Hierarchical control of DC micro-grid for photovoltaic EV charging station based on flywheel and battery energy storage system. *Electric Power Systems Research*. 2020; 179: 106079.

[12] Mohamed, Ahmed A.S., et al. Grid integration of a PV system supporting an EV charging station using Salp Swarm Optimization. *Solar Energy*. 2020; 205: 170–182.

[13] Chen, Zhiyong, et al. Adaptive sliding-mode voltage control for inverter operating in islanded mode in microgrid. *International Journal of Electrical Power & Energy Systems*. 2015; 66: 133–143.

[14] Saravanan, S., Karunanithi, K., Pragaspathy, S. A novel topology for bidirectional converter with high buck boost gain. *Journal of Circuits, Systems and Computers*. 2020 Nov 1; 29 (14): 2050222.

[15] Elgammal, A.A.A., Sharaf, A.M. Self-regulating particle swarm ptimized controller for (photovoltaic – fuel cell) battery charging of hybrid electric vehicles. *IET Electrical Systems in Transportation*. 2012; 2 (2): 77–89.

[16] Chakravarthi, B.N.C.H.V., Siva Krishna Rao, G.V. A high gain novel double-boost converter for DC microgrid applications. *Journal of Circuits, Systems and Computers*. 2020 Dec 15; 29 (15): 2050246.

[17] Moharil, R.M., Kulkarni, P.S. Reliability analysis of solar photovoltaic system using hourly mean solar radiation data. *Solar Energy*. 2010 Apr 1; 84 (4):691–702.

[18] Gulkowski, S., Zdyb, A., Dragan, P. Experimental efficiency analysis of a photovoltaic system with different module technologies under temperate climate conditions. *Applied Sciences*. 2019 Jan; 9 (1): 141.

[19] Chakravarthi, B.C., Rao, G.S. Impact of power quality issues in grid connected photovoltaic system. In *2020 4th International Conference on Electronics, Communication and Aerospace Technology (ICECA)*, 2020 Nov 5 (pp. 155–158). IEEE.

[20] Lakshmi, M., Hemamalini, S. Nonisolated high gain DC – DC converter for DC microgrids. *IEEE Transactions on Industrial Electronics*. 2017 Jul 31; 65 (2): 1205–1212.

[21] Patsamatla, H., Karthikeyan, V., Gupta, R. Universal maximum power point tracking in wind-solar hybrid system for battery storage application. In *2014 International Conference on Embedded Systems (ICES)*, 2014 Jul 3 (pp. 194–199). IEEE.

[22] Ishaque, K., Salam, Z., Lauss, G. The performance of perturb and observe and incremental conductance maximum power point tracking method under dynamic weather conditions. *Applied Energy*. 2014 Apr 15; 119: 228–236.

[23] Fathabadi, Hassan. Novel solar powered electric vehicle charging station with the capability of vehicle-to-grid. *Solar Energy*. 2017; 142: 136–143.

[24] Khan, Wajahat, Furkan Ahmad, Mohammad Saad Alam. Fast EV charging station integration with grid ensuring optimal and quality power exchange. *Engineering Science and Technology, an International Journal*. 2019; 22 (1): 143–152.

[25] Gautam, D.S., Musavi, F., Edington, M., Eberle, W., Dunford, W.G. An automotive onboard 3.3-kw battery charger for PHEV application. *IEEE Trans Vehicular Technol*. 2012; 61 (8): 3466–3474

[26] Hilton, George, et al. Dynamic charging algorithm for energy storage devices at high rate EV chargers for integration of solar energy. *Energy Procedia*. 2018; 151: 2–6.

[27] Chae, H.J., Moon, H.T., Lee, J.Y. On-board battery charger for PHEV without high-voltage electrolytic capacitor. *Electronics Letters*. 2010 Dec 9; 46 (25): 1691–1692.

[28] Praneeth, A.V., Williamson, S.S. Modeling, design, analysis, and control of a nonisolated universal on-board battery charger for electric transportation. *IEEE Transactions on Transportation Electrification*. 2019 May 27; 5 (4): 912–924.

[29] Kouka, Karima, et al. Dynamic energy management of an electric vehicle charging station using photovoltaic power. *Sustainable Energy, Grids and Networks*. 2020; 24: 100402.

[30] Berigai Ramaiaha, A., Maurya, R., Arya, S.R. Bidirectional converter for electric vehicle battery charging with power quality features. *International Transactions on Electrical Energy Systems*. 2018 Sep; 28 (9): 2589.

[31] Mohammadi, P., Moghani, J.S. Double-input high-gain bidirectional DC-DC converter for hybrid energy storage systems in DC-micro grid. In *2018 9th Annual Power Electronics, Drives Systems and Technologies Conference (PEDSTC)* 2018 Feb (pp. 312–317). IEEE.

[32] Lee, Y.J., Khaligh, A., Emadi, A. Advanced integrated bidirectional AC/DC and DC/DC converter for plug-in hybrid electric vehicles. *IEEE Transactions on Vehicular Technology*. 2009 Jul 21; 58 (8): 3970–3980.

23 Battery Packs in Electric Vehicles

Antonio Peršić

CONTENTS

23.1 Introduction 305
23.2 Electric Vehicle Configuration 305
23.3 High-Voltage Battery Pack Calculation 306
    23.3.1 Vehicle Parameters 307
    23.3.2 Measured Driving Profile 307
    23.3.3 Energy Consumption 308
    23.3.4 Battery Package Design 308
    23.3.5 Electric Vehicle Auxiliary System Consumption 309
    23.3.6 Electric Vehicle Efficiency 310
    23.3.7 Total Energy Consumption 310
    23.3.8 Battery Pack Architectures 310
    23.3.9 Battery Pack Configuration 311
23.4 Electric Vehicle Examples 313
23.5 Electric Vehicle Charging 313
    23.5.1 AC Charging Systems 313
    23.5.2 DC Charging Systems 313
23.6 Renewable Sources of Energy and Electric Vehicles 314
    23.6.1 Overview of Solar-Powered Electric Vehicle Power Train Configurations 314
    23.6.2 Solar-Powered Charging Station 316
    23.6.3 Solar Parking Lot Charging Stations 316
    23.6.4 Wind-Powered Charging Stations 316
        23.6.4.1 Off-Grid Wind Power Plants and EV Chargers 317
        23.6.4.2 Sanya Skypump 318
23.7 Hybrid-Powered EV Charging Stations 318
    23.7.1 Solar-Grid-Based Fast Charging System 318
    23.7.2 Solar-Wind-Based Charging Stations 318
    23.7.3 Solar-, Wind-, Air-, Hydrogen-, and Ammonia-Based Charging Station 318
23.8 Conclusion 320
23.9 References 320

23.1 INTRODUCTION

When talking about technological advances in wind and solar applications, it is important to mention electric vehicles and electric energy storage units.

It is a known fact that if all systems and devices that generate a substantial amount of pollution are transformed (if possible) into electrical devices (i.e., electrified) and all the electrical energy that powers those devices is generated from a renewable energy source, the result would be a significant decrease of CO_2 contributors. With the transportation sector generating a substantial amount of the world's carbon emission, this chapter is concentrated on the electrification process of vehicles by usage of electrical devices, thus transforming them into electrical vehicles.

This chapter is based on a battery package design procedure and its integration with renewable energy sources for the purpose of CO_2 reduction. Although this chapter focuses on electric vehicles, these ideas can be readily applied to other energy storage applications. In the end, the chapter will complete with an overview of wind and solar power implementation and present the possibility of a fully clean electric vehicle and how to achieve it.

23.2 ELECTRIC VEHICLE CONFIGURATION

The main component of any vehicle is its power train. It provides power to the vehicle, and it consists of a motor (engine) that generates power to achieve movement. The power train of an ICE (internal combustion engine) vehicle consists of a substantial amount of moving parts. Its main circuit consists of the engine that generates the power, then transmits it via the transmission gear system to the driveshaft. To achieve this, the ICE engine needs the following systems and modules, two or more axles, a differential, exhaust, an emissions control unit, an engine cooling and lubrication system, etc. [1].

Compared to the ICE's power train, the EV's (electric vehicle) power train has 60% fewer components. To achieve electrical propulsion, the electric vehicle consists of at least the following components:

- DC–AC converter – used to achieve highly efficient control of the electric motor. Also referred to as inverter, frequency converter, etc.
- Electric motor – converts electrical to mechanical energy, which is via transmission delivered to the wheels.
- High-voltage battery package – the main and only source of energy that the power train consumes to achieve propulsion. It consists of a serial-parallel combination of smaller battery cells to achieve the desired energy. The battery cells are usually based on the lithium-ion technology.
- Internal charger – a controllable AC/DC converter (rectifier) used for charging the battery pack. It has various charging functions, like the constant current and constant voltage charging functions.

Apart from the presented key elements, the EV's power train consists of multiple hardware and software components. Various electronic control units (ECUs) are distributed along the entire EV. Their main purpose is to control its accompanied system (like ABS, ESP, HVAC, brake system, airbag, seat belts, etc.) and to exchange data between the main (vehicle control unit, VCU) or any other controller. There are several small ECUs in an EV that perform specific functions. The communication between different ECUs in a vehicle is commonly carried over CAN protocol. More examples of core ECUs are:

- *Battery management system (BMS)*. A controller that is used for monitoring and control of the battery package. Its main concern is the battery pack's safety and voltage balancing functions as well as the state of charge, state of health, and depth of discharge estimation functions; this is to achieve maximum safety and efficiency. It is also connected to the main CAN bus in order to exchange data with other ECUs and the VCU. The BMS also controls the internal and external charging modules.
- *DC–DC converter*. A buck, boost, or buck-boost DC–DC converter is used to adapt the battery pack's voltage to any other auxiliary system (e.g., wipers, lights, infotainment system, mirror control, HVAC, ECUs, etc.) that needs power.
- *Thermal management system*. A cooling or heating system whose main purpose is to keep the system's modules temperature's within the rated range. It can consist of multiple small systems controlled with one controller, or it can be a fully integrated system.
- *Body control module (BCM)*. The main controller for the EV's body (e.g., power, window, mirrors, security, and vehicle access control), it supervises and controls the functions of electronic accessories.

The following sections discuss the configuration, assembly, components, and importance of the high-voltage battery package.

23.3 HIGH-VOLTAGE BATTERY PACK CALCULATION

The electric vehicle's battery package (i.e., battery pack) is the main source of power for all the onboard systems. It contains the energy to power the electrical propulsion system, main systems (inverter; VCU, or vehicle control unit; servo system; cooling system; lights; etc.) via the standard car battery, media system (infotainment), and secondary systems (HVAC, or heating, ventilation, and air-conditioning; cabin lights; charger; etc.). This underscores the battery package's importance and defines the need for it to be as efficient, reliable, safe, and environment-friendly as possible.

The lithium-ion battery cell is the building block of the battery pack, and it comes not only in different shapes and sizes (cylindrical, pouch, prismatic) but also in different chemical compositions ($LiMn_2O_4$, $LiNiMnCoO_2$, etc.). The most widely used battery cell is the lithium-ion-based battery cell, with its columbic efficiency of 99% [2]. When compared to other battery technologies, it has the most energy density per kilogram and the best value for money.

A battery pack is formed as a serial-parallel connection of identical battery cells. The configuration of a serial string or the number of parallels depends on the desired voltage and capacity that the battery pack needs to meet.

It is necessary to define certain terms used in this chapter because of the inconsistency in the terminology found in literature regarding electric vehicles. The following definitions are used in the following chapters:

- Cell – the smallest segment of a battery package that has the functionality of an electrochemical battery. This is the smallest increment that the voltage and capacity may increase by.
- Module – also known as a segment or a brick, is a bundle of electrically connected cells grouped in a solid encasing. It usually contains its own BMS slave modules along with thermal control components.
- Battery pack – a group of segments, modules, or bricks (cells) along with the BMS master module and the thermal control module, encased in a solid enclosure that provides protection for the battery pack and for the environment as well.
- Battery – a colloquial reference to any electrochemical energy storage device, including battery cells, bricks, segments, modules, UPS devices, high-voltage energy storage devices, battery packs.

The EV's battery pack is considered as one of the most (if not the most) important components of an electric vehicle. It is responsible for the following EV's parameters: road performance (acceleration and deceleration time), maximum range, and charging time. The major parameters required to design a battery package are the required average energy consumption and the required range of the electric vehicle.

23.3.1 Vehicle Parameters

The performance parameters are the starting point of any electric vehicle design procedure. The vehicle parameters are based on the type of use; this can be a bus, car, forklift, truck, etc. Usually, the main focus is on maximum velocity, ramp-up time from 0 to 100 (km/h), and maximum range.

In order to estimate the required parameters, a thorough market study and some statistic data should be analyzed. Afterwards, the parameters can be smoothened out and adapted; they will not affect the calculation by much. Table 23.1 contains an example of the calculation parameters.

23.3.2 Measured Driving Profile

Although there are government-approved drive cycles [3], the best drive cycle for battery pack calculation is the one that the vehicle is actually going to drive through. If possible, the calculation should be based on a specific route that the vehicle needs to drive on. The chosen route terrain should be driven with the desired velocity and distance and measured with a high-frequency accelerometer, gyroscope, and GPS module. With this data, a more accurate driving profile can be created for the vehicle, resulting with the calculated energy consumption that can fit the requirements even more.

The raw data is usually exported in a. CSV file for easier data processing, and they usually contain the following data:

- Acceleration X-axis [g]
- Acceleration Y-axis [g]
- Acceleration Z-axis [g]
- Gyroscope X-axis [°/s]
- Gyroscope Y-axis [°/s]
- Gyroscope Z-axis [°/s]
- Latitude [°]
- Longitude [°]
- Velocity [km/h]
- Altitude [m]
- Position [m]
- Time [DD.MM.YYYY HH:MM:ss.sss]

The goal is to calculate the necessary parameters to be able to calculate the forces and energy that the vehicle creates and is subjected to while driving the route. There are numerous ways to calculate these parameters, of which the following is the most common.

The **haversine formula** is used to calculate the traveled distance using the measured longitude and latitude.

$$ds = 2 \cdot r \cdot \sin^{-1} \left(\sqrt{\sin^2\left(\frac{\varphi_2 - \varphi_1}{2}\right) + \cos(\varphi_1) \cdot \sin^2\left(\frac{\lambda_2 - \lambda_1}{2}\right)} \right) \quad (23.1)$$

The symbols used in the previous equation are:

- ds [m] – distance difference between two points on the coordinate system
- r [m] – earth radius

TABLE 23.1
Vehicle Calculation Parameters

Type	Parameter [unit]	Symbol	Value
Battery Pack	Rated voltage [V]	Un	400
	Chemistry	-	Lithium-ion
Power Train	Electric motor	-	IPM sync.
Vehicle	Required max. range [km]	s_{req}	250
	Max. velocity [km/h]	v_{max}	90
	Time, 0–100 km/h [s]	t_{rup}	15
	Total mass [kg]	m_v	5,000
	Tire radius [m]	r_w	0,3647
	Tire friction coefficient (wheel–road) (no slip) [-]	μ_f	1
	Electric vehicle height [m]	h_v	2,8
	Electric vehicle width [m]	w_v	2,5
	Rated number of passengers [-]	n_{psg}	23
	Passenger average weight [kg]	m_{psg}	90
Other	Radius of earth [m]	r	6,378,000
	Gravitational acceleration constant [m/s^2]	g	9,81
	Air density [kg/m3]	ρ	1,225
	Rolling resistance coefficient rough estimate [-]	c_{rr}	0,015
	Aerodynamic drag coefficient rough estimate [-]	c_d	0, 6

- φ_1 [rad] – latitude of the first point
- φ_2 [rad] – latitude of the second point
- λ_1 [rad] – longitude of the first point
- λ_2 [rad] – longitude of the second point

To calculate the **total traveled distance** s [m], an integration is required.

$$s = \int ds \cdot dt \quad (23.2)$$

The following expressions are used to calculate **velocity** v [m/s]:

$$v = \frac{ds}{dt} \quad (23.3)$$

And **acceleration** a [m/s^2]:

$$a = \frac{dv}{dt} \quad (23.4)$$

The last parameter to be calculated is derived from the measured altitude. The following expression is for **the road incline** α [rad]:

$$\alpha = \tan^{-1}\left(\frac{dh}{ds}\right) \quad (23.5)$$

The symbol used in the previous equation is:

- dh [m] – altitude difference

With these basic movement parameters, the energy consumption can be calculated.

23.3.3 Energy Consumption

The energy consumption is calculated based on the road loads. The **total road load** F_{tot} [N] is the sum of the inertial force, road slope force, road load (friction) force, and aerodynamic drag force.

$$F_{tot} = F_d + F_s + F_r + F_a \quad (23.6)$$

The symbols used in the previous equation are:

- F_d [N] – driving force
- F_s [N] – climbing resistance
- F_r [N] – rolling resistance
- F_a [N] – aerodynamic drag

$$P_{tot} = F_{tot} \cdot v_v \quad (23.7)$$

The **absolute total power** P_{abs_tot} [W] represents the total power without the potential energy received from regenerative braking.

$$P_{abs\_tot} = |P_{tot}| \quad (23.8)$$

The **total energy consumed** E_{tot} [Wh] on the driven distance can be calculated:

$$E_{tot} = \int P_{tot} \cdot dt \quad (23.9)$$

The **absolute total energy consumption** E_{abs_tot} [Wh], is calculated from the absolute total power.

$$E_{abs\_tot} = \int P_{abs\_tot} \cdot dt \quad (23.10)$$

Figure 23.1 contains the calculated energy function. It can clearly be seen that if the total energy (red function) is subtracted by total absolute energy (blue function), the result would be the regenerative energy (yellow function). The calculation needs to be done without the regenerative component to produce the worst-case consumption scenario; this means that the total absolute energy is used.

The maximum value of the absolute energy consumption is the energy required to drive through the required route with the driving profile used.

If the maximum value of the absolute energy consumption is divided with the total distance, the result would be **the average energy consumption per kilometer** E_{avg} [kWh/km].

$$E_{avg} = \frac{\max(E_{abs\_tot}(t))}{\max(s(t))} \quad (23.11)$$

This value is important because it contains the driving profile and habits of the driver and performance of the vehicle, and it is scalable to any range.

The following expression is used to scale the average energy consumption per kilometer with the required distance s_{req} [m] to get **the scaled total consumed energy for the required distance** E_{con} [kWh]. This is the mechanical energy rated on the wheels of the vehicle.

$$E_{con} = E_{avg} \cdot s_{req} \quad (23.12)$$

Table 23.2 contains the calculated energy consumption and distance values covered in the previous expressions.

The measured route contains the driving profile, route topology, and vehicle dynamics; with a simple change of a few parameters, an accurate calculation can be derived for any vehicle.

It is necessary to point out that the calculated total energy consumption, scaled on a distance of sreq, is not the battery pack required energy; it is the energy that needs to be consumed on the wheels of the vehicle. The following chapter will focus on calculating the battery pack required energy.

23.3.4 Battery Package Design

The parameters of the battery pack have a major influence on the characteristics of the EV:

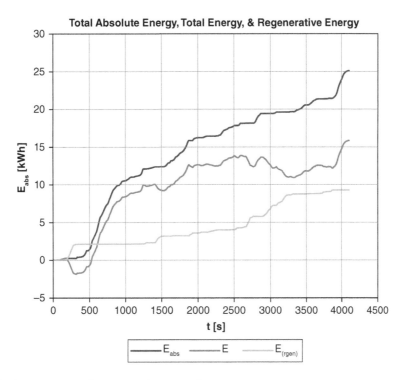

FIGURE 23.1 Total absolute energy, total energy, and regenerative energy.

TABLE 23.2
Propulsion Energy Consumption

Parameter [Unit]	Symbol	Value
Absolute total energy consumption [kWh]	$\max(E_{abs\_tot})$	16,08
Total distance [km]	s_{tot}	27,61
Average energy consumption for propulsion [Wh/km]	E_p	582,10
Required distance [km]	s_{req}	250
Consumed energy for the given distance [kWh]	E_{con}	145,53

- **Maximum traction motor torque** – depends on the maximum current that the electric motor can receive from the battery pack.
- **Maximum regeneration brake torque** – depends on the maximum charging current that the battery pack is allowed to take in.
- **Vehicle range** – depends on the battery pack capacity and continuous current that it can deliver.
- **Vehicle total mass** – the battery pack is known to form up to 50% of the electric vehicle's mass.
- **Vehicle price** – depends on the battery pack size and number of components. It usually comprises about 60% of the total electric vehicle construction price.

When talking about battery package design, there are four main components that need to be defined:

- **Battery chemistry** – the correct battery cell needs to be chosen for the desired application.

- **Voltage** – the battery pack voltage needs to be chosen. If the voltage is too high, the battery pack losses will decrease because of the lower current. Where the AC/DC inverter will be more expensive because the IGBTs will need to be rated for higher voltages, a balance needs to be made.
- **Average energy consumption** – this determines the battery pack's size. The energy, combined with the defined voltage, will lead to the required capacity.
- **Vehicle range** – another component that is necessary to determine the battery pack size.

23.3.5 Electric Vehicle Auxiliary System Consumption

The energy to achieve movement is not the only consumer of energy. Every vehicle has auxiliary systems which are fed from the battery pack; thus, the average auxiliary energy consumption E_{aux_avg} [Wh/km] needs to be added. These systems include low-voltage electrical systems, control voltage for the dashboard and controllers, the standard car battery, heating, cooling, etc. The typical power consumption data for these auxiliary systems can be found online [4]. The auxiliary electrical loads are divided into continuous loads P_{cont} [W] (controllers and background systems), prolonged loads P_p [W] (high- and low-beam headlights, HVAC, multimedia, etc.), and occasional loads P_i [W] (heaters, washing fluid pump, taillights, reverse lights, front and rear wipers, etc.).

The required energy consumed by the auxiliary system E_{aux} is scaled with the driving time t_{drv} [h]; also, the corresponding type of electrical load needs to be multiplied with its usage factor. The prolonged electrical load is roughly used 60% of the driving time, where the intermittent electrical load is used only 10% of the time.

$$E_{aux} = \left(P_{cont} + P_p \cdot c_p + P_i \cdot c_i \right) \cdot t_{drv} \quad (23.13)$$

This results in the following usage factors:

- Prolonged electrical load usage factor: $c_p = 0.6$ [-]
- Intermittent electrical load usage factor: $c_i = 0.1$ [-]

The average auxiliary systems electrical consumption E_{aux_avg} [Wh/km] can be calculated using the drive cycle distance s_{tot} [km] in the following expression:

$$E_{aux\_avg} = \frac{E_{aux}}{s_{tot}} \quad (23.14)$$

23.3.6 Electric Vehicle Efficiency

The power train efficiency η_p [-] needs to be considered when energy transformation from the electric motor to the wheels is achieved. It depends on the power train type and configuration, but the usual losses are located in the gearbox, transmission, electric motor, inverter or drive, and output of the battery package [5]. As per the following expression:

$$\eta_p = \eta_{gtr} \cdot \eta_{mot} \cdot \eta_{inv} \cdot \eta_{bp\_out} \quad (23.15)$$

Table 23.3 contains the calculation result of the efficiency of the power train that is usual for a standard electric vehicle.

23.3.7 Total Energy Consumption

With (23.15), the expression for the total average energy consumption E_{tot_avg} [Wh/km] of the electric vehicle can be derived.

$$E_{tot\_avg} = \left(\frac{E_p}{\eta_p} + \frac{E_{aux\_avg}}{\eta_{bp\_out}} \right) \quad (23.16)$$

The next expression is used to scale the total average energy consumption per kilometer with the required distance s_{req} [m] to get the **scaled total consumed energy for the required distance** E_{tot_con} [kWh].

$$E_{tot\_con} = E_{tot\_avg} \cdot s_{req} \quad (23.17)$$

This is the total electrical energy that needs to be consumed, rated on the battery pack output. Table 23.4 contains all the calculated final energy values in this example.

The value E_{tot_avg} contains the **total average consumption** of a vehicle with the parameters specified in Table 23.1, driven with the measured driving profile and route, with all the necessary auxiliary vehicle systems included. This parameter is useful because it represents the factorized energy per the unit of kilometer. When multiplied with the required vehicle range, the result is the **total consumed energy for the given distance**.

23.3.8 Battery Pack Architectures

As mentioned before, high-voltage battery packs consist of a number of modules that contain battery cells. A battery cell's voltage is the smallest division of the battery pack's total voltage. This means that the total voltage can only be incremented or decremented by the factor of the battery cell voltage.

The same stands for the battery cell's capacity in regard to the battery pack's total capacity.

A module is formed when the battery cells are connected in a serial-parallel topology. Later on, the aforementioned modules need to be connected in a similar serial-parallel topology to form the battery pack. The levels of modularity solely depend on the battery cell specifications and the battery pack's required size.

TABLE 23.3
Power Train Efficiency

Parameter [Unit]	Symbol	Value
Gearbox and transmission efficiency [-]	η_{gtr}	0,94
Electric motor efficiency [-]	η_{mot}	0,80
DC/AC inverter efficiency [-]	η_{inv}	0,9
Battery pack output efficiency [-]	η_{bp_out}	0,9
Total power train efficiency [-]	η_p	0,61

TABLE 23.4
Total Average Energy Consumption

Parameter [Unit]	Symbol	Value
Drive cycle time [h]	t_{cyc}	1,13
Prolonged electrical load usage factor [-]	c_p	0,60
Intermittent electrical load usage factor [-]	c_i	0,10
Continuous electrical load [W]	P_{cont}	2010,00
Prolonged electrical load [W]	P_p	1464,00
Intermittent electrical load [W]	P_i	172,00
Auxiliary energy consumption of the driving cycle [Wh]	E_{aux}	4.119,98
Average auxiliary systems energy consumption [Wh/km]	E_{aux_avg}	150,4
Average total energy consumption [Wh/km]	E_{tot_avg}	811,60
Total consumed energy for the given distance [kWh]	E_{tot_con}	202,00

A **serial connection** of the battery cells results in an increase of voltage, where the total voltage of a string is the sum of voltages of each battery cell.

A parallel connection of serial strings results in the increase not only of the total capacity but also of the total current that the battery pack can deliver. If the battery cells are of different model and maker and there is a difference of the battery cell capacity, this will affect the battery pack's capacity. In this case, the total capacity of a serial string will be equal to the lowest capacity of the battery cells within that string.

This is one of the reasons the battery pack needs to be formed out of battery cells of the same maker and model. Furthermore, the battery cells used in a battery pack should also be of the same production batch, to secure the most similar chemical properties possible. This is to ensure a longer battery pack life expectancy and a more reliable state of charge (SoC) calculation. Of course, nothing is ideal, but this is a favored precaution.

It is made out of 22 segments that are made from 4 prismatic battery cells that result in a total of 88 serially connected battery cells. With a single voltage increment of 3.7 V, the total voltage of the 88 battery cells is around 330 V.

It is built from 16 modules that consist of 74 cells connected in parallel and 6 in series, and it results in a total of 7,104 battery cells. The battery cell's voltage is 3,6 V, with the capacity of 3,400 mAh; this results in a total of 345,6 V and 251,6 Ah.

23.3.9 Battery Pack Configuration

The voltage level of the battery pack determines the maximum electrical power which can be delivered continuously. As always, the nominal power P_n [W] is calculated with the nominal voltage U_n [V] and the nominal current I_r [A]:

$$P_n = U_n \cdot I_r \quad (23.18)$$

Low voltages result in higher currents, and as the current rises, the thermal and voltage drop losses rise as well, which leads to greater wire diameters. While the current is limited not only by the battery cell maximum current but also by the connector and wiring losses, the voltage is only limited by the price of the inverters and converters connected to it. It is common practice to increase the battery pack's total voltage and keep the current at a lower value to increase the total power of the battery pack. The most common nominal voltage used in the electric vehicle industry is just under **400 V DC**, but more and more manufacturers have started to increase the voltage to **800 V DC** to decrease the electrical and thermal loses and still get the desired power.

When the correct battery cell is selected for the application, the total battery pack voltage selected and the required energy calculated, the battery pack parameter calculation can proceed.

The **battery pack's required energy** E_{bp} [Wh] is the average energy consumption E_{avg} [Wh/km] driven on the required range D_v [km].

$$E_{bp} = E_{avg} \cdot D_v \quad (23.19)$$

If the battery pack is designed as a **battery storage unit**, then the battery pack total energy is equal to the calculated energy demands of the device, household, industrial plant section, or the energy storage plant grid demand.

The total battery pack voltage U_{bp} [V] divided with the battery cell voltage U_{bc} [V] results in the **number of battery cells in a serial string** N_{cs} [-]. This integer division needs to be rounded up to a higher number.

$$N_{cs} = \frac{U_{bp}}{U_{bc}} \quad (23.20)$$

Then the **battery pack's total voltage** U_{bp} [V] needs to be recalculated to match the rounded-up battery cell number in a serial string. This is accomplished with rearranging the previous expression.

$$U_{bp} = N_{cs} \cdot U_{bc} \quad (23.21)$$

The **available energy of serial string** E_{bs} [Wh] is calculated by multiplying the number of battery cells in a serial string N_{cs} [-] with the energy of each cell E_{bc} [Wh].

$$E_{bs} = N_{cs} \cdot E_{bc} \quad (23.22)$$

The division of the battery pack's total energy E_{bp} [Wh] and the available energy of serial string E_{bs} [Wh] results in the **total number of cells connected in parallel** N_{cp} [-]. This integer division needs to be rounded up to a higher number.

$$N_{cp} = \frac{E_{bp}}{E_{bs}} \quad (23.23)$$

Then the **battery pack's total energy** E_{bp} [Wh] needs to be recalculated to match the rounded-up number of battery cells in parallel. This is accomplished with rearranging the previous expression.

$$E_{bp} = N_{cp} \cdot E_{bs} \quad (23.24)$$

The **total capacity of the battery pack** C_{bp} [Ah] can be seen in the following expression. With the number of battery cells in parallel N_{cp} [-] and the capacity of a single battery cell C_{bc} [Ah].

$$C_{bp} = N_{cp} \cdot C_{bc} \quad (23.25)$$

The **final number of cells** N_{cb} [-] can be seen in the following expression.

$$N_{bp} = N_{cp} \cdot N_{cs} \quad (23.26)$$

The **battery pack's size and mass** are imperative parameters when talking about battery pack design. Statistically, the battery package comprises from 30% to 50% (depending on the battery pack size) of the vehicle's total mass [6]; this is not that important when the battery package is a standstill energy storage unit.

Most of the battery pack's mass and volume come from the battery cells themselves, and the rest is in the wiring and cooling system (again, this depends on the hardware design of the battery package). To get a rough estimate of the battery pack's mass and volume, it is safe to use only the battery cell mass and dimension in consideration. A more accurate various circuits, wires, and encasement configurations should be taken into consideration.

The **total mass of the battery cells** m_{bp} [kg] can be found in the following expression, where m_{bc} [kg] is the mass of a single battery cell.

$$m_{bp} = N_{bp} \cdot m_{bc} \quad (23.27)$$

The **volume of all the battery cells gathered together** V_{bp} [m3] can be found in the following expression, where $V_{cc(pc)}$ [l] is the volume of a single cell (cylindrical or pouch).

$$V_{bp} = N_{bp} \cdot V_{cc(pc)} \quad (23.28)$$

The **maximum current of a serial string** I_{spc} [A] is equal to the capacity of a single battery cell C_{bc} [Ah] scaled via the corresponding C-rate factor $C\text{-}rate_{bcp}$ [-], given by the manufacturer.

$$I_{spc} = C\text{-}rate_{bcp} \cdot C_{bc} \quad (23.29)$$

The **total maximum current** I_{bpp} [A] of the battery pack is equal to the maximum current of a serial string multiplied by the number of parallels.

$$I_{bpp} = I_{spc} \cdot N_{cp} \quad (23.30)$$

The **total maximum power** P_{bpp} [W] of the battery pack is the product between the total maximum current I_{bpp} [A] and the total voltage U_{bp} [V].

$$P_{bpp} = I_{bpp} \cdot U_{bp} \quad (23.31)$$

The **nominal current of a serial string** I_{scc} [A] is equal to the capacity of a single battery cell C_{bc} [Ah] scaled via the corresponding C-rate factor $C\text{-}rate_{bcc}$ [-], given by the manufacturer.

$$I_{scc} = C\text{-}rate_{bcc} \cdot C_{bc} \quad (23.32)$$

The **total nominal current** I_{bpc} [A] of the battery pack is equal to the nominal current of a serial string multiplied by the number of parallels.

$$I_{bpc} = I_{scc} \cdot N_{cp} \quad (23.33)$$

The **total nominal power** P_{bpc} [W] of the battery pack is the product between the total nominal current I_{bpc} [A] and the total voltage U_{bp} [V].

$$P_{bpc} = I_{bpc} \cdot U_{bp} \quad (23.34)$$

Table 23.5 contains the calculated parameters of each battery cell example using the preceding expressions.

It can be seen in the previous table that the **pouch cells** contain a higher energy factor, have greater discharge and charge currents, and have greater capacity in comparison to the cylindrical ones, but the cylindrical cells can be more space- and mass-efficient.

The mass and dimensions of the battery pack are calculated only by taking the battery cell's dimensions and mass into consideration. The actual battery pack will be somewhat greater in mass and volume because of the additional essential components (wiring, battery management system, thermal management unit, encasing, etc.). Nevertheless,

TABLE 23.5
Battery Package for Different Battery Cells

Manufacturer	Panasonic	A123-Systems	Molicel	A123-Systems	Toshiba	Kokam
Serial connection [-]	112	122	109	122	174	112
Serial con. energy [Wh]	1291	1007	1050	7851	8003	6290
Parallel connections [-]	157	200	193	26	25	33
Battery pack energy [kWh]	202,57	219,60	196,91	222,67	313,20	207,57
Battery pack capacity [Ah]	502,40	500,00	501,80	507,00	500,00	514,80
# total cells [-]	17584	24400	21037	3172	4350	3696
Battery pack mass (cells) [kg]	863,69	1882,22	1068,20	1633,82	2307,24	1171,63
Battery pack volume (cells) [l]	312,42	854,25	378,29	867,38	1178,91	634,75
Battery pack peak current [A]	508,80	12180,00	1019,20	5265,00	520,00	1544,40
Battery pack peak power [kW]	205,15	4903,67	411,04	2119,69	208,10	622,70
Battery pack continuous current [A]	508,80	5075,00	509,60	526,50	520,00	1029,60
Battery pack continuous power [kW]	205,15	2043,20	205,52	211,97	208,10	415,13

this still is an accurate estimation for battery pack mass comparison.

When talking about which battery cell is best for the battery pack. If the vehicle is a racing vehicle, then cells with high current capacity need to be chosen; if the vehicle range needs to be emphasized, then the cells with greater capacity need to be chosen. In conclusion, the battery cells that need to be chosen are the ones that suit the application the best with the least compromise. A compromise needs to be made between mass, volume, energy/power density, and price. There is no such thing as a battery cell for all application.

23.4 ELECTRIC VEHICLE EXAMPLES

A few representatives of the electric vehicle industry are presented. **Porsche Taycan** is a good representative of a high-powered electric vehicle. The battery pack contains 33 modules with 12 cells each; this results in 396 pouch cells in total. Two synchronous motors of the total power of 390 kW get the average consumption up to 289 Wh/km. The total capacity is 93.4 kWh, with a voltage of 800 V; this is the first electric vehicle [7] with a high voltage of that size. The low energy is compensated with a sophisticated thermal management system to ensure the battery cells won't go into thermal getaway [8]. With the mass of 2.3 tons, it can drive in a maximum range of 416 km and a maximum velocity of 250 km/h, with an acceleration from 0 to 100 km/h in 4 s. The charging time in fast charging mode is from 5% to 80% in 22.5 minutes, but only with a charging station of 270 kW.

With their modular design and a wide range of components, the **Movitas E-bus** [9] and **Tribus E-bus** [10] are the best representatives of a compact, highly efficient electric minibus. Its feature of the modular design allows the customer to add or remove modules to increase or decrease its length. With a battery pack of 68 kWh, maximum speed up to 60 km/h, it can drive a range of 200 km. The integrated PV on the rooftop is to increase its efficiency even more, as well as an advanced thermal management system integrated with its HVAC along with floor heating.

23.5 ELECTRIC VEHICLE CHARGING

A battery pack of an EV can be charged via the embedded plug in two ways: via the AC voltage or via the DC voltage. When using the AC charging system, it is connected to the existing charger integrated within the EV. With the DC charging system, a direct connection from an outside charger to the battery pack is established. In both cases, the BMS sends the required charging parameters to the chargers, where only in the case of the DC charging system the connection is achieved with the CAN bus via the terminal embedded within the charging plug.

23.5.1 AC Charging Systems

As mentioned before for these charging systems, an AC/DC charger with a power factor correction module needs to be embedded within the EV.

There are various power levels that the chargers can be connected to: portable **low-power** chargers with a standard household power outlet (120 V AC, 16 A), **medium-power** chargers that are embedded within the EV (240 V AC, 60 A, up to 14.4 kW), and **high-power** chargers that are embedded within the EV (240 V AC, greater than 14.4 kW). **Fast chargers** that are embedded within the EV are classified as high-power chargers, but this does not mean that the high-power chargers are all fast chargers. A fast charger can be called that if it can charge the battery pack within 30 minutes or less. AC chargers need to be automotive-grade components that need to be robust and reliable.

23.5.2 DC Charging Systems

These types of chargers are an analogy of a gas station. They are usually located in a convenient location, like garages, parking lots, or even conventional gas stations. These chargers are high-powered charging systems, and they usually come with the fast charging option. Like previously mentioned, this charger needs to establish communication with the EV's BMS to receive the necessary charging parameters. There are various power levels that these chargers can deliver:

- Voltages from 200 to 450 VDC, with currents up to 80 A, and power 36 kW
- Voltages from 200 to 450 VDC, with currents up to 200 A, and power 90 kW
- Voltages from 200 to 600 VDC, with currents up to 400 A, and power 240 kW

23.6 RENEWABLE SOURCES OF ENERGY AND ELECTRIC VEHICLES

Renewable sources of energy can be utilized in countless combinations when talking about EVs. Along with regenerative braking, auxiliary sources of energy (alternators, generators, range extenders, solar panels, etc.) are used to charge the battery pack while in operation, which results in wider usability range and a reduction in charging frequency of the electric vehicle; it also extends its drive range.

For **onboard** applications, a solar panel installed on the roof is the only way to utilize renewable energy source in an EV. It doesn't seem very effective to use wind as an energy source on board an EV during operation.

For **off-board** applications, various combinations of EV chargers and renewable energy supplies can be used. In this chapter, examples of using renewable energy sources with EV charging are presented.

23.6.1 Overview of Solar-Powered Electric Vehicle Power Train Configurations

The main principle of this kind of electric vehicle is based on converting solar energy via solar panels to electrical energy, only to be stored in the battery system or to be used directly [11]. The following examples show different topologies available for solar-powered vehicles, but only the most important individuals are presented.

Figure 23.2 represents the classic solar-powered power train topology. The solar panels charge a battery package, which is directly used to power the electric motor via the frequency converter (DC/AC). Though this topology is simple to control and implement, it can drive a very limited range, lacks the power to accomplish fast changes in velocity, with low efficiency and a high sensitivity to weather conditions.

The following topology presented in Figure 23.3 is a modification of the aforementioned topology, with one exception: it utilizes a maximum power point tracking (MPPT) unit. This is done to increase the efficiency of the solar panels at different irradiation levels [11].

The MPPT is a controllable DC/DC converter that measures the generated voltage from the solar panel to charge a small energy storage unit (capacitor, battery cell). It then compares the voltage to the total voltage of the battery pack and delivers the voltage at peak power to the battery pack. Figure 23.4 contains the representation of the maximum power point shown on a current–voltage graph [11].

Figure 23.5 contains the representation of the maximum power point shown on a power–voltage graph. Alternatively, the MPPT can be used to directly power an electric motor, and the excess power can be used for battery pack charging.

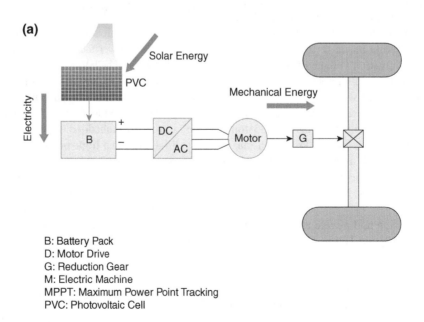

B: Battery Pack
D: Motor Drive
G: Reduction Gear
M: Electric Machine
MPPT: Maximum Power Point Tracking
PVC: Photovoltaic Cell

FIGURE 23.2 Solar-powered electric vehicle power train configuration: (a) conventional solar powered power train [11].

Battery Packs in Electric Vehicles

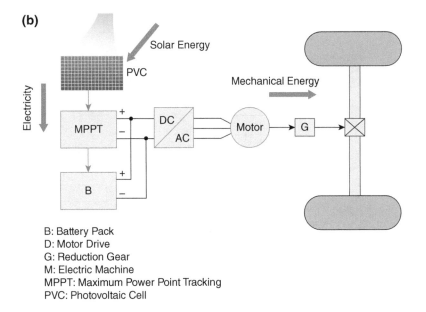

FIGURE 23.3 Solar-powered electric vehicle power train configuration: (b) solar-powered power train with maximum power point tracking [11].

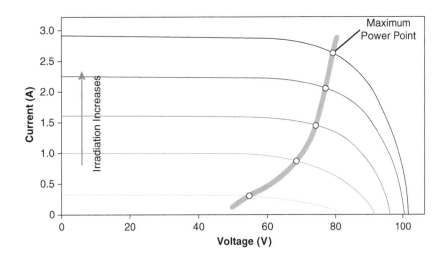

FIGURE 23.4 Photovoltaic array current – voltage different irradiance levels [11].

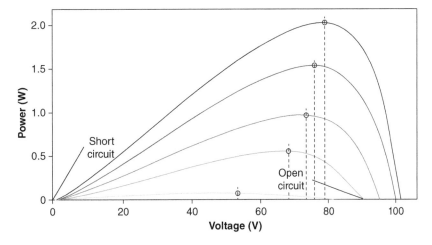

FIGURE 23.5 Power–voltage relationships at different irradiance levels [11].

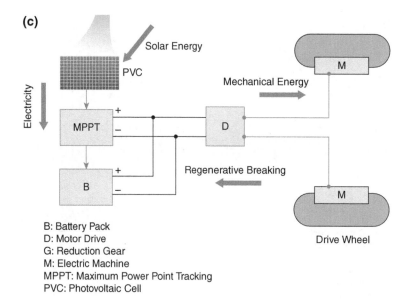

FIGURE 23.6 Solar-powered electric vehicle power train configuration: © solar-powered in-wheel drive with MPPT and regenerative braking [11].

Then the battery pack's energy is used to power the electric motor in times when the irradiation levels are low [11].

Figure 23.6 represents another power train configuration for solar-powered vehicles, by means of using an in-wheel motor system; this is done to remove the energy losses in the gearbox and differential. This way, the regenerative braking is also more efficient [11].

Another approach for extending the EV's range using solar panels from PV modules is by means of utilizing the active balancing functionality of the BMS. Rather than connecting a large MPPT to the entire battery pack, smaller MPPTs are connected directly to the balancing circuit of the battery cells. In this method, the MPPT's voltage needs to meet the smaller battery cell voltage. For lower voltages, the circuit gets better efficiency.

Figure 23.7 presents an example of the MPPT active balancing method. The balancing is done by means of charging each parallel connected in series separately, in regards to their voltage balance/imbalance. The balancing function is controlled by the BMS.

Fully PV-powered electric vehicles are still not efficient enough for mass production, only as concepts. They are mostly used to deliver small doses of energy to the battery pack, just to boost up efficiency for a few percent. This branch of EVs is waiting for improvements in nanotechnology for the purpose of boosting the PV's efficiency [11] because, currently, utilization of solar power (with the power generation efficiency of 0.7 to 0.85) to achieve electric propulsion is still too inefficient to be standardized.

23.6.2 Solar-Powered Charging Station

The continuous increase in the pollutant factors requires the implementation of renewable energy–powered chargers. EVs charged fully on renewable energy–based chargers are completely pollutant-free. Figure 23.8 represents a standard solar-powered charging station topology, where the charger uses a battery pack as an energy buffer to store energy while the charger is not being used. If there is a lack of sunlight, the alternative is a grid connection as a backup.

With this topology, the PVs can be mounted at charging stations, parking lots, gas stations, and even in homes, for the purpose of charging the EV. The following subsection contains examples of this application.

23.6.3 Solar Parking Lot Charging Stations

The introduction of solar panels on parking lots proved beneficial for renewable energy EV charging. The charging is controlled with a control unit to fairly distribute the renewable energy among the connected EVs. With the final results, in one charging period, 80% of the EVs are charged to at least 75%. The testing was done all along the year, with variable demand profile and weather conditions. The study proved that the grid-isolated solar charging parking lot is feasible when the PV system is proportional to the size of the parking lot. Only then can the energy be appropriately distributed to the EVs that are being charged on that parking lot [13].

23.6.4 Wind-Powered Charging Stations

The classic version of this charging station is the least common. The reason for this is that it's very difficult to find a suitable location which is abundant with wind and, at the same time, convenient for the customers to easily charge up and continue with their business. But this doesn't mean that they or their alternatives don't exist.

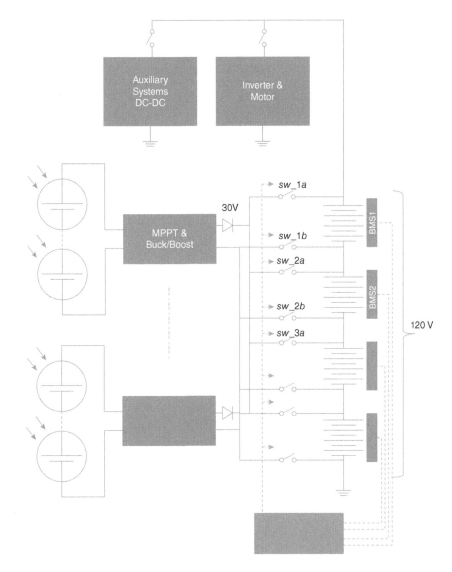

FIGURE 23.7 Solar-powered electric vehicle power train configuration: solar active balancing [12].

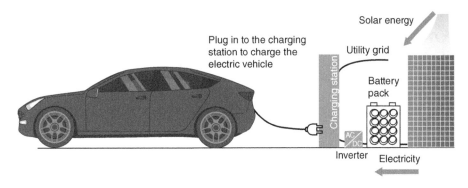

FIGURE 23.8 PV-based charging station [11].

23.6.4.1 Off-Grid Wind Power Plants and EV Chargers

With the problem of location and feasibility, a solution can be found in a classic off-grid wind farm and EV charging station combination, shown in the Figure 23.9. The problem with this is the expensive infrastructure that needs to be built, and the questionable feasibility. To be able to supply electrical energy off-grid is very complicated because the supply of electrical energy from the wind turbine needs to be balanced. The supply of electrical energy varies with wind speed and the electrical loads connected to it.

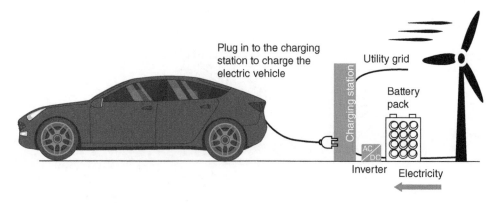

FIGURE 23.9 Wind-powered charging station.

However, this solution can be implemented with wind turbines under 10 kW, combined with energy storage devices (battery packs) to store the excess energy. This may help with high energy demands when the turbine is not sufficient enough to provide energy.

23.6.4.2 Sanya Skypump

Though these types of charging stations can't be found in abundance, some have actually been built and are in operation. The first ever fully wind-powered electric vehicle charging station was installed in Barcelona and across the United States by Urban Green Energy and General Electric. Sanya Skypump [14] gathers wind energy with a vertical wind turbine integrated with GE's standard charging systems for the purpose of achieving zero CO_2 emissions in EVs [15]. Tests needed to be conducted to gather all the necessary wind data, and they were done in two different locations in two different one-year periods. The result of the feasibility study concluded that the average wind velocity over a 3 min. period gives the best system performance; to achieve this, only nine turbines were used.

23.7 HYBRID-POWERED EV CHARGING STATIONS

Instead of the single renewable energy charging station, an integration of multiple renewable energies into a hybrid-powered charging station has proven to be highly efficient. The focus of this section is to present an overview of the possible configurations and topologies of electric charging stations.

23.7.1 Solar-Grid-Based Fast Charging System

Figure 23.10 contains a solar-grid-based DC fast charging system. It consists of a photovoltaic array, unidirectional step-up DC/DC converter, battery storage unit, and the electric vehicle fast charger. All the aforementioned components are integrated to an internal DC bus via converters. The bidirectional AC/DC converter is the link to the grid. The philosophy is based on the PV unit providing energy to the charger; when the energy is low, the battery storage unit is connected. If both of the sources are depleted, then the grid energy is transferred to the charger until one of the aforementioned sources is available [16].

23.7.2 Solar-Wind-Based Charging Stations

This is an **off-grid hybrid RE charging station** that includes wind turbines, a PV system, and a battery storage system. The concept was created to reduce the use of diesel-based generators in small isolated communities and was implemented in a small island in South Korea. The population is of 60 people, with 20 houses. The "eTuk" vehicles were selected for island transportation, to fill the people's transport needs to acces small and narrow spaces. An overview of the system is shown in the following block diagram in Figure 23.11.

The concept is based on replacing the discharged batteries with charged ones, where the EV can either pull up to the station for replacement or the batteries can be brought to the EV from the central station.

When talking about charging, the charging stations are located on various locations: within residential areas, around workspace areas, in market and community facilities, along landmarks and tourist hot spots, etc.

With the consumption of 1 kWh/km, the EVs are scheduled to charge between 6:00 and 7:00 a.m., where the batteries are charged in two hours. The island will not need more than five EVs with a 20% variation in both day-to-day charging amount and in time steps of the EV. A total of 368 or 130 kWh of battery modules was implemented as energy storage units, with a 30 kW charger. The share of energy generated from wind is 24%, where the solar is dominant at 76%. The total net present cost was minimized with optimization techniques, and an efficient and affordable hybrid system was created [17].

23.7.3 Solar-, Wind-, Air-, Hydrogen-, and Ammonia-Based Charging Station

This stand-alone hybrid charging station's philosophy is based on storing renewable generated energy in various

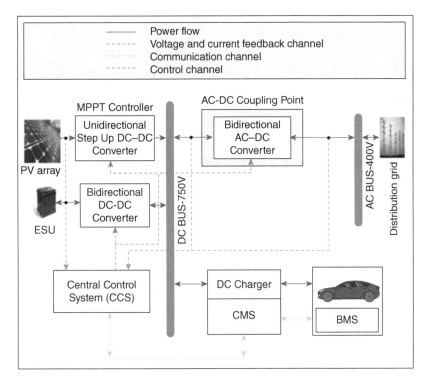

FIGURE 23.10 Solar-grid-based fast charging system topology [16].

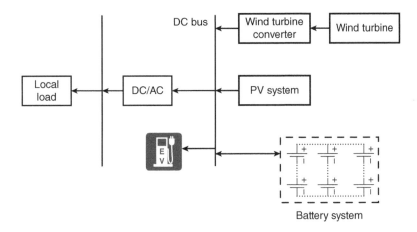

FIGURE 23.11 The topology of the solar-wind-based charging system [17].

storage units for the most efficient utilization with EV charging stations. The innovation lies in the chemical energy generated from air and electrolysis that is stored within the ammonia and hydrogen storage units in times when renewable energy is not available. This concept was implemented because it can provide a continuous feed of electrical energy at all times. Figure 23.12 contains the topology of this proposed charging station [18].

The photovoltaics primarily feed the electrolyzes unit via a DC/DC converter; the rest of the energy is used to supply the DC/AC inverter separately, then the NH3 and NH2 production, where the last priority is to charge the battery storage unit in times when the H2 supply is full and the inverter is not in use. The wind turbine's first priority is to supply the DC/AC inverter, and then the NH3 and NH2 production, and lastly, the battery storage charging, when the other two are not in use or are full. The concept of the electrolyzer unit is based on a proton-exchange membrane that recieves DC power from the photovoltaics units and decomposes the water into O2 and H2. Some of the H2 is used to fill the fuel cell, and the other for ammonia production. The ammonia production is based on two cycles, producing N2 from air, and then liquidizing it into NH3 with the use of the produced H2. The hydrogen and ammonia stored in their separated fuel cells can be converted to DC. The battery storage unit stores the excess energy from the wind turbine and the photovoltaic units to be used when the other units are unavailable to deliver energy.

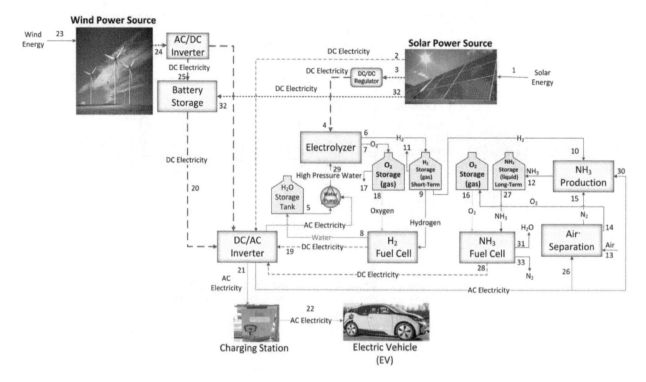

FIGURE 23.12 The topology of the hybrid charging station [18].

This charging station contains the photovoltaic units spread over 1,500 m2 surface area, along with a 250 kW wind turbine. The generated energy can easily charge 50 electric vehicles (35 kWh capacity) per day. The share of energy generated from wind is 49%, the solar 43%, and the fuel cells 8%. It needs to be noted that 23% of the fuel cell energy is repurposed back to the ammonia- and hydrogen-making process.

23.8 CONCLUSION

The main focus of these chapters is in the battery package itself. An electric vehicle battery package design procedure was given, but the procedure is adaptable to any type of battery storage unit; the only difference is in the energy consumption rate. The mentioned battery cells may be outdated in a few years because of the rapid improvements made in that industry, but the design philosophy in the battery package remains unchanged. Core elements of the battery package control modules and functions are presented, along with an introduction to supercapacitors and their power train integration topologies.

Electric vehicles and electric energy storage units play a major role in the battle against greenhouse gases and pollution. The rest of these chapters were dedicated to point out that the electric vehicle is not as clean as advertised in the media, to give a new point of view on the entire EV industry. Simply put, an electric vehicle is a vessel that is "filled" with electric energy and used to achieve movement. The "cleanliness" of the electric vehicle solely depends on the type of energy source the electric energy was generated from. Without this knowledge and any action, the electric vehicle would do more harm than help in the battle against pollution. It currently consumes the energy produced mostly by fossil fuels; this, added to the pollution share of the production of the electric vehicle itself, results in even greater pollution factor than ICE vehicles. For that cause, an overview of examples was given to prove the feasibility and importance on integrating renewable energy sources within the electric vehicle and its charging stations. The charging station examples given can be grouped by power supply hybrid, on- and off-grid. The most effective way to use renewable power sources in EVs is to pursue the tendency to introduce more renewable energy into the grid. All other solutions are isolated off-grid systems that are too expensive to build, or just proof of concepts that were never adopted.

23.9 REFERENCES

[1] 'EV Powertrain Components – Basics', Accessed on: January 22, 2021. [Online]. Available: https://evreporter.com/ev-powertrain-components/
[2] 'BU-808c: Coulombic and Energy Efficiency with the Battery', Accessed on: January 22, 2021. [Online]. Available: https://batteryuniversity.com/learn/article/bu_808c_coulombic_and_energy_efficiency_with_the_battery
[3] 'Worldwide Harmonized Light Vehicles Test Procedure', Accessed on: February 01, 2021. [Online]. Available: https://en.wikipedia.org/wiki/Worldwide_Harmonised_Light_Vehicles_Test_Procedure
[4] Tom Denton, *Automobile Electrical and Electronic Systems*, Third edition. Elsevier Butterworth-Heinemann, 2004, page 129.
[5] T. Hofman, C.H. Dai, [IEEE 2010 IEEE Vehicle Power and Propulsion Conference (VPPC) – Lille, France

(2010.09.1–2010.09.3)], *2010 IEEE Vehicle Power and Propulsion Conference,* 'Energy efficiency analysis and comparison of transmission technologies for an electric vehicle', (2010), 1–6. doi:10.1109/VPPC.2010.5729082

[6] Dainis Berjoza, Inara Jurgena, 'Influence of Batteries Weight on Electric Automobile Performance'. [Engineering for Rural Development, Jelgava, 24.-26.05.2017.]. Latvia University of Agriculture, 2017

[7] '161 MPH Taycan Revealed AS Porsche's First Electric Car', Accessed on: January 22, 2021. [Online]. Available: www.euronews.com/living/2019/09/04/161-mph-taycan-revealed-as-porsche-s-first-electric-car

[8] 'The Battery: Sophisticated Thermal Management', 800-volt System Voltage, Accessed on: January 22, 2021. [Online]. Available: https://newsroom.porsche.com/en/products/taycan/battery-18557.html

[9] 'Movitas 100% Electric Citybus Concept', Accessed on: January 22, 2021. [Online]. Available: https://siamagazin.com/movitas-100-electric-citybus-concept/

[10] 'Tribus, 100% Electric Citybus', Accessed on: January 22, 2021. [Online]. Available: www.tribus-group.com/newinnovation-movitas/

[11] Ali Emadi, Florence Berthold, *Advanced Electric Drive Vehicles.* US, CRC Press, 2017, page 510.

[12] Michelangelo Grosso, Davide Lena, Alberto Bocca, Alberto Macii, Salvatore Rinaudo, [IEEE 2016 IEEE 2nd International Forum on Research and Technologies for Society and Industry Leveraging a better tomorrow (RTSI) – Bologna, Italy (2016.9.7–2016.9.9)], *2016 IEEE 2nd International Forum on Research and Technologies for Society and Industry Leveraging a Better Tomorrow (RTSI),* 'Energy-efficient battery charging in electric vehicles with solar panels', (2016), 1–5. doi:10.1109/RTSI.2016.7740569

[13] Stephen Lee, Srinivasan Iyengar, David Irwin, Prashant Shenoy, [IEEE 2016 Seventh International Green and Sustainable Computing Conference (IGSC) – Hangzhou, China (2016.11.7–2016.11.9)], *2016 Seventh International Green and Sustainable Computing Conference (IGSC),* 'Shared solar-powered EV charging stations: Feasibility and benefits', (2016), 1–8. doi:10.1109/IGCC.2016.7892600

[14] 'Sanya Skypump: World's First Integrated Wind-Powered EV Charging Station Installed in Barcelona', Accessed on: January 22, 2021. [Online]. Available: https://inhabitat.com/worlds-first-integrated-wind-powered-ev-charging-station-unveiled-in-barcelona/sanya-skypump-in-use/

[15] 'World's First Wind Powered Electric Charging Station', Accessed on: January 22, 2021. [Online]. Available: www.designboom.com/technology/worlds-first-wind-powered-electric-charging-station/

[16] Ratil H. Ashique, Zainal Salam, Bin Abdul Aziz, Bhatti Mohd Junaidi, Abdul Rauf, 'Integrated photovoltaic-grid dc fast charging system for electric vehicle: A review of the architecture and control', *Renewable and Sustainable Energy Reviews,* 69 (2017), 1243–1257. doi:10.1016/j.rser.2016.11.245

[17] Amir Ahadi, Shrutidhara Sarma, Jae Moon, Sangkyun Kang, Jang-Ho Lee, 'A robust optimization for designing a charging station based on solar and wind energy for electric vehicles of a smart home in small villages', *Energies,* 11(7) (2018), 1728. doi:10.3390/en11071728

[18] Abdulla Al Wahedi, Yusuf Bicer, 'Assessment of a stand-alone hybrid solar and wind energy-based electric vehicle charging station with battery, hydrogen, and ammonia energy storages', *Energy Storage,* 1(5) (2019). doi:10.1002/est2.84

24 Alternative Wind Energy Turbines

Andrej Predin, Matej Fike, Marko Pezdevšek, and Gorazd Hren

CONTENTS

- 24.1 Introduction 323
  - 24.1.1 Betz Limit 323
  - 24.1.2 Conservation of Mass, Energy, and Momentum at Wind Turbines 324
    - 24.1.2.1 Conservation of Mass 324
    - 24.1.2.2 Conservation of Energy 324
    - 24.1.2.3 Conservation of Momentum 325
  - 24.1.3 Betz Limit Definition 325
    - 24.1.3.1 Power Coefficient Definition 326
- 24.2 Ducted Wind Turbines 326
- 24.3 Altered Geometry of the Wind Turbine Rotor 328
  - 24.3.1 Tubercles 328
- 24.4 Shark Skin 329
- 24.5 Alternative Bladeless Wind Turbine Designs 329
  - 24.5.1 Magnus 329
- 24.6 Saphonian Technology 330
- 24.7 O-Wind 331
- 24.8 Vortex Bladeless 331
- 24.9 Summary 331
- 24.10 Literature and sources 331

24.1 INTRODUCTION

Harnessing kinetic energy from the wind and its conversion to valuable types of energies is a process that has been known and used for centuries. Historians believe that the first windmills were invented almost 2000 years ago in Persia and China. A drawing of a Persian windmill is seen in Figure 24.1. They were used to grind corn and pump groundwater. In Europe, their use started much later, in the fourteenth century, when the Dutch originally developed windmills to drain their land and later used them to grind corn and produce electric energy in the nineteenth century [1, 2].

In America, Dani Halladay invented a windmill to pump water in 1854. Later, he established Halladay's Company – Engine & Pump Co. and became the largest manufacturer of windmills in America. Charles Francis Brush invented the first significant wind turbine for the generation of electricity in 1888. The oil crisis in 1973 increased interest in wind power, as attested by the rapid growth in electricity system production that led to the establishment of wind farms.

The potential of wind energy worldwide is enormous. Firstly, there is a significant interest in rural communities with limited access to grid electricity and where wind energy is an economically viable alternative to diesel engines and even thermal (coal-burning) power plants. This way, fuel would not have to be purchased and imported from other countries, and the luxury of "free fuel" could be enjoyed. Secondly, the linking of wind farms to national grids has been very successful. The next significant advancement will be the connection of wind farms to hydro power plant storage as a more effective form of energy storage of large amounts of energy when the wind is available and electric energy is not required.

The potential of wind as a renewable energy source has been recognized to reduce CO_2 emissions worldwide. The future looks very bright for the wind turbine industry.

24.1.1 Betz Limit

The energy yield of all commercial versions of freestanding wind turbines used currently is limited by the maximum kinetic energy of the wind – the Alfred Betz limit. The physical basis for the limit is the equality of wind velocities and pressure in front of and behind the windmill or the rotor of the wind turbine. Therefore, wind velocities behind the wind turbine rotor cannot, physically, be zero. Comparatively, such a state can only be achieved by a Pelton turbine, where the water flow from the nozzle emits all the available kinetic energy of the jet on the buckets of the Pelton wheel. In this case, the water jet falls freely (gravity) towards the lower water in the turbine housing as the remaining speed of the jet is zero.

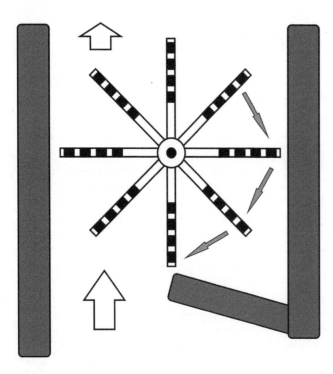

FIGURE 24.1 Persian windmill. Adapted from source [2].

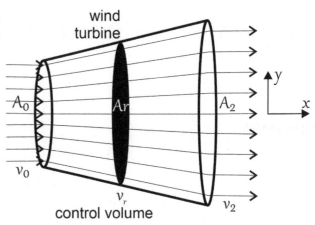

FIGURE 24.2 Typical control volume of HAWT.

The kinetic energy of the wind is:

$$E_{k,wind} = \frac{1}{2}mv^2, \quad (24.1)$$

Where m is mass and v is wind velocity. The energy is extracted from the air mass in the cylindrical volume of air that will flow through the rotor with the same diameter as is the rotor diameter of a horizontal axis wind turbine (HAWT). In this case, the rotor is put in the free stream of the wind flow.

The energy per unit of time (power) can be calculated as:

$$\dot{E} = \frac{1}{2}\dot{m}v^2 = \frac{1}{2}(\rho A v)v^2, \quad (24.2)$$

Where $\dot{m}$ is mass flow rate, ρ air density, A the turbine rotor cross-section area, that is, for HAWT, $A = \frac{\pi D^2}{4} = \pi r^2$ and v wind velocity. Available power in this case is:

$$P_A = \dot{E} = \frac{1}{2}\rho \pi r^2 v^3. \quad (24.3)$$

Three fundamental physics concepts are used for further study of energy extracted from the wind:

- Conservation of mass
- Conservation of energy
- Conservation of momentum

These are applied in a defined control volume which includes the wind turbine rotor. The typical control volume around a wind turbine rotor is shown in Figure 24.2.

24.1.2 Conservation of Mass, Energy, and Momentum at Wind Turbines

24.1.2.1 Conservation of Mass

To make it easier to understand [3], we introduce some assumptions:

- Air enters and exits so that the fluid flow is streamed so that there are no mass losses in the direction normal to the control volume.
- Because the flow velocities are low, the fluid can be considered incompressible in the passage of the control volume.

Under these assumptions, conservation of mass could be written as:

$$\dot{m} = \rho A_0 v_0 = \rho A_r v_r = \rho A_2 v_2, \quad (24.4)$$

Where v_0, v_r and v_2 are mean wind velocities across the section areas A_0, A_r and A_2, respectively. The wind's kinetic energy reduces flow velocities $v_0 > v_r > v_2$, while the wind turbine rotor extracts the energy from the air.

24.1.2.2 Conservation of Energy

The total flow energy could be simplified and considered as a sum of kinetic, pressure, and potential energy:

$$E_{tot} = E_{kin} + E_{press} + E_{pot}. \quad (24.5)$$

The value of the kinetic energy is greater than zero when flow motion is present. Pressure energy is the consequence of random particle (molecules) motion in the fluid. Potential energy is determined from the relative position of an object relative to other objects.

The assumptions previously mentioned, at mass conservation, must be expanded:

- Fluid is incompressible, which means the density is constant while pressure can change.

Alternative Wind Energy Turbines

- Fluid is inviscid, meaning, the equations are valid for the free stream flow, that is, outside the boundary layer flow.
- During transition through the control volume, no heat energy is exchanged.
- During transition through the control volume, there is no change of mass.
- Potential energy is considered a constant, while relative positions concerning the Earth's surface do not change.

The first assumption is defining an ideal fluid. Bernoulli's equation can be applied from the fluid direction to the wind turbine rotor. We can calculate the total energy per unit of the control volume as:

$$E_{tot} = \rho \frac{v^2}{2} + p = const., \qquad (24.6)$$

Where the first term on the right-hand side of the equation represents kinetic energy or dynamic pressure, and the second term represents the static pressure of the fluid flow. Bernoulli's equation can be applied from A_0 to A_2 but cannot be applied across the rotor disc that extracts energy. The equation constants in front of the rotor will be different from the ones behind.

24.1.2.3 Conservation of Momentum

A wind turbine must be considered as a machine that harnesses kinetic energy from the wind (airflow). That is why wind velocities will be reduced across the control volume in the flow direction. Since momentum is mass times velocity, there is a change in momentum. Considering Newton's second law, the rate of change of momentum in a control volume is equal to the sum of all acting forces. The same could be concluded for mass flow rate times velocity:

$$\dot{m}_0 v_0 - \dot{m}_2 v_2 = F. \qquad (24.7)$$

In order to simplify the problem (equations), the following assumptions [3] are required:

- There are no shear forces in the flow (x) direction.
- The pressure forces, caused by the surrounding pressure at boundary lines of cross-sections A_0 and A_2, are equal.
- There is no momentum loss or gain other than from cross-sections A_0 to A_2.
- The equation (7) is applied in the x-axis direction.

To change the momentum of an object, external forces must be acting on it. In this case, Newton's third law defined the external force in the sense of an acting (F_A) – reacting force (F_R). The force F that acts on the wind turbine rotor must be equal to the force exerted by the wind, with pressure difference across the rotor divided by the area of the rotor:

$$F_A = F_R = \frac{F}{A} = \Delta p. \qquad (24.8)$$

The wind turbine rotor blocks the airflow partly. This causes the pressure ($p_{r,0}$) to rise directly in front of the rotor, which is higher than the pressure of the free stream in front of the rotor (p_0). Directly behind the rotor is the flow pressure ($p_{r,2}$) below the free-stream pressure (p_0), as is shown in Figure 24.3.

24.1.3 Betz Limit Definition

Applying conservation of mass in the control volume (Figure 24.2) at cross-sections A_0, A_r, and A_2 with constant density, we can calculate the mass flow as:

$$\dot{m} = \rho A_0 v_0 = \rho A_r v_r = \rho A_2 v_2, \qquad (24.9)$$

Considering Newton's second law, the force exerted at the rotor by wind can be written as:

$$F = \dot{m} v_0 - \dot{m} v_2, \qquad (24.10)$$

$$F = \dot{m}(v_0 - v_2) = \rho A_r v_r (v_0 - v_2). \qquad (24.11)$$

The force exerted on the rotor is also because of the pressure difference acting across the rotor, meaning:

$$F = A_r (p_{r,0} - p_{r,2}). \qquad (24.12)$$

By equating equations (11) and (12) of force expressions, we get:

$$F = A_r (p_{r,0} - p_{r,2}) = \rho A_r v_r (v_0 - v_2). \qquad (24.13)$$

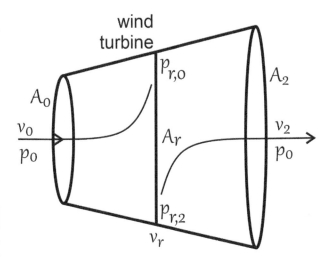

FIGURE 24.3 The wind pressure across the control volume.

Applying conservation of energy, or Bernoulli's law, on two control volumes (Figure 24.2) along the streamlines in the flow direction, the first from A_0 to A_r, and the second from A_r to A_2, the following energy equations can be written:

$$p_0 + \frac{1}{2}\rho v_0^2 = p_{r,0} + \frac{1}{2}\rho v_r^2, \quad (24.14)$$

$$p_{r,2} + \frac{1}{2}\rho v_r^2 = p_0 + \frac{1}{2}\rho v_2^2. \quad (24.15)$$

The pressure difference across the rotor can be expressed as:

$$p_{r,0} - p_{r,2} = \frac{1}{2}\rho\left(v_0^2 - v_2^2\right). \quad (24.16)$$

The pressure difference across the wind turbine rotor can also be obtained from the momentum equation, combining equations (11) and (13):

$$\frac{F}{A_r} = p_{r,0} - p_{r,2} = \rho v_r(v_0 - v_2) = \frac{\rho}{2}\left(v_0^2 - v_2^2\right), \quad (24.17)$$

$$\rho v_r(v_0 - v_2) = \frac{\rho}{2}\left(v_0^2 - v_2^2\right), \quad (24.18)$$

From where the velocity at the rotor is:

$$v_r = \frac{v_0 + v_2}{2}. \quad (24.19)$$

Equation (19) shows that the wind velocity v_r at the rotor is the average of the free-stream velocity at the front of the rotor and velocity at the rotor wake flow behind the rotor. The wind velocity in airflow wake (v_2) is where the pressure is equalized with free-stream pressure (v_0). This equation also implies that the first half of the wind velocity loss occurs in front of the rotor, and the second half after the rotor in the downstream direction. *Power* is defined as force times wind velocity at the rotor cross-section:

$$P = Fv_r = (p_{r,0} - p_{r,2})A_r v_r = \frac{\rho}{2}\left(v_0^2 - v_2^2\right)A_r v_r, \quad (24.20)$$

And:

$$P = \frac{1}{2}\rho A_r v_r (v_0 - v_2)(v_0 + v_2), \quad (24.21)$$

Or:

$$P = \frac{1}{2}\rho A_r v_r\left(v_0^2 - v_2^2\right) = \frac{1}{2}\dot{m}\left(v_0^2 - v_2^2\right), \quad (24.22)$$

That is the change in kinetic energy applied to the mass flow per unit time through the rotor. Meaning, the work done by force due to pressure difference is equal to the change in wind kinetic energy. Combining equation (24.21) in (24.19), one can write:

$$P = \rho A_r v_r^2(v_0 - v_2) = 2A_r v_r^2(v_0 - v_r). \quad (24.23)$$

The maximum power is realized when:

$$\frac{\partial P}{\partial v_r} = 0 = 2v_r v_0 - 3v_r^2, \quad (24.24)$$

Which yields the following expression:

$$v_r = \frac{2}{3}v_0, \quad (24.25)$$

That implies:

$$v_2 = \frac{1}{3}v_0. \quad (24.26)$$

Combining equations (23), (25), and (26), we get:

$$P = 2\rho A_r v_r^2(v_0 - v_r) = \rho A_r v_0^3\left(\frac{8}{27}\right). \quad (24.27)$$

24.1.3.1 Power Coefficient Definition

Estimation of usable power [4] from wind turbines is the ratio between the maximal power extracted from the wind and the calculated power (kinetic energy):

$$c_P = \frac{P}{P_A} = \frac{\rho A_r v_0^3\left(\frac{8}{27}\right)}{\frac{1}{2}\rho A_r v_0^3} = \frac{16}{27} = 0.593. \quad (24.28)$$

This represents the Betz limit and states that the maximum power an ideal rotor can extract from wind is 59.3%.

There are several ways to improve the efficiency of standard three-bladed wind turbines. Here, we will list three modifications that would, theoretically, improve the performance of a wind turbine.

24.2 DUCTED WIND TURBINES

The wind turbine's power increases with an increase in velocity in front of the rotor and with the largest possible surface of the rotor (large diameter) and with the highest possible air density. We cannot influence wind velocity and air density, and we are limited with the strength of modern materials when defining the rotor size.

The pressure difference on the wind turbine rotor and the velocity conditions indicates that the power on the wind turbine rotor will increase with the difference between the highest free-flow velocity in front of the rotor and the lowest possible velocity behind the rotor. We can achieve this with a ducted (coated) wind turbine. Ducted wind turbines (DWT) are HAWT with a diffuser (Figure 24.4).

Alternative Wind Energy Turbines

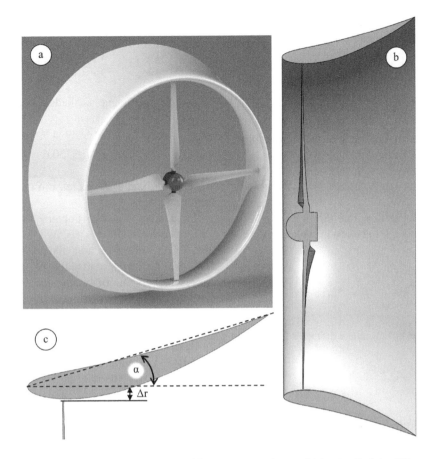

FIGURE 24.4 3D model of a (a) ducted wind turbine, (b) diffuser cross-section, and (c) a detail of the diffuser cross-section.

A ducted wind turbine takes advantage of the Venturi effect, in which the static pressure at the cross-section A_1 is bigger than the static pressure at the cross-section A_2. The converse can be said about the velocity, where the velocity at the cross-section A_1 is lower than the velocity at the cross-section A_2, as seen in Figure 24.5.

The following relations can be expressed in the Bernoulli equation:

$$\Delta p = \frac{1}{2}\rho\left(v_1^2 - v_2^2\right). \tag{24.29}$$

Let us consider the case of a HAWT with a diameter of $D_1 = 4\,\text{m}$, at a free-stream wind velocity of 10 m/s (v_0). Surrounding pressure is $p_B = p_0 = 1.013$ bar $= 1.013 \cdot 10^{-5}$ Pa, at a temperature $T_0 = 293$ K (20°C) and air density $\rho = 1.23$ kg/m3. Now we place the rotor in the middle of a symmetric nozzle (in the flow direction), as shown in Figure 24.6. The nozzle has an intake diameter and an exit diameter, 4.4 m (10% more than D_1).

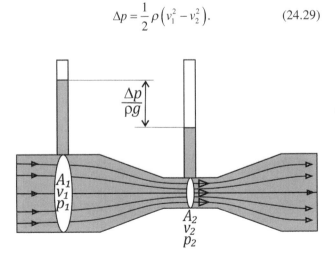

FIGURE 24.5 Venturi effect.

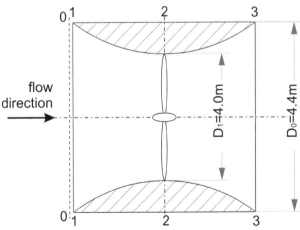

FIGURE 24.6 Wind turbine rotor at the middle of the symmetric nozzle.

From the energy equation that is written for cross-sections 0–1, Figure 24.3, one can calculate, according to equation (16), the pressure drops as:

$$\Delta p_{0-1} = \frac{1}{2}\rho\left(v_1^2 - v_0^2\right). \quad (24.30)$$

If we treat the flow through the nozzle as incompressible at a constant mass flow:

$$\dot{m} = \rho v A = \rho Q = \text{const.},$$

Where Q is the volumetric flow rate through the nozzle, from cross-sections 0 up to 3, the constant volumetric flow rate is:

$$Q = \pi \frac{D_0^2}{4} v_0 = \pi \frac{4.4^2}{4} 10 = 152.1 \, \text{m}^3/\text{s}.$$

The cross-flow area at the nozzle intake diameter $D_0 = D_3$ is:

$$A_0 = \pi \frac{D_0^2}{4} = \pi \frac{4.4^2}{4} = 15.2 \, \text{m}^2,$$

And the cross-flow area at the rotor diameter $D_r = D_1 = D_2$ is:

$$A_r = A_1 = \pi \frac{D_r^2}{4} = \pi \frac{4^2}{4} = 12.57 \, \text{m}^2.$$

According to Bernoulli's equation, the flow velocities can be calculated:

$$v_1 = \frac{Q}{A_1} = \frac{152.1}{12.57} = 12.1 \, \text{m/s}.$$

The pressure ratio at the first part of the nozzle (treated as a funnel part), in front of the rotor, according to equation (29), is now:

$$\Delta p_{0-1} = \frac{1}{2} 1.23 \left(12.1^2 - 10^2\right) = 46.4 \, \text{Pa},$$

Which means that this is the pressure difference that could be yielded from this velocity increase in the nozzle. With this pressure ratio, the exerted power (equation 20) is:

$$P = F v_r = \left(p_{r,0} - p_{r,1}\right) A_r v_r,$$

Where:

$$v_r = \frac{2}{3} v_1 = \frac{2}{3} 12.1 = 8.1 \, \text{m/s},$$

And:

$$P = 46.4 \times 12.57 \times 8.1 = 4724.3 \, \text{W}.$$

With available power (from the kinetic energy of the wind):

$$P_A = \frac{1}{2}\rho A_1 v_0^3 = \frac{1}{2} 1.23 \times 12.57 \times 10^3 = 7730.6 \, \text{W}.$$

The power coefficient is calculated as:

$$c_P = \frac{P}{P_A} = \frac{4724.3}{7730.6} = 0.611,$$

Which is around a 2% increase if compared with the Betz limit (0.593).

An increase in power is obtained when comparing DWT with open-rotor turbines (HAWT). R. Venters [5] reports that, for the same rotor area, the power output of the largest DWT was 66% greater than that of an open rotor.

Researchers in [6] conducted a study of an optimal diffuser design. The shape of the shroud is defined as the second-order polynomials, where the rotor geometry, namely, the twist and chord distributions, is approximated by Bezier curves. The conclusion was that it was possible to gain a 12% improvement in the power performance while decreasing the drag and thrust coefficients by 10% and 8.7%, but at the cost of increasing the diffuser and nozzle length.

24.3 ALTERED GEOMETRY OF THE WIND TURBINE ROTOR

24.3.1 Tubercles

The inspiration for this design is lifted from the fins of a humpback whale that has tubercles on one side, as seen in Figure 24.7(b). Some researchers have applied the design to the leading edge of a blade in the hope of injecting

FIGURE 24.7 Model of a tubercles blade section (a) and whale fins (b), adapted from [8].

momentum into the boundary layer, therefore delaying the flow separation (similar to vortex generators) [7]. A model of a tubercles blade is seen in Figure 24.7(a).

Other researchers have speculated that tubercles alter the pressure distribution on the wing so that separation of the boundary layer is delayed. That ultimately leads to a gradual onset of the stall and a larger stall angle [8].

Wind tunnel experiments have shown that stall is delayed from an angle of attack of 12° for the standard profile to 18° for the profile with tubercles. It was also noted that the drag was reduced by 32%, and lift increased by 8% [7, 8].

24.4 SHARK SKIN

The skin of a shark is covered in scales that are aligned with the direction of flow. The scales of a shark have a positive effect in reducing drag. In an attempt at replicating the effect of scales, researchers have studied different riblet designs and concluded that the drag reduction for different riblet shapes is between 5 and 10% [9]. A model of an experimental riblet is seen in Figure 24.8. The study [9] supported the theory that correctly designed riblets can reduce drag, and highlights two mechanisms of riblet drag reduction:

- Riblets impede the translation of the stream-wise vortices, which causes a reduction in vortex ejection and outer-layer turbulence.
- Riblets lift the vortices off the surface and reduce the amount of surface area exposed to the high-velocity flow.

The conclusion was that, by modifying the velocity distribution, riblets facilitate a net reduction in shear stress at the surface.

24.5 ALTERNATIVE BLADELESS WIND TURBINE DESIGNS

24.5.1 MAGNUS

The Magnus effect was named after Heinrich Magnus, a German physicist who described the effect in 1852 in his research in the deflection of projectiles from firearms. The Magnus effect is present with a rotating object moving through a fluid. A rotating object (in our case, a cylinder) experiences a deflection that can be explained by the difference in pressure of the fluid on opposite sides of the spinning object. The direction in which the deflection happens is related to the direction of rotation (clockwise or counter-clockwise). The forces that act on a rotating cylinder are seen in Figure 24.9; the *lift force* is defined as a force that is perpendicular to the free-stream velocity, while the drag force is parallel to the free-stream velocity.

German engineer Anton Flettner was the first to build a ship that attempted to use the Magnus effect for propulsion. In 1924, he constructed the ship *Backau*, which had two large cylinders, each 15 m in height and 3 m in diameter, driven by a 37 kW motor. The ship was later renamed *Baden Baden* and used for crossing the Atlantic Ocean [10]. In 2008, Enercon made a ship called *E-Ship* that used four cylinders with a diameter of 4 m and 25 m in height to help

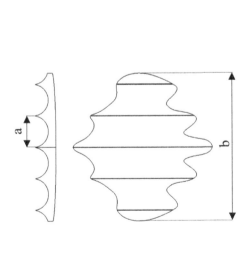

FIGURE 24.8 Experimental shark skin riblets.

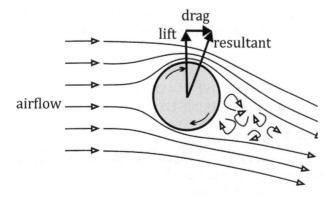

FIGURE 24.9 Forces acting on a rotating cylinder.

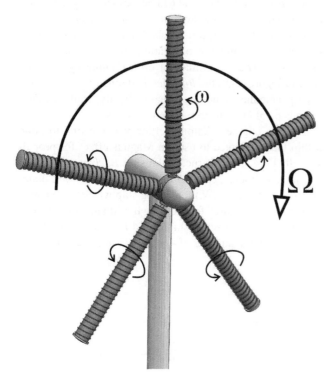

FIGURE 24.10 Magnus wind turbine.

raise the efficiency. The oversized Magnus rotors allegedly helped reduce fuel costs by 30%. The ship was used to transport wind turbine parts and other components.

Research into Magnus wind turbines has increased in recent years. Researchers have experimented with modifying the typical rotor design by adding end plates to the rotors. Data showed that, by adding this modification, the lift and drag coefficients increase significantly. In a paper written by Niel Lopez, numerical experimentations were conducted for different Magnus rotor types; investigation on the effect that spirals, bumps, and humps have on the lift and drag coefficients of the rotor was also done [11]. The authors concluded that the performance would be improved by using frustum cylinders and helical grooves. Figure 24.10 shows a model of a Magnus wind turbine with five rotors, end plates, and spirals.

Like every other technology, the Magnus wind turbine has some advantages and disadvantages. The advantages are:

- Compared to the complex design of classical wind turbine blades, the Magnus rotors are simplified and easier to manufacture.
- The wind turbine can be effective at lower wind speeds.
- Due to its simple design, the strength of the rotors is higher than the strength of conventional blades.

The disadvantages of a Magnus wind turbine are:

- There is a constant need for energy to power the rotating cylinders.
- The rotational speed of the cylinders is high.

24.6 SAPHONIAN TECHNOLOGY

A Saphonian bladeless wind turbine draws inspiration from the design of a ship's sails and converts the wind's kinetic energy into electricity. That involves channelling the wind in a back-and-forth motion until it is converted into mechanical energy using pistons. The pistons then produce hydraulic pressure that is converted to electricity via a hydraulic motor and a generator [12]. An illustration of a Saphonian wind turbine can be seen in Figure 24.11.

FIGURE 24.11 Saphonian wind turbine, adapted from [13].

Alternative Wind Energy Turbines

24.7 O-WIND

The O-Wind turbine is spherical, with a single axis of rotation going through it, as seen in Figure 24.12. The turbine makes use of Bernoulli's principle for its mechanical motion. The structure is lined up with vents that have large entrances and smaller exits for air. In the presence of wind, there is a pressure difference between the two terminals, causing the turbine to move. The vents are placed all across the sphere, making it receptive to wind from all directions in both vertical and horizontal planes [14].

24.8 VORTEX BLADELESS

Vortex Bladeless is a vortex-induced vibration resonant wind generator. It harnesses wind energy from a phenomenon of vorticity called vortex shedding. The cylinder oscillates on a wind range, which then generates electricity through an alternator system. The structure consists of two parts, the base, which includes the anchoring and the location of the power output system, and the cylindrical-shaped mast that is the oscillating element. The Vortex Bladeless is subjected strongly to fatigue effects [16]. A drawing of a Vortex Bladeless turbine can be seen in Figure 24.13.

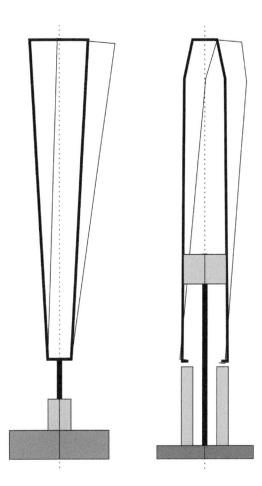

FIGURE 24.13 Vortex Bladeless.

24.9 SUMMARY

Finally, we can conclude that the presented alternative forms of wind turbines show the primary directions of further developments that could be divided into three large groups:

- Developments in the direction of developed accessories (guides) to classic designs
- Developments in the direction of changing the aerodynamics of geometry
- Developments based on other physical laws of fluid flow acting on solids in the flow

Moreover, we must not forget the development of new materials to enable different and more advanced solutions in wind turbine design.

24.10 LITERATURE AND SOURCES

[1] K. Farrokh, www.kavehfarrokh.com/arthurian-legends-and-iran-europe-links/the-windmill-and-the-contribution-of-persia. [3.3.2021].
[2] http://historicaliran.blogspot.com/2012/03/nashtifan-windmills.html [3.3.2021].
[3] Z. Warsi, Fluid Dynamics, *Theoretical and Computational Approaches*, Third Edition, CRC Press, 2006.

FIGURE 24.12 O-Wind turbine, adapted from [15].

[4] T. Burton, D. Sharpe, N. Jenkins, E. Bossanyi, *Wind Energy Handbook*, John Wiley & Sons Ltd, 2001.

[5] R. Venters, B. T. Helenbrook, K. D. Visser, "Ducted Wind Turbine Optimisation", *Journal of Solar Energy Engineering*, Vol. 140, 2018.

[6] T. A. Khamlaj, M. P. Rumpfkeil, "Analysis and Optimization of Ducted Wind Turbines", *Energy*, Vol. 162, pp. 1234–1252, November 2018.

[7] D. S. Miklosovic, M. M. Murray, L. E. Howle, F. E. Fish, "Leading-edge Tubercles Delay Stall on Humpback Whale", *Physics of Fluids*, Vol. 16, 2004.

[8] E. van Nierop, S. Alben, M. P. Brenner, "How Bumps on Whale Flippers Delay Stall: An Aerodynamic Model", *Physical Review Letter*, Vol. 100, 2008.

[9] B. Dean, B. Bhushan, "Shark-skin Surfaces for Fluid-drag Reduction in Turbulent Flow: A Review", *Philosophical Transactions of the Royal Society A*, Vol. 368, 2010.

[10] J. Seifert, "A review of the Magnus effect in aeronautics", *Progress in Aerospace Sciences*", *Izv.*, Vol. 55, pp. 17–45, 2012.

[11] N. Lopez, B. Mara, B. Mercado L. Mercado, M. Pascual, M. A. Promentilla, "Design of Modified Magnus Wind Rotor Using Computational Fluid Dynamics Simulation and Multi-response Optimization", *Journal of Renewable and Sustainable Energy*, 7, 2015.

[12] https://newatlas.com/saphonian-bladeless-wind-turbine/24890. [3.3.2021].

[13] https://newatlas.com/saphonian-bladeless-wind-turbine/24890/ [6.8.2021].

[14] www.jamesdysonaward.org/2018/project/o-wind-turbine [3.3.2021].

[15] www.jamesdysonaward.org/2018/project/o-wind-turbine/ [6.8.2021].

[16] https://vortexbladeless.com. [3.3.2021].

25 MPPT Controller for Partially Shaded Solar PV System

M. Subashini and M. Ramaswamy

CONTENTS

25.1 Introduction 333
25.2 Theory of Partial Shading 334
    25.2.1 Reasons for Partial Shading 334
    25.2.2 Effects of Partially Shaded PV Panel 335
25.3 MPPT Controller 335
    25.3.1 Central MPPT 336
    25.3.2 Modular MPPT 336
25.4 Design of MPPT Controller for Partially Shaded PV Systems 337
    25.4.1 Steady-State Analysis 338
    25.4.2 Transient Analysis 343
25.5 Hardware Implementation 343
    25.5.1 Hardware Implementation of GMPPT Control Algorithm 343
    25.5.2 Results and Discussion 343
25.6 Summary 348

25.1 INTRODUCTION

Energy appears to exist as a crucial element towards achieving sustainable development around the globe. While fossil fuels remain the major source, their depleting trend together with the challenges arising from the climatic change enforces a need to explore the use of the renewable form of energy resources.

The renewable energy sources (RES) that include solar, wind, and fuel cells among others offer to be a significant alternative in being able to contribute to the share of the energy demand. Solar energy enjoys the advantage of being environmentally friendly, offering lower carbon footprints, acclaiming to be cost-efficient, and generally remaining unlimited in availability. A large amount of solar energy available to meet ever-increasing energy needs makes solar photovoltaic (PV) systems a predominant source of renewable energy.

The power output from a PV panel or a module depends on the irradiation of the sun and the temperature of the panel. Despite the technological advancements, there arise challenges in the form of smaller conversion of efficiency, depending on the varying irradiation and temperature conditions. Figure 25.1 shows that the power generated from the PV panel varies with its terminal voltage and reveals that there occurs only one maximum power point (MPP) associated with each combination of the radiation and temperature.

When the PV arrays remain directly connected with the load, the potential power that can be extracted from the PV arrays may be wasted because the power output of the PV arrays mainly depends on the characteristic of the load. The efficiency of the PV system can, however, be increased by using a DC–DC converter along with a maximum power point controller.

The photovoltaic module engages a maximum power point tracking (MPPT) controller to ensure the extraction of the maximum power at any environmental condition (irradiance and temperature) from the PV modules by matching its P–V operating point with that of the corresponding DC–DC converter interface. It enforces the PV array to operate at the voltage corresponding to the MPP and, in turn, allows the increase in the energy produced substantially.

Due to the nonlinear I–V characteristics of the PV curve, the tracking of the maximum power point (MPP) at various environmental conditions can sometimes be a challenging task. The discrepancy that occurs in the electrical characteristics of PV panels interrupts the energy harvest of the PV system to a large extent. The method of manufacturing, the way of soldering, and the age of modules account for the mismatch in the electrical characteristics of the PV panel. The issue becomes more complicated when the entire PV array does not receive uniform irradiance – a condition known as partial shading that results in loss of energy and hence reduces the yield efficiency of the PV system.

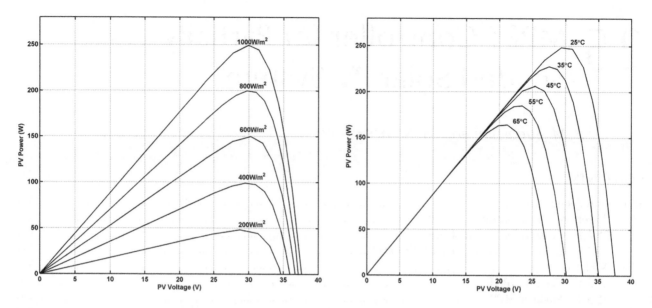

FIGURE 25.1 PV panel characteristics for various irradiation and temperature conditions.

25.2 THEORY OF PARTIAL SHADING

A solar panel is made up of several solar cells connected in series. Under uniform solar insolation, the output power of the PV panel is equal to the total output power of all solar cells. Nonuniform insolation condition causes a disproportionate loss in panel output or local overheating and PV panel destruction. This condition is referred to as partial shading.

A shaded cell causes a drop in PV panel current. The unshaded cells accommodate this drop by moving along their operating curves in the direction towards open-circuit voltage. The extra voltage generated in this way drives the voltage across the shaded cell in the opposite direction in an attempt to increase the current through it. With even small amounts of shading, the shaded cell will become reverse-biased and dissipate electrical power.

The loss in panel output resulting from such shading can be as high as that expected from the loss in incident solar radiation multiplied by the number of series-connected cells in the solar PV panel. Even more importantly, in the worst case, the power being dissipated in a single shaded cell can become as high as the entire generating capacity of the remainder of the PV panel. It is not surprising that this causes the rapid overheating of the cell and panel destruction.

The standard technique to protect against the destructive effects of partial shading is to connect bypass diodes across every string that includes series-connected cells. This restricts the reverse bias which can be generated across any shaded cell and hence the power which can be dissipated in it. Under the PS conditions, multiple maximum points appear on the PV curve, mentioned as local maxima (LM) and global maxima (GM). The partial shading condition mainly occurs due to clouds, rainy weather, dust, or shadow of something.

25.2.1 Reasons for Partial Shading

The main reasons for partial shading are as follows:

1) A tree that has grown tall enough to cast a significant amount of shade on the solar panel of a nearby roof.
2) Clouds are another source of potential shading. Clouds passing through the sky during the day may also result in fluctuations in system output, but these are unavoidable.
3) Bird droppings and animal remains that are not washed away also result in partial shading. The features of this kind of shadow are relative to a small area and of random shape and distribution. Thus, it is needed to check and clean the modules in PV power plants one by one to completely rule out the shading
4) The shadows of wire poles and power distribution buildings. Setting up electricity poles and wires is to collect and transport the electrical power produced by PV arrays. However, some shadings and shadows would be formed and fall on the PV arrays if the positions of wire poles and power distribution buildings were laid out inappropriately and too close to the PV modules.
5) The shadows of front-row PV arrays on the modules due to the improper design will greatly impact the performance of PV arrays. The shading caused by front rows of PV arrays appears when the solar altitude angle is relatively small in the early morning or late afternoon. This type of shading is due to the improper distance between the adjacent PV arrays.
7) Module irregularities caused by physical damage or aging.
8) Presence of cracks on the PV panel.

25.2.2 Effects of Partially Shaded PV Panel

The partial shading (PS) issue invites considerable interest due to its significance in influencing the energy yield of a PV system. The effect of partial shading is shown by using current–voltage (I–V) and power–voltage (P–V) curves when the generator is operated under specific operating conditions. When the PV array is operating under uniform insolation, the resulting P–V characteristics curve of the array exhibit a single maximum power point.

The shading effect is one of the influencing factors resulting in power output reduction of PV modules and arrays caused by the passing clouds, nearby trees or buildings, long-lasting dust, etc. To protect against hot spots emerging in the partially shaded PV modules, connecting a bypass diode with reverse polarity in parallel to a group of solar cells in a serial connection of the module is one of the most common strategies applied in the current commercial product. When the PV panels with bypass diodes are operated under partially shaded conditions, the characteristic of the P–V curve is transformed into a more complicated shape – characterized by several local and one global peak.

Currently, PV systems encounter many challenges to maximize their efficiency and output power under varying weather conditions. Under varying weather conditions and particularly low irradiance levels, it's difficult for PV systems to provide the maximum output, which results in low efficiency. The I–V and P–V curves have a unique maximum power point (MPP), at which the PV system operates with the maximum output and the highest efficiency. Although there have been many MPPT techniques, most of them consider constant irradiance conditions. Their performance is usually measured by complexity, tracking speed, maximum output energy, time to reach MPP, and the number of required input devices (sensors). However, the performance of conventional MPPT techniques may significantly degrade under the partial shaded condition. The effect of shading is the occurrence of multiple local maxima points, that is, the problem is transformed from single- to multimodal. The drawback with the conventional MPPT is that for a majority of the cases, the algorithm is likely to trap at the local peak simply because it could not differentiate the local with the GP. Consequently, it oscillates around the vicinity of the local peak and will remain in that location indefinitely, and output power is reduced. The power loss due to the PS can vary from 10 to 70% of the system yield, depending on the severity and type of shading pattern.

25.3 MPPT CONTROLLER

The V–I characteristics of PV systems are nonlinear, thereby making it difficult to be used to power a certain load. For extracting the maximum power from the cell, the operating voltage or current should be corresponding to the maximum power point under a given temperature and solar radiation. To increase efficiency, methods should be undertaken to match source and load properly. The MPPT operates the photovoltaic (PV) modules in a manner that allows the modules to produce all the power they are capable of. It is a fully electronic system that varies the electrical operating point of the modules so that the modules can deliver maximum available power.

The block diagram of the PV system with the MPPT technique is shown in Figure 25.2. It consists of a PV array, boost converter, MPPT block, and load. A combination of series and parallel solar cells constitute the PV array. The series connection of the solar cells boosts up the array voltage, and parallel connection increases the current. The system involves the use of a DC-to-DC converter to change the input resistance of the panel to match the load resistance (by varying the duty cycle).

The input of the boost converter is connected to the PV array, and the output is connected to the load. The MPPT block receives signals from the PV array. The output of the MPPT block is the series of pulses for the power switch in the boost converter. The converter works based on these pulses to make the PV system operate at the maximum power point (MPP).

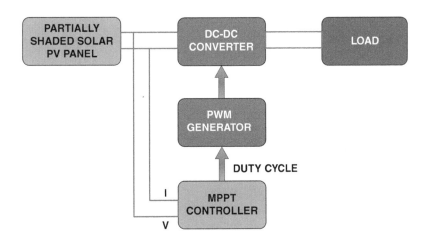

FIGURE 25.2 Block diagram of PV system with MPPT.

A PV module is composed of several solar cells connected in series and shielded with glass to protect against environmental changes. A typical PV generation system is composed of several such modules to meet the load power demand. The MPPT controller can be incorporated in such a PV system using two different approaches.

25.3.1 Central MPPT

In this configuration, all the modular PV systems are controlled by a single centralized MPPT controller, as shown in Figure 25.3. This configuration may not produce the desired results in the case of partial shading. Under partial shading conditions, some of the modules in the PV system may not generate the same amount of current. As all the PV modules are connected in series, the modules carry the least value of current generated from any module in the given PV system. The excess current generated by each module flows through the bypass diode attached in parallel to each module. This will create multiple peaks in the characteristics of the PV panel. If there are more modules, the characteristics under partial shading are complicated and may exhibit more local peaks. In such cases, it becomes difficult for the central MPPT controller to realize the MPPT using conventional methods. Even if it is possible to identify the global MPP, each module cannot be operated at the optimal condition, as their optimal current is inherently different at various levels of irradiation.

25.3.2 Modular MPPT

In this configuration, each modular PV system is provided with its MPPT controller, as shown in Figure 25.4. If the PV system is divided into submodules and each such module is controlled with its MPPT controller, the power loss due to partial shading can be minimized. However, this scheme requires more voltage and current sensors. To reduce the cost as well as the problems associated with the tracking scheme, the control circuitry must be simple and easy to implement, with a minimum number of sensors.

The partial shading conditions introduce a mismatch in the V–I characteristics that transform into a loss of power extraction from the PV systems. Mismatch losses are a serious problem in PV modules and arrays under some conditions because the output of the entire PV module under worst-case conditions is determined by the solar cell with the lowest output. For example, when one solar cell is shaded while the remainder in the module is not, the power being generated by the "good" solar cells can be dissipated by the lower-performance cell rather than powering the load.

For two cells connected in series, the current through the two cells is the same. The total voltage produced is the sum of the individual cell voltages. Since the current must be the same, a mismatch in current means that the total current from the configuration is equal to the lowest current. This type of mismatch is called a series mismatch.

For cells or modules in parallel, the total current from the combination is the sum of the currents in the individual cells or modules. Since the voltage across the parallel combination must always be the same, a mismatch in voltage means that the total voltage from the configuration is equal to the lowest voltage across the modules.

The mismatches in the PV system are also classified based on the objects causing the shading. The mismatches caused by shade from obstructions like a pole or nearby building are called a hard mismatch. Soft mismatch on modules accounts for small things like soiling (e.g., bird droppings) or vegetation (e.g., a weed). Soft mismatch causes create a series mismatch. While mismatch happens in both series and parallel types of connections, it is the mismatch in series connections that affects performance the most.

The parallel mismatch can be handled by connecting the MPPT controller in each string of the PV array. However, the string-level MPPT cannot handle the series mismatch in the modules. As a result, the soft mismatch persists until the array has module-level MPPT. Module-level MPPT has a ton of additional benefits beyond mismatch mitigation, including safety, design flexibility, electrical BOM benefits, and data visibility.

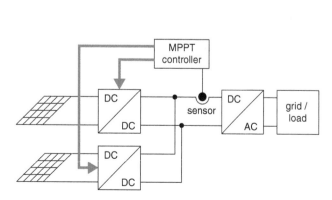

FIGURE 25.3 PV system with centralized MPPT controller.

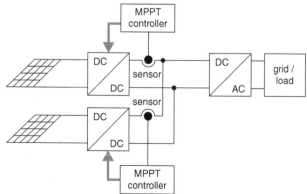

FIGURE 25.4 PV system with modular MPPT controller.

Panel mismatch resulting from shade or other factors can result in disproportionate power loss in solar panels, whereby as little as 10% of shading can lead to a 50% loss of energy harvested. Small variations in cell parameters can affect the system-level performance of the PV array. Partial shading will result in an array having multiple MPPs from different panels because of nonuniform parameters. With a centralized MPPT, this can lead to additional disproportionate losses. This is for two reasons: First, the centralized MPPT becomes confused, stopping on a local maximum point and settling in a suboptimal point of the voltage-to-power configuration. Second, the voltage point of the MPP can be very diverse due to irregular conditions, going beyond the scope and voltage range of the centralized MPPT. Because the variations between panels are significant, it is in these cases where the ability of module-level MPPT can enhance the performance of panels independently and boost performance.

Module-level MPPT offers the ability to maximize the energy extracted from every panel while maximizing the energy transfer in the PV system, recuperating up to 57% of energy lost to panel mismatch issues. Module-level MPPT addresses head-on the problems inherent with centralized systems by increasing total energy output by up to 37%, therefore mitigating successfully the panel mismatch problem.

The maximum power point tracking (MPPT) algorithm is crucial in attaining the maximal PV power, facilitating optimal PV cell performance. The conventional MPPT algorithms demonstrate excellent tracking efficiency in uniform insolation conditions. However, under partially shaded conditions, when the entire array does not receive uniform insolation, the PV characteristics become more complex, displaying multiple peaks, only one of which is the global peak (GP); the rest are local peaks. The occurrence of partially shaded conditions is quite common (e.g., due to clouds, trees, etc.) and echoes a need to develop special MPPT schemes that can track the GP under these conditions.

The main emphasis orients to explore the use of the maximum power from the solar PV system under the partial shading environment. It augurs the role of the closed-loop controller to vary the duty cycle of the converter interface and arrive at the delivery of the maximum power to the load. The exercise relates to modeling the solar PV system under a partially shaded environment and analyzing the performance for different shading patterns in the solar panel.

25.4 DESIGN OF MPPT CONTROLLER FOR PARTIALLY SHADED PV SYSTEMS

The proposed global maximum power point tracking (GMPPT) algorithm uses two stages to search for the GMPP. The PV voltage corresponding to the GMPP may be anywhere between zero to the open-circuit voltage of the PV array. The search area, therefore, becomes confined in the sense it requires to identify the PV voltage corresponding to the GMPP within the defined region. The duty cycle of the DC–DC converter decides the value of PV voltage at which the power transfer takes place from the source to the load. It necessitates the converter to operate at the value of the duty cycle corresponding to the PV voltage at the GMPP to ensure the transfer of maximum power to the load. Thus, the search of GMPP narrows down to the range of duty cycle of the DC–DC converter, and the complete range of variation of the duty cycle may be divided into various intervals. The first stage is to search for the global MPP location interval, and the traditional MPPT methods are used in the second stage to find the precise GMPP location. The first stage uses stepped duty cycle values between 0 and 1 for the MPPT boost converter and measures the corresponding power values from where it provides the duty cycle for obtaining the maximum power to the second stage of the search.

The second-stage search uses the direct duty cycle perturb and observe (P and O) MPPT technique with the initial duty cycle returned from the first-stage search. The advantage of this kind of method is its simplicity in the sense that it can easily be integrated with the traditional firmware (requiring only an additional global-biased search stage). The disadvantage of this kind of method is that when a small step is selected, the system requires relatively more time. A trade-off between the search step size and the time required must be addressed to use this method.

The MPPT control technique is required to detect whether the current system is exposed to uniform insolation or PSC and whether the system operating condition is changed. Therefore, an accurate understanding of the current operating status facilitates selecting suitable algorithms and avoiding extending the tracking time. The proposed method uses the change in current values between present and past iterations compared to a pre-set value to detect the partial shading condition.

The algorithm of the proposed system seen using the flowchart in Figure 25.5 and Figure 25.6 is as follows:

Step 1: Start GMPPT.
Step 2: Read n = no. of steps of duty cycle, and D_i = [$D_1 \ldots D_i \ldots D_n$].
Step 3: Read converter time constant t_c and normal operation current change ΔI_{spec}.
Step 4: Compute $\Delta I = I(k-1) - I(k)$.
Step 5: If $|\Delta I| > \Delta I_{spec.}$, go to step 6; else, go to step 14.
Step 6: Set i = 1.
Step 7: Return D_i and wait for the time t_c.
Step 8: Read P_i.
Step 9: Then i = i +1.
Step 10: If i > 3, go to step 11; else, go to step 7.
Step 11: P_{max} = max(P_i), i = 1, 2, 3 … n.
Step 12: $D_{best} = D_i$, such that $P_i = P_{max}$.
Step 13: Return D_{best} to P&O.
Step 14: Start P&O.

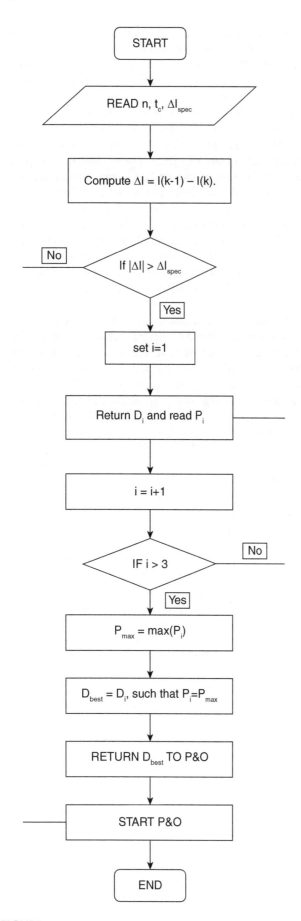

FIGURE 25.5 Flowchart of proposed MPPT controller.

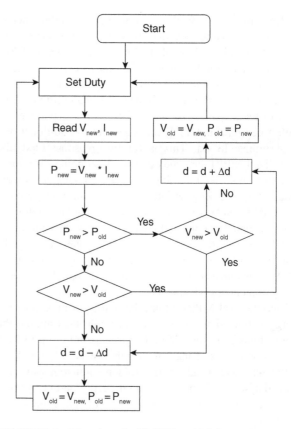

FIGURE 25.6 Flowchart for GMPPT and P&O algorithm.

TABLE 25.1
Parameters of Solar Panel under STC Conditions

Module Type	SRP-250–6PB
Rated Maximum Power (P_{max})	250 W
Current at P_{max} (I_{mp})	8.35 A
Voltage at P_{max} (V_{mp})	29.9 V
Short-Circuit Current (I_{sc})	8.92 A
Open-Circuit Voltage (V_{oc})	37.1 V
V_{oc} Temperature Coefficient	−0.32%/°C
I_{sc} Temperature Coefficient	+0.04%/°C

STC: Irradiance 1,000 W/m²; module temperature 25°C; AM 1.5; power measurement tolerance +/−3%.

25.4.1 Steady-State Analysis

The simulation study is based on a 250 W solar panel from Seraphim Solar, a leading PV manufacturer. The panel under study has 60 PV cells connected in series. The specifications of the solar panel are listed in Table 25.1. The PV cells are grouped into three PV strings in series with 20 series-connected PV cells in each string. Three bypass diodes are connected in parallel with the three strings respectively to prevent damage due to overheating. Each string is simulated with three irradiance levels of 200 W/m², 600 W/m², and 1,000 W/m² to study the partial shading effects on the performance of the panel. It is possible to study a diverse of

TABLE 25.2
Performance of Panel under Various Irradiation and Partial Shading Possibilities

Irradiation Condition Number	Irradiance Values (kW/m²)			Global MPP (Watts)
	String 1	String 2	String 3	
1	0.2	0.2	0.2	47.86
2	0.6	0.6	0.6	148.59
3	1	1	1	247.04
4	0.2	0.6	0.6	95.53
5	0.2	1	1	158.8
6	0.6	0.2	0.2	51.68
7	0.6	1	1	166.6
8	1	0.2	0.2	70.6
9	1	0.6	0.6	156.75
10	0.2	0.6	1	103.95
11	0.2	1	0.6	103.95
12	0.6	0.2	1	103.95
13	0.6	1	0.2	103.95
14	1	0.2	0.6	103.95
15	1	0.6	0.2	103.95

insolation conditions by covering a wide range of irradiance levels to which a PV system may be subjected from dawn to dusk. In a panel of three strings simulated for three irradiance levels, there are 15 possible combinations of irradiance values, as shown in Table 25.2.

The 15 possible combinations of irradiance conditions given in Table 25.2 can be grouped into four pattern categories according to the shading pattern characteristics produced, which are described in Table 25.3.

The simulation uses a panel with three strings of PV cells. It is thus obvious that the maximum number of possible peaks that will appear in its P–V characteristics is three. Therefore, three steps of duty cycle 0.1, 0.5, and 0.9, covering the entire range of 0–1, are used for the first stage of research. The time interval between the duty cycle steps is chosen to be 0.01 seconds, considering the time constant of the boost MPPT converter. The simulation of tracking of maximum power in four shading patterns 1 to 4 is given in Table 25.3. The proposed GMPPT, as seen from Figures 25.7, 25.9, 25.11, and 25.13, respectively, illustrates the performance for the four shading patterns 1 to 4 through the P–V and I–V characteristics shown in Figures 25.8, 25.10, 25.12, and 25.14.

TABLE 25.3
Categorization of Shading Pattern and Their Characteristics

Pattern No:	Irradiation Condition No:	Description
1	1, 2, 3	Uniform Irradiation with one peak
2	6, 7, 9	Partial shading with two peak values (Global peak at right)

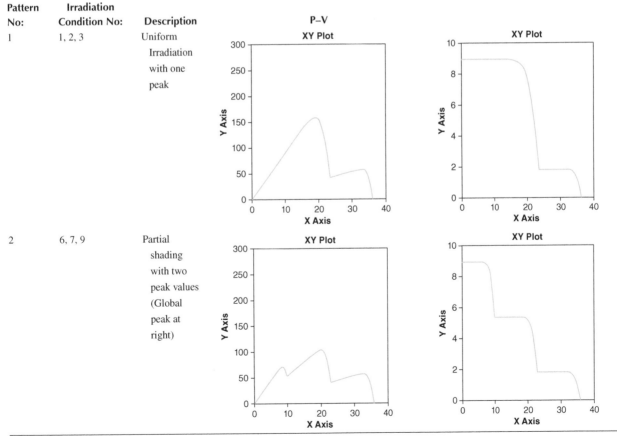

(Continued)

TABLE 25.3 (Continued)

Pattern No:	Irradiation Condition No:	Description	P–V	
3	4, 5, 8	Partial shading with two peak values (Global peak at right)		
4	10 to 15	Partial shading with three peak values		

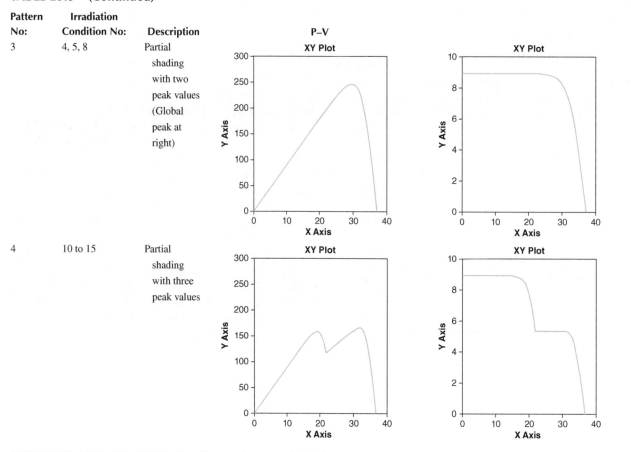

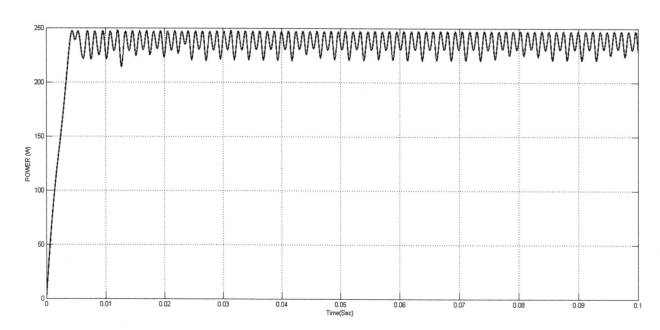

FIGURE 25.7 Tracking of maximum power in shading pattern 1.

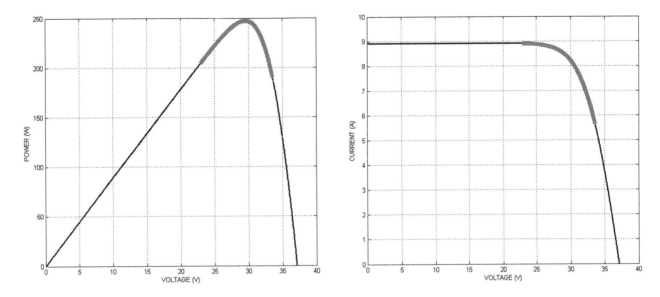

FIGURE 25.8 Performance of the proposed GMPPT in shading pattern 1.

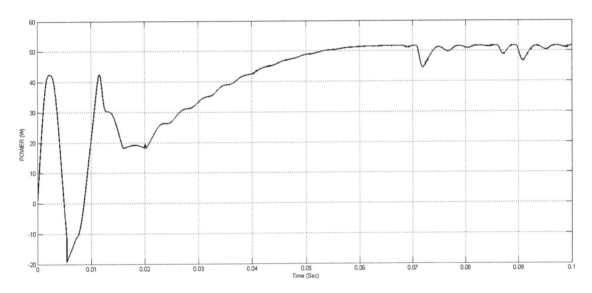

FIGURE 25.9 Tracking of maximum power in shading pattern 2.

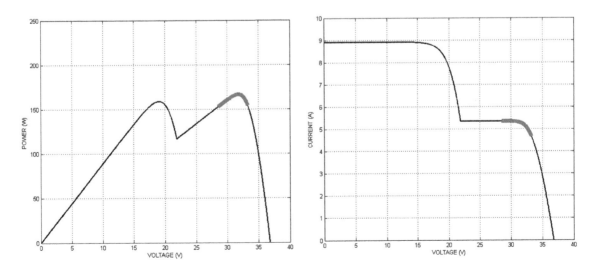

FIGURE 25.10 Performance of proposed GMPPT in shading pattern 2.

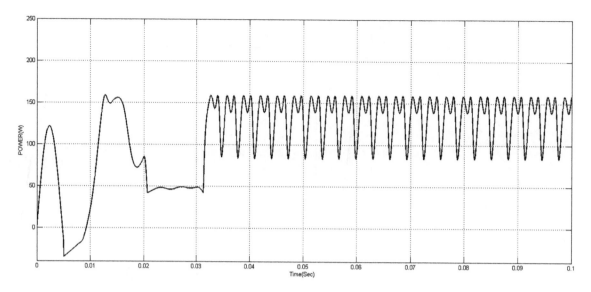

FIGURE 25.11 Tracking of maximum power in shading pattern 3.

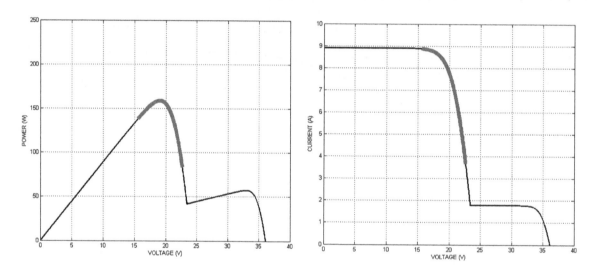

FIGURE 25.12 Performance of proposed GMPPT in shading pattern 3.

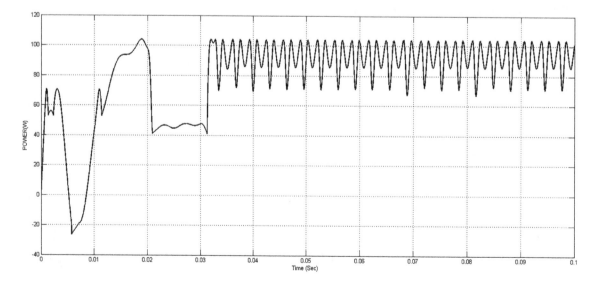

FIGURE 25.13 Tracking of maximum power in shading pattern 4.

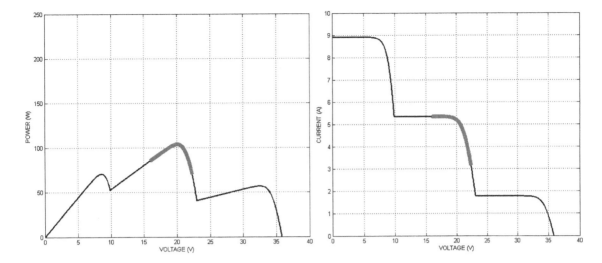

FIGURE 25.14 Performance of proposed GMPPT in shading pattern 4.

The proposed method of tracking GMPP is tested for all 15 possible irradiation conditions, and the performance of the controller is presented in Table 25.4.

25.4.2 Transient Analysis

The performance of the proposed GMPPT algorithm is tested for sudden variations of the shading patterns to prove its viability in rapid shading environments. The transient analysis simulation of tracking of maximum power with the proposed GMPPT is shown in Figure 25.15. The change in duty cycle to MPPT boost converter at the time of transient is also shown through the PWM pulses. The performance of the proposed GMPPT for the transient cases is exhibited through the P–V and I–V characteristics shown in Figure 25.16.

25.5 HARDWARE IMPLEMENTATION

The primary focus attempts to realize the GMPPT algorithm through the use of the dsPIC microcontroller for capturing the maximum power from the solar PV system under the partially shaded environment. The procedure involves the construction of a prototype for the converter interface and operates the solar PV system with the PWM pulses from the microcontroller and orients to examine the performance for different levels of irradiance.

The global maximum power point tracking (GMPPT) algorithm is implemented using the dsPIC30F4011 microcontroller. During partial shading, the parameters of the solar panel are sensed and given as an input to the microcontroller. This microcontroller includes an in-built PWM generation, which results in the production of PWM pulses. The pulses are used to trigger the switch in the converter.

The block diagram seen in Figure 25.17 depicts the methodology used in the hardware, through which it attempts to implement the GMPPT algorithm for tracking the maximum power from the solar panel.

25.5.1 Hardware Implementation of GMPPT Control Algorithm

The solar panel parameters like voltage and current are sensed and transferred as an input to the microcontroller for determining the variation in power. The voltage is sensed by a voltage divider circuit, and the current is sensed using a shunt. The GMPPT algorithm for tracking the maximum power point, depending on variation in solar intensity, is implemented using microcontroller dsPIC30F4011.

The microcontroller be-hives an in-built PWM generation, wherein the modulating signal is used for the generation of converter gate signals. To maintain the output of the converter, the signal is tapped from the load and stepped down to 5 V. It is multiplied with the modulation index using an analog multiplier to get the required modulating signal.

The triangular wave is generated using the combination of an operational amplifier operated as a square wave generator and integrator. The variation in the duty cycle given to the PWM generation generates gate pulses to trigger the power switches in the boost converter.

The isolator and driver circuit is used to provide the isolation between the power module and the control module. The converter used here is the boost converter, because it maintains the maximum constant voltage irrespective of the weather condition. The converter is then connected to the load. The flowchart shown in Figure 25.18 explains the steps relating to the functioning of the microcontroller, using which the scheme generates the PWM pulses following the GMPPT algorithm.

25.5.2 Results and Discussion

A 1 kW PV system consisting of four 250 W PV panels connected in series is experimented to study the partial shading effects on the output power of a PV system. The experiment is aimed at recording the output power from the PV system without MPPT, with P&O MPPT, and with the proposed global

TABLE 25.4
Performance of Proposed GMPPT Controller

	P&O MPPT		Proposed Two-Stage MPPT				
Irradiation Condition No.	Whether MPP-Tracked	Mean Max. Power	Whether MPP-Tracked	D_{best} Returned from the First Stage	Mean Max. Power	Time to Track MPP	MPPT Efficiency
1	Yes	47	Yes	NA	47	0.045	98.2
2	Yes	145	Yes	NA	145	0.008	97.6
3	Yes	236	Yes	NA	236	0.004	95.5
4	No	55	Yes	0.5	86.4	0.0322	90.4
5	No	57.5	Yes	0.5	134	0.0326	84.38
6	Yes	51	Yes	0.9	51	0.0663	98.7
7	Yes	150	Yes	0.5	150	0.0313	90.0
8	No	52	Yes	0.1	65	0.0355	92.06
9	Yes	152	Yes	0.5	152	0.0314	96.96
10	No	56	Yes	0.5	98	0.032	94.23
11	No	56	Yes	0.5	98	0.032	94.23
12	No	56	Yes	0.5	98	0.032	94.23
13	No	56	Yes	0.5	98	0.032	94.23
14	No	56	Yes	0.5	98	0.032	94.23
15	No	56	Yes	0.5	98	0.032	94.23

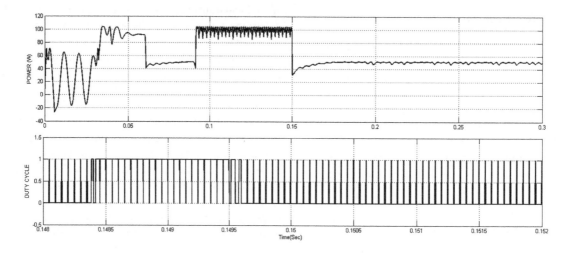

FIGURE 25.15 Transient analysis simulation of tracking of maximum power for case IV.

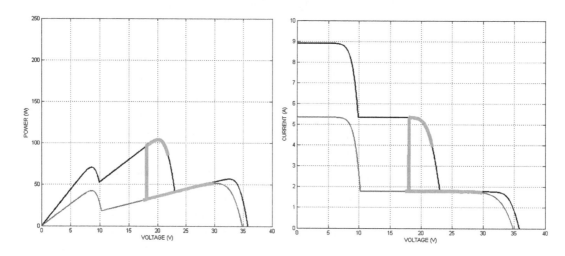

FIGURE 25.16 Performance of proposed GMPPT in transient case IV.

MPPT Controller for Partially Shaded Solar PV System

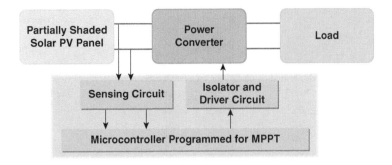

FIGURE 25.17 Block diagram of hardware implementation of proposed GMPPT.

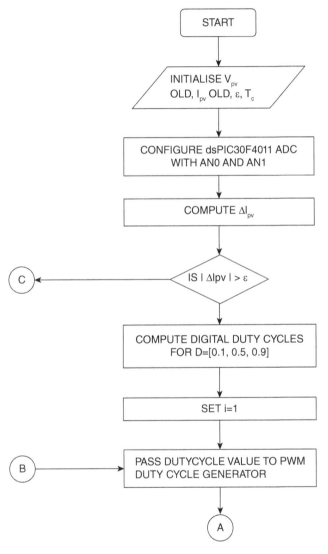

FIGURE 25.18 Flowchart of proposed MPPT control algorithm in dSPIC30f4011.

FIGURE 25.19 Experimental setup of proposed solar PV GMPPT controller system.

MPPT, with a view to explore the supremacy of the proposed tracking technique in utilizing the available PV power. The experimental setup with the control module, boost DC–DC converter for MPPT programmed with dSPIC30F4011 microcontroller, and the resistive load is shown in Figure 25.19.

The experiment is designed to study four cases of shading patterns which are explained in Table 25.5. Each case is tested for four different loading conditions. The performance of the PV system without MPPT, with P&O MPPT, and with the proposed global MPPT in cases I, II, III, and IV is shown in Figure 25.20 to Figure 25.23, respectively. The formulation expertise of the proposed global MPPT and the proficiency of implementing it in a real-time interface design are showcased through the coordinated readings in Table 25.6. The importance of having MPPT is clearly illustrated. The inability of P&O MPPT to track the maximum power in partial shading conditions is revealed in the results where the operation is struck around a local MPP. The aid of the proposed global MPPT to move the operating point to the global peak and to extract maximum power for a given load proves its effectiveness in a partially shaded PV system.

TABLE 25.5
Cases of Shading Patterns

S. No.	Case	No. of Cells Shaded in Panel 1	No. of Cells Shaded in Panel 2	No. of Cells Shaded in Panel 3	No. of Cells Shaded in Panel 4
1	I	Nil	Nil	Nil	Nil
2	II	Nil	20	Nil	Nil
3	III	Nil	20	30	Nil
4	IV	Nil	20	30	40

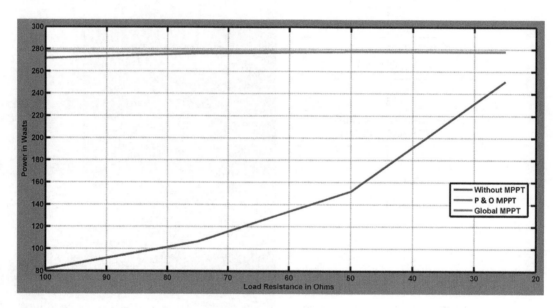

FIGURE 25.20 Performance of PV system under case I shading pattern.

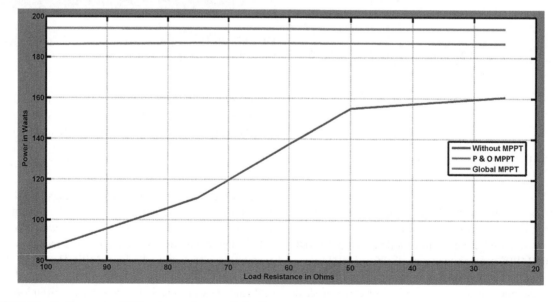

FIGURE 25.21 Performance of PV system under case II shading pattern.

MPPT Controller for Partially Shaded Solar PV System 347

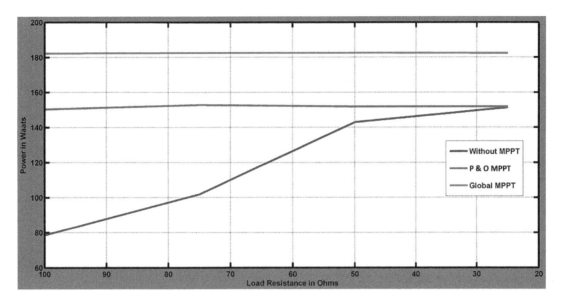

FIGURE 25.22 Performance of PV system under case III shading pattern.

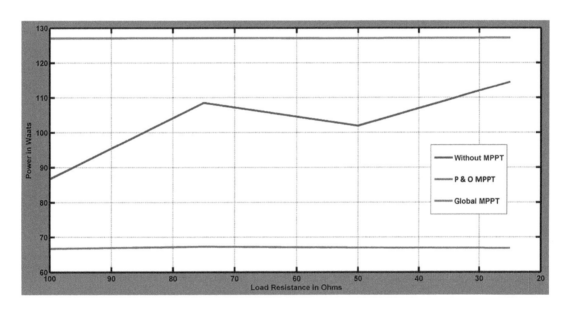

FIGURE 25.23 Performance of PV system under case IV shading pattern.

TABLE 25.6
Comparison of Simulated and Experimental Readings of Proposed Global MPPT

S. No.	Case I		Case II		Case III		Case IV	
	Sim.	Exp.	Sim.	Exp.	Sim.	Exp.	Sim.	Exp.
1	277.87	270	194.19	196	182.29	176	127.04	124
2	277.84	275	194.15	192	182.44	180	127.07	122
3	277.81	280	194.12	190	182.44	182	127	129
4	277.84	282	194.14	193	182.32	182	127.14	127

25.6 SUMMARY

The principal effort has been centered on the study of a partially shaded solar PV system. The system has been formed with a solar panel, a power electronic boost converter interface, and a suitable load. It has been operated through the etiquettes of a closed-loop MPPT controller. The investigations have been tailored to generate different shading patterns in a single solar panel. The influence of the variation of the irradiation has been sought to exhibit the occurrence of the maximum power. The main contribution has been oriented to the design of a global MPPT through a two-stage process. While the first stage has relied on incremental advancements in the search, the second stage hovers over the use of traditional methods to facilitate the capture of the maximum power.

The performance has been evaluated through MATLAB-based simulation in terms of its IV and PV characteristics. The results have been obtained with different shading patterns to illustrate the variation of the occurrence of the maximum power. The benefits of the developed algorithm have been brought out to foster the claim of the use of the GMPPT for a single panel in real-world applications.

The prototype that includes the boost converter along with a dsPIC controller has been constructed. The controller has been programmed to function as a PI controller and create the reference for the PWM pulse. It generates the pulse for the switch following the design of the GMPPT strategy.

The variation of the irradiation that occurs on the panel has been reflected in the location of the maximum power point. The study has been belied through the ability of the dsPIC microcontroller in being able to implement the GMPPT strategy.

The model has been examined for different shading patterns to elucidate the changes in the occurrence of the maximum power. The results have been portrayed to bring out the suitability of the prototype for measuring the maximum power and its viability for use in practice.

26 Adaptive Control of Smart Microgrid Using AI Techniques

Krishna Degavath and Mallesham Gaddam

CONTENTS

- 26.1 Introduction: Background and Motivation 350
- 26.2 Modeling of Smart Microgrid 351
  - 26.2.1 Renewable Energy Sources (RES) 352
    - 26.2.1.1 Wind Energy Conversion System (WECS) 352
    - 26.2.1.2 Solar Panel Modeling under Maximum Power Point Tracking (MPPT) Operation 352
  - 26.2.2 Secondary Energy Sources in Microgrid 353
  - 26.2.3 The Smart Microgrid 354
  - 26.2.4 Working Principle of Microgrid Central Controller (MGCC) 354
  - 26.2.5 The Communication Channel for Microgrid 355
- 26.3 Design of Reinforced Learning MLP-Based Controller for the Microgrid ... 356
  - 26.3.1 The Architecture of the MLP Controller 356
- 26.4 Simulation Results and Analysis 358
  - 26.4.1 Increase in Load by 5% at 10th Second 358
  - 26.4.2 Solar Power Increased by 3% at 10th Second 359
- 26.5 Conclusions 360
- 26.6 Appendix 360
- 26.7 References 360

NOMENCLATURE

P_G	Power generation (kW)
P_L	Load power (kW)
P_W	Power from wind source (kW)
P_S	Power from solar PV panels (kW)
P_{dg}	Power from diesel generator (kW)
P_{fc}	Power from fuel cell (kW)
P_{ae}	Power from aqua-electrolyzer (kW)
H_{2s}	Hydrogen stored in tank (Moles/s)
P_b	Power from battery power (kW)
Q_b	State of the charge of the battery (kWh)
ΔP	Power deviation (kW)
Δf	Frequency deviation (Hz)
f_{sys}	System frequency (Hz)
B_i	Frequency bias (MW/Hz)
R_i	P-f droop-in (MW/Hz)
T_{sim}	P-f droop-in (MW/Hz)
J	Performance index based on ITSE (integral time squared error) criteria
T_m	Mechanical torque (N-m)
w_s	Wind speed (m/s)
w_r	Rotor speed (rad/sec)
β	Pitch angle (degrees)
C_p	Power coefficient
n_p	Number of PV strings in parallel
C	DC link capacitance (F)
S	Solar insolation
V_{dc}	DC link (PV array) voltage (V)
I_{rs}	Reverse saturation current (A)

26.1 INTRODUCTION: BACKGROUND AND MOTIVATION

Energy is essential to everyone's life, no matter when and where they are. It is especially true in this new century, where people keep pursuing a higher quality of life. Among different energy types, electric energy is one of the most important that people need every day. World energy consumption is expected to grow about 49% from 2007 to 2035, and there is strong growth of global energy demand in the next two decades [1, 2]. From the report [3], it is clear that a large part of the total energy is provided by fossil fuels (about 86%). The future of global economic growth is highly dependent on whether the ever-increasing energy demand can be met or not. Fossil fuels are not evenly distributed worldwide, and regional or global conflicts may arise from energy crises if national economy is still heavily dependent on them.

Moreover, during generating and using electrical energy with today's conventional technologies, the global environment has already been significantly affected, and the environment of some regions has been damaged severely. Therefore, it is a big challenge for the whole world to figure out how to generate the needed amount of energy and efficient coordination. So the world is moving towards the utilization of renewable energy sources, such as wind power and solar power.

The distributed generation system (DGS) having small-scale energy sources is called a microgrid or active distribution network [4]. These sources are small in capacity and are mostly connected at the distribution voltage level, indirectly reducing transmission and distribution losses since the sources are around the load. However, still, these microgrid sources can pose both positive and negative impacts on the existing power systems. These new issues have led DGS operation to be an important research area.

The microgrid generally operates in a grid-connected mode. However, circumstances such as fault, voltage sag, and large frequency oscillations in the main grid may force the active distribution network to be disconnected from the grid and operate as an isolated microgrid. During this isolation or change in load, there will be a change in power output from the controllable microgrid sources and the need to be regulated properly to have a stable operation concerning balance power and operation frequency, which is to be properly addressed. In this isolated microgrid scenario, there will be changes in power output from the controllable microgrid sources which need to be regulated properly to balance demand and supply within the isolated microgrid. The diesel generator, combined-cycle gas turbine–based system, fuel cell, aqua-electrolyzer, and battery can be considered controllable sources in the microgrid.

It is well-known that when there is a mismatch in demand and supply power, the frequency of the system changes. Therefore, in this chapter, the controllable sources' power output is regulated based on the frequency deviation of the microgrid. This regulation will not be smooth due to inherent time delay and ramp rate limit/generation rate constraint (GRC) associated with the sources. Besides, for proper load sharing among the microgrid sources, we introduce a power–frequency (P-f) droop (R) for the power-generating sources participating in frequency control [5, 6]. Similarly, in conventional control, each controllable source can have a control signal to regulate output power, proportional to frequency deviation, termed as frequency bias (B), which is to be properly selected.

The time–domain analysis of small-signal stability of a hybrid power system is given in [7]. Authors in [7] have proposed the different configurations with different power sources for the frequency control due to changes in the source and load.

In [8], a hybrid power system uses many wind turbine generators with fluctuating power output in small isolated islands, leading to the system's fluctuating frequency. In order to solve this problem, they have proposed additional sources, like aqua-electrolyzer and fuel cell. The controllers, which are tuned by the trial-and-error method, are used for each device.

In [9], an automatic generation control (AGC) to DGS is given. The gains of the controllers are tuned by the trial-and-error method. This approach gives performances far from optimality, so proper optimization techniques are necessary for tuning. The more realistic power system model would have been taken with an automatic voltage regulator (AVR) and power system stabilizer (PSS), which is not dealt with in this analysis.

In the aforementioned references, the authors have modeled the microgrid and studied the frequency control without the nonlinear elements generation rate constraint (GRC), power–frequency (P–f) droop. The controller gains and frequency bias are decided through a trial-and-error approach. Furthermore, the microgrid is not studied for the management of power from the sources in the microgrid. So the present research work emphasizes the aforementioned issues.

In most of the literature, research has been done to select proportional integral and derivative (PID) controller gains based on the integral squared error (ISE) technique, which is a trial-and-error method. When the number of controllers' gains to be optimized simultaneously is more based on ISE, it is extremely time-consuming and may result in a suboptimal solution.

In our previous work [10–14], modeling of the microgrid and the frequency control using conventional proportional and integral (PI) controller with tuning methods – Ziegler's-Nichol method [10, 11] and gradient descent method [12] – have been presented. In [13], population-based optimization techniques like genetic algorithms (GA), particle swarm optimization (PSO), and bacterial foraging optimization (BFO) based parameter tuning of controllers for the microgrid has been done. However, in these heuristic methods, the parameters are tuned by the offline method and then used for the controllers. It involves excessive numerical iterations, resulting in huge computations. Moreover, many

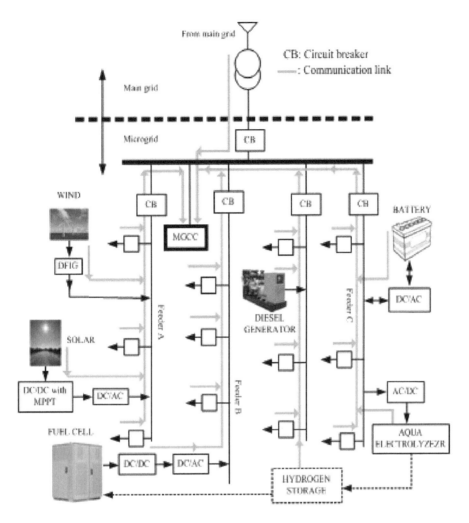

FIGURE 26.1 General configuration of a smart microgrid.

times, these offline-based tuning of controllers may give suboptimal performance when implemented practically.

In the literature, the tuning of controller parameters has been carried by using a fuzzy logic controller [15], artificial neural network (ANN) controller [16]. All these methods are offline-based and need sufficient knowledge about the system.

We have also identified the new approach for microgrid, online frequency control of microgrid using reinforced lean multilayer perceptron (MLPNN) [17]. This method's performance is compared with the previous method, where PI controllers are tuned with BFO and established its advantages.

In this book chapter, an efficient energy management strategy as well as keeping proper control coordination among the controllers using the concept of multiagent systems (MAS), treating each load and source as an agent, is provided and discussed [18]. Each agent communicates with each other following user datagram protocol (UDP/IP) over the internet [19].

The layout of the chapter is as follows. In Section 26.2, element-wise description and mathematical formulation of constituents of the microgrid are given. Also, the working principle of the microgrid central controller is briefly explained in the same section. Section 26.3 presents the methodology used in solving the problem statement. Case studies and relevant simulation results are showcased in Section 26.4. Conclusive remarks and Appendix are indicated in Section 26.5 and Section 26.6, respectively.

26.2 MODELING OF SMART MICROGRID

In this chapter, a self-sufficient isolated microgrid with the total rating of the controllable sources being equal to that of the load is considered. As shown in Figure 26.1, a smart microgrid consists of renewable energy sources (RES)/primary energy sources – wind and solar – and secondary energy sources – diesel generator, fuel cell, aqua-electrolyzer, and battery. The power from renewable energy–based sources fluctuates. When the renewable power is more than the load, to store this precious excess energy, aqua-electrolyzer is used in the microgrid, which can dissociate water into hydrogen and oxygen. The hydrogen so produced can be stored in a storage tank and further used by the fuel cell when required. On the other extreme, when the generation from the renewable sources becomes zero, the total power required by the load needs to be catered by the controllable sources; otherwise, load shedding needs to be carried out. The battery is considered to improve frequency excursion damping following each disturbance by regulating its power during the transient period.

26.2.1 Renewable Energy Sources (RES)

26.2.1.1 Wind Energy Conversion System (WECS)

From [20], for the power generating sources used for the regulation service, irrespective of continuous frequency control or contingency, a certain amount of generating power is kept as reserve to accommodate the governor's actions. The same idea has been utilized, as shown in Figure 26.2, for the wind turbines and is used to provide frequency control for which the deloaded operation is necessary.

The power coefficient (Cp) value of a wind turbine power coefficient is a function of the fourth-order polynomial of λ (tip speed ratio), and β is given here in (26.1).

$$C_p(\lambda,\beta) = \sum_{i=0}^{4}\sum_{j=0}^{4}\alpha_{i,j}\beta^i\lambda^j \quad (26.1)$$

The values of the coefficients $\alpha_{i,j}$ are given in [12]. The expression for λ is defined as:

$$\lambda = w_0 R w_r / w_s \quad (26.2)$$

Where w_r is rotor speed in pu, w_s is the wind speed in m/s, w_0 is rotor base speed in rad/s, and R is rotor radius in meter.

From Figure 26.2, the mechanical torque (T_m) is a complex function of wind speed (w_s), rotor speed (w_r), and pitch angle (β). When the output power exceeds the machine's rating, the pitch angle control is initiated in the case of a wind turbine. In this case, it is assumed that the wind speed is such that the power output is always less than its rating, and hence the pitch angle control is absent.

In this book chapter, two cases are studied, namely, (1) WECS under maximum power point tracking (WECS-MPPT) and (2) WECS with a deloaded operation (WECS-D).

26.2.1.2 Solar Panel Modeling under Maximum Power Point Tracking (MPPT) Operation

The power output from the solar photovoltaic (PV) panel [20] is given by (26.3):

$$P_{pv} = n_p CSV_{dc} - n_p I_{rs} V_{dc}\left(e^{\alpha V_{dc}} - 1\right) \quad (26.3)$$

The change in PV panel power for a small perturbation in V_{dc} can be written as:

$$dP_{pv}/dV_{dc} = n_p CS - n_p I_{rs}\left(e^{\alpha V_{dc}} + \alpha V_{dc} e^{\alpha V_{dc}} - 1\right) \quad (26.4)$$

Equating $dP_{pv}/dV_{dc} = 0$, we obtain (26.5), from which the required DC voltage (V_{dc0}) to extract maximum power from the PV panel at any given insolation S_0 can be calculated.

$$CS/I_{rs} = \left(e^{\alpha V_{dc}} + \alpha V_{dc} e^{\alpha V_{dc}} - 1\right) \quad (26.5)$$

Now, perturbing (26.5) around V_{dc0} and S_0, we arrive at (26.6).

$$\Rightarrow \Delta V_{dc} = C\Delta S /\left(2\alpha e^{\alpha V_{dc0}} + \alpha^2 V_{dc0} e^{\alpha V_{dc0}}\right)I_{rs} \quad (26.6)$$

Similarly, perturbing (26.3) around V_{dc0} and S_0 results to (26.7).

$$\therefore \Delta P_{pv} = \left(n_p C\Delta SV_{dc0} + n_p C\Delta V_{dc} S_0 - n_p I_{rs}\Delta V_{dc} e^{\alpha V_{dc0}} - n_p I_{rs} V_{dc0}\alpha e^{\alpha V_{dc0}}\Delta V_{dc} + n_p I_{rs}\Delta V_{dc}\right) \quad (26.7)$$

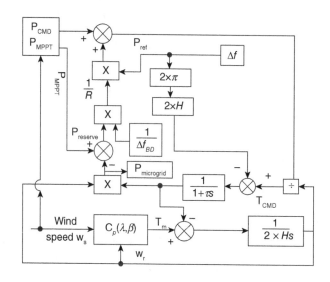

FIGURE 26.2 WECS model for frequency control using deloaded operation.

By substituting ΔV_{dc} given from (26.6) in (26.7), we get:

$$\Rightarrow \frac{\Delta P_{pv}}{\Delta S} = K_s = n_p C V_{dc0}$$
$$+ \left(CS_0 - I_{rs} e^{\alpha V_{dc0}} - I_{rs} V_{dc0} \alpha e^{\alpha V_{dc0}} + I_{rs} \right)$$
$$\times n_p C / \left[\left(2\alpha e^{\alpha V_{dc0}} + \alpha^2 V_{dc0} e^{\alpha V_{dc0}} \right) I_{rs} \right] \quad (26.8)$$

Now, the challenge here is how to find the operating point which must be perturbed so that the gain K_s will be valid for a wide range of isolation values. The following experimentation has been carried out to overcome this challenge.

Firstly, calculate K_s at the 20% insolation. Keeping the K_s unchanged varies S (say, $S = 0.3$, and hence $\Delta S = 0.3 - 0.2$) and gets ΔP_{pv}. Add this ΔP_{pv} to the MPPT power obtained at 20% insolation level. Let this power be termed as estimated power (P_{est}) at $S = 0.3$. Calculate the actual power (P_{act}) at this $S = 0.3$ by first obtaining the MPPT V_{dc} from (26.5) and then substituting that into (26.3). Now the % error at $S = 0.3$ can be defined as:

$$err_s = \frac{100 \left(P_{est} - P_{act} \right)}{P_{act}} \quad (26.9)$$

A similar exercise is carried out for other S (such that it covers the whole range from 10% to 90%), and % errors are calculated based on (26.9). The histogram corresponding to % error and insolation level is presented in Figure 26.3.

Looking at Figure 26.3, the maximum % error introduced due to the assumption of constant K_s is recorded. Now the whole procedure is repeated for another insolation, say, $S = 0.1$, and maximum % error is obtained. The histogram of maximum % error against solar insolation in the range 0.1 to 0.9 insolation level is depicted in Figure 26.4.

Looking carefully at Figure 26.4, it may be concluded that if K_s is calculated at $S = 0.2$, the maximum % error in estimating the P_{pv} is just around 5% in the whole operation range. Therefore, in this manuscript, K_s value corresponding to $S = 0.2$ is considered for all the simulation purpose.

Since the PV panel is interfaced to the microgrid via a voltage source converter (VSC), its input power can be translated to its output by the PWM controller in two to three cycles. Therefore, the transfer function of a PV panel under MPPT can be represented as in (26.10).

$$\Delta P_s / \Delta P_{pv} = K_s / (1 + sT_s) \quad (26.10)$$

Where T_s is the time constant of the converter controller and P_s is the electrical power output from the solar system.

26.2.2 Secondary Energy Sources in Microgrid

In this chapter, the secondary energy sources considered, and their transfer functions, are for governor, turbine of diesel generator, fuel cell, aqua-electrolyzer, and battery [7, 8].

$$\left. \begin{array}{l} G_{dgg}(s) = 1/(1 + sT_{dgg}) \\ G_{dgt}(s) = \dfrac{1}{1 + sT_{dgt}} \\ G_{ae}(s) = \dfrac{1}{1 + sT_{ae}} \\ G_{fc}(s) = 1/(1 + sT_{fc}) \\ G_b(s) = 1/(1 + sT_b) \end{array} \right\} \quad (26.11)$$

T_{dgg}, T_{dgt}, T_{ae}, and T_{fc} are the time constants of the governor, turbine of diesel generator, fuel cell generator, and aqua-electrolyzer, respectively. The values of all the parameters are given in the Appendix.

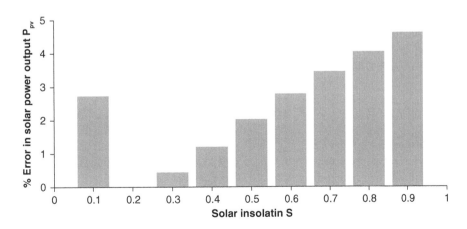

FIGURE 26.3 Percentage error in estimating maximum P_{pv} when K_s is calculated at $S = 0.2$.

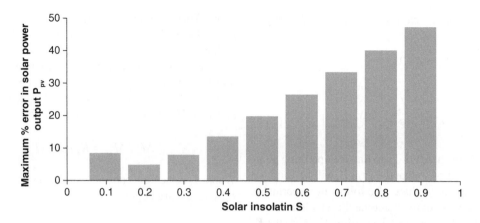

FIGURE 26.4 The maximum percentage error in estimating maximum P_{pv} when K_S is calculated for different solar insolation range of 0.1 to 0.9.

The swing equation for a synchronous generator [8] is depicted in (26.12):

$$\Delta f = f_{sys} / 2Hs \times [\Delta P_G - \Delta P_e] \quad (26.12)$$

Where:

$$\left. \begin{array}{l} P_G = P_W + P_s + P_{dg} + P_{fc} - P_{ae} \pm P_b \\ \Delta P_e = \Delta P_L + D\Delta f \end{array} \right\} \quad (26.13)$$

Therefore, the transfer function for system frequency variation to per unit power deviation is given by (26.14).

$$\begin{aligned} G_{sys}(s) &= \Delta f / (\Delta P_G - \Delta P_L) = 1 / \left[D + (2H / f_{sys}) s \right] \\ &= K_{ps} / (1 + sT_{ps}) \end{aligned} \quad (26.14)$$

Where K_{ps} and T_{ps} are $1/D$ and $2H / (Df_{sys})$, respectively.

It is to be noted here that (26.12) is valid only when there is a synchronous machine in the microgrid. Therefore, the researchers should be careful in using (26.14) when they are simulating the microgrid.

There are three sources in this smart microgrid: diesel generator, fuel cell, and aqua-electrolyze with P–f droop. The P–f droop characteristics are required when multiple power sources are connected in parallel, like in a microgrid [4]. The individual power generators are responsible for maintaining the frequency. Conventionally, it is implemented as in (26.15):

$$m_2 / m_1 = P_{1rated} / P_{2rated} \quad (26.15)$$

Where m_1, m_2 are P–f droop coefficients and P_{1rated}, P_{2rated} are power ratings of generating sources in the microgrid. Since the droops are related to the sources' rating, there is a need to calculate for at least one source at the outset and then examine for the others. The P–f droop calculations considered in this research have been derived from [13] based on the Bode plot stability criterion.

26.2.3 THE SMART MICROGRID

To implement the frequency control strategy and for efficient use of available resources, the microgrid is made smart using the concept of multiagent systems (MAS), as proposed in [18]. In this scheme, each load and source will be called an agent and assigned with an IP address. Using the IP address, the MGCC acts as a server and will fetch the agents' status. Based on this information, the MGCC will take a decision and generate a control signal transmitted to the respective sources being identified by their IP addresses. The two-way transmission of information in a digital domain is carried out using UDP/IP. The control variable, real power produced by different sources and power consumed by loads, is sensed through different transducers. As the transducers' outputs are analog by nature, an analog-to-digital conversion (A/D) is required before it gets transmitted to the communication network.

Similarly, a digital-to-analog (D/A) converter will be required to decode the instruction sent by the MGCC to the agents. Apart from the control unit, MGCC also maintains the states of each source and load and can act as a monitoring system. The simplified block diagram of the smart microgrid, neglecting real power loss, from the perspective of real power exchange and, consequently, the frequency deviation, is shown in Figure 26.5.

26.2.4 WORKING PRINCIPLE OF MICROGRID CENTRAL CONTROLLER (MGCC)

As shown in Figure 26.5, the status of all the agents is obtained at MGCC. Based on the information on power generation, storage, and load, MGCC will decide and give on/off signals to the circuit breakers (CB) via UDP/IP. The MGCC produces the signals based on renewable energy sources and load.

When the contribution of renewable energy sources (wind P_w and solar P_s) (RES) is higher than the load (P_{L0}) in a steady state, the aqua-electrolyzer will function, whereas the fuel cell and diesel generator will not contribute any power

Adaptive Control of Smart Microgrid Using AI Techniques

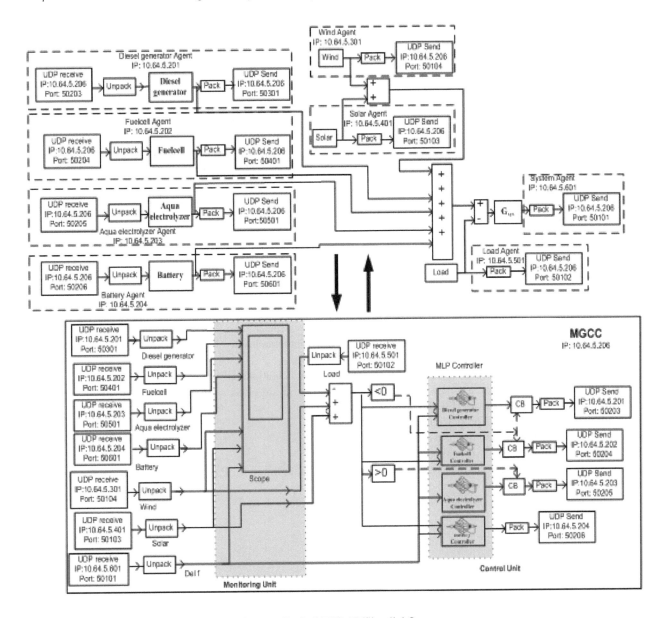

FIGURE 26.5 The smart microgrid with MLP controller in MATLAB/Simulink®.

to the microgrid. This rule is framed because when there is excess power from the renewable sources in microgrid, it should be converted and be stored in the form of hydrogen. Similarly, if the contribution of renewable energy (windP_w and solarP_s) is lower than the load (P_{L0}) in a steady state, the aqua-electrolyzer should not function. The fuel cell and diesel generator should start contributing power to the microgrid. When the hydrogen stored in the tank comes to a level such that the pressure is not sufficient for the fuel cell to work satisfactorily, then the MGCC should give an alarm.

26.2.5 THE COMMUNICATION CHANNEL FOR MICROGRID

The loads and sources present in the microgrid are treated as agents, and MAS is formulated. The sources in microgrid – wind, solar, diesel generator, fuel cell, aqua-electrolyzer, and battery – are considered as agents, and MGCC is modeled as shown in Figure 26.5, and the exchange of the data between these agents and MGCC is achieved by using UDP/IP.

For each agent, the UDP send and UDP receive blocks are used to send and receive the data to and from the agents. UDP send block has one input, and the parameters to be specified in this block are the IP address of the receiving agent and the port address through which the data will enter the receiving agent. Similarly, the UDP receive block has one input, and the parameters to be given are the IP address of the sending agent and the port address through which the data is to be received by the receiving agent. For example, as shown in Figure 26.5, the diesel generator agent is assigned with IP address 10.64.5.201, and the status of diesel generator power output is transmitted to MGCC, defined with IP address

10.64.5.206, through the port number 50301. Similarly, the diesel generator's control input in MGCC is sent back to the diesel generator agent via port number 50203. This IP-based two-way transmission of information in a digital domain is carried out using UDP/IP through a GSM (global system for mobile communications) technology. Proper sensors/transducers are used to fetch the desired information. As the transducers' outputs are analog by nature, an analog-to-digital conversion (A/D) is used before transmission to the communication network. Similarly, a digital-to-analog (D/A) converter will be required to decode the instruction sent by the MGCC to the agents. The effect of data loss and wrong sequence (packets) in smart microgrid with the UDP/IP has been discussed in [13] and shows that there is no change in frequency deviation.

26.3 DESIGN OF REINFORCED LEARNING MLP-BASED CONTROLLER FOR THE MICROGRID

In this section, an MLP controller using reinforcement learning has been presented for a smart microgrid frequency control. In our previous work [13], the bacterial foraging optimization (BFO) based proportional and integral (PI) controllers were implemented for the frequency control of the microgrid. In this chapter, this method has been compared with the MLP-based microgrid control. The MLP controllers are in MGCC, and based on the microgrid's output, MGCC generates control signals to the secondary power sources.

26.3.1 THE ARCHITECTURE OF THE MLP CONTROLLER

As shown in Figure 26.6, the details of MLP are:

1) *Number of MLPs: six.*
   Each secondary power source is provided with an MLP controller, which generates a control signal and then required power to compensate for the disturbance in the smart microgrid.
2) *Number of input neurons: one.*
   The input to each MLP is $x = \Delta P + B \times \Delta f$, which goes to the hidden neurons as input.
3) *Number of hidden neurons: six.*
   There are six neurons in the hidden layer. The nonlinearity in this layer is log-sigmoidal.
4) *Number of output neurons: one.*

   One neuron in the output layer is considered, and the transfer function for this neuron is purely linear.

The mathematics involved in the MLP neural network is as follows.

The deviations in the frequency and power of the smart microgrid are given as input to the MLP and are given by (26.16):

$$x = \Delta P + B \times \Delta f \qquad (26.16)$$

The input to the MLP of the nth secondary energy source is given by:

$$x_n = \Delta P + B_n \times \Delta f \qquad (26.17)$$

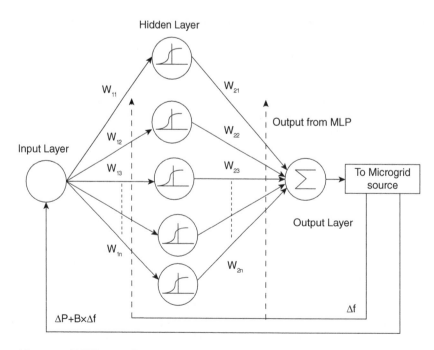

FIGURE 26.6 The architecture of MLP controller with smart microgrid.

The aforementioned signal is transmitted to all hidden neurons by multiplying the input layer weights W_{ji} (26.18). The resulting signal I_{ji} passes through the log-sigmoidal function of the hidden layer, resulting in output signal of h_{ij} given in (26.19):

$$I_{ji} = x_n \times W_{ji}; i = 1, j = 1....N_h \quad (26.18)$$

$$h_{ji} = \frac{1}{1+e^{-I_{ji}}} \quad (26.19)$$

The h_{ji} signal multiplies with the output weights W_{kj}, and summation at the output neuron results into a control signal to the secondary sources, given in (26.20):

$$y_k = \sum_{j=1}^{N_k} h_{ji} \times W_{kj}; K = 1, j = 1....N_h \quad (26.20)$$

5) *Updating weights in MLP.*

When the smart microgrid's frequency deviation is not reaching zero, then based on this signal, weights in the output layer and input layer are updated. Initially, the weights are assumed to be zero in the hidden and output layer. The weights W_{kj} in the output layer are updated through the least mean square (LMS), given in (26.21):

$$\Delta W_{kj} = -\mu \times \Delta f \times h_{kj} + \alpha \times \Delta W_{kj(prev)}; \\ K = 1, j = 1....N_h \quad (26.21)$$

Whereas the frequency deviation in the microgrid is back-propagated to the hidden layer from the output layer, presented in (26.22):

$$\Delta_{kj} = \Delta f \times W_{kj}; k = 1, j = 1.... N_h \quad (26.22)$$

The derivative of output of a log-sigmoidal function with respect to its associated input weights is given by (26.23):

$$\Delta_{ji} = h_{ji} \times (1 - h_{ji}) \times \Delta_{kj}; j = 1....N_h \quad (26.23)$$

From (26.23), the weights in hidden layer neurons are updated and given in (26.24):

$$\Delta W_{ji} = -\eta \times \Delta_{ji} \times x_i + \alpha \times \Delta W_{ji(prev)}; j = 1....N_h \quad (26.24)$$

Where the constants η and μ are the learning rate and α is the momentum constant. The training algorithm of MLP is shown in Figure 26.7 in the form of a flowchart.

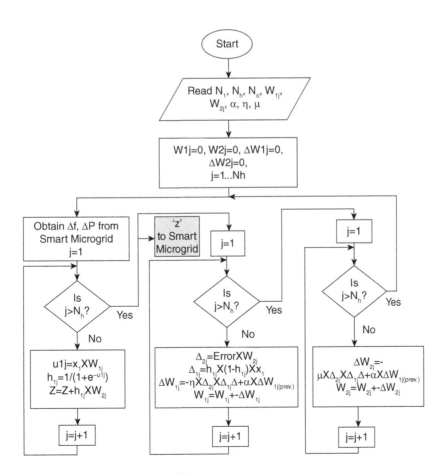

FIGURE 26.7 The flowchart of the MLP training algorithm.

In general, at least one integral controller is used in the frequency control of a power system to bring the frequency fluctuations to zero in steady state [13]. But this is achieved by MLP controller without using any integral controllers. The hidden layer's activation function is log-sigmoidal and gives 0.5 when the input to the controller is zero. This signal from the hidden layer is multiplied by the output neurons weights such that the summation of all the signals results in zero output.

26.4 SIMULATION RESULTS AND ANALYSIS

The frequency control of a smart microgrid with PI controllers (tuned with BFO) [13] and MLP controllers is simulated in MATLAB. Detailed descriptions of the microgrid parameters are given in the Appendix. The performance analysis of the proposed MLP is compared with the PI controller tuned with BFO (PI-BFO).

26.4.1 Increase in Load by 5% at 10th Second

Initially, the load in the smart microgrid is 626.5 kW, and after 10 seconds, there is a sudden increase of 5%, which is 657.5 kW. Two possible case studies are examined. One is when the WECS is under maximum power point tracking (WECS-MPPT), that is, 307.3 kW, and in the second case, the WECS is participating frequency control by keeping 10% MPPT power as margin (deloading operation) (WECS-D).

In Figure 26.8, the simulation results for frequency control of a smart microgrid are given with (1) BFO-tuned PI controllers (PI-BFO), (2) MLP with WECS under MPPT (MLP-WECS-MPPT), and (3) MLP with deloaded WECS (MLP-WECS-D).

In Figure 26.8(a) and Figure 26.8(b), the diesel generator and fuel cell's power output are given. The time taken to reach a steady state for these sources is the same except for the peak overshoot for PI-BFO and WECS-MPPT, which is not required.

Also, for the same sources, the power output under MLP-WECS-D reaches steady state without oscillations. Figure 26.8(c) and Figure 26.8(d) show that the power output and state of the charge of the battery are reduced under MLP-WECS-D. This reduction is advantageous in terms of the cost and life span of the battery.

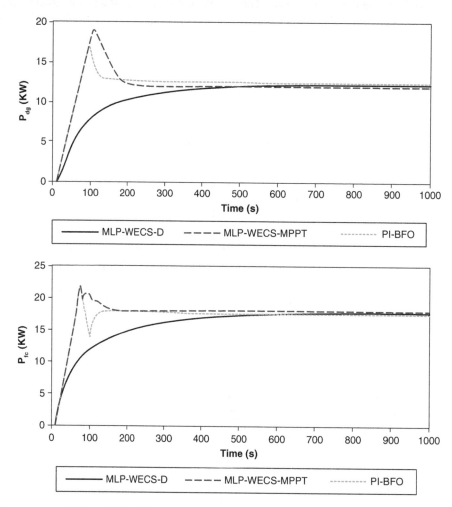

FIGURE 26.8 The power supply form (a) diesel generator P_{dg}, (b) fuel cell P_{fc}, (c) battery P_b, (d) battery state of the charge Q_b, and (e) frequency deviation of a smart microgrid Δf.

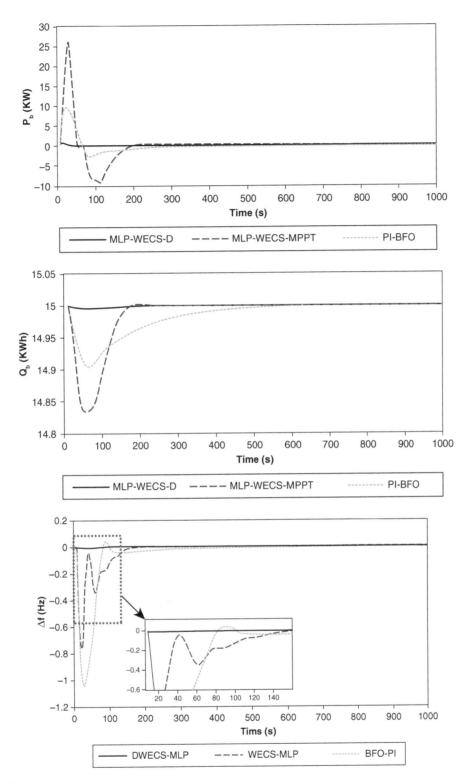

FIGURE 26.8 (Continued)

In Figure 26.8(e), it can be observed that the frequency deviation with MLP-WECS-D is very small compared to the MLP-WECS-MPPT and BFO-PI. From this it is evident that MLP-based controller is advantageous for frequency control of a smart microgrid, and if possible, the deloaded WECS operation is much more useful.

26.4.2 Solar Power Increased by 3% at 10th Second

In this case, solar power is varied from 319.2 kW to 328.8 kW at 10 s, that is, 3%. The simulation results are shown in Figure 26.9. All the secondary sources,

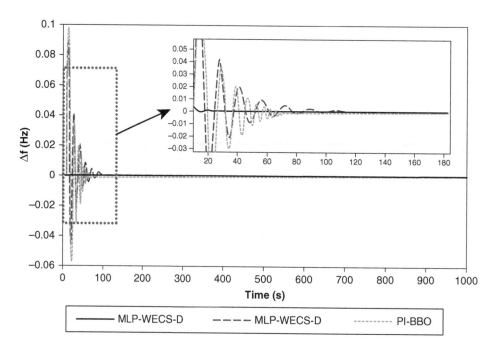

FIGURE 26.9 The frequency deviation of power systems Δf due to the 3% increase in solar power.

aqua-electrolyzer and battery, respond to the microgrid's disturbance in the transient period. In the steady-state period, the aqua-electrolyzer takes the excess power from solar and converts it into hydrogen, then stores it in the hydrogen storage tank.

It is observed that the MLP-WECS-D performance is much better when compared to the other two methods shown in Figure 26.9.

26.5 CONCLUSIONS

A state-of-the-art MLP has been successfully implemented for the frequency control of a smart microgrid. The performance analysis of the proposed MLP is compared with the BFO-PI, and it is found that the MLP is efficient.

It is observed that when the WECS participates in frequency control, the usage of the battery reduces, resulting in increasing the lifetime and decreasing the capacity of the battery. For the optimal usage of the microgrid resources, the concept of the smart microgrid is proposed based on MAS and UDP/IP and is successfully implemented.

26.6 APPENDIX

A) *Parameters for solar power*
   $I_{rs} = 1.2e^{-7}$; $I_{scr} = 8.03$; $T_r = 300$; $T = 320$; $K_t = 0.0017$; $A = 1.92$; $n_p = 176$; $n_s = 1500$; $K = 1.3805e^{-23}$; $q = 1.610e^{-19}$; $S = 0.2$; $V_{dc0} = 542.2651V$; $K_s = 1.7018$; $T_s = 0.05$ s.

B) *Nominal parameters of the smart microgrid*
   $f_{sys} = 50Hz$; Base Power = 1 MVA; $D = 0.012$ MW/Hz; $H = 5s$; $T_{dgg} = 2s$; $T_{fc} = 4s$; $T_b = 0.1s$; $T_{ae} = 0.2s$; $T_{dgt} = 20s$; $GRC_{dg} = 3\%$; $GRC_{ae} = 10\%$; $GRC_{fc} = 10\%$; $GRC_b = 30\%$.

26.7 REFERENCES

[1] International Energy Outlook 2005, Energy Information Administration (EIA), www.eia.doe.gov/iea.
[2] S. R. Bull, "Renewable energy today and tomorrow", *Proceedings of IEEE*, vol. 89, no. 8, pp. 1216–1221, Aug. 2001.
[3] Renewables and Alternate Fuels, Energy Information Administration (EIA), W. S. Fyfe, M. A. Powell, B. R. Hart, and B. Ratanasthien, "A global crisis: Energy in the future", *Nonrenewable Resources*, vol. 2, pp. 187–195, 1993.
[4] R. H. Lasseter, "Extended CERTS microgrid", *Proceedings of the IEEE Power Energy Society General Meeting*, Jul. 2008, pp. 1–5.
[5] Gayadhar Panda, Sidharrtha Panda, and Cemal Ardil, "Automatic generation control of interconnected power system with generation rate constraints by hybrid neuro fuzzy approach", *World Academy of Science, Engineering and Technology*, vol. 52, 2009.
[6] R. Majumder, B. Chaudhuri, A. Ghosh, R. Majumder, G. Ledwich, and FiruzZare, "Improvement of stability and load sharing in an autonomous microgrid using supplementary droop control loop", *IEEE Transactions on Power Systems*, vol. 25, no. 2, pp. 796–808, May 2010.
[7] D. Lee and Li Wang, "Small signal stability analysis of an autonomous hybrid renewable energy power generation/energy storage system time domain simulations", *IEEE Transactions on Energy Converters*, vol. 23, no. 1, pp. 311–320, Mar. 2008.
[8] T. Senjyo, T. Nakaji, K. Uezato, and T. Funabashi, "A hybrid power system using alternative energy facilities in isolated island", *IEEE Transactions on Energy Converters*, vol. 20, no. 2, pp. 406–414, Jun. 2005.

[9] B. S. Kumar, S. Mishra, and N. Senroy, "AGC for distribution generation", *Proceedings of International Conference on Sustainable Energy Technologies*, pp. 89–94, 2008.

[10] S. Mishra, G. Mallesham, and A. N. Jha, "Ziegler-Nichols based controller parameters tuning for load frequency control in a microgrid", *Proceedings of the IEEE Sponsored International Conference on Energy, Automation, and Signals (ICEAS-2011)*, Bhubaneswar, pp. 1–8, Dec. 2011.

[11] G. Mallesham, S. Mishra, and A. N. Jha, "Optimization of control parameters in AGC of microgrid using gradient descent method", *16th National Power Systems Conf. (NPSC-2010)*, Hyderabad, pp. 37–42, 2010.

[12] G. Mallesham, S. Mishra, and A. N. Jha, "Automatic generation control of microgrid using artificial intelligence methods", *Proceedings of 2012 IEEE Power & Energy Society General Meeting*, San Diego, pp. 1–8, July 2012.

[13] S. Mishra, G. Mallesham, and A. N. Jha, "Design of controller and communication for frequency regulation of a smart microgrid", *IET Renewable Power Generation*, vol. 6, no. 4, pp. 248–258, July 2012.

[14] G. Mallesham and Krishna Degavath, "Optimal frequency control of a microgrid", *International Conference on Recent innovations and Electrical and Electronic Engineering-2017 (ICRIEEE-2017)*, Helix, vol. 8, no. 2, pp. 3079–3086, Feb. 28, 2018.

[15] G. A. Chown and R. C. Hartman, "Design & experience of fuzzy logic controller for automatic generation control (AGC)", *IEEE Transactions on Power Systems*, vol. 13, no. 3, pp. 965–970, Aug. 1998.

[16] F. Beaufays, Y. Abdel-Magid, and B. Widrow, "Application of neural network to load frequency control in power systems", *Neural Networks*, vol. 7, no. 1, pp. 183–194, 1994.

[17] Lalit Chandra Saikia, Sukumar Mishra, Nidul Sinha, and J. Nanda, "Automatic generation control of a multi area hydrothermal system using reinforced learning neural network controller", *International Journal of Electrical Power & Energy Systems*, vol. 33, no. 4, pp. 1101–1108, May 2011.

[18] M. Pipattanasomporn, H. Feroze, and S. Rahman, "Multi-agent systems in a distributed smart grid: Design and implementation", *IEEE PES Power Systems Conference and Exposition*, Mar. 15–18, 2009.

[19] K. J. P. Macken, K. Vanthournout, J. Van Den Keybus, G. Deconinck, and R. J. M. Belmans, "Distributed control of renewable generation units with integrated active filter", *IEEE Transactions on Power Electronics*, vol. 19, no. 5, pp. 1353–1360, Sep. 2004.

[20] A. Yazdani and P. P. Dash, "A control methodology and characterization of dynamics for a photovoltaic (PV) system interfaced with a distribution network", *IEEE Transactions on Power Delivery*, vol. 24, no. 3, pp. 1538–1551, July 2009.

Index

A

absolute total energy consumption, 308
adaptive P&O algorithm, 291
aerodynamic drag force, 64
aesthetic design of lighting, 190
afore said hybrid system, 198
alliance, 49
ambient temperature, 210
anaerobic digestion, 113
anemometer, 30
aqua-electrolyzer, 350
arbitrary, 54
artificial illumination, 111
artificial network (AN), 18
artificial night light, 189
attributes, 47
auto regressive moving average (ARMA), 18
auto-synchronization, 172
auxiliary inductance, 183
axillary power supply, 70

B

back propagation, 41
band width, 38
battery chemistry, 309
battery electric vehicle, 60
battery pack, 69
battery SDC, 95
battery storage unit, 311
Bernoulli's law, 326
Betz limit, 328
Bezier curves, 328
bibliography, 79
bidding, 283
bidirectional power, 295
bifurcation, 281
bio-mass, 109
bird draft pings, 336
blackout, 227
blade-swept area, 37
boost converter, 178
bridge voltage, 181
buck converter, 82
buoyant, 213
buzzword, 77

C

California, 79
capacitor banks, 260
cardless phones, 62
ChadeMO, 72
Chevrolet Spark, 71
city matrix, 191
clamping, 151
closed-loop converter, 290
CMI, 17
CMLI topology, 14
CMV-differential mode voltage, 155
CMV evaluation, 157
Coda, 71
combo, 73
common mode current, 153

compatible, 72
competitiveness, 170
COMSAT Laboratories, 62
concealing, 47
conservation of mass, 32
control volume, 325
consumer-grade nickel, 63
conventional vehicle, 60
conversion ratio, 180
COP, 223
correlation, 194
cost inadequate, 52
custom power task, 41

D

DAB power, 298
DAB topologies, 176
DCHE, 223
DC link voltage, 301
DC paradigm, 197
decentralized, 48
decision making block, 127
decision mode, 149
deducting, 285
degrade, 89
dehumidification, 223
demonstrate, 47
density function, 03
depleted, 137
depreciation, 285
desalination, 216
design of thermosyphon coding system, 104
DG sets, 170
discrete, 43
disruptive effects, 334
distribution function, 02
DNI, 223
double full-bridge topology, 177
drag coefficient, 66
drastically, 283
drastic shift, 198
drivetrain, 20
droop control, 199
ducted wind turbines, 326
dynamic response, 251

E

E-bus, 313
effectiveness, 345
electric vehicle efficiency, 310
electrolyzers, 316
elucidation, 254
e-mobility, 46
emphasis orients, 337
endowed, 172
energy harvest, 333
enhancing, 173
EPS, 181
era, 61
ESS converter, 234
ETC tubes, 216
EV charger, 267

evolution of BEV, 292
extremophile species, 112

F

FACTS, 229
feasible, 171, 197
fermentation, 110
FET analysis, 148
flow formation, 265
fluctuates, 351
fluctuating, 37
foreseen, 138
fossil fuels, 109
freewheeling period, 154
frustrum cylinders, 330
fuel-based electrical energy system, 45
fuel cell system, 149
full bridge DC-AC converter, 129
full bridge DC-AC inverter, 126
full bridge inverter, 124
fuzzification, 51
fuzzy logic controller (FLL), 14
fuzzy logic controller suboptimal, 351

G

galvanic isolation, 293
gas chromatography (GC), 117
gas chromatography mass spectroscopy (GCMS), 118
gasification, 113
genetic algorithm, 252
global carbon emission, 45
globalization, 01
global peak, 48, 345
Goodenough's, 63
GPREC, 171
graphene, 77
green energy, 196
greenhouse gasses, 109
grid, 171
grid integration, 148
grid interactive battery, 233
grid interconnection, 46
GSM, 356
Guobiao, 75

H

hardware module, 30
harmonic compensation, 277
haversine formula, 307
HAWT, 324
H-bridge, 152
HBZVR with DC link capacitor, 152
hidden neurons, 356
hierarchical coordination, 204
hill climbing, 65
holistic whole, 196
hybrid electric vehicle, 60
hybrid grid, 199
hybrid system, 300
hybrid system design consideration, 105
hysteresis logic, 84

hysteresis scheme, 98
Hyundai, 78

I

IBDABC, 294
illumination, 50
illumination intensity, 194
illustrate, 347
infotainment, 306
insolation, 56, 339
integrated, 46
interleaved configuration, 294
international space station, 62
inverter-to-grid, 253
inverter types, 151
IRDEA, 172
irradiance, 51
irradiation conditions, 49
islanded mode, 202

J

jet, 323

L

leakage inductance, 182
likability, 196
load hormones, 271
load requirement, 277
local peaks, 48
log-sigmoidal, 356
loss distribution, 160
losses comparison, 168
low switching count, 154
LSC-H5, 156
LSC-HERICS5, 156

M

maloperation, 254
margin, 89
master slave control, 199
mean absolute error (MAE), 19
methodology, 53
microalgae slurry, 116
micro controller, 343
micro grid, 288, 350
minimum fuel space price (MFSP), 115
mitigate, 42, 169, 213
mitigation, 99
MLP transmitted algorithm, 357
module irregularities, 334
monetary, 171
monocrystalline, 289
Moore-Penrose method, 19
MPPT technique, 145
multiagent systems, 354

N

nano technology, 76
negligible oscillations, 299
Nernst equation, 141
non detection zone, 237
non hydro renewable energy, 137
non-MPPT mode, 127
non shaded lighting element, 190
nozzle, 323

nullified, 43
nutrient recycling, 115

O

off board, 314
off-peak, 284
on board, 314
onboard charger, 69
opted, 170
optimization, 55
overwhelm, 169
O-Wind, 331

P

parking lots, 192
pasteurization, 208
penetration, 284
persistent extreme learning machine (PELM), 17
perturb, 139
perturbation process, 141
phase locked loop (PLL), 268
phase shift modulation, 177
phasor measurement unit, 228
phospho olivine, 64
photo bioreactor tubular system, 112
photo voltaic system, 265
photo voltaic, 49
photosynthesis radiation, 111
photovoltaic(s), 58, 114, 151
photovoltaic module engages, 333
PI controller, 133, 358
plug-in-installations, 35
PMSG, 14
PMSG dynamic modeling, 15
polycrystalline, 289
polygeneration, 36
Porsche Taycan, 313
post distribution analysis, 231
potential grid, 198
pouch cells, 312
power gener, 58
power system oscillator, 239
power system stabilizers, 230
principal of diversity, 191
proliferation, 13
proposed system, 200
propulsion, 306
prosumers, 174
protocol, 75, 306
PV cell array, 184
PV integrated, 267
PV module current, 291
PV module simulation, 140
PWM controller, 353
PWM pulse generation, 16
PWM with LSC-TLIT'S, 155
pyrolysis, 114

Q

quad demodulation, 236
quadrillion BTU, 138

R

ramp, 285
rapidity, 54

Rayleigh probability, 03
reactive power support (RPS), 267
recurrent neural network (RNN), 242
re-enacts, 54
reliability, 39
remote-area power system, 13
renewable energy system, 198
renewable integration, 301
residential load, 275
ripple energy, 83
robust, 286
robustness, 91
ROCOF measurements, 241
rolling resistance, 65

S

sacred building, 193
sanction, 283
SAPE control strategy, 16
self-sabotaging, 61
serial connection, 311
settling time, 91
shark skin, 329
shift control, 179
signal-processing device, 14
simulation model, 128
simulation time, 20
single bias, 38
single phase shift, 179
smart grid, 288
SOFC cell, 141
SOGI, 268
solar cell, 102, 262
solar irradiance, 275
solar panel, 316, 338
solar radiations, 335
solar thermal collector, 102
squirrel cage, 253
state estimation accuracy, 236
steady state operation in MPPT MODE, 134
streamlining, 53
stringent, 139
substantial, 305
sustainability coefficient, 193
Swagelok pipes, 118
switching frequency, 182
switching losses, 296
switching states, 296
switch-mode rectifier MPPT algorithm, 16
symmetric nozzle, 327
synchronization, 202
synchronizing PLL, 14
synchro phasor measurement, 234
system dynamics, 140

T

techno economic analysis, 114
thermochemical methods, 110
thermosyphon, 210
thyristor, 36
time of arrival (TOA), 243
TLC configuration, 157
TLI, 157
tolerance limit, 257
topical, 97

Index

Toyota, 78
tracking, 84
track sinusoidal, 87
transducers, 354
transesterification, 110
transient angle instability, 229
transmission state estimation, 235
tremendous, 198
triggering, 39
turbine generating system, 265

U

unconstrained, 173
uninterruptible supply, 278
unpredictable, 50

unwavering, 52
utility grid, 200

V

vehicle range, 309
Venturi effect, 327
versatile, 52
viability, 347
vicinity, 335
villain, 67
Volta, Alessandro, 61
voltage source oscillator, 239
voltage unbalance, 254
Vortex blade less, 331
vortex generator, 329

W

war of currents, 287
WASP algorithm, 04
WEGS, 14
Weibull parameters, 02
Weibull probability, 02
well-regulated, 83
wide area measurement system, 227
wide area monitoring system, 228
wide-bandgap, 185
wind power density, 05

Z

zero current switching (ZCS), 178
ZVS, 177

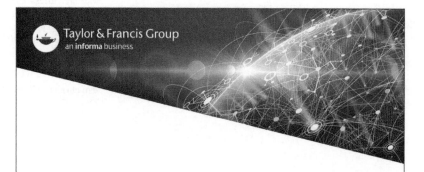